More Than Just a Textbook

Internet Resources

Step 1   **Connect to** Math Online   **macmillanmh.com**

Step 2   **Connect to online resources by using** *QuickPass* **codes.**
You can connect directly to the chapter you want.

MC7320c1

Enter this code with the appropriate chapter number.

For Students

Connect to the student edition *eBook* that contains all of the following online assets. You don't need to take your textbook home every night.

- Personal Tutor
- Extra Examples
- Self-Check Quizzes
- Multilingual eGlossary
- Concepts in Motion
- Chapter Test Practice
- Test Practice
- Study to Go
- Math Adventures with Dot and Ray
- Math Tool Chest
- Math Songs

For Teachers

Connect to professional development content at **macmillanmh.com** and the *eBook Advance Tracker* at **AdvanceTracker.com**

For Parents

Connect to **macmillanmh.com** for access to the *eBook* and all the resources for students and teachers that are listed above.

T60104

Macmillan McGraw-Hill

Math Connects

3

Admission $2

6 feet

2 feet

Authors

Altieri • Balka • Day • Gonsalves • Grace • Krulik
Malloy • Molix-Bailey • Moseley • Mowry • Myren
Price • Reynosa • Santa Cruz • Silbey • Vielhaber

Mc
Graw
Hill **Macmillan
McGraw-Hill**

About the Cover

Multiplication and division are featured topics in Third grade. The cover shows different representations of multiplication such as the array of buttons on the snowman and the design on penguin's swimsuit. Have students use multiplication to find the price of admission for 5, 6, and 7 penguins as well as estimate the total number of dots on the penguin's inner-tube.

The **McGraw·Hill** Companies

 Macmillan/McGraw-Hill

Copyright © 2009 by The McGraw-Hill Companies, Inc. All rights reserved. Except as permitted under the United States Copyright Act, no part of this publication may be reproduced or distributed in any form or by any means, or stored in a database or retrieval system, without prior permission of the publisher.

Send all inquiries to:
Macmillan/McGraw-Hill
8787 Orion Place
Columbus, OH 43240-4027

ISBN: 978-0-02-105732-0
MHID: 0-02-105732-X

Math Connects, Grade 3

Printed in the United States of America.

5 6 7 8 9 10 WDQ/LEH 16 15 14 13 12 11 10

Contents in Brief

Focal Points

The Curriculum Focal Points identify key mathematical ideas for this grade. They are not discrete topics or a checklist to be mastered; rather, they provide a framework for the majority of instruction at a particular grade level and the foundation for future mathematics study. The complete document may be viewed at www.nctm.org/focalpoints.

KEY

G3-FP1
Grade 3 Focal Point 1

G3-FP2
Grade 3 Focal Point 2

G3-FP3
Grade 3 Focal Point 3

G3-FP4C
Grade 3 Focal Point 4
Connection

G3-FP5C
Grade 3 Focal Point 5
Connection

G3-FP6C
Grade 3 Focal Point 6
Connection

G3-FP7C
Grade 3 Focal Point 7
Connection

G3-FP1 *Number and Operations* and *Algebra:* **Developing understandings of multiplication and division and strategies for basic multiplication facts and related division facts**

Students understand the meanings of multiplication and division of whole numbers through the use of representations (e.g., equal-sized groups, arrays, area models, and equal "jumps" on number lines for multiplication, and successive subtraction, partitioning, and sharing for division). They use properties of addition and multiplication (e.g., commutativity, associativity, and the distributive property) to multiply whole numbers and apply increasingly sophisticated strategies based on these properties to solve multiplication and division problems involving basic facts. By comparing a variety of solution strategies, students relate multiplication and division as inverse operations.

G3-FP2 *Number and Operations:* **Developing an understanding of fractions and fraction equivalence**

Students develop an understanding of the meanings and uses of fractions to represent parts of a whole, parts of a set, or points or distances on a number line. They understand that the size of a fractional part is relative to the size of the whole, and they use fractions to represent numbers that are equal to, less than, or greater than 1. They solve problems that involve comparing and ordering fractions by using models, benchmark fractions, or common numerators or denominators. They understand and use models, including the number line, to identify equivalent fractions.

G3-FP3 *Geometry:* **Describing and analyzing properties of two-dimensional shapes**

Students describe, analyze, compare, and classify two-dimensional shapes by their sides and angles and connect these attributes to definitions of shapes. Students investigate, describe, and reason about decomposing, combining, and transforming polygons to make other polygons. Through building, drawing, and analyzing two-dimensional shapes, students understand attributes and properties of two-dimensional space and the use of those attributes and properties in solving problems, including applications involving congruence and symmetry.

G3-FP4C *Algebra:* Understanding properties of multiplication and the relationship between multiplication and division is a part of algebra readiness that develops at grade 3. The creation and analysis of patterns and relationships involving multiplication and division should occur at this grade level. Students build a foundation for later understanding of functional relationships by describing relationships in context with such statements as, "The number of legs is 4 times the number of chairs."

G3-FP5C *Measurement:* Students in grade 3 strengthen their understanding of fractions as they confront problems in linear measurement that call for more precision than the whole unit allowed them in their work in grade 2. They develop their facility in measuring with fractional parts of linear units. Students develop measurement concepts and skills through experiences in analyzing attributes and properties of two dimensional objects. They form an understanding of perimeter as a measurable attribute and select appropriate units, strategies, and tools to solve problems involving perimeter.

G3-FP6C *Data Analysis:* Addition, subtraction, multiplication, and division of whole numbers come into play as students construct and analyze frequency tables, bar graphs, picture graphs, and line plots and use them to solve problems.

G3-FP7C *Number and Operations:* Building on their work in grade 2, students extend their understanding of place value to numbers up to 10,000 in various contexts. Students also apply this understanding to the task of representing numbers in different equivalent forms (e.g., expanded notation). They develop their understanding of numbers by building their facility with mental computation (addition and subtraction in special cases, such as 2,500 + 6,000 and 9,000 − 5,000), by using computational estimation, and by performing paper-and-pencil computations.

Mary Behr Altieri
Putnam/Northern
    Westchester BOCES
Yorktown Heights,
    New York

Don S. Balka
Professor Emeritus
Saint Mary's College
Notre Dame, Indiana

Roger Day, Ph.D.
Mathematics Department Chair
Pontiac Township High School
Pontiac, Illinois

Philip D. Gonsalves
Mathematics Coordinator
Alameda County Office
    of Education and
    California State
    University East Bay
Hayward, California

Ellen C. Grace
Consultant
Albuquerque,
    New Mexico

Stephen Krulik
Professor Emeritus
    Mathematics Education
Temple University
Cherry Hill, New Jersey

Carol E. Malloy, Ph. D.
Associate Professor of
    Mathematics Education
University of North
    Carolina at Chapel Hill
Chapel Hill, North
    Carolina

Rhonda J. Molix-Bailey
Mathematics Consultant
Mathematics by Design
Desoto, Texas

Lois Gordon Moseley
Staff Developer
NUMBERS: Mathematics
    Professional
    Development
Houston, Texas

Brian Mowry
Independent Math Educational
    Consultant/Part-Time Pre-K
    Instructional Specialist
Austin Independent School District
Austin, Texas

Christina L. Myren
Consultant Teacher
Conejo Valley Unified
    School District
Thousand Oaks, California

Jack Price, Ed. D.
Professor Emeritus
California State
    Polytechnic University
Pomona, California

Mary Esther Reynosa
Instructional Specialist for
    Elementary Mathematics
Northside Independent
    School District
San Antonio, Texas

Rafaela M. Santa Cruz
SDSU/CGU Doctoral
    Program in Education
San Diego State University
San Diego, California

Robyn Silbey
Math Content Coach
Montgomery County
    Public Schools
Gaithersburg, Maryland

Kathleen Vielhaber
Mathematics Consultant
St. Louis, Missouri

Contributing Authors

Donna J. Long
Mathematics Consultant
Indianapolis, Indiana

FOLDABLES **Dinah Zike**
Educational Consultant
Dinah-Might Activities, Inc.
San Antonio, Texas

Consultants

Macmillan/McGraw-Hill wishes to thank the following professionals for their feedback. They were instrumental in providing valuable input toward the development of this program in these specific areas.

Mathematical Content

Viken Hovsepian
Professor of Mathematics
Rio Hondo College
Whittier, California

Grant A. Fraser, Ph.D.
Professor of Mathematics
California State University, Los Angeles
Los Angeles, California

Arthur K. Wayman, Ph.D.
Professor of Mathematics Emeritus
California State University, Long Beach
Long Beach, California

Assessment

Jane D. Gawronski, Ph.D.
Director of Assessment and Outreach
San Diego State University
San Diego, California

Cognitive Guided Instruction

Susan B. Empson, Ph.D.
Associate Professor of Mathematics
  and Science Education
University of Texas at Austin
Austin, Texas

English Learners

Cheryl Avalos
Mathematics Consultant
Los Angeles County Office of Education, Retired
Hacienda Heights, California

Kathryn Heinze
Graduate School of Education
Hamline University
St. Paul, Minnesota

Family Involvement

Paul Giganti, Jr.
Mathematics Education Consultant
Albany, California

Literature

David M. Schwartz
Children's Author, Speaker, Storyteller
Oakland, California

Vertical Alignment

Berchie Holliday
National Educational Consultant
Silver Spring, Maryland

Deborah A. Hutchens, Ed.D.
Principal
Norfolk Highlands Elementary
Chesapeake, Virginia

Reviewers

Each Reviewer reviewed at least two chapters of the Student Edition, giving feedback and suggestions for improving the effectiveness of the mathematics instruction.

Ernestine D. Austin
Facilitating Teacher/Basic
  Skills Teacher
LORE School
Ewing, NJ

Susie Bellah
Kindergarten Teacher
Lakeland Elementary
Humble, TX

Megan Bennett
Elementary Math Coordinator
Hartford Public Schools
Hartford, CT

Susan T. Blankenship
5th Grade Teacher – Math
Stanford Elementary School
Stanford, KY

Wendy Buchanan
3rd Grade Teacher
The Classical Center at Vial
Garland, Texas

Sandra Signorelli Coelho
Associate Director for
  Mathematics
PIMMS at Wesleyan University
Middletown, CT

Joanne DeMizio
Asst. Supt., Math and
  Science Curriculum
Archdiocese of New York
New York, NY

Anthony Dentino
Supervisor of Mathematics
Brick Township Schools
Brick, NJ

Lorrie L. Drennon
Math Teacher
Collins Middle School
Corsicana, TX

Ethel A. Edwards
Director of Curriculum and
  Instruction
Topeka Public Schools
Topeka, KS

Carolyn Elender
District Elementary Math
  Instructional Specialist
Pasadena ISD
Pasadena, Texas

Monica Engel
Educator Second Grade
Pioneer Elementary School
Bolingbrook, IL

Anna Dahinden Flynn
Math Teacher
Coulson Tough K-6 Elementary
The Woodlands, TX

Brenda M. Foxx
Principal
University Park Elementary
University Park, MD

Katherine A. Frontier
Elementary Teacher
Laidlaw
Western Springs, IL

Susan J. Furphy
5th Grade Teacher
Nisley Elementary
Grand Jct., CO

Peter Gatz
Student Services Coordinator
Brooks Elementary
Aurora, IL

Amber Gregersen
Teacher – 2nd Grade
Nisley Elementary
Grand Junction, CO

Roberta Grindle
Math and Language Arts
  Academic Intervention
  Service Provider
Cumberland Head
  Elementary School
Plattsburgh, NY

Sr. Helen Lucille Habig, RSM
Assistant Superintendent/
  Mathematics
Archdiocese of Cincinnati
Cincinnati, OH

Holly L. Hepp
Math Facilitator
Barringer Academic Center
Charlotte, NC

Martha J. Hickman
2nd Grade Teacher
Dr. James Craik Elementary
  School
Pomfret, MD

Margie Hill
District Coordinating Teacher
  for Mathematics, K-12
Blue Valley USD 229
Overland Park, KS

Carol H. Joyce
5th Grade Teacher
Nathanael Greene Elementary
Liberty, NC

Stella K. Kostante
Curriculum Coach
Roosevelt Elementary
Pittsburgh, PA

Pamela Fleming Lowe
Fourth Grade eMINTS Teacher
O'Neal Elementary
Poplar B0luff, MO

Lauren May, NBCT
4th Grade Teacher
May Watts Elementary School
Naperville, IL

Lorraine Moore
Grade 3 Math Teacher
Cowpens Elementary School
Cowpens, SC

Shannon L. Moorhead
4th Grade Teacher
Centerville Elementary
Anderson, SC

Gina M. Musselman, M.Ed
Kindergarten Teacher
Padeo Verde Elementary
Peoria, AZ

Jen Neufeld
3rd Grade Teacher
Kendall
Naperville, IL

Cathie Osiecki
K-5 Mathematics Coordinator
Middletown Public Schools
Middletown, CT

Phyllis L. Pacilli
Elementary Education Teacher
Fullerton Elementary
Addison, IL

Cindy Pearson
4th/5th Grade Teacher
John D. Spicer Elementary
Haltom City, TX

Herminio M. Planas
Mathematics Curriculum
  Specialist
Administrative Offices-
  Bridgeport Public Schools
Bridgeport, CT

Jo J. Puree
Educator
Lackamas Elementary
Yelm, WA

Teresa M. Reynolds
Third Grade Teacher
Forrest View Elementary
Everett, WA

Dr. John A. Rhodes
Director of Mathematics
Indian Prairie SD #204
Aurora, IL

Amy Romm
1st Grade Teacher
Starline Elementary
Lake Havasu, AZ

Delores M. Rushing
Numeracy Coach
Dept. of Academic Services-
  Mathematics Department
Washington, DC

Daniel L. Scudder
Mathematics/Technology
  Specialist
Boone Elementary
Houston, TX

Laura Seymour
Resource Teacher Leader –
  Elementary Math & Science,
  Retired
Dearborn Public Schools
Dearborn, MI

Petra Siprian
Teacher
Army Trail Elementary School
Addison, IL

Sandra Stein
K-5 Mathematics Consultant
St. Clair County Regional
  Educational Service Agency
Marysville, MI

Barb Stoflet
Curriculum Specialist
Roseville Area Schools
Roseville, MN

Kim Summers
Principal
Dynard Elementary
Chaptico, MD

Ann C. Teater
4th Grade Teacher
Lancaster Elementary
Lancaster, KY

Anne E. Tunney
Teacher
City of Erie School District
Erie, PA

Joylien Weathers
1st Grade Teacher
Mesa View Elementary
Grand Junction, CO

Christine F. Weiss
Third Grade Teacher
Robert C. Hill Elementary
  School
Romeoville, IL

Contents

Start Smart

H.O.T. Problems

WRITING IN ▶MATH 3, 5, 7, 9, 11, 13

CHAPTER 1
Use Place Value to Communicate

Focal Points and Connections
See page iv for key.

G3-FP7C Number and Operations
G3-FP1 Number and Operations and Algebra

Test Practice 27, 31, 37, 41, 51, 55, 63, 64, 65

H.O.T. Problems
Higher Order Thinking 19, 27, 30, 37, 41, 46, 51, 55

WRITING IN ▸MATH 19, 21, 23, 27, 30, 31, 33, 37, 41, 46, 51, 55, 63

Contents

CHAPTER 2 — Add to Solve Problems

Focal Points and Connections
See page iv for key.

G3-FP7C Number and Operations

Test Practice 71, 77, 81, 85, 99, 105, 106, 107

H.O.T. Problems
Higher Order Thinking 71, 77, 80, 85, 94, 99

WRITING IN ►MATH 71, 73, 77, 80, 81, 85, 89, 91, 99, 105

CHAPTER 3 Subtract to Solve Problems

Focal Points and Connections
See page iv for key.

G3-FP7C Number and Operations
Test Practice 117, 121, 131, 141, 151, 152, 153

H.O.T. Problems
Higher Order Thinking 113, 117, 120, 131, 136, 141, 143

WRITING IN MATH 113, 117, 120, 121, 125, 127, 131, 133, 136, 141, 143, 151

Contents

CHAPTER 4
Develop Multiplication Concepts and Facts

Focal Points and Connections
See page iv for key.

G3-FP1 Number and Operations and Algebra
G3-FP4C Algebra

Test Practice 167, 170, 181, 188, 195, 196, 197

H.O.T. Problems
Higher Order Thinking 161, 164, 170, 176, 181, 188

WRITING IN ►MATH 158, 161, 164, 167, 170, 173, 176, 181, 185, 188, 195

CHAPTER 5

Develop More Multiplication Facts

Focal Points and Connections
See page iv for key.

G3-FP1 Number and Operations and Algebra
G3-FP4C Algebra

Contents

Focal Points and Connections
See page iv for key.

G3-FP1 Number and Operations and Algebra
G3-FP4C Algebra

Test Practice 261, 269, 273, 280, 289, 290, 291

H.O.T. Problems
Higher Order Thinking 255, 261, 266, 273, 279, 283

WRITING IN ►MATH 252, 255, 257, 261, 263, 266, 269, 273, 277, 279, 283, 289

CHAPTER 7

Develop More Division Facts

Focal Points and Connections
See page iv for key.

G3-FP1 Number and Operations and Algebra
G3-FP4C Algebra

Test Practice 303, 309, 319, 327, 328, 329

H.O.T. Problems
Higher Order Thinking 299, 303, 308, 314, 319

WRITING IN ▸MATH 296, 299, 303, 305, 308, 309, 314, 319, 321, 327

Contents

CHAPTER 8
Use Patterns and Algebraic Thinking

Focal Points and Connections
See page iv for key.

G3-FP1 Number and Operations and Algebra
G3-FP4C Algebra

Test Practice 341, 347, 353, 367, 368, 369

H.O.T. Problems
Higher Order Thinking 335, 341, 347, 351, 359

WRITING IN ►MATH 335, 337, 341, 343, 347, 351, 353, 355, 359, 367

CHAPTER 9
Measure Length, Area, and Temperature

Focal Points and Connections
See page iv for key.

G3-FP5C Measurement

Test Practice 381, 389, 391, 395, 401, 411, 417, 418, 419

H.O.T. Problems
Higher Order Thinking 377, 381, 389, 395, 400, 411

WRITING IN ►MATH 374, 377, 381, 383, 385, 389, 391, 395, 397, 400, 403, 407, 411, 417

Contents

CHAPTER 10 Measure Capacity, Weight, Volume, and Time

Focal Points and Connections
See page iv for key.

G3-FP5C Measurement

Test Practice 428, 435, 441, 447, 453, 455, 461, 462, 463

H.O.T. Problems
Higher Order Thinking 427, 434, 441, 447, 453, 455

WRITING IN ►MATH 424, 427, 431, 434, 435, 437, 441, 447, 449, 453, 455, 461

CHAPTER 11 Identify Geometric Figures and Spatial Reasoning

Focal Points and Connections
See page iv for key.

G3-FP5C Measurement
G3-FP3 Geometry

Test Practice 471, 481, 483, 497, 507, 508, 509

H.O.T. Problems
Higher Order Thinking 470, 475, 481, 485, 490, 493, 496

WRITING IN ►MATH 470, 475, 477, 481, 483, 485, 487, 490, 493, 496, 507

Contents

CHAPTER 12
Organize, Display, and Interpret Data

Focal Points and Connections
See page iv for key.

G3-FP6C Data Analysis

Test Practice 521, 525, 531, 535, 539, 545, 553, 554, 555

H.O.T. Problems
Higher Order Thinking 517, 521, 530, 535, 538, 545

WRITING IN ▶**MATH** 514, 517, 521, 523, 525, 527, 530, 535, 538, 545, 547, 553

CHAPTER 13 Develop Fractions

**Focal Points
and Connections**
See page iv for key.

G3-FP2 Number and Operations
G3-FP5C Measurement

Test Practice 567, 574, 577, 583, 587, 595, 596, 597

H.O.T. Problems
Higher Order Thinking 563, 567, 574, 583, 587

(WRITING IN ►MATH) 560, 563, 567, 569, 571, 574, 577, 579, 583, 587, 595

Contents

CHAPTER 14 — Understand Fractions and Decimals

Focal Points and Connections
See page iv for key.

G3-FP2 Number and Operations
G3-FP5C Measurement

Test Practice 606, 611, 621, 629, 630, 631

H.O.T. Problems
Higher Order Thinking 606, 610, 621

WRITING IN ▸MATH 602, 606, 610, 611, 615, 617, 621, 623, 629

CHAPTER 15 Multiply by One-Digit Numbers

Focal Points and Connections
See page iv for key.

G3-FP1 Number and Operations and Algebra
G3-FP4C Algebra

Test Practice 643, 647, 655, 667, 668, 669

H.O.T. Problems
Higher Order Thinking 637, 642, 647, 655, 658

WRITING IN ►MATH 637, 639, 642, 643, 647, 649, 651, 655, 658, 667

Contents

Looking Ahead

Problem-Solving Projects

Student Handbook

Built-In Workbook

Reference

To the Student

As you gear up to study mathematics, you are probably wondering, "What will I learn this year?"

- **Number and Operations** and **Algebra:** Multiply and divide whole numbers and relate multiplication and division.

- **Number and Operations:** Understand fractions and equivalent fractions.

- **Geometry:** Describe and analyze properties of two-dimensional figures.

Along the way, you'll learn more about problem solving, how to use the tools and language of mathematics, and how to THINK mathematically.

How to Use Your Math Book

Have you ever been in class and not understood all of what was being presented? Or, you understood everything in class, but got stuck on how to solve some of the homework problems? Don't worry. You can find answers in your math book!

- **Read** the MAIN IDEA at the beginning of the lesson.

- **Find** the New Vocabulary words, highlighted in yellow, and read their definitions.

- **Review** the EXAMPLE problems, solved step-by-step, to remind you of the day's material.

- **Refer** to the EXTRA PRACTICE boxes that show you where you can find extra exercises to practice a concept.

- **Go** to Math Online where you can find extra examples to coach you through difficult problems.

- **Review** the notes you've taken on your FOLDABLES.

- **Refer** to the Remember boxes for information that may help you with your examples and homework practice.

TREASURE HUNT

CHAPTER 1

Let's Get Started

Use the Treasure Hunt below to learn where things are located in each chapter.

1 What is the title of Chapter 1?

2 What is the Main Idea of Lesson 1-1?

3 How do you know which words are vocabulary words?

4 What are the vocabulary words for Lesson 1-3?

5 What is the key concept shown in Lesson 1-9?

6 How many Examples are presented in Lesson 1-4?

7 What is the web address where you could find extra examples?

8 On page 25, there is a Remember tip box. How does the Remember tip help you?

9 How many exercises are there in Lesson 1-6?

10 Suppose you need more practice on a concept. Where can you go for Extra Practice?

11 Suppose you're doing your homework on page 36 and you get stuck on Exercise 18. Where could you find help?

12 What is the web address that would allow you to take a self-check quiz to be sure you understand the lesson?

13 On what pages will you find the Chapter 1 Study Guide and Review?

14 Suppose you can't figure out how to do Exercise 7 in the Study Guide and Review on page 57. Where could you find help?

MATH? SYMBOLS?

Start Smart

Let's Review!

Bald Eagle

Quite Deep and Quite Tall

South Carolina's state tree is the Sabal palmetto. It is also known as the Cabbage Palm. It can grow to 82 feet tall. The roots of the Sabal palmetto can be 20 feet deep.

What is the total length of a sabal palmetto?

You can use the four-step problem-solving plan to solve math problems. The four steps are Understand, Plan, Solve, and Check.

Understand

- **Read the problem carefully.**
- **What facts do you know?**
- **What do you need to find?**

You know the height of a Sabal palmetto above ground and the depth of its roots. You need to find the total length of the Sabal palmetto.

Plan

- **Think about how the facts relate to each other.**
- **Plan a strategy to solve the problem.**

To find the total length of the tree, use addition.

Solve

Use your plan to solve the problem.

Add the height of a sabal palmetto above ground, and the length of its roots.

$$\begin{array}{ll} 82 \text{ feet} & \text{height above ground} \\ \underline{+20 \text{ feet}} & \text{depth of roots} \\ 102 \text{ feet} \end{array}$$

The total length is 102 feet long.

Check

- **Look back.**
- **Does your answer make sense?**
- **If not, solve the problem another way.**

Use subtraction to check the addition. $102 - 20 = 82$. This is the height of the tree above ground. So, the answer is correct.

CHECK What You Know

1. List the four steps of the four-step problem-solving plan.

2. Describe each step of the four-step problem-solving plan.

3. **WRITING IN ►MATH** Use the tree facts on page 3 to write a real-world problem. Ask a classmate to solve.

An Exciting Two Minutes!

The Kentucky Derby is the most famous horse race in the United States. Real Quiet, one of the most recent winners, placed first in the 124th Kentucky Derby.

 **CHECK What You Know** Place Value

The model shows the value of each digit in 124.

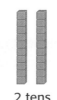

1 hundred   2 tens   4 ones

1. Copy and complete the place value chart.

Number	Hundreds	Tens	Ones
124	1	2	4
134	▦	▦	▦
120	▦	▦	▦
903	▦	▦	▦

2. Write each number in Exercise 1 in word form.

The world's largest bat (120 feet long) is found in Louisville, Kentucky. It is modeled after Babe Ruth's bat which was actually 34 inches long. Since 120 is greater than 34, you can write 120 > 34.

Compare. Use <, >, or =.

3. 87 ● 108

4. 425 ● 45

5. 307 ● 307

6. 333 ● 313

7. 580 ● 508

8. 919 ● 929

Did you Know

Mammoth Cave Park has over 367 miles of explored caves to visit. It is thought to be the longest cave system in the world.

The table shows how long visitors could expect to spend on some Kentucky cave tours.

Cave Tours	Grand Avenue	Trog	Wild Cave	Great Onyx
Time (min)	270	150	360	135

9. Which tour is the longest?

10. Which tour is the shortest?

11. Order these times from greatest to least.

12. **WRITING IN ►MATH** Roll a number cube 3 times to make a three-digit number. Record the numbers. Repeat three times to make 4 three-digit numbers. Explain how to order the numbers from least to greatest.

At the Aquarium

The New York Aquarium in Brooklyn is home to many types of marine life from all over the world. Among these are alien stingers, sharks, seahorses, and sea lions.

 **CHECK What You Know** **Addition Properties** · · · · · · · · · · · · · · ·

Sally sees 3 stingrays and 4 sharks in the Marine Explorers program. Tomas sees 4 sharks and 3 stingrays. The sums $3 + 4$ and $4 + 3$ are the same since the order in which numbers are added does not change the answer.

1. Fernando sees 5 stingers and 8 sea lions. Draw a picture using circles to show that $5 + 8 = 8 + 5$.

2. Draw a picture to show that $3 + 2 + 4 = 2 + 4 + 3$.

Tell what number is missing in each number sentence.

3. $3 + \blacksquare = 2 + 3$

4. $\blacksquare + 6 = 6 + 5$

5. $10 + 7 + 1 = 1 + 7 + \blacksquare$

6. $9 + 12 + 3 = 3 + \blacksquare + 9$

A crab has 10 legs. What is the total number of legs for 5 crabs?

The table shows that for every crab added, 10 more legs are added.

So, there would be 40 + 10, or 50 legs if there were 5 crabs in a tank.

Crab legs	
Number of Crabs	Total Legs
1	10
2	20
3	30
4	40
5	■

+10
+10
+10
+10

Did you Know

There are over 350 different types of sharks. Some only grow to about 7 inches long.

Copy and complete each table.

7.

Dolphin Dorsal Fins	
Dolphins	Dorsal Fins
1	1
2	2
3	■

8.

Stingray Eyes	
Stingrays	Eyes
1	2
2	4
3	■
4	8

9. **WRITING IN ►MATH** Lydia adds 7 + 5 and then adds 10 to the result. Will she get the same number if she adds 7 to the result of 5 + 10? Explain.

Measure Away

Illinois' state quarter was the first quarter released in 2003. It shows Abraham Lincoln, the 16th president of the United States.

CHECK What You Know

Customary Units

One customary unit of measurement is the inch. One Illinois state quarter has a length of 1 inch.

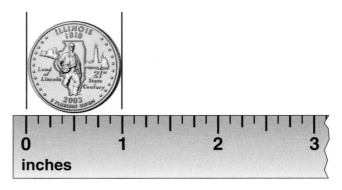

Estimate the length of each set of coins to the nearest inch. Then use a quarter to measure the length.

1.

2.

3. How many pennies in a row would measure about 5 quarters?

CHECK What You Know · Temperature

Coins can be melted at very, very hot temperatures.

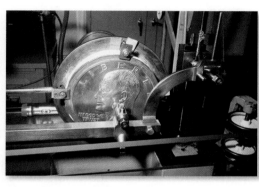

Tell which thermometer shows the temperature described.

4. Water boils at 212 °F.

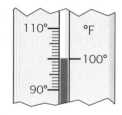 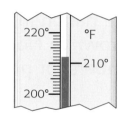  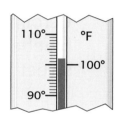

5. The temperature in the classroom was 75°F.

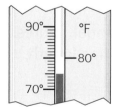  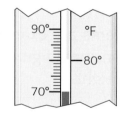 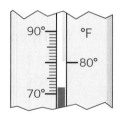

6. **WRITING IN ►MATH** What happens to the liquid in the thermometer as the temperature increases?

Geometry

Food and Geometry

Blueberries grow all over North Carolina. In fact, North Carolina ranks fourth in the nation in the production of blueberries. Other crops grown in North Carolina include sweet potatoes and cucumbers for pickles.

CHECK What You Know **Three-Dimensional Figures** · · · · · · · · · ·

Many food items are examples of three-dimensional figures. A three-dimensional figure is a solid figure that has length, width, and height.

1. What three-dimensional figure is a blueberry?

   A cylinder   **C** cube

   B sphere   **D** pyramid

2. What three-dimensional figure is a jar of pickles?

   A cone   **C** sphere

   B cylinder   **D** cube

CHECK What You Know   Two-Dimensional Figures ···········

A two-dimensional figure has length and width. A circle and a square are examples of two-dimensional figures.

Copy and complete the table.

	Figure	Name	Sides
3.			
4.			
5.			
6.			

7. **WRITING IN ►MATH** How is a square different than a cube? How are they alike?

How Fast can You Go?

Pennsylvania's Kennywood Park is home to one of the fastest roller coasters in the world. In fact, The Phantom's Revenge sends riders traveling faster than 85 miles per hour.

CHECK **What You Know** **Picture Graph** ·

Justin surveyed his friends to name their favorite amusement park ride. He put the data in a picture graph.

Favorite Amusement Park Rides					
Roller coaster	😊	😊	😊	😊	
Ferris wheel	😊	😊			
Water rides	😊	😊	😊	😊	😊
Merry-go-round	😊	😊			
Key: Each face = 2 votes					

1. How many friends did Justin survey?

2. Which kind of ride is most liked?

3. How many more friends like roller coasters than Ferris wheels?

✓ **CHECK What You Know**   **Tally Chart and Bar Graph** ··········

Flora asked her friends how many times they rode a roller coaster last summer. She recorded the results in a tally chart.

Ride a Roller Coaster?		
Name	**Tally**	**Total**
Val	⌗⌗ ||	■
Jerome	|||	■
Rosa	||	■
Aaron	⌗⌗ ⌗⌗	■

Did you Know?

The Steel Phantom was the first ride with a record-breaking *second* 225-foot drop.

4. Copy and complete the tally chart.

5. How many friends did Flora survey?

6. How many times did they ride a roller coaster in all?

7. How many more times did Aaron ride a roller coaster than Rosa?

8. Copy the table below. Use the data above to make a bar graph. The data for Val is shown.

Ride a Roller Coaster?									
Val									
Jerome									
Rosa									
Aaron									

7. **WRITING IN ►MATH** Explain the difference between a picture graph and a tally chart.

Use Place Value to Communicate

BIG Idea What is the place value of a digit in a number?

Place value is the value given to a digit by its place in a number.

Example The state cat of Massachusetts is the Tabby cat. Every 5 seconds, a cat purrs about 125 times. The number 125 is read *one hundred twenty-five.*

Hundreds	Tens	Ones
1	2	5

$\uparrow$ 100 $\uparrow$ 20 $\uparrow$ 5

What will I learn in this chapter?

- Count, read, and write whole numbers.
- Identify place value of whole numbers.
- Compare and order whole numbers.
- Round numbers to the nearest ten, hundred, and thousand.
- Use the four-step problem-solving plan.

Key Vocabulary

pattern

place value

equal to

round

Math Online > **Student Study Tools** at **macmillanmh.com**

FOLDABLES®
Study Organizer

Make this foldable to help you organize information about place value. Begin with one sheet of 11″ × 17″ paper.

1 **Fold** the paper in half as shown.

2 **Unfold** and fold one side up 5 inches.

3 **Fold** again so the pockets are inside. Glue the edges.

4 **Label** as shown. Record what you learn.

Place Value Through 1,000

Place Value Through 10,000

ARE YOU READY for Chapter 1?

You have two ways to check prerequisite skills for this chapter.

Option 2

Math Online Take the Chapter Readiness Quiz at macmillanmh.com.

Option 1

Complete the Quick Check below.

QUICK Check

Write each number. (Prior grade)

1.

Hundreds	Tens	Ones
	1	4

2.

Hundreds	Tens	Ones
	3	3

3.

Hundreds	Tens	Ones
1	1	0

4. 1 ten 5 ones

5. 1 hundred 2 ones

6. twenty-four

7. one hundred thirty-eight

Write the number of tens and ones in the following numbers. (Prior grade)

8. 12　　　　**9.** 26　　　　**10.** 31　　　　**11.** 85

12. Manuel and his family went to the circus. They spent a total of $65. Name how many tens and ones are in 65.

Algebra Find a pattern. Write the next two numbers.

(Prior grade)

13. 2, 4, 6, 8, ■, ■

14. 1, 3, 5, 7, ■, ■

15. 5, 10, 15, 20, ■, ■

16. 10, 20, 30, 40, ■, ■

17. Sonja read 4 pages the first day, she read 8 pages the second day, and she read 12 pages the third day. If the pattern continued, how many pages did she read on the fourth day?

Algebra: Number Patterns

GET READY to Learn

Number patterns are everywhere. Look at the speed limit signs shown. What number pattern do you see?

SPEED **15** SPEED **25** SPEED **35** SPEED **45**

MAIN IDEA

I will find patterns in numbers.

New Vocabulary

pattern

Math Online

macmillanmh.com
• Extra Examples
• Personal Tutor
• Self-Check Quiz

A **pattern** is a series of numbers or figures that follow a rule. A hundred chart shows many number patterns.

1	2	3	4	5	6	7	8	9	10
11	12	13	14	15	16	17	18	19	20
21	22	23	24	25	26	27	28	29	30
31	32	33	34	35	36	37	38	39	40
41	42	43	44	45	46	47	48	49	50
51	52	53	54	55	56	57	58	59	60
61	62	63	64	65	66	67	68	69	70
71	72	73	74	75	76	77	78	79	80
81	82	83	84	85	86	87	88	89	90
91	92	93	94	95	96	97	98	99	100

EXAMPLE   Find and Extend a Number Pattern

① **Identify a pattern in 15, 25, 35, 45, ▢. Then find the missing number.**

The pattern shows that 10 is added to each number.

15,   25,   35,   45,   ▢
   +10   +10   +10   +10

The missing number is 55.

2 **READING** Lakeisha is reading a book. If the pattern continues, how many pages will she read on Saturday?

Pages Lakeisha Read

Day	Pages
Monday	3
Tuesday	6
Wednesday	9
Thursday	12
Friday	15
Saturday	▨

Each day, Lakeisha reads 3 more pages than the day before.

3,   6,   9,   12,   15,   18

+3   +3   +3   +3   +3

So, Lakeisha will read 18 pages on Saturday.

Remember

When looking for a pattern, see how the next number changes from the one before it.

3 **SPORTS** Mia's bowling scores are 150, 145, 140, ▨, 130, ▨. Find the missing numbers in the pattern.

Notice that 5 is subtracted from each number.

150,   145,   140,   135,   130,   125

−5   −5   −5   −5   −5

The missing numbers are 135 and 125.

CHECK What You Know

Identify a pattern. Then find the missing numbers. See Examples 1–3 (pp. 17–18)

1. 10, 12, 14, 16, ▨, 20

2. 5, 10, 15, 20, ▨, 30

3. 20, ▨, 40, 50, ▨, 70

4. 110, 107, ▨, 101, 98, ▨

5. A track team runs 4 laps on Day 1, 6 laps on Day 2, and 8 laps on Day 3. The pattern continues. How many will they run on Day 5?

See Example 3 (p. 18)

6. **Talk About It** Suppose you start at 20 and skip count to 36. Is the pattern skip counting by 3s? Explain.

Identify a pattern. Then find the missing numbers. See Examples 1–3 (pp. 17–18)

7. 10, 14, 18, ▩, 26, 30

8. 13, 18, 23, ▩, 33, 38

9. 28, 24, 20, ▩, 12, 8

10. 63, 60, ▩, 54, 51, 48

11. 34, 36, ▩, 40, ▩, 44

12. 71, 76, 81, ▩, 91, ▩

13. 105, 100, ▩, 90, ▩, 80

14. 100, 110, 120, ▩, ▩

15. Each soccer player below has a number. If the pattern continues, what is Takisha's number?

Soccer Players' Numbers	
Name	**Number**
Kisho	3
Kayla	5
Michael	7
Lenora	9
Takisha	▩

16. Dillon is saving his allowance. How much money will he have saved at week 5? at week 10?

Dillon's Savings	
Week	**Total Saved**
1	$4
2	$8
3	$12
4	$16
5	▩

17. Elki draws 6 stars, 10 stars, 14 stars, and then 18 stars. If he continues the pattern, how many stars will he draw next?

18. Measurement A school bell rings at 8:15, 8:45, 9:15, and 9:45. If the pattern continues, when will the bell ring next?

H.O.T. Problems

19. OPEN ENDED Create a number pattern. Explain your pattern.

NUMBER SENSE Copy and complete. Use a hundred chart if needed.

20.
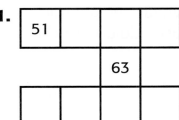

21.

22.

23. **WRITING IN ►MATH** Describe the pattern that can produce the numbers 104, 99, 94, 89, ... What number is next?

Problem-Solving Skill

MAIN IDEA I will use the four-step plan to solve problems.

Dina's family went to a zoo. They learned that a roadrunner is 1 foot tall. An African elephant is 12 feet tall. How much taller is an African elephant than a roadrunner?

Roadrunner

Understand	**What facts do you know?** • The roadrunner is 1 foot tall. • The African elephant is 12 feet tall. **What do you need to find?** • You need to find how much taller an African elephant is than a roadrunner.
Plan	To find out how much taller an African elephant is than a roadrunner, subtract.
Solve	$\begin{array}{r}12 \\ -\ 1 \\ \hline 11\end{array}$ ← height of elephant ← height of roadrunner So, the elephant is 11 feet taller than the roadrunner.
Check	Since addition and subtraction are inverse operations, you can use addition to check the subtraction. $\begin{array}{r}11 \\ +\ 1 \\ \hline 12\end{array}\qquad\begin{array}{r}12 \\ -\ 1 \\ \hline 11\end{array}$ So, the answer is correct.

Refer to the problem on the previous page.

1. Explain why you subtract 1 from 12 to find how much taller an elephant is than a roadrunner.

2. Suppose an elephant is 8 feet tall. How much shorter is a roadrunner?

3. Suppose a roadrunner is 3 feet tall. How much taller would an elephant be than the roadrunner?

4. Look back at Exercise 3. Check your answer. How do you know that it is correct? Explain.

PRACTICE the Skill

EXTRA **PRACTICE**
See page R2.

Solve. Use the four-step plan.

5. Cameron and Mara walk 2 blocks and then they turned a corner and walk 4 blocks. How many blocks do they need to walk to return to their starting place?

6. **Algebra** Find the missing numbers.

Input	16	■	24	28	32
Output	18	22	■	■	34

7. Rachel sold 4 glasses of lemonade. How much money did she make?

LEMONADE
25¢ per glass

8. Lola read a book that has 24 more pages than the book Fran read. Fran's book has 12 pages. How many pages are there in Lola's book?

9. **Algebra** If the pattern continues, what number will be the 6th and 7th number in the pattern?

2, 5, 8, 11, 14

10. Cortez bought a loaf of wheat, a loaf of rye, and a loaf of white bread. Gloria bought a loaf of raisin bread, a loaf of cinnamon, and a loaf of rye. How many different loaves of bread did they buy?

11. **Algebra** Darnell drew 4 pictures Monday. He drew 8 pictures Tuesday and 12 on Wednesday. If the pattern continues, how many pictures will he draw on Thursday?

12. **WRITING IN** ►**MATH** Explain how the four-step plan helps you solve a problem.

A **digit** is any symbol used to write whole numbers. The numbers (0, 1, 2, 3, 4, 5, 6, 7, 8, 9) are all digits. The **place value** of a digit tells what value it has in a number. Base-ten blocks can be used to explore place value.

MAIN IDEA

I will use models to explore place value through thousands.

You Will Need
base-ten blocks

Math Online

macmillanmh.com
• Concepts in Motion

ACTIVITY

1 **Use base-ten blocks to model 142 in two ways.**

One Way   **Use hundreds, tens, and ones.**

1 hundred   4 tens   2 ones

Another Way   **Use tens and ones.**

14 tens   2 ones

ACTIVITY

2 **Use base-ten blocks to model 1,025 in two ways.**

One Way **Use thousands, hundreds, tens, and ones.**

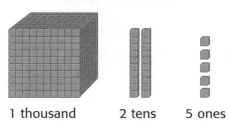

1 thousand   2 tens   5 ones

Another Way **Use hundreds, tens, and ones.**

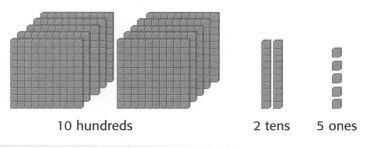

10 hundreds   2 tens   5 ones

Think About It

1. Why can you use different combinations of thousands, hundreds, tens, and ones to model the same number?

CHECK What You Know

Use base-ten blocks to model each number in two ways.

2. 135   **3.** 304   **4.** 1,283   **5.** 1,890

Write each number modeled.

6.
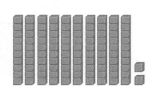

7.
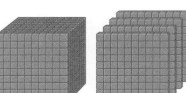

8. **WRITING IN** ►**MATH** Explain how base-ten blocks are helpful in understanding numbers.

MAIN IDEA

I will read, write, and identify place value of whole numbers through thousands.

New Vocabulary

digit
place value
standard form
expanded form
word form

Math Online

macmillanmh.com

• Extra Examples
• Personal Tutor
• Self-Check Quiz

GET READY to Learn

The Statue of Liberty recently celebrated her 120th birthday. The height from the top of the base to the top of the torch is 1,813 inches.

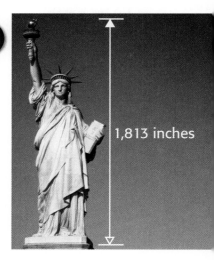

1,813 inches

The number 1,813 has four **digits**. A digit is any symbol used to write whole numbers. The **place value** of a digit tells what value it has in a number.

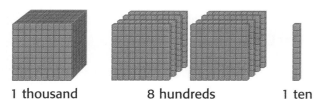

1 thousand   8 hundreds   1 ten   3 ones

A place-value chart can help you understand place value.

EXAMPLE   Identify Place Value

1. **Identify the place of the underlined digit in** <u>1</u>,813. **Then write the value of the digit.**

Thousands	Hundreds	Tens	Ones
1	8	1	3

↑ The value of the 1 is 1,000. $1 \times 1,000$

↑ The value of the 8 is 800. 8×100

↑ The value of the 1 is 10. 1×10

↑ The value of the 3 is 3. 3×1

The underlined digit, 1, is in the thousands place. Its value is 1,000.

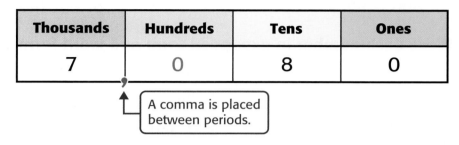
2 **STATUES** If ten people climb the stairs to the top of the Statue of Liberty and back down, they will have walked 7,080 steps. Identify the place of the underlined digit in 7,<u>0</u>80. Then write its value.

The place-value chart shows 7,080.

Thousands	Hundreds	Tens	Ones
7	0	8	0

A comma is placed between periods.

The underlined digit, 0, is in the hundreds place. Its value is zero. There are no hundreds. When 0 is used in a number, it is sometimes called a place holder.

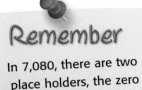

Remember

In 7,080, there are two place holders, the zero in the hundreds place and the zero in the ones place.

Numbers can be written in different ways. **Standard form** shows only the digits. **Expanded form** shows the sum of the value of the digits. **Word form** uses words.

Real-World **EXAMPLE**   Write Numbers

3 **TRAVEL** It is 1,215 miles from Mobile, Alabama to the Statue of Liberty in New York City. Write 1,215 in three ways.

The place-value chart shows 1,215.

Thousands	Hundreds	Tens	Ones
1	2	1	5

Standard Form: 1,215

Expanded Form: 1,000 + 200 + 10 + 5

Word Form: one thousand, two hundred fifteen

Write the place of the underlined digit. Then write the value of the digit. See Examples 1 and 2 (pp. 24–25)

1. 8̲70

2. 2̲,312

3. 7,5̲09

Write each number in standard form. See Example 3 (p. 25)

4. 800 + 50 + 6

5. one thousand, six hundred four

Write each number in expanded form and word form.
See Example 3 (p. 25)

6. 375

7. 5,230

8. 9,909

9. Lindsey uses each digit 3, 8, 0, and 1 once. Find the greatest whole number she can make.

10. (Talk About It) How do you tell the place value of each digit when given a number?

▶ Practice and Problem Solving

EXTRA PRACTICE
See page R2.

Write the place of the underlined digit. Then write the value of the digit. See Examples 1 and 2 (pp. 24–25)

11. 5̲01

12. 5,77̲2

13. 1,02̲0

14. 4,81̲0

15. 3̲,176

16. 8̲04

Write each number in standard form. See Example 3 (p. 25)

17. 4,000 + 600 + 70 + 8

18. 3,000 + 20 + 1

19. seven thousand, six hundred forty-one

20. eight thousand, seven hundred sixty

Write each number in expanded form and word form.
See Example 3 (p. 25)

21. 4,332

22. 1,324

23. 6,219

24. 6,503

25. 8,150

26. 1,001

27. A motorcycle costs $3,124. What is the value of each digit?

28. Write all of the three-digit numbers that have 3 in the tens place and a 5 in the hundreds place.

H.O.T. Problems

29. CHALLENGE Carlos is thinking of a four-digit number. The thousands digit is double the ones digit. The sum of these two digits is 9. What is the number? Explain your work.

30. FIND THE ERROR Julio and Caitlyn are writing 2,013 in word form. Who is correct? Explain your reasoning.

Julio
Two hundred thirteen

Caitlyn
Two thousand, thirteen

31. **WRITING IN MATH** Explain why a zero needs to be used when writing the number four thousand, sixty-eight in standard form.

TEST Practice

32. Marcus has 1,270 baseball cards. Which of these equals 1,270? (Lesson 1-3)

A $1 + 2 + 7 + 0$

B $100 + 20 + 7$

C $100 + 200 + 70$

D $1,000 + 200 + 70$

33. Which number is 7 more than 1,097? (Lesson 1-2)

F 1,004

G 1,104

H 2,004

J 2,007

Spiral Review

34. Josefina had 15 math problems for homework on Monday night. On Tuesday night, she had 18. How many math problems did she have in all? (Lesson 1-2)

Algebra Identify a pattern. Then find the missing numbers. (Lesson 1-1)

35. 19, ■, 23, ■, 27

36. 145, ■, 165, ■, ■, 195

1-4 Place Value through 10,000

GET READY to Learn

Scientists found that a gooney bird once traveled 24,983 miles in just 90 days. That is almost the distance around Earth.

Gooney bird

MAIN IDEA

I will read, write, and identify place value of whole numbers through ten thousands.

New Vocabulary

period

Math Online

macmillanmh.com
• Extra Examples
• Personal Tutor
• Self-Check Quiz

A place-value chart can be used to help read large numbers. A group of 3 digits is called a **period**. Commas separate the periods. At each comma, say the name of the period.

EXAMPLES Place Value

1 Identify the place of the underlined digit in <u>2</u>4,983. Then write its value.

The place-value chart shows 24,983.

Thousands Period			Ones Period		
hundreds	tens	ones	hundreds	tens	ones
	2	4	9	8	3

The underlined digit, 2, is in the ten thousands place. So, its value is 20,000.

2 Write 24,983 in three ways.

Standard Form: 24,983

Expanded Form: 20,000 + 4,000 + 900 + 80 + 3

Word Form: twenty-four thousand, nine hundred eighty-three

PLANETS While studying planets, Mario found a chart comparing the width of the three largest planets in our solar system.

The Solar System's Largest Planets

Saturn   72,368 miles
Uranus   31,518 miles
Jupiter   86,822 miles

Source: *The World Almanac*

Remember

Place a comma between the thousands and hundreds place.

③ Write the width of Uranus in expanded form.

31,518 = 30,000 + 1,000 + 500 + 10 + 8

④ Write the width of Jupiter in word form.

eighty-six thousand, eight hundred twenty-two

CHECK What You Know

Write the place of the underlined digit. Then write its value. See Example 1 (p. 28)

1. 62,57<u>4</u> **2.** 38,<u>0</u>35 **3.** <u>5</u>3,456 **4.** 1<u>2</u>,345

Write each number in standard form. See Example 2 (p. 28)

5. 50,000 + 1,000 + 300 + 3 **6.** twelve thousand, four

Write each number in expanded form and word form.

See Examples 2–4 (pp. 28–29)

7. 23,472 **8.** 49,602 **9.** 52,220 **10.** 71,002

11. A car's mileage is thirty-six thousand, five hundred twenty-three miles. Write this number in standard and expanded form.

12. **Talk About It** Dominic said that the number 61,903 is the same as 60,000 + 1,000 + 90 + 3. Is he correct? Explain.

Write the place of each underlined digit. Then write its value. See Example 1 (p. 28)

13. 15,3<u>8</u>8

14. 1<u>9</u>,756

15. 3<u>0</u>,654

16. <u>4</u>3,543

17. 57,08<u>1</u>

18. <u>6</u>9,003

19. 70,00<u>0</u>

20. 86,0<u>6</u>0

Write each number in standard form. See Example 2 (p. 28)

21. 20,000 + 4,000 + 200 + 20 + 2

22. 10,000 + 1,000 + 100 + 10 + 1

23. forty thousand, three hundred eighty

24. thirty-two thousand, twenty-five

Write each number in expanded form and word form.

See Examples 2–4 (pp. 28–29)

25. 12,194

26. 28,451

27. 39,234

28. 51,160

29. 60,371

30. 73,100

31. 81,001

32. 99,027

Data File

The table lists the location and altitude of the world's largest telescopes.

33. Which altitudes have a digit in the ten thousands place?

34. Write the altitude of the Palomar Mountain observatory in word form.

35. Which observatory's altitude has a digit with a value of 700?

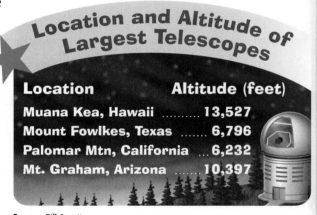

Location and Altitude of Largest Telescopes

Location	Altitude (feet)
Muana Kea, Hawaii	13,527
Mount Fowlkes, Texas	6,796
Palomar Mtn, California	6,232
Mt. Graham, Arizona	10,397

Source: Bill Arnett

H.O.T. Problems

36. OPEN ENDED Write three different numbers that have 5 in the thousands place.

37. **WRITING IN ►MATH** Explain the difference between standard form and expanded form.

Algebra Identify a pattern. Then find the missing numbers. (Lesson 1-1)

1. 20, ■, 60, 80, ■

2. 5, 15, ■, 35, ■

3. Hong has saved $37. He spends $19 on school clothes. He earns $15 for mowing the neighbor's yard. How much money does Hong have now? Use the four-step plan. (Lesson 1-2)

Write the place of each underlined digit. Then write the value of the digit. (Lesson 1-3)

4. 5<u>4</u>9

5. 3,<u>5</u>20

6. **MULTIPLE CHOICE** How is five thousand, three hundred nineteen written in standard form? (Lesson 1-3)

   A 5,193

   B 5,309

   C 5,319

   D 5,391

7. **Measurement** A hippopotamus at a zoo weighs 3,525 pounds. Write this number in expanded and word form. (Lesson 1-3)

Write the place of each underlined digit. Then write the value of the digit. (Lesson 1-4)

8. <u>1</u>6,846

9. <u>2</u>8,950

Write each number in standard form. (Lesson 1-4)

10. twenty-three thousand, seven hundred forty-two

11. 60,000 + 4,000 + 8

Write each number in expanded form. (Lesson 1-4)

12. Jennifer hopes to read 10,240 pages this summer.

13. Forty-five thousand, sixty-seven people.

14. **MULTIPLE CHOICE** Which digit is in the ten thousands place in the number 92,108? (Lesson 1-4)

   F 0   H 2

   G 1   J 9

15. **WRITING IN ►MATH** Describe a pattern that can produce the numbers shown below. What number is next? (Lesson 1-1)

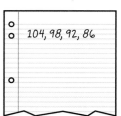

104, 98, 92, 86

Problem-Solving Investigation

MAIN IDEA I will use the four-step plan to solve a problem.

P.S.I. TEAM +

DERRICK: My sister gave me drawing paper for my birthday. There were 32 sheets. I want to make it last 8 days.

YOUR MISSION: Find how many sheets Derrick can use each day if he uses the same number of sheets each day.

Understand	There are 32 sheets of paper to last for 8 days. Find how many sheets he can use each day.
Plan	You know the total number of sheets of paper and how many days they need to last. You can show this using counters.
Solve	Use 32 counters to represent the 32 sheets of paper. Make 8 equal groups, placing the counters one at a time into each group until gone. 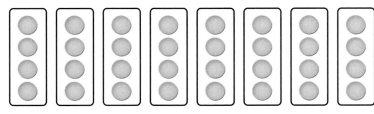 day 1 day 2 day 3 day 4 day 5 day 6 day 7 day 8 So, he can use 4 sheets of paper each day.
Check	Look back. Use addition to check the answer. $4 + 4 + 4 + 4 + 4 + 4 + 4 + 4 = 32$ So, the answer is correct.

Use the four-step plan to solve each problem.

1. **Measurement** Juan exercised 20 minutes yesterday. Today he is going to exercise twice as long. How long does Juan plan to exercise today? Explain.

2. **Algebra** What is the next figure in the pattern shown?

3. Mr. Hernandez brought pretzels to share with his 36 students. Only 18 of his students ate the pretzels. How many students did not eat the pretzels?

4. Marjorie baked 48 pancakes for the school breakfast. Elian ate some of the pancakes, and now Marjorie only has 43 pancakes. How many pancakes did Elian eat?

5. Gabriela buys the following items. She gives the cashier $20. How much change will she receive?

6. **Measurement** Joshua gets up at 8:30 A.M. He needs to leave for school by 9:00 A.M. How many minutes does he have to get ready?

7. **Geometry** Uncle Ramos is putting up a fence in the shape of a triangle. How much fencing is needed?

Side A	Side B	Side C
36 feet	half of side A	same as side A

8. Austin's garden has 5 rows of 6 plants. How many plants does Austin have in his garden?

9. Look at the table. How many pens do Cesar and Pamela have in all? How many more pens does Carmen have than Pamela?

Name	Pens
Pamela	7
Cesar	9
Carmen	20

10. Mrs. Reinhart read her students one book each day for 2 weeks. If there are 5 days in each school week, how many books did she read in all? Explain your reasoning.

11. **WRITING IN ►MATH** How is the plan step is different than the solve step in the four-step plan.

1-6 Compare Numbers

MAIN IDEA

I will compare numbers through ten thousands.

New Vocabulary

is less than (<)
is greater than (>)
is equal to (=)

Math Online

macmillanmh.com
• Extra Examples
• Personal Tutor
• Self-Check Quiz

The table lists the maximum speeds of two kinds of go-carts. Which go-cart is faster?

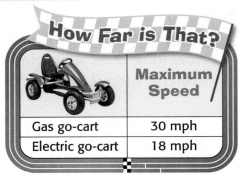

How Far is That?

	Maximum Speed
Gas go-cart	30 mph
Electric go-cart	18 mph

When comparing two numbers, the first number is either **less than**, **greater than**, or **equal to** the second number.

Symbol	Meaning
<	is less than
>	is greater than
=	is equal to

Real-World EXAMPLE   Use a Number Line

1 **MEASUREMENT** **Which go-cart is faster, the gas go-cart or the electric go-cart?**

You can use a number line to compare 30 and 18.

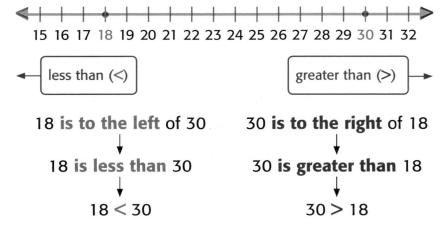

15 16 17 18 19 20 21 22 23 24 25 26 27 28 29 30 31 32

less than (<)   greater than (>)

18 is to the left of 30   30 is to the right of 18

18 is less than 30   30 is greater than 18

18 < 30   30 > 18

So, the gas go-cart is faster than the electric go-cart.

2 **MEASUREMENT** The Tyee family is planning a road trip to the Grand Canyon. One route from Laverne, Oklahoma is about 840 miles. A second route is about 835 miles. Which route is shorter?

Compare 835 and 840 to see which route is shorter.

Remember

Always line up the numbers by their place value and start to compare from the left.

Step 1 Line up the numbers by place-value.

Hundreds	Tens	Ones
8	3	5
8	4	0

Step 2 Compare. Start with the greatest place-value position.

Hundreds	Tens	Ones
8	3	5
8	4	0

same   different 3 tens < 4 tens

Since 3 is less than 4, the number 835 is less than 840. So, 835 < 840. The second route is shorter.

830 831 832 833 834 835 836 837 838 839 840

3 **MONEY** Which is greater, 987 coins or 1,400 coins?

You need to compare 1,400 and 987. Line up the numbers. Then compare.

Thousands	Hundreds	Tens	Ones
1	4	0	0
	9	8	7

1,400 has 1 thousand but 987 has 0 thousands

1 thousand is greater than 0 thousands. So, 1,400 > 987.

Compare. Use >, <, or =. See Examples 1–3 (pp. 34–35)

1. 60 ● 59

2. 88 ● 98

3. 100 ● 85

4. 64 ● 46

5. 1,000 ● 1,000

6. 27,345 ● 27,357

7. The Flips Gymnastics Club has 131 members. The Tumblers have 113 members. Which club has more members? Explain.

8. **Talk About It** Why is it not necessary to compare the ones digits in the numbers 4,365 and 4,378?

Practice and Problem Solving

EXTRA PRACTICE See page R3.

Compare. Use >, <, or =. See Examples 1–3 (pp. 34–35)

9. 55 ● 72

10. 99 ● 99

11. 70 ● 80

12. 93 ● 83

13. 121 ● 112

14. 657 ● 765

15. 303 ● 330

16. 998 ● 989

17. 8,008 ● 8,080

18. 23,753 ● 23,735

19. 7,654 ● 7,654

20. 9,999 ● 1,000

Algebra Compare. Use >, <, or =.

21. 65 ● 62 + 3

22. 35 + 4 ● 39

23. 209 ● 200 + 90

24. The table shows the number of tickets sold for a movie. Which showing sold more tickets?

Revenge of Dinosaurs	
Showing	**Tickets sold**
5:00 P.M.	235
7:00 P.M.	253

25. **Measurement** Which day was warmer in the desert, Tuesday or Wednesday?

Desert Temperature	
Day	**Temperature**
Tuesday	119°F
Wednesday	109°F

26. There are 165 students in the third grade. There are 35 students in each of the 5 classes in the second grade. Which has more students? Explain.

27. Keith's family bought a computer for $1,200. Margareta's family bought a computer for $1,002. Which computer costs less? Explain.

H.O.T. Problems

28. OPEN-ENDED Write the largest and smallest number you can make using the numerals 3, 6, 7, and 9.

29. WHICH ONE DOESN'T BELONG? Identify the number that is not more than 4,259.

| 4,295 | 4,260 | 4,300 | 4,209 |

30. WRITING IN ►MATH Explain the first step to comparing 2,032 and 203. Which number is greater? Explain.

TEST Practice

31. Which number will make the number sentence true? (Lesson 1-6)

$$1,426 > \blacksquare$$

A 1,425   C 1,452

B 1,426   D 1,524

32. Mrs. Phillips' class is having a pizza party. There are 30 students. Each pizza is cut into 10 pieces. If each student gets one piece, how many pizzas are there? (Lesson 1-5)

F 3   H 7

G 5   J 10

Spiral Review

Write each number in expanded form and word form. (Lesson 1-3)

33. 982   **34.** 2,045   **35.** 1,900

Write the place of the underlined digit. Then write the value of the digit. (Lesson 1-3)

36. 2<u>4</u>,981   **37.** 6,<u>0</u>79   **38.** 2,76<u>1</u>

39. Ricardo said the word form of 60,287 is six thousand, two hundred eighty-seven. Is this correct? Explain. (Lesson 1-2)

40. Identify the pattern of Byron's stamp collection. (Lesson 1-1)

1-7

Order Numbers

MAIN IDEA

I will use a number line and place value to order numbers through ten thousands.

Math Online

macmillanmh.com
• Extra Examples
• Personal Tutor
• Self-Check Quiz

GET READY to Learn

The table shows the length of three whales. Which whale is the longest? Which is the shortest?

Average Length of Whales

Whale	Length (inches)
Orca Whale	264
Blue Whale	1,128
Humpback Whale	744

Source: Advanced Technologies Academy

Comparing numbers can help you to order numbers.

Real-World EXAMPLE   Order Least to Greatest

1 **MEASUREMENT** Order the lengths from least to greatest.

One Way: Use a Number Line

$$264 < 744 < 1{,}128$$

Another Way: Use a Place-Value Chart

Line up the numbers by their place value. Compare from the left.

Thousands	Hundreds	Tens	Ones
	2	6	4
1	1	2	8
	7	4	4

1 thousand is the greatest number.

7 hundreds > 2 hundreds

The order is 264 inches, 744 inches, and 1,128 inches.

2 **MEASUREMENT** The table shows the distances whales travel to feed in the summertime. This is called migration. Order these distances from greatest to least.

Whale Migration	
Whale	Distance (miles)
Humpback Whale	3,500
Gray Whale	12,000
Orca Whale	900

Source: Island Marine Institute

Use the place value chart to line up the numbers by their place value. Compare from the left.

Ten Thousands	Thousands	Hundreds	Tens	Ones
	3	5	0	0
1	2	0	0	0
		9	0	0

12,000 is the greatest number.

3 thousands > no thousands so 3,500 is the next greatest number.

The order from greatest to least is 12,000 miles, 3,500 miles, and 900 miles.

Remember

When you move to the left on a number line, the numbers get smaller.

CHECK What You Know

Order the numbers from least to greatest. See Example 1 (p. 38)

1. 39; 32; 68

2. 224; 124; 441

3. 202; 2,202; 220

Order the numbers from greatest to least. See Example 2 (p. 39)

4. 231; 136; 178

5. 1,500; 150; 15

6. 9,009; 909; 6,999

7. Team A won 19 games, Team B won 40 games, and Team C won 22 games during the season. What place did each team earn for the season?

8. **Talk About It** Order these numbers from the greatest to the least: 435; 345; 3,453. Explain how you can tell which number is the greatest.

Order the numbers from least to greatest. See Example 1 (p. 38)

9. 303; 30; 3,003

10. 4,404; 4,044; 4,040

11. 39; 78; 123

12. 1,234; 998; 2,134

13. 598; 521; 3,789

14. 2,673; 2,787; 2,900

Order the numbers from greatest to least. See Example 2 (p. 39)

15. 60; 600; 6,006

16. 288; 209; 2,899

17. 49; 43; 60

18. 3,587; 875; 2,435

19. 451; 409; 415

20. 999; 1,342; 2,000

21. Carra's dad bought the three items shown below. Which item costs the most?

22. Kurt wants to buy a parrot, lizard, or hamster. Order the animals from the least to the most expensive.

23. Three elementary schools have 2,500 students, 3,002 students, and 2,536 students. Which is the least number of students?

24. In a set of numbers, 59 is the least number and 10,000 is the greatest. Write 4 ordered numbers that could come between these numbers.

Real-World PROBLEM SOLVING

Animals The lengths of three different whales are shown.

25. Order the lengths from greatest to least.

26. Which whale is the longest?

27. How much longer is the humpback whale than the orca whale?

Orca
← 30 ft →

Humpback
← 48 ft →

Gray
← 45 ft →

Source: World Wide Whale

H.O.T. Problems

28. FIND THE ERROR Juliana and Alex are ordering a set of numbers from least to greatest. Who is correct? Explain.

Alex
1,268
1,264
1,168

Juliana
1,168
1,264
1,268

29. NUMBER SENSE Between which two numbers will 567 be placed if we are placing the numbers 467; 980; 745 in order from greatest to least?

30. **WRITING IN ►MATH** Write a real-world problem that asks to order numbers from least to greatest.

TEST Practice

31. Which number sentence is false? (Lesson 1-6)

A 227 > 232

B 368 < 386

C 958 > 887

D 1,587 > 1,547

32. Which set of numbers is in order from greatest to least? (Lesson 1-7)

F 2,587; 3,610; 5,846; 8,745

G 1,587; 567; 987; 1,453

H 362; 542; 464; 558

J 268; 251; 158; 119

Spiral Review

Compare. Use >, <, or =. (Lesson 1-6)

33. 29 ● 38 **34.** 69 ● 58 **35.** 98 ● 85

36. Measurement Mrs. Garrison needs the longest string. Whose string does she need? (Lesson 1-5)

Student	Tracy	Nichelle	Collin
String	24 inches	36 inches	28 inches

Problem Solving in Geography

THE MIGHTY MISSISSIPPI

The Mississippi River is part of the largest river system in North America. The river begins in Minnesota and empties into the Gulf of Mexico. The Mississippi River system extends from the Rocky Mountains in the western United States to the Appalachian Mountains in the east.

The Mississippi River is about 2,340 miles long. The shallowest point is 3 feet. The deepest point is 198 feet. It's no wonder that the Mississippi River is called the "Mighty Mississippi."

MAJOR RIVERS OF THE MISSISSIPPI RIVER SYSTEM

River	Length (miles)
Arkansas	1,469
Mississippi	2,340
Missouri	2,540
Ohio	1,310
Red	1,290

Source: United States Geological Survey

Real-World Math

Use the information on page 42 to solve each problem.

1 Which river is the longest?

2 Which river lengths have the same value for the hundreds place? What is that value?

3 Write the length of the Arkansas River in expanded form.

4 The total of the lengths of the Missouri River and Mississippi River is 4,880 miles. How is this number written in words?

5 How does the length of the Red River compare to the lengths of the other 4 rivers? Use >, <, or = for each comparison.

6 Which is the third longest river?

7 Is the total length of the two longest rivers less than, greater than, or equal to the total length of the three shortest rivers?

8 Write the length of the Ohio River in word form.

9 What is the difference in the depths of the Mississippi from its shallowest point to its deepest point?

10 The Amazon River in South America is 3,920 miles long. Which river is longer, the Amazon or the Missouri?

Round to the Nearest Ten and Hundred

GET READY to Learn

Cassandra used 62 minutes of time on her family's cell phone plan. Her brother Matao used 186 minutes of time. About how many minutes did each person use?

To **round** is to change the value of a number to one that is easier to work with. You can use a number line to round.

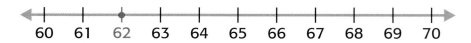

Real-World EXAMPLES Round to the Nearest Ten

1 **TECHNOLOGY** About how many minutes did Cassandra use? Round to the nearest ten.

The closest ten *less than* 62 is 60. The closest ten *greater than* 62 is 70. Use a number line from 60 to 70.

60 61 62 63 64 65 66 67 68 69 70

Since 62 is closer to 60 than to 70, round 62 to 60.

2 **TECHNOLOGY** About how many minutes did Matao use? Round to the nearest ten.

The closest ten *less than* 186 is 180. The closest ten *greater than* 186 is 190. Use a number line from 180 to 190.

180 181 182 183 184 185 186 187 188 189 190

Since 186 is closer to 190 than to 180, round 186 to 190.

You can also round numbers to the nearest hundred.

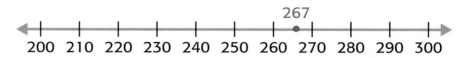
Real-World EXAMPLES

Vocabulary Link

round

Everyday Use in the form of a circle

Math Use to change the value of a number to one that is easier to work with

3 **BOOKS** **Bruno read a book that was 267 pages long. To the nearest hundred, how many pages did he read?**

The closest hundred *less than* 267 is 200. The closest hundred *greater than* 267 is 300.

```
        267
◄─┼──┼──┼──┼──┼──┼──┼─•┼──┼──┼──┼─►
 200 210 220 230 240 250 260 270 280 290 300
```

267 is closer to 300 than to 200. Round 267 to 300.

4 **SHELLS** **Olivia collected shells. To the nearest hundred, how many seashells did she collect?**

The closest hundred *less than* 1,423 is 1,400. The closest hundred *greater than* 1,423 is 1,500.

OLIVIA'S SEASHELL COLLECTION
1,423

```
        1,423
◄─┼────•┼───────┼───────┼───────┼─►
1,400   1,425   1,450   1,475   1,500
```

Since 1,423 is closer to 1,400 than to 1,500, round 1,423 to 1,400.

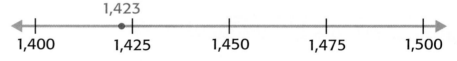
CHECK What You Know

Round to the nearest ten. See Examples 1 and 2 (p. 44)

1. 58   **2.** 62   **3.** 685   **4.** 552

Round to the nearest hundred. See Examples 3 and 4 (p. 45)

5. 449   **6.** 473   **7.** 415   **8.** 1,450

9. Kayla has to read 67 pages for homework tonight. To the nearest ten, how many pages does she need to read?

10. **Talk About It** What should you do to round a number that is exactly halfway between two numbers?

Round to the nearest ten. See Examples 1 and 2 (p. 44)

11. 77 **12.** 67 **13.** 13 **14.** 21

15. 285 **16.** 195 **17.** 157 **18.** 679

Round to the nearest hundred. See Examples 3 and 4 (p. 45)

19. 123 **20.** 244 **21.** 749 **22.** 750

23. 353 **24.** 850 **25.** 1,568 **26.** 4,829

27. Myron has 179 baseball cards. He says he has about 200 cards. Did he round the number of cards to the nearest ten or hundred? Explain.

28. Measurement A passenger train traveled 1,687 miles. To the nearest hundred, how many miles did the train travel?

29. Coco collected 528 cans of food for the school food drive. If she collects 25 more cans, what will the number of cans be, rounded to the nearest hundred?

30. Mrs. Boggs ran for mayor. She received 1,486 votes. Mrs. Swain received 1,252 votes. What is the difference in the number of votes to the nearest ten?

Real-World PROBLEM SOLVING

Sports Danilo is practicing bowling. The table shows his scores for one week.

31. Round all scores to the nearest hundred. Which days were the scores about 300?

32. To the nearest ten, what was the score on Tuesday?

33. Which day's score rounds to 250?

BOWLING SCORES

Monday	252
Tuesday	83
Wednesday	164
Thursday	256
Friday	290
Saturday	283
Sunday	173

H.O.T. Problems

34. OPEN ENDED I am thinking of a number that when it is rounded to the nearest hundred is 400. What is the number? Explain.

35. **WRITING IN ►MATH** Explain why 238 can be rounded to 240 or 200.

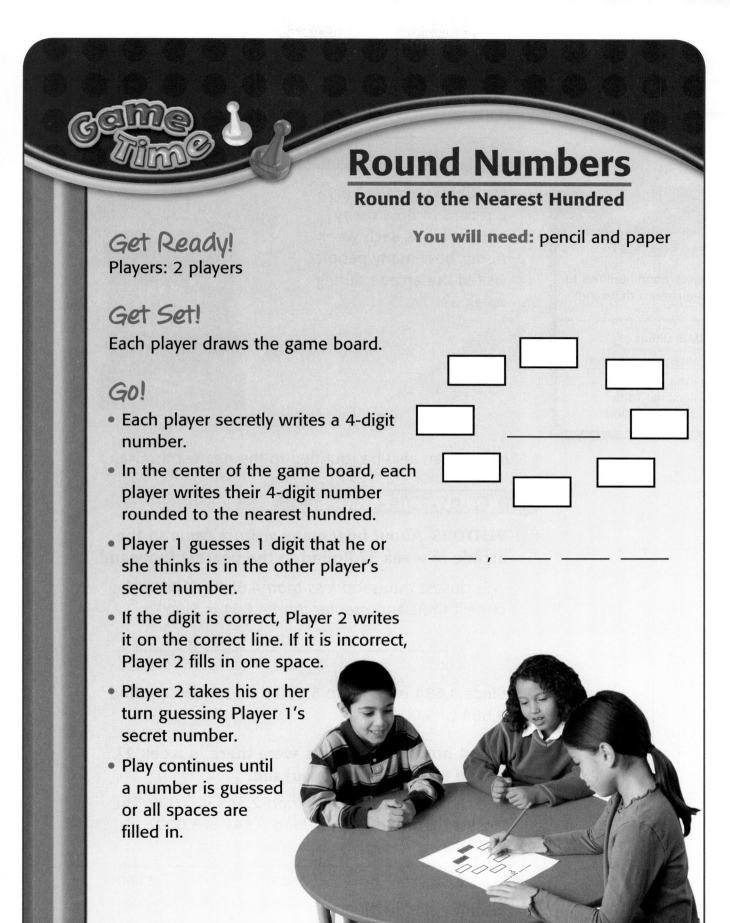

Round Numbers
Round to the Nearest Hundred

Get Ready!
Players: 2 players

You will need: pencil and paper

Get Set!
Each player draws the game board.

Go!
- Each player secretly writes a 4-digit number.

- In the center of the game board, each player writes their 4-digit number rounded to the nearest hundred.

- Player 1 guesses 1 digit that he or she thinks is in the other player's secret number.

- If the digit is correct, Player 2 writes it on the correct line. If it is incorrect, Player 2 fills in one space.

- Player 2 takes his or her turn guessing Player 1's secret number.

- Play continues until a number is guessed or all spaces are filled in.

1-9 Round to the Nearest Thousand

MAIN IDEA

I will round numbers to the nearest thousand.

Math Online

macmillanmh.com
- Extra Examples
- Personal Tutor
- Self-Check Quiz

GET READY to Learn

Mr. Chou's Arcade keeps a record of how many visitors it has each week. About how many people visited the arcade during week 3?

Mr. Chou's Arcade

Week	Number of Visitors
1	1,258
2	2,341
3	4,684
4	2,500
5	3,499

Numbers can also be rounded to the nearest thousand.

Real-World EXAMPLES Use a Number Line

1 **VISITORS About how many visitors came to the arcade in week 3? Round to the nearest thousand.**

The closest thousand *less than* 4,684 is 4,000. The closest thousand *greater than* 4,684 is 5,000.

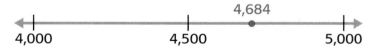

Since 4,684 is closer to 5,000 than to 4,000, round 4,684 to 5,000.

2 **About how many visitors were there in week 2? Round to the nearest thousand.**

The closest thousand *less than* 2,341 is 2,000. The closest thousand *greater than* 2,341 is 3,000.

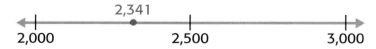

Round 2,341 to 2,000.

You can use rounding rules to round a number.

Rounding Whole Numbers — Key Concept

Step 1	Underline the digit to be rounded.
Step 2	Look at the digit to the right of the place being rounded.
Step 3	If the digit is 4 or less; do not change the underlined digit. If the digit is 5 or greater, add 1 to the underlined digit.
Step 4	Replace all digits after the underlined digit with zeros.

Real-World EXAMPLE   Use Rounding Rules

Remember

Use rounding rules for rounding in *all* place values.

③ ZOO A zoo had 5,499 visitors last week. To the nearest thousand, about how many people visited the zoo?

You need to round 5,499 to the nearest thousand.

Step 1 Underline the digit in the place to be rounded. In this case, the 5 is in the thousands place.   <u>5</u>,499

Step 2 Look at the 4, the digit to the right of the underlined digit.   <u>5</u>,499

Step 3 This digit is less than 5, so do not change the underlined digit.   <u>5</u>,499

Step 4 Replace all digits after the underlined digit with zeros.   <u>5</u>,000

To the nearest thousand, 5,499 rounds to 5,000.

```
        5,499
  ←——+————————•———————————+——→
  5,000        5,500        6,000
```

Round to the nearest thousand. See Examples 1–3 (pp. 48–49)

1. 3,922

2. 2,798

3. 7,099

4. 1,499

5. 2,500

6. 3,601

7. There are 1,250 houses in our city. Round the number of houses to the nearest thousand.

8. **Talk About It** Explain how you would use the rounding rules to round 5,299 to the nearest thousand.

Practice and Problem Solving

EXTRA PRACTICE
See page R4.

Round to the nearest thousand. See Examples 1–3 (pp. 48–49)

9. 8,611

10. 3,651

11. 1,099

12. 4,243

13. 2,698

14. 1,503

15. 1,257

16. 5,598

17. 5,299

18. 1,500

19. 2,400

20. 3,789

21. The fourth-grade class read a total of 2,389 pages this week. Round the number of pages read to the nearest thousand.

22. The attendance at a recent high school football game was 1,989. What is the attendance rounded to the nearest thousand?

23. To the nearest thousand, what will the cost be for the third grade to take a trip to the zoo?

24. Irene's scores on her favorite video game got better each day. What is her score on Wednesday rounded to the nearest thousand?

Third Grade Trip to The Zoo

$1,855

Video Game Scores	
Day	**Score**
Monday	1,735
Tuesday	2,200
Wednesday	2,585

25. Alton and his friends collected 1,683 rocks. How many rocks is that rounded to the nearest thousand?

26. **Measurement** Chong rode a train 2,156 miles one way. To the nearest thousand how many miles did he ride the train both ways?

H.O.T. Problems

27. NUMBER SENSE Describe all the four digit numbers that when rounded to the nearest thousand the result is 8,000.

28. WHICH ONE DOESN'T BELONG? Identify the number that is not rounded correctly to the nearest thousand. Explain.

| 2,184 → 2,000 | 5,500 → 5,000 | 3,344 → 3,000 | 8,456 → 8,000 |

29. WRITING IN ►MATH Round 499 to the nearest hundred. Then round 499 to the nearest ten. Are the two answers the same? Explain.

TEST Practice

30. Which number is 549 rounded to the nearest ten? (Lesson 1-8)

   A 500   **C** 540

   B 600   **D** 550

31. Margo rounded the number of beads in her craft set to 4,000. What number could be the exact number of beads? (Lesson 1-9)

   F 2,989   **H** 4,576

   G 3,576   **J** 5,004

Spiral Review

Round to the nearest ten. (Lesson 1-8)

32. 89   **33.** 319   **34.** 5,568   **35.** 8,728

Order the numbers from greatest to the least. (Lesson 1-7)

36. 1,234; 998; 2,134   **37.** 598; 521; 3,789   **38.** 2,673; 2,787; 2,900

39. Elias purchased the following items. He also bought a book about sports for $8. How much did he spend in all? (Lesson 1-5)

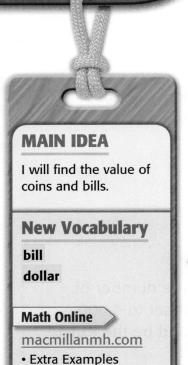

MAIN IDEA

I will find the value of coins and bills.

New Vocabulary

bill

dollar

Math Online

macmillanmh.com
• Extra Examples
• Personal Tutor
• Self-Check Quiz

GET READY to Learn

Allison used the coins shown to buy a salad. How much did she spend?

In the United States, money includes coins and bills.

Real-World EXAMPLE   **Find the Value of Coins**

1 **MONEY** How much did Allison spend on her salad?

Value of Coins			Key Concept
Penny			1¢ or $0.01
Nickel			5¢ or $0.05
Dime			10¢ or $0.10
Quarter			25¢ or $0.25
Half-Dollar			50¢ or $0.50

Allison used 2 quarters, 1 dime, 1 nickel, and 5 pennies.

Add the value of each coin. Start with the greatest value.

$25 + 25 + 10 + 5 + 1 + 1 + 1 + 1 + 1 = 70¢$

So, Allison spent 70 cents, 70¢, or $0.70. ¢ is read *cents*

You can also use the dollar sign ($) to write amounts of money. One **dollar** = 100¢ or 100 cents. It is also written as $1.00.

Real-World EXAMPLE Find the Value of Coins

2 **MONEY** **What is the value of the coins shown?**

There are 4 nickels, 2 quarters, 2 pennies, and 4 dimes.

First, put the coins in order from greatest to least in value. Use skip counting to add the values of each coin.

$0.25 $0.50 $0.60 $0.70 $0.80 $0.90 $0.95 $1.00 $1.05 $1.10 $1.11 $1.12

So, the value of the coins shown is $1.12.

Another name for paper money is **bill**. The unit is dollar ($).

Real-World EXAMPLE Find the Value of Bills and Coins

3 **MONEY** **What is the value of the money shown?**

The value of each bill is shown in the corners.

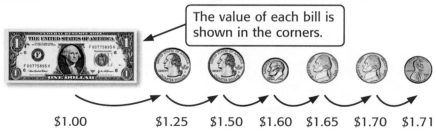

$1.00 $1.25 $1.50 $1.60 $1.65 $1.70 $1.71

$1.00 + $0.25 + $0.25 + $0.10 + $0.05 + $0.05 + $0.01 = $1.71

The value of the money shown is 1 dollar and 71 cents or $1.71.

Remember

Put the coins in order from the greatest to the least before you count the money.

Find the value of the coins. See Examples 1 and 2 (p. 52–53)

1.

2.

Find the value of the bills and coins. See Example 3 (p. 53)

3.

4.

5. Monique has 95¢. Which coins could she have?

6. 💬 **Talk About It** Is there more than one way to make 4¢? Explain.

Practice and Problem Solving

EXTRA PRACTICE
See page R5.

Find the value of the coins. See Examples 1 and 2 (p. 52–53)

7.

8.

9.

Find the value of the bills and coins. See Example 3 (p. 53)

10.

11.

12.

13.

14. Tara has 7 coins that equal $1.25. Which coins does she have?

15. Todd has 3 nickels and 2 dimes. Will he have enough money to buy a snack for $0.25? Explain.

16. Which coins could be used to make $2.00?

17. Which bills could be used to make $74.00?

H.O.T. Problems

18. OPEN ENDED Describe three combinations of coins and bills to come up with the same amount of money.

19. Challenge What is the least number of coins needed to make 99¢? Which coins are used?

20.  **MATH** Explain why a person might trade a dollar bill for four quarters.

**Practice**

21. What is the value of the coins?

    A $0.164   **C** $1.16

    B $0.86   **D** $1.21

22. What numeral means the same as 30,000 + 4,000 + 200 + 8?

    F 30,208   **H** 34,208

    G 30,280   **J** 34,280

Spiral Review

Round to the nearest ten. (Lesson 1-8)

23. 48   **24.** 82   **25.** 692

Order the numbers from greatest to least. (Lesson 1-7)

26. 902; 962; 692   **27.** 444; 333; 555   **28.** 1,645; 1,564; 1,465

Write each number in word, standard, and expanded form. (Lesson 1-4)

29. 937,026

30. 200,000 + 80,000 + 1,000 + 600 + 50 + 4

31. eight hundred thirty two thousand, six hundred one.

32. Algebra On day one, Kara did one cartwheel. On day two, she did three cartwheels. On day three, she did six. On day four, she did ten. On day five, she did fifteen. If the pattern continues, how many cartwheels will Karly do on day eight? (Lesson 1-1)

Study Guide and Review

FOLDABLES
Study Organizer **GET READY to Study**

Be sure the following Key Vocabulary words and Key Concepts are written in your Foldable.

Place Value Through 1,000   Place Value Through 10,000

Key Concepts

- A **pattern** is a sequence of numbers, or figures that follows a rule. (p. 17)

  5,   15,   25,   35,   45,   55
  + 10   + 10   + 10   + 10   + 10

- **Place value** is the value given to a digit by its place in a number. (p. 24)

Hundreds	Tens	Ones
9	8	3

- To compare numbers use **is less than** <, **is greater than** >, or **is equal to** =. (p. 34)

  46 < 50   46 is less than 50.

  125 > 89   125 is greater than 89.

  60 = 60   60 is equal to 60.

Key Vocabulary

bill (p. 52)

dollar (p.52)

is equal to (p. 34)

pattern (p. 17)

place value (p. 24)

round (p. 44)

Vocabulary Check

Choose the vocabulary word that completes each sentence.

1. One _____ is the same as 100¢.

2. Another name for paper money is _____.

3. The number 887 ___?___ eight hundred eighty-seven.

4. When you ___?___ 87 to the nearest 10, you get 90.

5. The order of the numbers 50, 60, 70, and 80 is an example of a ___?___.

6. The value of a digit in a number is its ___?___.

Lesson-by-Lesson Review

1-1 Algebra: Number Patterns (pp. 17–19)

Example 1
Identify a pattern in 140, 135, 130, ▪, 120, ▪. Then find the missing numbers.

The pattern shows that 5 is subtracted from each number.

140, 135, 130, ▪, 120, ▪
  −5 −5 −5 −5 −5

The missing numbers are 125 and 115.

Identify a pattern. Then find the missing numbers.

7. 85, ▪, 105, 115, ▪, 135

8. 200, 400, ▪, ▪, 1,000; 1,200

9. 120, 110, ▪, 90, 80, ▪

10. The first four numbers in a pattern are 27, 30, 33, and 36. If the pattern continues, what are the next four numbers?

1-2 Problem-Solving Skill: The Four-Step Plan (pp. 20–21)

Example 2

Estella runs 1 mile the first week, then doubles her miles each week after that. How many weeks will it take for her to run 8 miles?

The first week Estella ran 1 mile. She doubles that each week. Find how many weeks it will take to run 8 miles.

You can start with 1 and keep doubling it until you reach 8.

1 mile	Week 1
1 + 1 = 2 miles	Week 2
2 + 2 = 4 miles	Week 3
4 + 4 = 8 miles	Week 4

She reached 8 miles by week 4.

Solve each problem.

11. Raini wants a bike that costs $65. Raini's father will match any amount of money Raini saved. Raini has $30. With his father's help, can he buy the bike? Explain.

12. Vincent played soccer for 3 seasons. Mitchell has played for 3 years. If there are 2 seasons each year, who has played more seasons? Explain.

13. Bo brought 25 pencils to school the first week. He used 5 the first week and 7 the second week. How many are still unused?

1-3 **Place Value Through 1,000** (pp. 24–27)

Example 3
Write 3,456 in expanded form and word form.

The number 3,456 is written in standard form.

Standard Form: 3,456

Expanded Form: 3,000 + 400 + 50 + 6

Word Form: three thousand, four hundred fifty-six

Write each number in expanded form and word form.

14. 4,013 **15.** 6,047

Write each number in standard form.

16. 7,000 + 600 + 20 + 2

17. 4,000 + 50 + 6

18. one thousand, two hundred three

19. two thousand eight hundred seventy-five

1-4 **Place Value Through 10,000** (pp. 28–30)

Example 4
Write the place of the underlined digit in 23,456. Then write its value.

The place-value chart shows 23,456

Thousands Period			Ones Period		
hundreds	tens	ones	hundreds	tens	ones
	2	3	4	5	6

The underlined digit, 2, is in the ten thousands place. So, its value is 20,000.

Write the place of each underlined digit. Then write its value.

20. 46,887 **21.** 63,004

Write each number in expanded form and word form.

22. 60,457 **23.** 54,010

Write each number in standard form.

24. 80,000 + 7,000 + 400 + 3

25. forty-seven thousand, nine hundred seventy-one

1-5 Problem-Solving Investigation: Use the Four-Step Plan (pp. 32–33)

Example 5
Bart lives 30 miles from a water park. Clint lives 25 miles more than Bart from the same water park. How many miles does Clint live from the water park.

Understand
You know that Bart lives 30 miles from the park. Clint lives 25 miles further than Bart. You want to find out how many miles Clint lives from the water park.

Plan
You can use addition to find the total.

Solve
Add the distance Bart lives from the water park and how much further Clint lives from the park.

$$\begin{array}{r} 30 \\ + \ 25 \\ \hline 55 \end{array}$$ distance Bart lives
distance further Clint lives
total distance Clint lives

So, Clint lives 55 miles away from the water park.

Check
Look back at the problem. Check by subtracting.

$$\begin{array}{r} 55 \\ - \ 25 \\ \hline 30 \end{array}$$

The answer is correct.

Use the four-step plan to solve each problem.

26. **Algebra** Garrett has twice as many coins as Luke. Luke has 12. How many coins do they have?

27. For each coupon book Julie sells, she earns 100 points. If she sold 4 books last week and 5 this week, does she have enough points for an 800-point prize? Explain.

28. **Measurement** Mr. Jonas needs to put a fence around part of his yard for his dog. How many feet of fence will he need?

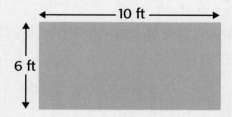

29. Mrs. Cassady made tea for her grandchildren. She used 3 tea bags for one pitcher. If she makes 4 pitchers, how many tea bags will she use?

30. Star gave each of her 6 friends 5 pieces of paper. She kept the rest of the paper. The pack now has 70 pieces of paper left. How much did she have to begin with?

1-6 **Compare Numbers** (pp. 34–37)

Example 6
Compare 679 ● 686.

You can use a number line.

<----+----+----+----+----+----+----+---->
679 680 681 682 683 684 685 686

less than   greater than

679 is to the left of 686

679 is less than 686

679 < 686

Compare. Use >, <, or =.

31. 2,045 ● 2,405

32. 201 ● 1,020

33. 10,567 ● 10,657

34. 5,801 ● 8,051

35. A school sold 235 tickets for the 3rd grade play. There were 253 tickets sold for the 4th grade play. Which play had more people? Explain.

1-7 **Order Numbers** (pp. 38–41)

Example 7
Order the numbers from least to greatest.

7,541; 5,300; 6,108

Use a place-value chart to compare.

Thousands	Hundreds	Tens	Ones
7	5	4	1
5	3	0	0
6	1	0	8

5,300 < 6,108 < 7,541

Order the numbers from least to greatest.

36. 36,201; 35,201; 36,102

37. 450; 540; 405

Order the numbers from greatest to least.

38. 89,554; 98,554; 87,554

39. 603; 630; 306

40. Explain how you know which number is the greatest without comparing the value of the digits.

535; 354; 4,435

Round to the Nearest Ten and Hundred (pp. 44–46)

Example 8
MEASUREMENT The students played on the playground for 78 minutes. To the nearest ten, about how many minutes is this?

The closest ten *less than* 78 is 70. The closest ten *greater than* 78 is 80.

Since 78 is closer to 80 than 70, round 78 to 80.

Example 9
Jordan has 236 toy cars. To the nearest hundred, about how many toy cars does he have?

The closest hundred *less than* 236 is 200. The closest hundred *greater than* 236 is 300.

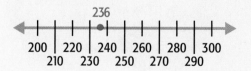

Since 236 is closer to 200 than 300, round 236 to 200.

Round to the nearest ten.

41. 56   **42.** 32

43. 801   **44.** 652

Round to the nearest hundred.

45. 569   **46.** 858

47. 1,593   **48.** 1,263

49. Coty has 465 marbles. Rounded to the nearest hundred, how many does he have?

50. Anita read 39 pages. Lisa read 33 pages. Rounded to the nearest ten, how many did they each read?

51. Raul says he has 200 army men when he rounds the total to the nearest hundred. How many men could Raul have? Explain how you know.

52. This is May's rock collection. What is the total number of rocks in her collection to the nearest ten?

1-9 **Round to the Nearest Thousand** (pp. 48–51)

Example 10
Round 5,256 to the nearest thousand.

The closest thousand less than 5,256 is 5,000. The closest thousand greater than 5,256 is 6,000.

5,256

5,000   5,500   6,000

Since 5,256 is closer to 5,000 than to 6,000, round 5,254 to 5,000.

Round to the nearest thousand.

53. 4,509   **54.** 4,905

55. 3,980   **56.** 3,089

57. Gayla found the receipt below. What is the total amount spent to the nearest thousand?

The Sports Store	
Treadmill	$ 2,500
Weight set	$ 2,000
Volleyball set	$ 150
Total	$ 4,650
CUSTOMER COPY	

1-10 **Value of Coins and Bills** (pp. 52–55)

Example 11
Catherine used the money shown to buy a journal. How much did she spend?

Add the value of each coin. Start with the greatest value.

$1.00 + $0.25 + $0.25 + $0.25 + $0.25 + $0.10 + $0.10 + $0.05 = $2.25

So, Catherine spent $2.25, or 2 dollars and 25 cents.

Find the value of the bills and coins.

58.

59.

60. Kathy has 8 coins that equal $2. Which coins does she have?

CHAPTER 1

Chapter Test

For Exercises 1 and 2, tell whether each statement is *true* or *false*.

1. The number 3,578 is written in standard form.

2. Expanded form is a way to write a number in words.

Algebra Identify a pattern. Then find the missing number.

3. 30, ▪, 50, 60, ▪

4. 5, 10, ▪, 20, ▪

Identify the place of the underlined digit. Then write its value.

5. <u>3</u>,720

6. 5<u>2</u>9

7. **Measurement** Darlene noticed that the meter on her family's new car showed they have driven two thousand, eight hundred eighteen miles so far. How is that number written in standard form?

Write each number in expanded form and word form.

8. 6,191

9. 19,804

10. **MULTIPLE CHOICE** How is four thousand, three hundred twenty-one written in standard form?

A 3,421

C 4,231

B 4,021

D 4,321

11. Find the value of the coins.

Compare. Use >, <, or =.

12. 8,415 ● 8,541

13. 500 + 80 + 9 ● 589

14. Order from least to the greatest.

4,804; 4,408; 8,440

15. Order the number of baskets from least to greatest.

Career Baskets	
Player	**Baskets**
Roz	2,308
Marquez	2,803
Amada	2,083

Round each number to the nearest ten, hundred, and thousand.

16. 2,942

17. 9,267

18. **MULTIPLE CHOICE** Which digit is in the thousands place in the number 92,108?

F 1

H 8

G 2

J 9

19. **WRITING IN** ►**MATH** Give an example of when it would be appropriate to round numbers.

TEST Example

A pet shop sold 1,372 turtles. Which of these equals 1,372?

A $1 + 3 + 7 + 2$

B $1 + 30 + 70 + 2000$

C $100 + 300 + 70 + 2$

D $1000 + 300 + 70 + 2$

TEST-TAKING TIP

You can use a number line to help you find numbers and compare numbers.

Read the Question

You need to find which equals 1,372.

Solve the Question

You can use a place-value chart to find the value of each digit in 1,372.

$1,372 = 1000 + 300 + 70 + 2$.
So, the answer is D.

Thousands	Hundreds	Tens	Ones
1	3	7	2

PART 1 Multiple Choice

Read each question. Then fill in the correct answer on the answer sheet provided by your teacher or on a sheet of paper.

1. Which point represents 415?

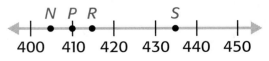

A N

B P

C R

D S

2. Which of the following is three hundred forty-two?

F 234

G three hundred twenty-four

H 342

J $300 + 40 + 2$

3. What number is 3,737 rounded to the nearest ten?

A 3,740   **C** 3,800

B 3,780   **D** 4,000

4. What is this number in standard form?

Thousands Period			Ones Period		
hundreds	tens	ones	hundreds	tens	ones
		1	3	4	2

F 1,432   **H** 1,234

G 1,342   **J** 132

5. What is the next number in the pattern 4, 10, 16, 22, 28, 34, ____?

A 38   **C** 42

B 40   **D** 44

6. What is the value of the money?

F $1.60   **H** $1.70

G $1.65   **J** $1.75

PART 2   Short Response

Record your answer on the answer sheet provided by your teacher or on a sheet of paper.

7. Leonardo has 158 baseball cards in his collection. Write 158 in expanded form.

8. What is 6,639 rounded to the nearest thousand?

9. Tell the value of each digit in the number 17,523.

PART 3   Extended Response

Record your answer on the answer sheet provided by your teacher or on a sheet of paper.

10. Bertram played with 5 toy sailboats in the pool. He gave 2 to his friend to use. How many sailboats does Bertram have now? Explain how you would use the four-step plan to solve this problem.

NEED EXTRA HELP?										
If You Missed Question...	1	2	3	4	5	6	7	8	9	10
Go to Lesson...	1-7	1-3	1-8	1-3	1-1	1-10	1-3	1-9	1-4	1-2

Add To Solve Problems

BIG Idea **When will I use addition?**

Addition is helpful when you want to buy something.

Example Renato wants to buy the skateboard items shown. What is the total cost of the items?

What will I learn in this chapter?

- Use the properties of addition.
- Add money.
- Estimate sums.
- Add two-digit and three-digit numbers.
- Solve problems by deciding if an estimate or exact answer is needed.

Key Vocabulary

Commutative Property of Addition

Identity Property of Addition

Associative Property of Addition

regroup

estimate

Math Online **Student Study Tools at** macmillanmh.com

FOLDABLES
Study Organizer

Make this Foldable to help you organize information about addition. Begin with one sheet of $8\frac{1}{2}'' \times 11''$ paper.

1. **Fold** the sheet of paper as shown.

2. **Fold** again in half as shown.

3. **Unfold** and cut along the two inside valley folds.

4. **Label** as shown. Record what you learn.

| Two-Digit Addition | Add Money |
| Estimate Sums | Three-Digit Addition |

You have two ways to check prerequisite skills for this chapter.

Option 2

Math Online ⟩ Take the Chapter Readiness Quiz at macmillanmh.com.

Option 1

Complete the Quick Check below.

QUICK Check

Add. (Prior grade)

1. 5
   + 4

2. 6
   + 7

3. 3
   + 9

4. 7
   + 7

5. 9 + 2

6. 4 + 6

7. 8 + 3

8. 9 + 8

9. Percy swam 8 laps today and 4 laps yesterday. How many laps did he swim in 2 days?

Find each sum. (Prior grade)

10.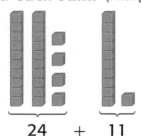

   24 + 11

11.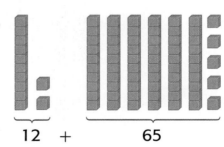

   12 + 65

Round to the nearest ten. (Lesson 1-8)

12. 72

13. 19

14. 59

15. 89

Round to the nearest hundred. (Lesson 1-8)

16. 470

17. 771

18. 301

19. 149

20. 99

21. 505

22. 77

23. 661

Algebra: Addition Properties

GET READY to Learn

Sal has 2 pieces of quartz and 3 pieces of granite. Ruby has 3 pieces of quartz and 2 pieces of granite. They both have the same number of rocks.

$$2 + 3 = 3 + 2$$

MAIN IDEA

I will use addition properties to add whole numbers.

New Vocabulary

Commutative Property of Addition

Identity Property of Addition

Associative Property of Addition

Math Online

macmillanmh.com
• Extra Examples
• Personal Tutor
• Self-Check Quiz

In math, properties are rules you can use with numbers.

Addition Properties	Key Concept

Models

$$\boxed{\ } + \boxed{\ } = \boxed{\ }$$

Examples $2 + 3 = 5$ $3 + 2 = 5$

Words **Commutative Property of Addition** The order in which the numbers are added does not change the sum.

Examples $3 + 0 = 3$ $0 + 3 = 3$

Words **Identity Property of Addition** The sum of any number and zero is the number.

Examples $(3 + 2) + 4 =$ $3 + (2 + 4) =$
  $5 + 4 =$ $3 + 6 =$
  9 9

Words **Associative Property of Addition** The way the addends are grouped does not change the sum.

EXAMPLE Use Properties to Add

1 Find the sum 4 + 5 = ■ and 5 + 4 = ■.
Identify the property.

4 + 5 = ■ and 5 + 4 = ■ ◀ | The sum is 9. The order in which the numbers are added does not change the sum.

This is the Commutative Property of Addition.

Real-World EXAMPLE Use the Associative Property to Add

Remember
Parentheses tell you which numbers to add first.

2 ANIMALS A zoo has 4 owl chicks, 2 cheetah cubs, and 6 lion cubs. How many baby animals are at the zoo?

You need to find 4 + 2 + 6. Rearrange the numbers so they are easier to add.

4 + 2 + 6

= 2 + 4 + 6 ◀ | Commutative Property of Addition.

= 2 + (4 + 6) ◀ | Associative Property of Addition. The grouping of the addends does not change the sum.

= 2 + 10

= 12

So, there are 12 baby animals.

CHECK What You Know

Find each sum. Identify the property. See Examples 1 and 2 (p. 70)

1. 6 + 5 = ■
5 + 6 = ■

2. (5 + 7) + 3 = ■
5 + (7 + 3) = ■

3. 0 + 12 = ■

4. Algebra Write a number sentence to show how many shells were collected. Which property did you use?

Seashell Collection			
Day	Friday	Saturday	Sunday
Shells	6	7	4

5. **Talk About It** Describe how you can use the Commutative and Associative Properties of Addition to add 7, 8, and 3.

Find each sum. Identify the property. See Examples 1 and 2 (p. 70)

6. $0 + 9 = \blacksquare$

7. $9 + 2 = \blacksquare$
$2 + 9 = \blacksquare$

8. $(2 + 5) + 8 = \blacksquare$
$2 + (5 + 8) = \blacksquare$

9. $2 + 8 = \blacksquare$
$8 + 2 = \blacksquare$

10. $100 + 0 = \blacksquare$

11. $4 + (6 + 3) = \blacksquare$
$(4 + 6) + 3 = \blacksquare$

Algebra **Find each missing number. Identify the property.**

12. $6 + \blacksquare = 6$

13. $(7 + 9) + 3 = (9 + \blacksquare) + 3$

14. $9 + \blacksquare = 2 + 9$

15. $(8 + 3) + \blacksquare = 8 + (3 + 2)$

Find each sum mentally.

16.
$\begin{array}{r} 1 \\ 7 \\ + 9 \\ \hline \end{array}$

17.
$\begin{array}{r} 5 \\ 7 \\ + 5 \\ \hline \end{array}$

18.
$\begin{array}{r} 4 \\ 2 \\ 6 \\ + 2 \\ \hline \end{array}$

19.
$\begin{array}{r} 2 \\ 1 \\ 9 \\ + 3 \\ \hline \end{array}$

Solve.

20. Necie has 3 dogs. Simona has 5 fish and 6 birds. Peyton has 1 snake. How many pets do the children have?

21. Luis drew the picture below. Write two number sentences that would be examples of the Associative Property of Addition.

22. Mrs. Jackson bought 6 blue, 2 red, and 2 yellow notebooks. There are 7 notebooks left on the store's shelf. How many were there to begin with?

H.O.T. Problems

23. OPEN ENDED Describe three different ways to find the sum of $7 + 9 + 3$. Which properties of addition did you use? Explain the way that you find easiest.

24. **WRITING IN** ►**MATH** Is there a Commutative Property of Subtraction? Explain.

Problem-Solving Skill

MAIN IDEA I will decide whether an estimate or an exact answer is needed to solve a problem.

To celebrate Arbor Day, a town planted trees one weekend. On Saturday, 53 trees were planted. Another 38 trees were planted on Sunday. About how many trees were planted in all?

Understand	**What facts do you know?** • On Saturday, 53 trees were planted. • On Sunday, 38 trees were planted. **What do you need to find?** • Find *about* how many trees were planted in all.
Plan	You need to decide whether to estimate or find an exact answer. Since the question asks *about* how many trees were planted, you need to estimate.
Solve	• First, find about how many trees were planted each day. Estimate by rounding to the closest ten. 53 → 50 ← [Round 53 to 50.] 38 → 40 ← [Round 38 to 40.] • Then, add. $$\begin{array}{r} 50 \\ +40 \\ \hline 90 \end{array}$$ So, about 90 trees were planted in all.
Check	Look back. If the question asked for an exact answer you would find $53 + 38 = 91$. The estimate is close to the exact answer. So, the estimate makes sense.

Refer to the problem on the previous page.

1. How do you know when to find an estimate or an exact answer?

2. Describe a situation when an exact answer is needed.

3. Would under-estimating ever cause a problem? Explain.

4. Explain one reason why only an estimate is needed for the number of trees planted.

PRACTICE the Skill

EXTRA PRACTICE
See page R5.

Tell whether an estimate or an exact answer is needed. Then solve.

5. During a career day, the students gave an author stories they wrote. How many stories were written?

Student Stories	
2nd grade	26
3rd grade	35

6. **Measurement** Kishi cut 2 lengths of rope. One was 32 inches long. The other was 49 inches long. Will he have enough rope for a project that needs 47 inches and 29 inches of rope? Explain.

7. There are enough seats for 60 students on the bus. Can all 32 boys and 26 girls ride the bus? Explain.

8. The number 7 septillion has 24 zeros after it. The number 7 octillion has 27 zeros after it. How many zeros is that altogether?

9. **Measurement** If each tablespoon of mix makes 1 glass of lemonade, will 96 ounces be enough for 15 glasses of lemonade? Explain.

Lemonade Directions	
Water	**Mix**
32 ounces	4 tablespoons
64 ounces	8 tablespoons
96 ounces	12 tablespoons

10. The directions on a treasure map told Rosaline to walk 33 paces forward. Then she was to turn right and walk 15 paces. How many paces does she need to walk altogether?

11. **WRITING IN ►MATH** Write two real-world problems. One should involve estimation and the other should involve an exact answer.

Estimate Sums

GET READY to Learn

The students at Glenwood Elementary School had an art show. The number of visitors is shown. About how many people visited the art show over the two days?

Art Show

Visitors
Friday 47
Saturday 34

MAIN IDEA

I will estimate sums using rounding and compatible numbers.

New Vocabulary

estimate

compatible numbers

Math Online

macmillanmh.com
• Extra Examples
• Personal Tutor
• Self-Check Quiz

The word *about* means that you do not need an exact answer. You can estimate. When you **estimate**, you find an answer that is close to the exact answer. You can use rounding to estimate.

Real-World EXAMPLE Estimate by Rounding

① **SCHOOL About how many people in all visited the art show on Friday and Saturday?**

To find the total, find 47 + 34. Since, the question says *about* how many people, estimate 47 + 34.

Step 1 Round each number to the nearest ten.

$$47 \longrightarrow 50$$
$$34 \longrightarrow 30$$

Round 47 to 50.
Round 34 to 30.

Step 2 Add.

$$47 \longrightarrow 50$$
$$+ 34 \longrightarrow + 30$$
$$\overline{ 80}$$

Use mental math to add. Since 5 + 3 = 8, 50 + 30 = 80.

So, *about* 80 people visited the art show.

Remember

See Lesson 1-8 to review rounding of whole numbers.

 **Real-World EXAMPLE**   Estimate by Rounding

2 **BAGELS** Mrs. Cruz bought 36 bagels and 32. **About how many bagels did Mrs. Cruz buy?**

You need to estimate 36 + 32.

$$
\begin{array}{rll}
36 & \longrightarrow & 40 \\
32 & \longrightarrow & +\ 30 \\
\hline
& & 70
\end{array}
$$

Round 36 to 40.
Round 32 to 30.

So, Mrs. Cruz bought *about* 70 bagels.

You can use **compatible numbers** to estimate. Compatible numbers are numbers that are easy to add.

EXAMPLES   Estimate by Using Compatible Numbers

3 **Estimate 12 + 39.**

Numbers ending in 0 are easy to add.

12 ⟶ 10
39 ⟶ 40

10 + 40 = 50

So, 12 + 39 is *about* 50.

4 **Estimate 73 + 23.**

The numbers 25, 50, 75, and 100 are easy to add.

73 ⟶ 75
23 ⟶ 25

 = $1

    75   + 25 = 100

So, 73 + 23 is *about* 100.

CHECK What You Know

Estimate each sum using rounding. See Examples 1 and 2 (pp. 74–75)

1.   31
    + 57

2.   38
    + 59

3.   35
    + 28

Estimate each sum using compatible numbers. See Examples 3 and 4 (p. 75)

4. 43 + 56   **5.** 91 + 94   **6.** 52 + 17

7. A movie theater will show 53 movies this week and 45 movies next week. About how many movies will they show in the two weeks?

8. **Talk About It** Look back at Exercise 7. How could it be rewritten so an exact answer is needed?

Estimate each sum using rounding. See Examples 1 and 2 (pp. 74–75)

9. 64
     + 34

10. 75
     + 11

11. 56
     + 22

12. 13
     + 39

13. 81
     + 10

14. 23
     + 25

15. 11 + 72

16. 49 + 20

17. 18 + 41

Estimate each sum using compatible numbers.

See Examples 3 and 4 (p. 75)

18. 23
     + 28

19. 94
     + 14

20. 80
     + 15

21. 33 + 37

22. 80 + 89

23. 11 + 72

24. 48 + 29

25. 91 + 14

26. 13 + 31

27. About how many racers were in the Summer Fun Race?

Summer Fun Race		
Start time	**Group**	**Entrants**
9:00 A.M.	runners	79
10:00 A.M.	race walkers	51

FINISH

28. What would be a reasonable estimate for attendance at the school fair?

School Fair Attendance	
Saturday	**Sunday**
62	92

ADMIT ONE   ADMIT ONE

29. Noshie made 2 bunches of balloons. One bunch had 9 blue and 12 yellow balloons. The other bunch had 14 red and 16 yellow balloons. About how many yellow balloons were there in all?

30. Look at the table below. Team A has 112 soccer players. Team B has 74 soccer players. Match the number of boys and girls to each team.

Team A and Team B Players	
Boys	**Girls**
55	33
41	57

31. Measurement Two walls of a room measure 21 feet each, and the other two measure 26 feet each. Estimate the total length of all four walls.

H.O.T. Problems

32. OPEN ENDED Using the digits 1, 2, 3, and 4 once, write two 2-digit numbers whose estimated sum is less than 50.

33. FIND THE ERROR Ed and Jayden are estimating 26 + 47. Who is correct? Explain your reasoning.

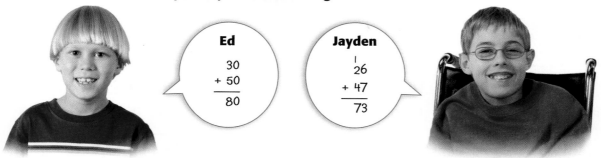

Ed

$$\begin{array}{r} 30 \\ + 50 \\ \hline 80 \end{array}$$

Jayden

$$\begin{array}{r} \overset{1}{2}6 \\ + 47 \\ \hline 73 \end{array}$$

34. WRITING IN ▸MATH Write about a real-world situation where using estimation would not be a good way to find a sum.

TEST Practice

35. For the party, Evelina made 39 celery sticks and 58 egg rolls. About how many snacks does she have for the party? (Lesson 2-2)

  A 60   **C** 90

  B 70   **D** 100

36. Mr. Moseley will plant 11 flowers in each of his 4 window boxes. About how many flowers does he need to buy? (Lesson 2-3)

  F 11   **H** 30

  G 20   **J** 40

Spiral Review

Algebra Find each missing number. Identify the property. (Lesson 2-1)

37. $(8 + 4) + 7 = \blacksquare$
   $\blacksquare + (4 + 7) = 19$

38. $25 + \blacksquare = 25$

39. $9 + \blacksquare = 16$
   $7 + \blacksquare = 16$

40. Describe two combinations of coins that equal 83¢. (Lesson 1-10)

41. Measurement Miss Sylvia drove 7 miles from home to the store. Then, she drove another 16 miles to work. At the end of the day, she drove the same path home. How many miles did she drive that day? (Lesson 1-2)

2-4 Two-Digit Addition

Math Online
macmillanmh.com

MAIN IDEA

I will regroup ones to add two-digit numbers.

New Vocabulary

regroup

Math Online

macmillanmh.com
• Extra Examples
• Personal Tutor
• Self-Check Quiz

GET READY to Learn

 Hands-On Mini Activity

Step 1
Use base-ten blocks to model 28 + 7.

Tens	Ones

Step 2
Add the ones. Regroup 10 ones as 1 ten.

Tens	Ones

1. What is 28 + 7?
2. Explain when you need to regroup.
3. How would you use regrouping to find 13 + 9?

When you add, you sometimes need to regroup. **Regroup** means to rename a number using place value.

Real-World EXAMPLE Add With Regrouping

1. **GAMES** Gaspar has 8 game tokens. His brother has 24 tokens. How many tokens do they have in all?

You need to find the sum of 24 and 8.

Estimate 24 + 8 ⟶ 20 + 10 = 30

Step 1 Add the ones.

$$\begin{array}{r} 1 \\ 24 \\ + 8 \\ \hline 2 \end{array}$$

4 ones + 8 ones = 12 ones
12 ones = 1 ten and 2 ones

Step 2 Add the tens.

$$\begin{array}{r} 1 \\ 24 \\ + 8 \\ \hline 32 \end{array}$$

1 ten + 2 tens = 3 tens

Check for Reasonableness

Compare 32 to the estimate. The answer is reasonable. ✔

2 DOGS There are 26 golden retrievers and 17 beagles. What is the total number of dogs?

You need to add 26 and 17.

One Way: Partial Sums	Another Way: Decompose Numbers
26 + 17 ——— 30 Add the tens. + 13 Add the ones. ——— 43 Add the partial sums.	Decompose, or take apart numbers to find the sum. 26 + 17 20 + 6 + 10 + 7 Rearrange the numbers so it is easier to add. 20 + 10 + 7 + 6 = 43

So, 26 + 17 = 43.

Sometimes you do not need to regroup.

EXAMPLE Add Without Regrouping

3 Find 51 + 23.

$$\begin{array}{r} 51 \\ + 23 \\ \hline 74 \end{array}$$ 1 one + 3 ones = 4 ones
5 tens + 2 tens = 7 tens

So, 51 + 23 = 74.

✓ CHECK What You Know

Add. Use models if needed. Check for reasonableness. See Examples 1–3 (pp. 78–79)

1. 27
+ 2

2. 42
+ 9

3. 17
+ 26

4. 20 + 79

5. In the park, 13 children are riding bikes, and 18 children are on skateboards. How many children are on bikes and skateboards?

6. **Talk About It** When adding, why do you need to line up the columns for the ones and tens digits?

Add. Use models if needed. Check for reasonableness. See Examples 1–3 (pp. 78–79)

7. 44
   $+ 5$

8. 62
   $+ 3$

9. 43
   $+ 7$

10. 57
   $+ 7$

11. 75
   $+ 12$

12. 72
   $+ 13$

13. 26
   $+ 34$

14. 61
   $+ 19$

15. $22 + 7$

16. $32 + 8$

17. $78 + 12$

18. $53 + 25$

19. There were 25 words on last week's spelling list. This week's list has 19 words. What is the two week total?

20. In two hours, 47 trucks and 49 cars passed through a tunnel. How many vehicles is this altogether?

21. Stuart and his father picked 38 red apples and 18 yellow apples at an orchard. They used 11 of them in pies. How many apples are there now?

22. One tray makes 24 ice cubes. Another tray makes 36. Are there enough ice cubes for 25 cups if each cup gets 2 ice cubes? Explain.

Data File

Rodeos have a great tradition in Oklahoma. A rodeo has several events. The person with the fastest time wins.

23. What is the combined time for steer wrestling and barrel racing?

24. What is the combined time for all four events?

Guyman Rodeo

Event	Winning Time (seconds)
Steer wrestling	18
Team roping	41
Tie down	40
Barrel racing	50

Source: Professional Rodeo Cowboys Association

H.O.T. Problems

25. OPEN ENDED Explain how to find $33 + 59$ mentally.

26. **WRITING IN ▶MATH** Miki had 60 minutes before her swim lesson. It took 45 minutes to do her homework and 18 minutes to eat a snack. Did she get to her lesson on time? Explain.

Find each sum. Identify the property. (Lesson 2-1)

1. $9 + 0 = \blacksquare$

2. $(3 + 4) + 2 = \blacksquare$
$3 + (4 + 2) = \blacksquare$

Algebra Find each missing number. Identify the property. (Lesson 2-1)

3. $2 + (7 + \blacksquare) = (2 + 7) + 3$

4. $\blacksquare + 4 = 4 + 7$

5. $6 + \blacksquare = 6$

6. MULTIPLE CHOICE Look at the number sentence below.

$$(7 + 2) + 9 = \blacksquare$$

Which number will make the number sentence true? (Lesson 2-1)

A 18   **C** 81

B 23   **D** 126

7. The worker washed 41 windows today and 54 yesterday. How many windows were washed in the two days? (Lesson 2-2)

8. Fina bought 8 daffodils and 13 daisies for her mother. About how many flowers did she buy? (Lesson 2-3)

Estimate each sum using compatible numbers. (Lesson 2-3)

9. 45
       $\underline{+\ 37}$

10. 12
        $\underline{+\ 46}$

11. Mrs. Barnes bought supplies for the classroom. Estimate the total number of items by rounding. (Lesson 2-3)

Add. Use models if needed.
(Lesson 2-4)

12. 58
      $\underline{+\ 3}$

13. 73
       $\underline{+12}$

14. 26
       $\underline{+\ 37}$

15. 39
       $\underline{+\ 18}$

16. MULTIPLE CHOICE There are a total of 38 second graders and 59 third graders at a school. How many students are in the second and third grades? (Lesson 2-4)

F 87   **H** 107

G 97   **J** 151

17. **WRITING IN ▸MATH** Explain what it means to regroup. Give an example. (Lesson 2-4)

GET READY to Learn

Claudio paid Rey 35¢ for one goldfish and 50¢ for one angelfish. How much money did Claudio pay for the two fish?

MAIN IDEA

I will add money as dollars or cents.

New Vocabulary

cent sign (¢)

Math Online

macmillanmh.com

• Extra Examples
• Personal Tutor
• Self-Check Quiz

Adding cents is like adding whole numbers. You place a **cent sign (¢)** *after* the sum.

Real-World EXAMPLES   Add Cents

1 **MONEY How much money did Claudio pay for the two fish?**

You need to find the sum of 35¢ + 50¢.

One Way: Use Models

Use coins to model 35¢ + 50¢.

  +

Another Way: Use Numbers and Words

$$\begin{array}{r} 35¢ \\ +\ 50¢ \\ \hline 85¢ \end{array}$$ Add the ones.

Add the tens.

Place cents sign *after* the sum.

So, Claudio paid 85¢ for the two fish.

You can also add dollars.

Real-World EXAMPLE   Add Dollars

2 **TICKETS** Ted spent $25 for a ticket to a baseball game, and $17 for a ticket to a basketball game. How much did Ted spend for the two tickets?

You need to add $25 and $17.
Use bills to model $25 + $17.

$25 $17

Order from greatest to least. Then skip count.

Read: 20, 30, 35, 40, 41, 42

So, Ted spent $42 for the tickets.

CHECK What You Know

Add. Use models if needed. See Examples 1 and 2 (p. 82–83)

1.	86¢ + 11¢	2.	$12 + $78	3.	$39 + $18	4.	19¢ + 30¢

5. 59¢ + 20¢   **6.** 42¢ + 37¢   **7.** $17 + $9   **8.** $66 + $14

9. Rory earns $6 allowance each week. Aida earns $4 each week. If Rory and Aida put 2 weeks of their allowance together, what three different toys could they buy to spend as much of their combined allowance as possible?

10. (**Talk About It**) Explain why you could trade 2 quarters for 5 dimes.

Add. Use models if needed. See Examples 1 and 2 (pp. 82–83)

11. 12¢
+ 23¢

12. 30¢
+ 38¢

13. 49¢
+ 19¢

14. 36¢
+ 19¢

15. $21
+ $38

16. $53
+ $45

17. $17
+ $26

18. $69
+ $13

19. 25¢ + 4¢

20. 21¢ + 2¢

21. 68¢ + 6¢

22. 8¢ + 74¢

23. $27 + $71

24. $55 + $41

25. $34 + $8

26. $7 + $66

27. A market sells oranges for 37¢ each. How much would 2 oranges cost?

28. A computer program is $19. The guidebook is $15. You have $25. Do you have enough money to buy both? Explain.

For Exercises 29–31, use the poster.

29. Beverly spent 85¢ at the store today. What 2 items did she buy?

30. Which list would cost Beverly more to buy? Explain.

31. What is the greatest number of items Beverly can buy without spending more than $1? Explain.

Real-World PROBLEM SOLVING

History The cost for many items has increased over the years. Find each sum.

32. a stamp and milk in 1900

33. bread, milk, and a stamp in 1940

Prices Then and Now			
Year	Bread	Milk	Stamp
1900	3¢	30¢	2¢
1940	8¢	51¢	3¢
1980	48¢	$2	15¢
2000	$2	$3	34¢

Source: Somethingtoremembermeby.org

H.O.T. Problems

34. CHALLENGE Chuck has $36. He wants to buy screwdrivers and a hammer. Kali has $42 and wants to buy wrenches and a tool box. Who has enough money to buy the items they need? Explain.

35. WRITING IN ►MATH Write about a time when you needed to know how to find sums of money.

TEST Practice

36. Look at the number sentence below.

$$79 + 13 = \blacksquare$$

Which sum would make it true? (Lesson 2-4)

A 96

B 93

C 92

D 90

37. Alexandra buys the two items.

What was the total cost? (Lesson 2-5)

F $32   **H** $42

G $35   **J** $51

Spiral Review

Add. Use models if needed. Check for reasonableness. (Lesson 2-4)

38. 22 + 68   **39.** 75 + 13   **40.** 79 + 87

41. Algebra Booker took the first tray of cookies out of the oven at 1:08. If he continues this pattern, what time will he take the fourth and fifth trays out?

(Lesson 1-1)

Time to Take Cookies Out	
Tray	**Time**
1	1:08
2	1:16
3	1:24

Problem Solving in Geography

A Walk in the PARK

Yellowstone National Park is located in Montana, Wyoming, and Idaho. It is home to hot springs, bubbling mud holes, and fountains of steaming water. These fountains shoot water more than 100 feet into the air.

Yellowstone is also home to 290 species of birds, 50 species of mammals, 6 species of reptiles, and 18 species of fish. What a place to visit! A seven-day visitor's pass to the park costs $10 for one person to hike, or $20 for a carload of people.

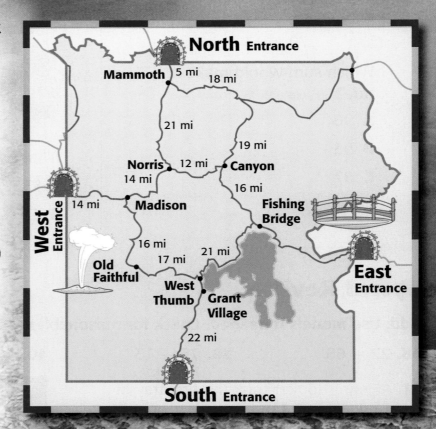

North Entrance
Mammoth   5 mi   18 mi
21 mi
19 mi
Norris   12 mi   Canyon
14 mi
West Entrance   14 mi   Madison   16 mi
Fishing Bridge
Old Faithful   16 mi   21 mi
17 mi   West Thumb   Grant Village   East Entrance
22 mi
South Entrance

 # Real-World Math

Use the information and map on page 86 to solve each problem.

1 What is the total number of species of birds, mammals, fish, and reptiles in Yellowstone

2 If 5 people travel by car, they need two cars. Does it cost less for them to drive or hike? Explain.

3 Suppose you walk from the Fishing Bridge to West Thumb and then to Old Faithful. How many miles will you walk?

4 The Diaz family drives from Mammoth to the Fishing Bridge. If their trip was 49 miles long, what route did they take?

5 What is the shortest distance you can walk from the North Entrance of the park to Canyon?

6 A group of visitors travels from the West Entrance to Norris and then to Canyon. Another group travels from the West Entrance to Madison and then to Canyon. Did they travel the same distance? Explain.

7 A group of visitors wants to travel to Norris from the South Entrance. They want to pass the lake. About how long is this route?

Problem-Solving Investigation

MAIN IDEA I will use the four-step plan to solve a problem.

P.S.I. TEAM +

KIRI: My father and I needed to catch at least 10 fish. During the first hour, we caught 9 fish but threw 4 back. The second hour we caught 16 fish and threw 9 back.

YOUR MISSION: Find if they caught and kept at least 10 fish.

Understand	You know how many fish they caught and how many they threw back. Find if they caught and kept at least 10 fish.
Plan	You need to find an exact answer. Use addition and subtraction and write number sentences.
Solve	First, subtract to find out how many fish they kept.

Solve (continued)

Hour One: 9   −   4   =   5

                fish caught   threw back   total each hour

Hour Two: 16   −   9   =   7

Next, add the total for each hour.

  first hour   second hour   total
    5   +   7   =   12

Kiri and her father caught and kept 12 fish.

Check	Look back at the problem. Yes, Kiri and her father caught and kept at least 10 fish. They caught 12.

Use the four-step plan to solve each problem.

1. Neva has a hamster and Sherita has a turtle. If they each have a $5-bill, how much change will each receive when they buy food for their pets?

Pet Food	
Hamster	Turtle
$3	$4

2. It takes one hour to make 4 pizzas. How many pizzas can be made in 4 hours and 30 minutes?

3. **Measurement** Rudy left for the New Jersey shore at 5 A.M. If the trip takes 10 hours, will he be at his destination by 3 P.M.? Explain.

4. **Geometry** Blaine built a cube staircase. How many cubes in all are needed to build 6 steps?

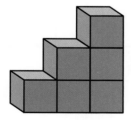

5. There are 4 snack-packs of yogurt. Each snack-pack has 6 yogurts. How many more snack-packs are needed for a total of 30 yogurts?

6. There are 3 children in line. Cami is right after Brock. Bo is third. What place is each child in line?

7. There are 37 stores on the first floor of a 2-floor mall. The second floor has 29 stores. About how many stores are at this mall?

8. The space museum gift shop opened at 10:30 A.M. In the morning, 15 model spaceships were sold. During the afternoon, 23 models were sold. How many spaceship models were sold that day?

9. At one campsite there are 3 tents with 5 people in each tent. Another campsite has 3 tents with 4 people in each. How many campers are there in all?

10. **Measurement** Mom's watering can holds 2 gallons. Each day she waters 12 large flower pots and 10 small pots. How many times will she need to fill her watering can?

Pots that can be watered with 2 gallons	
Large pots	4
Small pots	5

11. **WRITING IN ►MATH** The children in Mr. Robinson's class are designing a flag. The flag's background can be red or green with a blue or a purple stripe. How many flags can they design? Explain how you solved the problem.

Math Activity for 2-7
Add Three-Digit Numbers

Use base-ten blocks to model three-digit addition.

ACTIVITY **Find 148 + 153.**

Step 1 Model 148 and 153.

MAIN IDEA

I will use models to explore adding three-digit numbers.

You Will Need
base-ten blocks

Math Online
macmillanmh.com
• Concepts in Motion

Hundreds	Tens	Ones	
			148
			153

Step 2 Add the ones.

Hundreds	Tens	Ones

8 ones + 3 ones = 11 ones
Regroup 11 ones as
1 ten and 1 one.

Step 3 Add the tens.

Hundreds	Tens	Ones

5 tens + 5 tens = 10 tens

Regroup 10 tens as 1 hundred and 0 tens.

Step 4 Add the hundreds.

1 hundred + 1 hundred + 1 hundred = 3 hundreds

So, 148 + 153 = 301.

Think About It

1. Describe the sum of the digits that needed to be regrouped.

2. Why were the ones and the tens regrouped?

3. Does changing the order of the addends make a difference in whether you need to regroup? Explain.

CHECK What You Know

Add. Use base-ten blocks to model if needed.

4. 259 + 162　　　**5.** 138 + 371　　　**6.** 362 + 172

7. 541 + 169　　　**8.** 261 + 139　　　**9.** 285 + 75

10. **WRITING IN ►MATH** Write a rule that would explain when to regroup.

Three-Digit Addition

GET READY to Learn

During the annual backyard bird count, birdwatchers reported sighting 127 wrens and 68 eagles. How many birds is that in all?

In the Explore lesson, you used base-ten blocks to add three-digit numbers. You can also use paper and pencil.

Real-World EXAMPLE   Add with Regrouping

1 BIRDS How many wrens and eagles did the bird watchers report?

You need to add 127 + 68.

Estimate

$$127 \rightarrow 130$$
$$+\ 68 \rightarrow +\ 70$$
$$\overline{200}$$

Step 1 Add the ones.

$$\overset{1}{}$$
$$127$$
$$+\ 68$$
$$\overline{5}$$

7 ones + 8 ones = 15 ones
Regroup 15 ones as 1 ten and 5 ones.

Step 2 Add the tens and hundreds.

$$\overset{1}{}$$
$$127$$
$$+\ 68$$
$$\overline{195}$$

1 ten + 2 tens + 6 tens = 9 tens
Bring the 1 hundred down.

So, the birdwatchers reported 195 wrens and eagles.

Check for Reasonableness

195 is close to the estimate of 200. The answer is reasonable. ✔

2 **MONEY** A store received two boxes of butterfly nets. One box costs $175, and the other costs $225. How much did the nets cost altogether?

Find $175 + $225 to determine the total cost of the nets.

Remember

When adding 3-digit numbers, be sure to align the ones column, tens column, and hundreds column.

Step 1 Add the ones.

$$\begin{array}{r} {\scriptstyle 1} \\ \$175 \\ + \$225 \\ \hline 0 \end{array}$$

5 ones + 5 ones = 10 ones
Regroup 10 ones as 1 ten and 0 ones.

Step 2 Add the tens.

$$\begin{array}{r} {\scriptstyle 1\,1} \\ \$175 \\ + \$225 \\ \hline 00 \end{array}$$

1 ten + 7 tens + 2 tens = 10 tens
Regroup 10 tens as 1 hundred + 0 tens.

Step 3 Add the hundreds.

$$\begin{array}{r} {\scriptstyle 1\,1} \\ \$175 \\ + \$225 \\ \hline \$400 \end{array}$$

1 hundred + 1 hundred + 2 hundreds = 4 hundreds

Together, the two boxes of butterfly nets cost $400.

CHECK What You Know

Add. Check for reasonableness. See Examples 1 and 2 (pp. 92–93)

1. 164 + 17

2. 156 + 255

3.
$$\begin{array}{r} 468 \\ + 35 \\ \hline \end{array}$$

4.
$$\begin{array}{r} 227 \\ + 26 \\ \hline \end{array}$$

5.
$$\begin{array}{r} \$355 \\ + \$156 \\ \hline \end{array}$$

6.
$$\begin{array}{r} \$272 \\ + \$148 \\ \hline \end{array}$$

7. Chase has 176 video games. Estaban has 238 games. What is the total number of games they have?

8. **Talk About It** Why is it important to check for reasonableness?

Add. Check for reasonableness.

See Examples 1 and 2 (pp. 92–93)

9. 759
+ 19

10. 445
+ 26

11. $345
+ $93

12. $427
+ $217

13. 597
+ 51

14. 599
+ 59

15. $298
+ $408

16. $287
+ $453

17. 43 + 217

18. 607 + 27

19. $173 + $591

20. $108 + $589

21. 635 + 285

22. 398 + 355

23. $797 + $185

24. $490 + $288

25. A 10-speed bike is on sale for $199, and a 12-speed racing bike is on sale for $458. How much do the two bikes cost altogether?

26. Measurement Russell's bean stalk grew 24 inches the first month and 27 inches the second month. How tall was Russell's bean stalk after two months?

27. Measurement Use the map at the right. What is the total distance from the entrance of the park to Leonora's house and back to the park again?

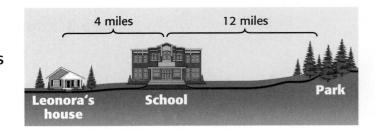

Algebra Find each missing number. Identify the property.

28. 240 + 679 = ■ + 240

29. (13 + 24) + 6 = ■ + (24 + 6)

30. 989 + ■ = 989

31. (565 + 6) + 39 = 565 + (■ + 39)

H.O.T. Problems

32. OPEN ENDED Write an addition problem whose sum is between 450 and 500.

33. CHALLENGE Use the digits 3, 5, and 7 to make two three-digit numbers. Use each digit one time in each number. Write an addition problem that would make the greatest sum possible.

How Low Can You Go?

Add Three-Digit Numbers

You will need: spinner

Get Ready!

Players: 2 players

Get Set!

- Divide and label a spinner as shown.

- Make two game sheets.

Go!

- Player 1 spins the spinner and records the digit in any box on his or her game sheet.

- Player 2 spins the spinner and records the digit in any box on his or her game sheet.

- Players repeat taking turns and recording numbers until all boxes are filled.

- Players find the sums of their numbers. The least sum wins.

2-8 Add Greater Numbers

GET READY to Learn

In the United States, 869 radio stations play rock music. There are 2,179 news/talk radio stations. How many radio stations play these two formats?

MAIN IDEA

I will add three- and four-digit numbers with regrouping.

Math Online

macmillanmh.com
• Extra Examples
• Personal Tutor
• Self-Check Quiz

Add 2,179 and 869 just like you add two-digit numbers.

Real-World EXAMPLE Add Numbers

1. **RADIO** Add 2,179 and 869 to find how many radio stations play these two formats.

   Step 1 Add the ones.

   $$
   \begin{array}{r}
   1 \\
   2,179 \\
   + 869 \\
   \hline
   8
   \end{array}
   $$
   9 ones + 9 ones = 18 ones
   Regroup as 1 ten and 8 ones.

   Step 2 Add the tens.

   $$
   \begin{array}{r}
   1\,1 \\
   2,179 \\
   + 869 \\
   \hline
   48
   \end{array}
   $$
   1 ten + 7 tens + 6 tens = 14 tens
   Regroup as 1 hundred and 4 tens.

   Step 3 Add the hundreds.

   $$
   \begin{array}{r}
   1\,1\,1 \\
   2,179 \\
   + 869 \\
   \hline
   048
   \end{array}
   $$
   1 hundred + 1 hundred + 8 hundreds = 10 hundreds
   Regroup as 1 thousand and 0 hundreds.

   Step 4 Add the thousands.

   $$
   \begin{array}{r}
   1\,1\,1 \\
   2,179 \\
   + 869 \\
   \hline
   3,048
   \end{array}
   $$
   1 thousand + 2 thousands = 3 thousands

So, 3,048 radio stations play the two formats.

96   **Chapter 2** Add To Solve Problems

2 **PLANES** The world's fastest plane can fly 2,139 miles in 32 minutes. What is the total distance if it flew another 2,314 miles?

You need to find 2,139 + 2,314.

Estimate

$$
\begin{array}{r}
2{,}139 \\
+2{,}314 \\
\end{array}
\longrightarrow
\begin{array}{r}
2{,}100 \\
+2{,}300 \\
\hline
4{,}400 \\
\end{array}
$$

Remember

To check if your answer makes sense, estimate first. Then compare the answer to the estimate.

One Way: Partial Sums		**Another Way:** Expanded Form
2,139		2,139 = 2,000 + 100 + 30 + 9
+ 2,314		+ 2,314 = 2,000 + 300 + 10 + 4
13	Add ones.	4,000 + 400 + 40 + 13
40	Add tens.	400
400	Add hundreds.	40
+ 4,000	Add thousands.	+ 13
4,453		4,453

Check for Reasonableness

4,453 is close to 4,400. So, the answer is reasonable. ✔

To add money, add as you would with whole numbers.

3 **SPORTS** Last year $3,295 was spent on a skate park. This year $3,999 was spent. How much money was spent over the two years?

Estimate $3,295 + $3,999 $\longrightarrow$ $3,000 + $4,000 = $7,000

Find $3,295 + $3,999.

$$
\begin{array}{r}
\scriptstyle 1\ 1\ 1 \\
\$3{,}295 \\
+\$3{,}999 \\
\hline
\$7{,}294 \\
\end{array}
$$

Add.

Place the dollar sign in front of the dollars.

Check for Reasonableness

$7,294 is close to $7,000. So, the answer is reasonable. ✔

CHECK What You Know

Find each sum. Use estimation to check for reasonableness.

See Examples 1–3 (p. 96–97)

1. 3,345
    + 654

2. 4,234
    + 500

3. $3,205
    + $1,709

4. 678 + 4,789

5. $3,445 + $6,547

6. $9,299 + $701

7. Lou's dad's car uses 1,688 gallons of gas a year. His mom's car uses 1,297 gallons. Find the total gallons used.

8.  **Talk About It** How is finding the sum of 4-digit numbers like finding the sum of 3-digit numbers?

Practice and Problem Solving

EXTRA **PRACTICE**
See page R7.

Find each sum. Use estimation to check for reasonableness.

See Examples 1–3 (p. 96–97)

9. 6,999
    + 543

10. $1,998
    + $300

11. $2,507
    +$2,899

12. $8,285
    +$1,456

13. $2,390
    +$3,490

14. 5,555
    +3,555

15. 2,865 + 5,522

16. 3,075 + 5,640

17. $1,603 + $3,509

18. $5,788 + $2,550

19. $3,999 + $4,800

20. 1,250 + 1,520

21. **Algebra** Write a number sentence to represent the total number of minutes each group read this month. Use > or <.

Time Spent Reading	
Group	**Minutes**
A	2,600
B	2,574

22. How many people were surveyed?

Favorite Summer Place
Beach   2,311
Amusement park 2,962

23. The Appalachian Trail is 2,174 miles long. How many miles is it if you hike it one way and then back?

24. Mr. Roth's car costs $7,681. Mr. Randalls car costs $1,406 more than Mr. Roth's. How much do their cars cost altogether?

H.O.T. Problems

25. CHALLENGE Use the digits 2, 3, 4, 5, 6, 7, 8, and 9 to create two four-digit numbers whose sum is greater than 10,000. Use each digit once.

26. FIND THE ERROR Selina and Elliott have found the sum for 4,573 + 2,986. Who is correct? Explain.

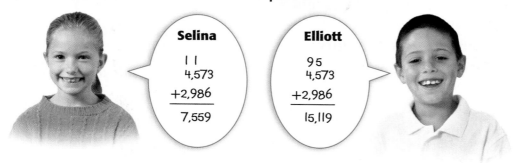

Selina

```
  1 1
  4,573
+2,986
───────
  7,559
```

Elliott

```
  9 5
  4,573
+2,986
───────
 15,119
```

27. **WRITING IN** ►**MATH** Explain what it means to check for reasonableness.

> **TEST Practice**

28. Yesterday's attendance at a baseball game was 4,237. Today's attendance was 3,176. What is the attendance for the two days? (Lesson 2-8)

   A 7,313

   B 7,403

   C 7,413

   D 7,611

29. What is the total cost? (Lesson 2-8)

$11 $4 $1

   F $12   **H** $16

   G $15   **J** $17

Spiral Review

Find each sum. (Lesson 2-5)

30. 78 + 19   **31.** 24¢ + 53¢   **32.** $46 + $46

33. Mrs. Vallez bought 2 packages of 127 stickers each. She added these to the 219 stickers she already has. How many stickers does she have now? (Lesson 2-2)

FOLDABLES™ Study Organizer GET READY to Study

Be sure the following Key Vocabulary words and Key Concepts are written in your Foldable.

Two-Digit Addition	Add Money
Estimate Sums	Three-Digit Addition

Key Concepts

Addition Properties (p. 69)

- **Commutative Property of Addition**

$$4 + 3 = 7 \qquad 3 + 4 = 7$$

- **Identity Property of Addition**

$$6 + 0 = 6 \qquad 0 + 6 = 6$$

- **Associative Property of Addition**

$$(2 + 5) + 1 = 2 + (5 + 1)$$

Estimate Sums (p. 74)

56	→	60	← Round 56 to 60.
+ 21	→	+ 20	← Round 21 to 20.
		80	

Add with Regrouping (p. 78)

1 1	4 ones + 7 ones = 11 ones
474	Regroup 11 ones as 1 ten and 1 one.
+ 237	1 ten + 7 tens + 3 tens = 11 tens
711	

Key Vocabulary

Associative Property of Addition (p. 69)

Commutative Property of Addition (p. 69)

Identity Property of Addition (p. 69)

estimate (p. 74)

regroup (p. 78)

Vocabulary Check

Choose the vocabulary word that completes each sentence.

1. When you _____?_____ , you find an answer that is close to the exact answer.

2. The _____?_____ states that the sum of any number and zero is the number.

3. To rename a number using place value is to _____?_____ .

4. The _____?_____ states that grouping the addends does not change the sum.

5. 2 + 4 = 6 and 4 + 2 = 6 is an example of the _____?_____ .

Lesson-by-Lesson Review

2-1   Algebra: Addition Properties (pp. 69–71)

Example 1
Find the sum in $2 + 7 = \blacksquare$ and $7 + 2 = \blacksquare$. Identify the property.

The sum is 9. The order does not change the sum. This is the Commutative Property of Addition.

Example 2
Find $8 + 0 = \blacksquare$. Identify the property.

The sum is 8. The sum of any number and 0 is that number. This is the Identity Property of Addition.

Find each sum. Identify the property.

6. $0 + 10 = \blacksquare$

7. $11 + 2 = \blacksquare$
$2 + 11 = \blacksquare$

8. Algebra Find the missing number. Identify the property.

$$(6 + 9) + 5 = 6 + (\blacksquare + 5)$$

Find each sum mentally.

9.
$\begin{array}{r} 4 \\ 6 \\ + \ 0 \\ \hline \end{array}$

10.
$\begin{array}{r} 8 \\ 6 \\ + \ 2 \\ \hline \end{array}$

2-2   Problem-Solving Skill: Estimate or Exact Answer (pp. 72–73)

Example 3
LaVonne saved \$36. Her sister saved \$29. About how much money do the two girls have in all?

You need to estimate \$36 + \$29. Round each amount to the nearest ten. Then add.

$\begin{array}{rcl} \$36 & \longrightarrow & \$40 \\ + \ \$29 & \longrightarrow & + \ \$30 \\ \hline & & \$70 \end{array}$

The girls have about \$70.

Tell whether an estimate or exact answer is needed. Then solve.

11. To enter a dance competition, Viviana needs \$35 for an entrance fee and \$75 for her costume. How much money does Viviana need?

12. Cole practices violin for 47 minutes three nights a week. Two other nights he has soccer practice for 29 minutes. About how many minutes does Cole spend practicing each week?

2-3 Estimate Sums (pp. 74–77)

Example 4

An artist created a piece of art with 66 round glass beads. There are 17 more beads in the shape of a heart. About how many beads are there altogether?

You need to estimate 66 + 17. Round each number to the nearest ten. Then add.

$$
\begin{array}{rcr}
66 & \rightarrow & 70 \\
+\,17 & \rightarrow & +\,20 \\
\hline
& & 90
\end{array}
$$

So, there are about 90 beads.

Estimate each sum using rounding.

13. $\begin{array}{r} 76 \\ +\,12 \\ \hline \end{array}$

14. $\begin{array}{r} 52 \\ +\,21 \\ \hline \end{array}$

Estimate using compatible numbers.

15. $\begin{array}{r} 33 \\ +\,58 \\ \hline \end{array}$

16. $\begin{array}{r} 28 \\ +\,47 \\ \hline \end{array}$

17. 31 + 68

18. 97 + 28

19. There are about 2,000 earthquakes each year in Yellowstone National Park. Is this an estimate or an exact number? Explain.

2-4 Two-Digit Addition (pp. 78–80)

Example 5

Find 25 + 3.

$\begin{array}{r} 25 \\ +\,3 \\ \hline 28 \end{array}$ 5 ones + 3 ones = 8 ones

2 tens + 0 tens = 2 tens

So, 25 + 3 = 28.

Example 6

Find 32 + 9.

$\begin{array}{r} 1 \\ 32 \\ +\,9 \\ \hline 41 \end{array}$ 2 ones + 9 ones = 11 ones

11 ones + 1 ten and 1 one

1 ten + 3 tens = 4 tens

So, 32 + 9 = 41.

Add. Use models if needed. Check for reasonableness.

20. $\begin{array}{r} 32 \\ +\,4 \\ \hline \end{array}$

21. $\begin{array}{r} 19 \\ +\,3 \\ \hline \end{array}$

22. Karly's math class has 21 students. Landon's math class has 29 students. How many math students are there in the two classes?

23. There are 18 juice boxes in the refrigerator and 12 on the shelf. How many juice boxes are there altogether?

2-5 Add Money (pp. 82–85)

Example 7

At school, Vito bought yogurt for 65¢ and milk for 25¢. How much money did he spend?

Find 65¢ + 25¢. To add money, add as you would with whole numbers.

$$\begin{array}{r} 1 \\ 65¢ \\ + \ 25¢ \\ \hline 90¢ \end{array}$$ Place the cents sign *after* the sum.

Example 8

Find $39 + $45.

$$\begin{array}{r} 1 \\ \$39 \\ + \ \$45 \\ \hline \$84 \end{array}$$ Place the dollar sign *before* the sum.

Add. Use models if needed.

24. $\begin{array}{r} 13¢ \\ + \ 43¢ \end{array}$
25. $\begin{array}{r} \$54 \\ + \ \$35 \end{array}$

26. $\begin{array}{r} 74¢ \\ + \ 6¢ \end{array}$
27. $\begin{array}{r} \$90 \\ + \ \$19 \end{array}$

28. Rondell's piggy bank has $23. For his birthday, his grandmother gave him $15. How much money does he have now?

29. Edgardo found 36¢ on the sidewalk. If he combines this with the 27¢ he has left from his lunch money, how much will he have altogether?

2-6 Problem-Solving Investigation: The Four-Step Plan (pp. 88–89)

Example 9

A music group bought two new guitars. Each guitar cost $488. How much did the guitars cost?

Since each guitar cost the same, add $488 two times.

$$\begin{array}{r} 11 \\ \$488 \\ + \ \$488 \\ \hline \$976 \end{array}$$

So, the guitars cost $976.

Use the four-step plan to solve each problem.

30. While fishing in Chesapeake Bay, Augusto and his uncle caught 17 catfish, 21 trout, and 6 bass. About how many fish did they catch?

31. Carolyn walked 2 blocks south to meet Tiffany. They walked 3 blocks east and 1 block north to Quincy's house. Does Quincy live on Carolyn's street? Explain.

2-7 Three-Digit Addition (pp. 92–94)

Example 10
Joel read one book with 175 pages and another with 409 pages. How many pages did Joel read total?

Find 175 + 409. Add the ones.

$$
\begin{array}{r}
1 \\
175 \\
+\ 409 \\
\hline
4
\end{array}
$$
5 ones + 9 ones = 14 ones
14 ones = 1 ten + 4 ones

Add the tens. Then hundreds.

$$
\begin{array}{r}
1 \\
175 \\
+\ 409 \\
\hline
584
\end{array}
$$
1 ten + 7 tens = 8 tens
1 hundred + 4 hundreds =
5 hundreds

So, 175 + 409 = 584.

Add. Check for reasonableness.

32. 377 + 26   **33.** 657 + 245

34. $67
      +$25

35. $325
      +$256

36. Flavio and Felix bought airline tickets for $213. Their rental car and hotel are $378. How much will they spend on these vacation expenses?

37. Measurement Last year Ithaca had a record snowfall of 124 inches. This year it snowed 117 inches. What was the total snowfall for the two years?

2-8 Add Greater Numbers (pp. 96–99)

Example 11
Find the total number of students surveyed.

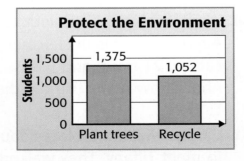

1,375
+ 1,052
2,427

Find each sum. Use estimation to check for reasonableness.

38. 1,003 + 7,927

39. 3,042 + 4,517

40. $8,385
      + $1,476

41. $2,190
      + $3,790

42. A chess set costs $24. The collector's edition, costs $12 more. How much will the collector's edition cost?

Chapter Test

Find each sum. Identify the property.

1. $5 + 3 = \blacksquare + 5$

2. $\blacksquare + 2 = 2$

3. $(1 + 2) + 3 = 1 + (\blacksquare + 3)$

Estimate each sum using rounding.

4. 54
   + 29

5. 18
   + 23

6. **MULTIPLE CHOICE** At a movie theater, 64 bags of popcorn were sold before the movie. During the movie, 29 bags of popcorn were sold. How many bags of popcorn were sold?

   A 90   C 103

   B 93   D 113

7. How many tickets were purchased the last two weeks?

Dolphin Watching Tickets

Week	Child Tickets	Adult Tickets
1	173	106
2	121	115

Add. Check for reasonableness.

8. $281 + 674$

9. $\$313 + \$731$

10. 2,103
    + 879

11.   6,545
    + 2,345

For Exercises 12 and 13, tell whether an estimate or an exact answer is needed. Then solve.

12. Toshi wants to buy new school supplies. She has $5. Does she have enough money? Explain.

School Supplies		
Paper	Crayons	Pencils
$1	$2	$1

13. There are 3 office buildings on a block. About how many offices are in the three buildings? Explain.

Number of Offices by Building		
A	B	C
2,114	3,112	2,295

14. **MULTIPLE CHOICE** Gen spent $378 at the mall. Her sister spent $291. About how much did the sisters spend together?

   F $700   H $600

   G $669   J $400

15. Abby's bird club spotted 328 birds. Nita's bird club spotted 576 birds. Did the two bird clubs spot more than 915 birds? Explain.

16. **WRITING IN ►MATH** How do you know when you need to regroup when adding? Give an example.

PART 1 Multiple Choice

Read each question. Then fill in the correct answer on the answer sheet provided by your teacher or on a sheet of paper.

1. Tyra bought a notebook that cost $3, a marker for $2, and a folder for $6. What was the total cost of these three items?

   A $2   **C** $10

   B $5   **D** $11

2. What is the total number of people in the park on Friday?

People in the Park	
Activity	**Number of People**
Biking	12
Walking	22
Running	45
Reading	18

   F 86   **H** 93

   G 87   **J** 97

3. What number makes this number sentence true?

   $$2 + 5 + 8 = 2 + 8 + \blacksquare$$

   A 2   **C** 8

   B 5   **D** 15

4. Which point on the number line represents 174?

   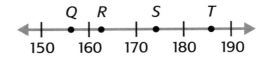

   F Q   **H** S

   G R   **J** T

5. How is two thousand, one hundred forty-six written in standard form?

   A 2,416   **C** 1,246

   B 2,146   **D** 2,14

6. What is 64 + 128 rounded to the nearest ten?

   F 200   **H** 180

   G 190   **J** 150

7. Fran has 41 stickers, Nina has 62 stickers, and Amy has 57 stickers. How can you find the total number of stickers?

   A 41 + 62   **C** 41 + 62 − 57

   B 41 − 62 − 57   **D** 41 + 62 + 57

8. Which symbol makes a true number sentence below?

561 ⬤ 559

F $<$ **H** $=$

G $>$ **J** $+$

---

9. Colleen and Vicky are selling ornaments for a fundraiser. Colleen has sold 82 ornaments, and Vicky has sold 47. About how many ornaments have they sold altogether?

A 110   **C** 130

B 120   **D** 140

---

10. Reed and his brother have 562 marbles. Which of these equals 562?

F $500 + 60 + 2$

G $500 + 6 + 2$

H $5 + 60 + 2$

J $5 + 6 + 2$

PART 2   Short Response

Record your answers on the answer sheet provided by your teacher or on a sheet of paper.

11. Kurt and his brother have 7,834 marbles. Write 7,834 in expanded form?

---

12. The River School sold 3,428 banners. The Gibson School sold 4,636 banners. How many banners were sold in all? Show your work.

PART 3   Extended Response

Record your answers on the answer sheet provided by your teacher or on a sheet of paper.

13. Cecelia used all the money shown below to buy some pencils. How much money did Cecelia spend on pencils? Explain how you found your answer.

NEED EXTRA HELP?													
If You Missed Question...	1	2	3	4	5	6	7	8	9	10	11	12	13
Go to Lesson...	2-5	2-4	2-1	1-7	1-3	2-3	2-4	1-6	2-4	1-3	1-3	2-8	1-10

Subtract To Solve Problems

BIG Idea What is subtraction?

Subtraction is an operation that tells the difference when some or all are taken away.

Example New York is one of the top apple-producing states. Tanisha has red and green apples. There are 30 apples in all. If 17 of the apples are green, 30 − 17 or 13 apples are red.

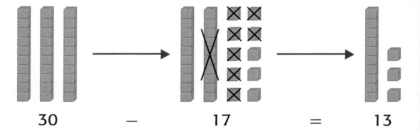

30   −   17   =   13

What will I learn in this chapter?

- Subtract two- and three-digit numbers.
- Estimate differences.
- Subtract with regrouping.
- Decide whether an answer is reasonable.

Key Vocabulary

difference

estimate

regroup

Math Online > **Student Study Tools**
at **macmillanmh.com**

FOLDABLES®
Study Organizer

Make this Foldable to help you organize information about subtraction. Begin with four sheets of $8\frac{1}{2}'' \times 11''$ paper.

1 **Stack** 4 sheets of paper as shown.

2 **Fold** upward so all layers are the same distance apart.

3 **Crease** well. Open and glue together as shown.

4 **Label** with the lesson titles. Record what you learn.

Subtraction
Two-Digit Subtraction
Estimate Differences
Subtract Money
Regrouping in Subtraction
Subtract Across Zeros
Subtract 3-digit and 4-digit Numbers
Addition or Subtraction

ARE YOU READY for Chapter 3?

You have two ways to check prerequisite skills for this chapter.

Option 2

Math Online > Take the Chapter Readiness Quiz at macmillanmh.com.

Option 1

Complete the Quick Check below.

QUICK Check

Subtract. (Prior grade)

1. 15
    − 9

2. 12
    − 4

3. 13
    − 6

4. 17
    − 9

5. 50
    − 20

6. 70
    − 10

7. 25
    − 15

8. 61
    − 31

9. Dalila had a package of 36 pens. She gave 14 to her friends. How many pens does she have left?

10. Abram took 27 magazine orders. He needs 49 orders in all. How many more does he need to get?

Round to the nearest ten. (Lesson 1-8)

11. 76

12. 57

13. 32

14. 99

Round to the nearest hundred. (Lesson 1-8)

15. 273

16. 923

17. 166

18. 501

Estimate. (Prior grade)

19. 52 − 42

20. 49 − 18

21. 67 − 28

22. 88 − 61

Two-Digit Subtraction

GET READY to Learn

The table shows that a tiger sleeps 16 hours a day. A cat sleeps 12 hours each day. Find the difference to show how much longer a tiger sleeps than a cat.

Hours of Sleep Each Day

Animal	Time (h)
python	18
tiger	16
cat	12
horse	3

You can use subtraction to solve the problem. The **difference** is the answer to a subtraction problem.

Real-World EXAMPLE   Subtract with No Regrouping

1 **ANIMALS How much longer does a tiger sleep than a cat?**

You need to find $16 - 12$. Use base-ten blocks to model the problem.

Step 1 Subtract ones.

$$\begin{array}{r} 16 \\ -\,12 \\ \hline 4 \end{array}$$ 6 ones − 2 ones = 4 ones

Step 2 Subtract tens.

$$\begin{array}{r} 16 \\ -\,12 \\ \hline 4 \end{array}$$ 1 ten − 1 ten = 0 tens

The difference is 4.

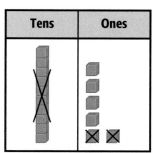

Tens	Ones

So, a tiger sleeps 4 hours more than a cat.

Check You can use addition to check your answer.

```
        ┌── same ──┐
    16     ↗ 4
  − 12   ╱ + 12
    4      16  ←── So, the answer is correct. ✔
```

Sometimes when you subtract, there are not enough ones to subtract from. In this case, you need to regroup.

Real-World EXAMPLE **Subtract with Regrouping**

Review Vocabulary

regroup
To use place value to exchange equal amounts when renaming a number. (Lesson 2-4)

② **CARS** At one time, Preston had 54 toy cars. He lost 18. How many does he have now?

You need to find 54 − 18.

Step 1 Subtract ones.

$$\begin{array}{r} \overset{4\,14}{\cancel{54}} \\ -\ 18 \\ \hline 6 \end{array}$$

You cannot take 8 ones from 4 ones.
Regroup 1 ten as 10 ones.
4 ones + 10 ones = 14 ones
14 ones − 8 ones = 6 ones

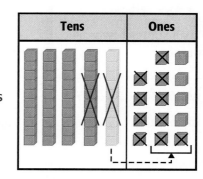

Tens	Ones

Step 2 Subtract tens.

$$\begin{array}{r} \overset{4\,14}{\cancel{54}} \\ -\ 18 \\ \hline 36 \end{array}$$

4 tens − 1 ten = 3 tens

Check You can use addition to check your answer.

— same —

$$\begin{array}{r} 54 \\ -\ 18 \\ \hline 36 \end{array} \qquad \begin{array}{r} 36 \\ +\ 18 \\ \hline 54 \end{array}$$

So, the answer is correct. ✔

CHECK What You Know

Subtract. Use models if needed. Check your answer.

See Examples 1 and 2 (pp. 111–112)

1. $\begin{array}{r} 39 \\ -\ 14 \end{array}$

2. $\begin{array}{r} 79 \\ -\ 18 \end{array}$

3. $\begin{array}{r} 94 \\ -\ 25 \end{array}$

4. $\begin{array}{r} 63 \\ -\ 46 \end{array}$

5. Nell brought lemonade and juice to the picnic. Of the 26 drinks, 8 were lemonade. How many were juice?

6. **Talk About It** Why do you start subtracting with the ones?

EXTRA PRACTICE
See page R8.

Subtract. Use models if needed. Check your answer.

See Examples 1 and 2 (pp. 111–112)

7.   28
    − 16

8.   74
    − 13

9.   45
    − 28

10.   54
    − 15

11.   99
    − 35

12.   34
    − 21

13.   32
    − 27

14.   41
    − 15

15. 70 − 48

16. 30 − 14

17. 96 − 68

18. 57 − 39

19. Measurement Howie is 43 inches tall. His brother is 51 inches tall. Find the difference in their heights.

20. There are 28 days left of summer vacation. If there were 90 days total, how many days have passed?

21. Elijah had 42 pieces of chalk. He gave 13 to Amado and 15 to Wapi. How many were left?

22. Ramona made 28 bracelets. Some are red and some are blue. If 17 are blue, how many are red?

Real-World PROBLEM SOLVING

Animals For Exercises 23–26, use the table, which shows speeds of animals in miles per hour (mph).

23. How much faster does a lion run than the fastest human?

24. What is the difference between the fastest animal and the slowest?

25. Which animal is 38 miles per hour slower than the lion?

26. Name two animals whose speeds have a difference of 7 miles per hour.

FAST MOVERS

Animal	Speed (mph)
Cheetah	70
Lion	50
Greyhound	39
Dragonfly	36
Rabbit	35
Giraffe	32
Fastest human	28
Elephant	25
Squirrel	12

Source: The World Almanac for Kids

H.O.T. Problems

27. NUMBER SENSE Without subtracting, how do you know if 31 − 19 is greater than 20?

28. **WRITING IN ►MATH** Refer to the table in Exercises 23–26. Write a real-world subtraction problem about the animal data for which the answer is 34.

GET READY to Learn

Toya had a choice to buy a 72-ounce bag of cat food or a 48-ounce bag of cat food. About how much more food is in the larger bag of cat food?

MAIN IDEA

I will estimate differences using rounding.

Math Online

macmillanmh.com
• Extra Examples
• Personal Tutor
• Self-Check Quiz

An exact answer is not needed. You can use rounding or compatible numbers to make an estimate that is close to the exact answer.

Real-World EXAMPLE   Estimate Differences

1 **PET FOOD** **About how much more food is in the larger bag of cat food?**

You need to estimate $72 - 48$.

One Way: Round	Another Way: Compatible Numbers
Step 1 Round to the nearest ten.	**Step 1** Change to compatible numbers.
72 $\longrightarrow$ 70 48 $\longrightarrow$ 50	72 $\longrightarrow$ 75 48 $\longrightarrow$ 50
Step 2 Subtract. $70 - 50 = 20$	**Step 2** Subtract. $75 - 50 = 25$

So, there are *about* 20 to 25 ounces more in the larger bag.

You can round to the nearest hundred when estimating.

 **Real-World EXAMPLE** **Estimate Differences**

2 **TRAVEL** The Cape Cod Lighthouse in Massachusetts is 66 feet tall. In South Carolina, the Morris Island Lighthouse is 161 feet tall. What is the difference between the heights of the two lighthouses?

66 ft

161 FT

You need to estimate 161 − 66.

Step 1 Round each number to the nearest hundred.

$$161 \longrightarrow 200$$
$$66 \longrightarrow 100$$

Round 161 to 200.
Round 66 to 100.

161
100 125 150 175 200

66
0 25 50 75 100

Step 2 Subtract.

$$161 \longrightarrow 200$$
$$\underline{-66} \longrightarrow \underline{-100}$$
$$100$$

So, Morris Island Lighthouse is *about* 100 feet taller than Cape Cod Lighthouse.

CHECK What You Know

Estimate. Round to the nearest ten or use compatible numbers.

See Example 1 (p. 114)

1.   84
      − 61

2.   91
      − 37

3.   46
      − 23

Estimate. Round to the nearest hundred. See Example 2 (p. 115)

4. 176
      − 64

5. 341
      − 183

6. 365
      − 119

7. Pedita invited 112 friends to a party. Of those, 37 could not come. About how many people will come?

8. **Talk About It** Explain the steps you would take to round 789 to the nearest hundred.

EXTRA PRACTICE
See page R8.

Estimate. Round to the nearest ten or use compatible numbers.

See Example 1 (p. 114)

9.   55
       − 37

10.   91
        − 73

11.   72
        − 49

12. 88 − 32

13. 86 − 68

14. 57 − 41

Estimate. Round to the nearest hundred. See Example 2 (p. 115)

15.   901
        − 260

16.   775
        − 191

17.   381
        − 265

18. 880 − 114

19. 322 − 199

20. 671 − 156

Data File

The American Robin is the state bird of Michigan. It can grow to a length of 28 centimeters. While building a nest, the female makes about 180 trips a day to gather materials.

21. Measurement What is the length of the robin rounded to the nearest 10 centimeters?

22. Round to the nearest ten to find the difference between the number of trips a robin can make to gather nesting materials, and a robin who made 157 trips.

Source: Cornell Lab of Ornithology

23. Measurement A gale has a wind speed up to 54 miles per hour. A breeze has a wind speed of 18 miles per hour. Estimate the difference between the two wind speeds.

24. The students want to buy 78 books for the school library. So far, they have bought 49 books. Estimate how many more books they need to buy to reach their goal.

25. Measurement Bernard is on a trip 62 miles one way. He will return. About how many miles are left in his trip if he has traveled 56 miles so far?

26. Margaret ordered 275 school T-shirts for the first day of school. Estimate the number of T-shirts left if she sold 183.

H.O.T. Problems

27. FIND THE ERROR Flor and Kenny estimated the difference of 78 and 45. Who is correct? Explain.

Flor

$78 \rightarrow 70$
$45 \rightarrow -50$
$\overline{20}$

Kenny

$78 \rightarrow 80$
$45 \rightarrow -50$
$\overline{30}$

28. **WRITING IN ►MATH** Write a real-world problem about a situation when you would use estimation.

TEST Practice

29. The temperature this morning was 59 degrees. This afternoon it warmed up to 87 degrees. What is the difference in temperature? (Lesson 3-1)

A 28   C 38

B 32   D 146

30. There are 92 pumpkins and 38 cornstalks in a field. About how many more pumpkins are there than cornstalks? (Lesson 3-2)

F 40   H 50

G 44   J 60

Spiral Review

Subtract. Use models if needed. Check your answer. (Lesson 3-1)

31. $45 - 28$ **32.** $51 - 16$ **33.** $37 - 9$

34. Mr. Sanchez has $450. Does he have enough money to buy the sofa and bed? Explain. (Lesson 2-5)

Sofa _____ $245

Bed _____ $207

Chair _____ $95

35. About how much money would Kele need to buy the items in the shop window? (Lesson 2-3)

$9.00

$4.00

$6.00

$10.00

Subtract Money

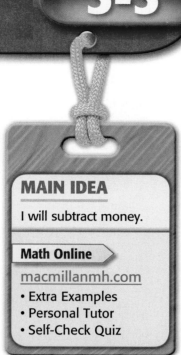

MAIN IDEA

I will subtract money.

Math Online

macmillanmh.com
• Extra Examples
• Personal Tutor
• Self-Check Quiz

GET READY to Learn

Large posters sell for 94¢, and small posters sell for 42¢. What is the difference in price?

Subtracting money is just like subtracting whole numbers.

Real-World EXAMPLE Subtract Cents

1 POSTERS **What is the difference in the price of the large and small posters?**

You need to find the difference between 94¢ and 42¢.

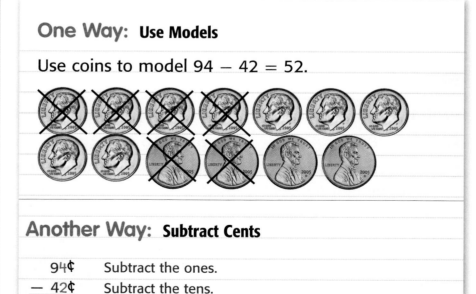

One Way: Use Models

Use coins to model 94 − 42 = 52.

Another Way: Subtract Cents

94¢	Subtract the ones.
− 42¢	Subtract the tens.
52¢	Place the cents sign *after* the difference.

So, the difference in price is 52¢.

When you subtract money, you often need to regroup.

Real-World EXAMPLE Subtract Dollars

2 **SKATING** Carson paid $59 for in-line skates. Mia paid $75. How much more did Mia pay for her skates?

You need to find the difference between $75 and $59.

Estimate $75 − $59 ⟶ $80 − $60 = $20

$$
\begin{array}{r}
6\,15 \\
\$ \cancel{75} \\
-\$59 \\
\hline
\$16
\end{array}
$$

Regroup 1 ten as 10 ones.
Subtract the ones.
Subtract the tens.
Place the dollar sign *before* the difference.

So, Mia paid $16 more than Carson.

Check for Reasonableness

Since $16 is close to the estimate, the answer is reasonable. ✓

Check Use addition to check your subtraction.

```
     ┌─── same ───┐
     ↓            │
  $75          $16
 −$59    ╱    +$59
 ────   ╱     ────
  $16  ╱      $75◄──
```

So, the answer is correct. ✓

The cent sign (¢) is placed *after* the money amount. The dollar sign ($) is placed *before* the money amount.

 CHECK **What You Know**

Subtract. Use models if needed. Check your answer.

See Examples 1 and 2 (pp. 118–119)

1. 29¢
   − 23¢

2. $77
   − $45

3. 45¢
   − 16¢

4. 75¢
   −6¢

5. $32
   − $27

6. 56¢
   − 48¢

7. The Westville Garden Club wants to make $65 on their flower sale. So far they have made $29. How much more money do they need?

8. **Talk About It** Explain how you know when you need to regroup a ten for 10 ones in a subtraction problem.

Subtract. Use models if needed. Check your answer.

See Examples 1 and 2 (pp. 118–119)

9. $52
− $41

10. $78
− $37

11. 66¢
− 25¢

12. 67¢
−9¢

13. 74¢
−7¢

14. 52¢
− 13¢

15. 93¢ − 42¢

16. 63¢ − 42¢

17. 73¢ − 31¢

18. 81¢ − 56¢

19. $28 − $19

20. 32¢ − 16¢

21. Cooper wants to buy 2 rubber snakes. Each snake is 39¢, and he has 95¢ to spend. Does he have enough money to buy a rubber spider also, for 15¢? Explain.

22. Enzo and his grandfather each paid $17 to go deep sea fishing. How much change did they receive from $40?

For Exercises 23–26, use the picture at the right. See Example 2 (p. 119)

23. Angel has $20. How much change will she get when she buys the baseball cap?

24. How much more does the tennis racket cost than the baseball bat?

25. Isabel has $60. She buys the tennis racket. Bruce has $40. He buys the baseball bat. Who gets more change back? Explain.

26. Coach paid for three things with four $20-bills. He got $10 in change back. What were the three things he bought?

H.O.T. Problems

27. OPEN ENDED A subtraction problem has an answer of 23. What could be the subtraction problem?

28. **WRITING IN** ▶**MATH** Write about a real-world situation when knowing how to make change will be helpful.

Subtract. Check your answer. (Lesson 3-1)

1. 28
      − 3

2. 37
      − 5

3. 70 − 19

4. 99 − 69

5. Devan had 38 ribbons for the art project. He gave 14 to Leon and 12 to Brendan. How many did he have left for himself? (Lesson 3-1)

6. Rodney had 23 baseball cards. He gave 6 to his best friend. How many cards does Rodney have now? (Lesson 3-1)

7. MULTIPLE CHOICE What is the difference of 97 and 65? (Lesson 3-1)

A 23   **C** 33

B 32   **D** 172

Estimate. Round to the nearest ten or use compatible numbers. (Lesson 3-2)

8. 83
      −62

9. 38
      −18

10. 63 − 28

11. 46 − 12

Estimate. Round to the nearest hundred. (Lesson 3-2)

12. 742
       −614

13. 567
       −113

14. 889 − 279

15. 335 − 142

16. Kaitlyn purchased a new blouse and skirt. The blouse was $19 and the skirt was $7. She paid with two $20 bills. How much change did she receive? (Lesson 3-3)

Subtract. Check your answer. (Lesson 3-3)

17. 34¢
       −14¢

18. $69
       −$35

19. MULTIPLE CHOICE Vanesa compared the prices of two cartons of milk. The table shows the prices.

Brand	Cost
A	99¢
B	81¢

How much more does Brand A cost than Brand B? (Lesson 3-3)

F 17¢

G 18¢

H 53¢

J 70¢

Subtract. Check your answer. (Lesson 3-3)

20. $64 − $6

21. 73¢ − 52¢

22. **WRITING IN ►MATH** Explain how subtracting money is like subtracting whole numbers and how it is different. (Lesson 3-3)

The Sounds of the Symphony

The Boston Philharmonic is a popular symphony orchestra. The orchestra has four instrument families— woodwinds, strings, brass, and percussion.

The musicians sometimes practice four times a week before a performance.

The Boston Philharmonic plays at Jordan Hall. Ticket prices are based on how close the seats are to the orchestra.

Boston Philharmonic	
Instrument Family	Number of Musicians
Percussion	6
Brass	18
Woodwinds	21
Strings	70

Weekend Series	
Seats	Ticket Price
A	$76
B	$58
C	$43
D	$29

Source: Boston Philharmonic

Real-World Math

Use the information on page 122 to solve each problem.

1. How many more woodwind musicians are there than percussion musicians?

2. Thirty-four of the string musicians play the violin. How many string musicians are not violinists?

3. The total number of musicians is 115. How many musicians do not play brass instuments?

4. On weeknights, the cost of a ticket in section D is $16. How much money do you save by going to the orchestra on Monday instead of Saturday?

5. Estimate the difference in the price of a ticket in section A and a ticket for a seat in section D.

6. If you paid for a ticket in section B with a $100-bill, how much change would you get?

7. How many musicians do not play brass or woodwind instruments?

8. Estimate the cost of each ticket to the nearest ten. Would it cost less to buy a ticket in section A and a ticket in section D or a ticket in section B and a ticket in section C?

Problem-Solving Skill

MAIN IDEA I will decide whether an answer to a problem is reasonable.

Brad bought a box of 85 straws of 3 different colors.
He found that 53 straws were blue and green.
Kenji thinks that about 30 straws are pink.
Is this a reasonable answer?

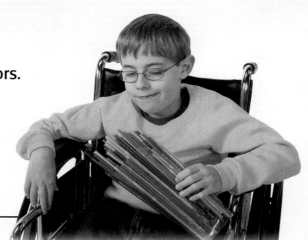

Understand	**What facts do you know?**
	• There are 85 straws.
	• There are 3 colors of straws.
	• There are 53 blue and green straws.
	What do you need to find?
	• Decide whether 30 is a reasonable amount of pink straws.
Plan	Use subtraction to find the number of pink straws. Then compare the answer to 30.

Solve	Subtract the number of blue and green straws from the total number of straws.
	
	  85
	−53
	———
	  32
	
	Since 32 is close to 30, it is reasonable to say that 30 of the straws are pink.

Check	Look back at the problem. Estimate by rounding.
	
	  85 ⟶ 90
	−53 ⟶ −50
	————
	      40
	
	So, the answer makes sense for the problem.

Refer to the problem on the previous page.

1. How do you know if the answer to a problem is reasonable?

2. Explain why you would ask yourself if an answer is reasonable.

3. If there are two colors of straws, and 57 are blue, about how many are green?

4. Explain why your answer to Exercise 3 is reasonable.

EXTRA PRACTICE
See page R9.

Solve.

5. Is 400 a reasonable estimate for the difference in attendance on Monday and Wednesday?

County Fair Attendance	
Monday	395
Tuesday	247
Wednesday	834

6. Anson swam 28 laps last week and 24 this week. He says he needs to swim about two more weeks to swim a total of 100 laps. Is this a reasonable estimate? Explain.

7. Aubrey's class earned tokens for good behavior. The tally table shows their votes for a reward.

Reward	Tally
Extra recess	卌 I
Game time	III
Pizza treat	卌 III
Read aloud time	卌 卌 卌

Is it reasonable to say about half voted for a read aloud time?

8. Mrs. Kinney's class of 30 students will play a game. Each child needs 3 cubes. Alfeo says that 100 cubes will be enough for the class to play the game. Is that reasonable? Explain.

9. Mr. Gonzalez made a table of the books he has collected. He says he has more than 50 books. Is this a reasonable estimate? Explain.

Book Collection	
Mystery	13
Gardening	25
Biography	8
Fiction	15

10. Julina estimated that she needs to make 100 favors for the family reunion. Is this a reasonable estimate if 67 relatives will come on Friday and 42 will come on Saturday? Explain your reasoning.

11. **WRITING IN ►MATH** Explain a situation when you would determine a reasonable answer to solve the problem.

Subtract Three-Digit Numbers with Regrouping

MAIN IDEA

I will model subtraction with regrouping.

You Will Need:
base-ten blocks

Math Online

macmillanmh.com
• Concepts in Motion

You can use models to regroup tens and hundreds.

ACTIVITY   Find 244 − 137.

Step 1   Use models.

244
− 137

Hundreds	Tens	Ones

Step 2   Subtract ones.

$$\begin{array}{r} {}^{3}\,{}^{14} \\ 2\,\cancel{4}\,\cancel{4} \\ -1\,3\,7 \\ \hline 7 \end{array}$$

You cannot take 7 ones from 4 ones.
Regroup 1 ten as 10 ones.
4 ones + 10 ones = 14 ones
Subtract. 14 ones − 7 ones = 7 ones

Hundreds	Tens	Ones

Step 3   Subtract tens.

$$\begin{array}{r} {}^{3}\,{}^{14} \\ 2\,\cancel{4}\,\cancel{4} \\ -1\,3\,7 \\ \hline 0\,7 \end{array}$$ 3 tens − 3 tens = 0 tens

Hundreds	Tens	Ones

Step 4 Subtract hundreds.

$$2\overset{3\ 14}{\cancel{4}\cancel{4}} \quad \text{2 hundreds} - \text{1 hundred} = \text{1 hundred}$$
$$-137$$
$$\overline{107}$$

Hundreds	Tens	Ones

So, 244 − 137 = 107.

Think About It

1. In Step 2, why did you regroup 1 ten as 10 ones?

2. What did you notice about the tens in Step 3 when you subtracted them?

3. Why do you sometimes have to regroup more than once?

CHECK What You Know

Use models to subtract.

4. 181 − 93

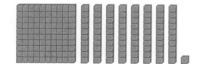

5. 322 − 148

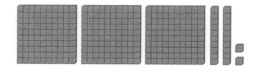

6. 342 − 179

7. 212 − 123

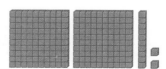

8. 328
  − 19

9. 308
  −125

10. 437
  − 243

11. 513
  − 155

12. **WRITING IN ►MATH** Explain when to regroup in subtraction.

Subtract Three-Digit Numbers with Regrouping

3-5

MAIN IDEA

I will subtract three-digit numbers with regrouping.

Math Online

macmillanmh.com
• Extra Examples
• Personal Tutor
• Self-Check Quiz

GET READY to Learn

Liseta, Will, and Alano each have craft paper. How many more sheets of paper of paper does Will have than Liseta?

Craft Paper	
Name	**Sheets**
Liseta	79
Will	265
Alano	128

In the Explore Activity, you learned to regroup tens. Regrouping hundreds works the same way.

Real-World EXAMPLE

1 **How many more sheets does Will have than Liseta?**

You need to find 265 − 79.

Step 1 Subtract ones.

$$\begin{array}{r} {\scriptstyle 5\ 15} \\ 2\cancel{6}\cancel{5} \\ -\ 79 \\ \hline 6 \end{array}$$

You cannot take 9 ones from 5 ones.
Regroup 1 ten as 10 ones.
5 ones + 10 ones = 15 ones
Subtract. 15 ones − 9 ones = 6 ones

Step 2 Subtract tens.

$$\begin{array}{r} {\scriptstyle 15} \\ {\scriptstyle 1\ 5\ 15} \\ \cancel{2}\cancel{6}5 \\ -\ 79 \\ \hline 86 \end{array}$$

You cannot take 7 tens from 5 tens.
Regroup 1 hundred as 10 tens.
5 tens + 10 tens = 15 tens
Subtract. 15 tens − 7 tens = 8 tens

Step 3 Subtract hundreds.

$$\begin{array}{r} {\scriptstyle 15} \\ {\scriptstyle 1\ 5\ 15} \\ \cancel{2}65 \\ -\ 79 \\ \hline 186 \end{array}$$

Subtract. 1 hundred − 0 hundreds = 1 hundred

So, 265 − 79 = 186.

2 **AIRPLANE** Denzel wants to buy a remote control airplane for $179. He has $350. How much money will he have left?

$179

You need to find $350 − $179.

Step 1 Subtract ones.

```
  4 10
$3 5̸0̸        You cannot take $9 from $0.
$179         Regroup $50 as $40 + $10.
    1        Subtract. $10 − $9 = $1
```

Step 2 Subtract tens.

```
    14
  2 4̸ 10
$3̸ 5̸ 0̸       You cannot take $70 from $40.
− $179       Regroup $300 as $200 + $100.
   71        Subtract. $140 − $70 = $70
```

Step 3 Subtract hundreds.

```
    14
  2 4̸ 10
$3̸ 5̸ 0̸       $200 + $100 = $100
− $179       Place the dollar sign before the difference.
$171
```

So, Denzel will have $171 left.

✓ CHECK What You Know

Subtract. Check your answer. See Examples 1 and 2 (pp. 128–129)

1. $764
   − $132

2. 458
   − 121

3. $614
   − $457

4. 391 − 178

5. 542 − 167

6. 317 − 198

7. This year, the third grade raised $342 for a dog shelter. Last year, they raised $279. How much more money did they raise this year than last year?

8. **Talk About It** What happens to the tens when you have to regroup twice?

Subtract. Check your answer. See Examples 1 and 2 (pp. 128–129)

9. $687
     − $353

10. $197
     − $94

11. 293
     − 172

12. 884
     − 63

13. $843
     − $187

14. $728
     − $359

15. 267
     − 178

16. 728
     − 259

17. 92¢ − 83¢

18. 58¢ − 27¢

19. 856 − 637

20. 531 − 499

21. Greta and Mulan are eating at the Good Eats Diner. Greta buys veggies and water. Mulan buys a piece of pizza and a salad. How much more does Mulan spend than Greta?

22. Measurement The Bank of America building in Charlotte is 871 feet tall. The One Liberty Place building in Philadelphia is 945 feet tall. How much taller is the One Liberty Place building?

Good Eats Diner	
Item	**Cost**
Pizza	$2
Salad	$2
Apple	$1
Veggies	$1
Water	$2
Smoothie	$3

Real-World PROBLEM SOLVING

Use the bar graph for Exercises 23–25.

23. How many more 3rd graders than 4th graders are buying their lunch?

24. What is the total number of students buying their lunch?

25. The lunchroom holds 150 students at one time. Name two classes that can eat at the same time. Explain.

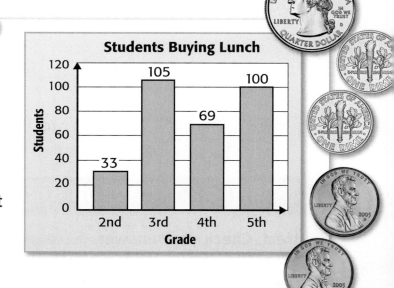

Algebra Find each missing digit.

26. 61■
      − 417
      ■02

27. ■99
      − 1■9
       750

28. 798
      − ■97
      4■1

29. 989
      − 77■
      ■18

H.O.T. Problems

30. NUMBER SENSE When Federico subtracted 308 from 785, he got 477. To check his answer he added 308 and 785. What did he do wrong?

31. FIND THE ERROR Odell and Liz are finding $566 − $347. Who is correct? Explain.

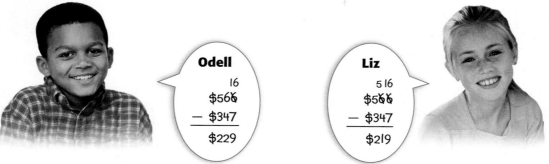

Odell

$$\begin{array}{r} {}^{1\,6} \\ \$5\,6\,6 \\ -\ \ \$3\,4\,7 \\ \hline \$2\,2\,9 \end{array}$$

Liz

$$\begin{array}{r} {}^{5\ 1\,6} \\ \$5\,6\,6 \\ -\ \ \$3\,4\,7 \\ \hline \$2\,1\,9 \end{array}$$

32. WRITING IN MATH Explain how to use addition to check the answer of a subtraction problem.

TEST Practice

33. Irene wrote this pattern.

24, 26, 28, 30, ___

What would be a reasonable answer for the next number? (Lesson 3-4)

A 28   **C** 35

B 32   **D** 40

34. Ty will hike 281 yards to get to the end of the trail. His friend is 187 yards from the end. What is the difference in the distance the boys have yet to walk? (Lesson 3-5)

F 94 yards   **H** 106 yards

G 104 yards   **J** 194 yards

Spiral Review

35. Baxter made $7 this week helping his dad around the house. Last week he made $8. Is it reasonable to say that he made at least $15 the last two weeks? Explain. (Lesson 3-4)

36. How much more does the bike cost than the skates? (Lesson 3-3)

$29

$53

Write the place value of the underlined digit. Then write the value of the digit. (Lesson 1-4)

37. 6<u>4</u>,284   **38.** 20,<u>0</u>02

MAIN IDEA I will choose the best strategy to solve a problem.

P.S.I. TEAM +

MIRANDA: For a class project, my teacher needs 155 paper towel rolls. So far, Marissa collected 24, Stan collected 32, and I collected 18.

YOUR MISSION: **Find out how many more cardboard rolls are needed.**

Understand	You know that 155 cardboard rolls are needed. You also know that three students have already collected 24, 32, and 18 cardboard rolls. Find how many more rolls are needed.
Plan	First, add to find the number of rolls collected. Then, subtract to find the amount still needed.
Solve	$\begin{array}{r} \overset{1}{}24 \\ 32 \\ +\ 18 \\ \hline 74 \end{array}$ So, 74 rolls have been collected.

Subtract 74 from 155 to find how many rolls are still needed.

$\begin{array}{r} \overset{0\ 15}{\cancel{1}\cancel{5}5} \\ -\ 74 \\ \hline 81 \end{array}$ So, 81 cardboard rolls are needed. |
| **Check** | Look back at the problem. You can check by adding. Since $81 + 74 = 155$, the answer is correct. |

Solve. Tell what strategy you used.

1. If 6 cans of tennis balls come in a box, about how much will the box of tennis balls cost? Show your work.

$3

2. There are 113 people riding a train. At the first stop, 32 get off. After the second stop, there are 14 people left. How many people got off at the second stop?

3. Mr. White bought 7 daisy plants for $28. He paid with two $20-bills. How much change did he get back?

4. Mrs. Carpenter received a bill for $134 for car repairs. Should she pay an estimated amount or should she pay the exact amount? Explain your reasoning.

Car Repairs
$134.00

5. **Measurement** It took Keisha 1 hour and 37 minutes to ride from her aunt's house to her grandmother's. Then she rode 3 hours and 14 minutes to her mom's house. About how many minutes were spent riding in the car?

6. The library received 155 new books today. If there are now 784 books, how many were there before the new books arrived?

7. Some children took part in a penny hunt. Use the table below to tell about how many more pennies Pat found than each of his two friends.

Penny Hunt	
Cynthia	133
Pat	182
Garcia	125

8. Hale has $20. He buys himself and two friends each a slice of pizza, a small salad, and water. How much money did he have left?

LUNCH MENU

SMALL SALAD	$ 1
LARGE SALAD	$ 1
SLICE OF PIZZA	$ 2
GRILLED CHEESE	$ 2
MILK	$ 1
WATER	$ 1

9. **WRITING IN ►MATH** Look back at Exercise 8. Give an example of an answer that is not reasonable. Explain your reasoning.

3-7 Subtract Greater Numbers

MAIN IDEA

I will learn to subtract three and four-digit numbers.

Math Online

macmillanmh.com
• Extra Examples
• Personal Tutor
• Self-Check Quiz

GET READY to Learn

The table shows the height of four different waterfalls. What is the difference in height between Ribbon Falls and Kalambo Falls?

Waterfalls

Name	Height (ft)
Ribbon	1,612
Angel	3,212
Yosemite	2,425
Kalambo	726

In this lesson, you will subtract greater numbers.

Real-World EXAMPLE

1 MEASUREMENT What is the height difference between Ribbon Falls and Kalambo Falls? Find 1,612 − 726.

Step 1 Subtract ones.

$$
\begin{array}{r}
0\,1\,2 \\
1,6\,1\,2 \\
-\ \ 726 \\
\hline
6
\end{array}
$$

You cannot take 6 ones from 2 ones.
Regroup 1 ten as 10 ones.
2 ones + 10 ones = 12 ones
12 ones − 6 ones = 6 ones

Step 2 Subtract tens.

$$
\begin{array}{r}
10 \\
5\,0\,1\,2 \\
1,6\,1\,2 \\
-\ \ 726 \\
\hline
86
\end{array}
$$

You cannot take 2 tens from 0 tens.
Regroup 1 hundred as 10 tens.
0 tens + 10 tens = 10 tens
10 tens − 2 tens = 8 tens

Step 3 Subtract hundreds and thousands.

$$
\begin{array}{r}
15\,10 \\
0\,5\,0\,1\,2 \\
1,6\,1\,2 \\
-\ \ 726 \\
\hline
886
\end{array}
$$

You cannot take 7 hundreds from 5 hundreds.
Regroup 1 thousand as 10 hundreds.
5 hundreds + 10 hundreds = 15 hundreds
15 hundreds − 7 hundreds = 8 hundreds
0 thousands − 0 thousands = 0 thousands

So, the difference in height is 886 feet.

2 BIKING The bar graph shows the length of two popular cross-country bike routes. How much longer is Route B than Route A?

You need to find 3,159 − 1,579.

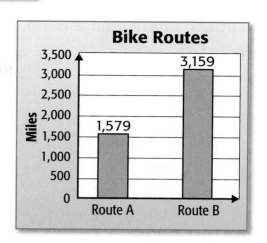

Bike Routes

Remember

When subtracting whole numbers, line up the digits in the ones place.

Step 1 Subtract ones.

$$\begin{array}{r} 3,15\mathbf{9} \\ -\ 1,57\mathbf{9} \\ \hline \mathbf{0} \end{array}$$

Step 2 Subtract tens.

$$\begin{array}{r} {\scriptstyle 0\ 15} \\ 3,\cancel{1}5\cancel{9} \\ -\ 1,5\mathbf{7}9 \\ \hline \mathbf{80} \end{array}$$

Step 3 Subtract hundreds and thousands.

$$\begin{array}{r} {\scriptstyle 10} \\ {\scriptstyle 2\ \cancel{0}15} \\ \cancel{3},\cancel{1}59 \\ -\ \mathbf{1,5}79 \\ \hline \mathbf{1,580} \end{array}$$

Check Check your answer.

same

$$\begin{array}{r} 3,159 \\ -\ 1,579 \\ \hline 1,580 \end{array} \qquad \begin{array}{r} 1,580 \\ +\ 1,579 \\ \hline 3,159 \end{array}$$

So, the answer is correct. ✓

So, 3,159 − 1,579 = 1,580.

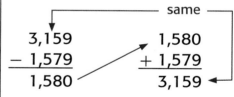 **What You Know**

Subtract. Check your answer. See Examples 1 and 2 (pp. 134–135)

1. $\begin{array}{r} \$7,371 \\ -\ \$365 \\ \hline \end{array}$

2. $\begin{array}{r} \$3,457 \\ -\ \$649 \\ \hline \end{array}$

3. $\begin{array}{r} 2,421 \\ -\ 865 \\ \hline \end{array}$

4. $\begin{array}{r} 7,234 \\ -\ 6,487 \\ \hline \end{array}$

5. Cell phones were invented in 1983. The TV was invented 56 years before that. What year was the TV invented?

6. **Talk About It** Explain the steps to find 8,422 − 5,995.

Subtract. Check your answer. See Examples 1 and 2 (pp. 134–135)

7. 1,392
      − 238

8. 3,298
      − 858

9. 3,475
      − 1,267

10. 3,665
       − 1,643

11. $3,421
       − $1,049

12. $5,452
       − $1,187

13. $4,875
       − $3,168

14. $6,182
       − $581

15. 6,340
       − 3,451

16. 5,123
       − 2,736

17. $1,856
       − $969

18. $4,137
       − $1,562

19. Of the 2,159 pre-sold county fair tickets, only 1,947 tickets were collected at the gate. How many tickets were not collected?

20. Measurement The distance around a rectangular swimming pool is 300 yards. What are the measurements of the remaining sides?

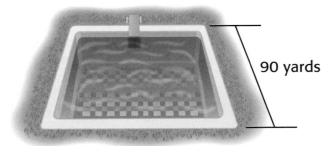

90 yards

21. To earn money for a trip, band members had to sell 1,590 boxes of popcorn. So far they have sold 779 boxes. How many more boxes do they have to sell?

Algebra Compare. Use >, <, or =.

22. 1,543 − 984 ● 5,193 − 4,893

23. 2,006 − 781 ● 5,224 − 3,999

24. 8,937 − 3,038 ● 3,598 − 1,084

25. 5,070 − 2,345 ● 8,765 − 1,965

H.O.T. Problems

26. OPEN ENDED Write a real-world subtraction problem whose difference is 1,379.

27. CHALLENGE Write a subtraction problem in which the answer is 1,735.

28. WRITING IN ►MATH Explain how subtracting four-digit numbers is like subtracting three-digit numbers.

Do Not Zero Out

Find Differences

Get Ready!

Players: 2 players

Get Set!

- Label the blank number cube 4–9.

- Each player writes 999 at the top of his or her paper.

Go!

- Player 1 rolls the number cubes and writes the two-digit number under 999. Subtract.

- Player 2 rolls the number cubes, makes a two-digit number, writes the number under 999 on his or her paper, and subtracts.

- Players continue, subtracting from the lowest number.

You will need:
one 0–5 number cube, one blank number cube

```
  999
-  74
  925

  925
-  32
  893
```

- When a player thinks the difference is as low as possible, he or she may stop. The other player may continue taking turns.

- If a player rolls a number that takes the difference below zero, the game is over. The other player wins. Otherwise, the player with the smallest difference wins.

Subtract Across Zeros

MAIN IDEA

I will learn how to subtract across zeros.

Math Online

macmillanmh.com
- Extra Examples
- Personal Tutor
- Self-Check Quiz

GET READY to Learn

A large box of watermelons weighs 300 pounds. A smaller box weighs 134 pounds. What is the difference in the weights?

Sometimes, before you can begin subtracting, you have to regroup more than one time.

Real-World EXAMPLE   Subtract Across Zeros

1 What is the difference in the weight of the two boxes?

You need to find 300 − 134.

Step 1 Regroup.

```
  210
  3̶0̶0
−134
```

You cannot take 4 ones from 0 ones.

Regroup.

There are no tens to regroup.

Regroup 3 hundreds as 2 hundreds and 10 tens.

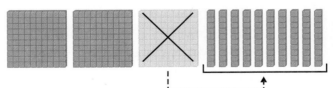

Step 2 Regroup.

```
      9
  2̶1̶010
  3̶0̶0̶
 −134
```

Regroup 10 tens as 9 tens and 10 ones.

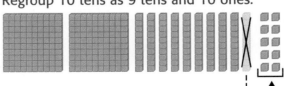

Step 3 Subtract.

```
      9
  2̶1̶010
  3̶0̶0̶
 − 134
   166
```

Subtract the ones, tens, and hundreds.

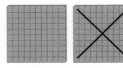

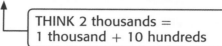

2 **TOYS** Brandy made 2,000 hops on her pogo stick last week. Roque hopped 253 times. How many more hops did Brandy make?

You need to find 2,000 − 253.

Step 1 Regroup.

$$\begin{array}{r} {\scriptstyle 1\ 10} \\ \cancel{2},\cancel{0}00 \\ -\ \ 253 \end{array}$$

You cannot take 3 ones from 0 ones. Regroup. There are no tens or hundreds.

Regroup 1 thousand as 10 hundreds.

THINK 2 thousands = 1 thousand + 10 hundreds

Step 2 Regroup.

$$\begin{array}{r} {\scriptstyle 9} \\ {\scriptstyle 1\,\cancel{10}\,10} \\ \cancel{2},\cancel{0}\cancel{0}0 \\ -\ \ 253 \end{array}$$

Regroup again.

Regroup 1 hundreds as 10 tens.

THINK 2 thousands = 1 thousand + 9 hundreds + 10 tens

Step 3 Regroup and subtract.

$$\begin{array}{r} {\scriptstyle 9\ \ 9} \\ {\scriptstyle 1\,\cancel{10}\,\cancel{10}\,10} \\ \cancel{2},\cancel{0}\cancel{0}\cancel{0} \\ -\ \ 253 \\ \hline 1,747 \end{array}$$

Regroup 1 ten as 10 ones.
Subtract.
10 ones − 3 ones = 7 ones, 9 tens − 5 tens = 4 tens,
9 hundreds − 2 hundreds = 7 hundreds,
1 thousand − 0 thousands = 1 thousand

So, Brandy hopped 1,747 times more than Roque.

Check Add up to check.

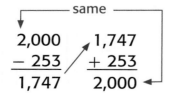

$$\begin{array}{cc} 2{,}000 & 1{,}747 \\ -\ \ 253 & +\ \ 253 \\ \hline 1{,}747 & 2{,}000 \end{array}$$

same

So, the answer is correct. ✔

Subtract. Check your answer. See Examples 1 and 2 (pp. 138–139)

1. 208
− 68

2. 802
− 77

3. $500
− $317

4. $3,300
−$226

5. There are 200 pennies in a jar. If 27 are removed, how many are left?

6. (Talk About It) Explain the steps to find 503 − 366.

Practice and Problem Solving

EXTRA PRACTICE
See page R10.

Subtract. Check your answer. See Examples 1 and 2 (pp. 138–139)

7. 401
−37

8. 902
−84

9. 300
− 217

10. 400
− 256

11. $500
− $388

12. $800
− $685

13. $7,400
−$1,211

14. $4,003
−$227

15. 2,001 − 132

16. 5,006 − 1,257

17. 9,000 − 6,652

18. 6,000 − 168

19. Darnell paid for his TV set with four $50-bills. How much did his TV set cost if he received $42 in change?

20. A farmer picked 2,008 oranges. Of these, 32 were thrown away and 1,469 were sold. How many oranges are left?

21. Hanako wants to earn 200 points for turning her homework in on time this year. How many more points does she need if she has 137 so far?

22. Measurement A playground measures 400 feet around its outside. Three sides together measure 293 feet. How many feet is the last side of the playground?

Use the bar graph for Exercises 23–25.

23. How many more 5th graders than 4th graders went to the safety assembly?

24. How many students in 2nd, 3rd and 4th grade went to the safty assembly?

25. The gym holds 175 students. Name two classes that could attend the assembly at the same time. Explain.

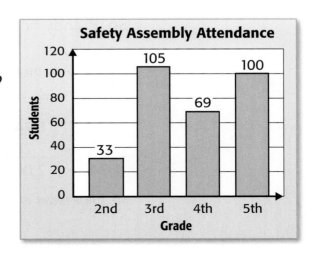

140 Chapter 3 Subtract To Solve Problems

H.O.T. Problems

26. WHICH ONE DOESN'T BELONG? Identify the problem with the incorrect answer. Explain your reasoning.

$2,017 - 1,713 = 304$

$8,500 - 4,764 = 4,836$

$1,109 - 768 = 341$

$5,000 - 3,574 = 1,426$

27. **WRITING IN** ►**MATH** Find $3,004 - 1,238$. Explain the steps you follow as you find the difference.

TEST Practice

28. Which problem could be used to check $3,624 - 1,896 = 1,728$?

(Lesson 3-7)

A $1,896 + 1,728 = $ ▇

B $1,896 - 1,728 = $ ▇

C $1,896 \times 1,728 = $ ▇

D $1,896 \div 1,728 = $ ▇

29. Which number is 8 less than 2,002? (Lesson 3-8)

F 1,046

G 1,054

H 1,994

J 2,044

Spiral Review

Subtract. Check your answer. (Lesson 3-7)

30.
$$\begin{array}{r} 1,951 \\ -\ \ 563 \\ \hline \end{array}$$

31.
$$\begin{array}{r} 3,298 \\ -1,699 \\ \hline \end{array}$$

32.
$$\begin{array}{r} \$3,679 \\ -\ \$2,789 \\ \hline \end{array}$$

33. Measurement Juwan jumped 32 inches. Grant jumped 3 feet 1 inch. Juwan said he jumped farther. Is his answer reasonable? Explain. (Lesson 3-4)

Compare. Use $>$, $<$, or $=$. (Lesson 1-6)

34. 475 ● 478

35. 3,392 ● 3,299

36. 2,381 ● 12,000

Select Addition or Subtraction

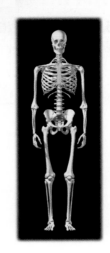

GET READY to Learn

A baby can have 300 bones.
An adult has 206 bones.
How many more bones does
a baby have?

MAIN IDEA

I will select addition or subtraction to solve problems.

Math Online

macmillanmh.com
• Extra Examples
• Personal Tutor
• Self-Check Quiz

In this lesson, you will select addition or subtraction to solve problems.

Real-World EXAMPLE

1. **SCIENCE** How many more bones do babies have than adults?

   Decide whether to use addition or subtraction to solve.

   The words *how many more* mean subtraction is needed.

   $$\begin{array}{r} \overset{9}{} \\ 2\,\overset{1}{\cancel{0}}\,\overset{10}{\cancel{0}} \\ \cancel{3\,0\,0} \\ -\,2\,0\,6 \\ \hline 9\,4 \end{array}$$

   THINK When subtracting across zeros, remember to regroup.

   So, a baby has 94 more bones than an adult.

Real-World EXAMPLE

2. **MONEY** Maya spent 45¢ on an apple and 52¢ on a banana. How much did she spend in all?

   The words *in all* show that you will need to add.

   $$\begin{array}{r} 45¢ \\ +\,52¢ \\ \hline 97¢ \end{array}$$ So, Maya spent 97¢.

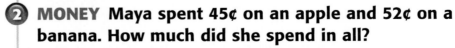

Select addition or subtraction to solve. See Examples 1 and 2 (p. 142)

1. Dale has 174 acorns. If Dale is trying to collect 225 acorns, how many more acorns must he collect?

2. Frankie raked 711 leaves. Lea raked 25 more leaves than Frankie. How many leaves did she rake in all?

3. Patsy has jumped rope 284 times in a row. To tie the record, she needs to jump 150 more times. What is the record?

4. **Talk About It** List two words or phrases that can be used to show that addition is needed to solve a problem.

Practice and Problem Solving

EXTRA PRACTICE
See page R10.

Select addition or subtraction to solve. See Examples 1 and 2 (p. 142)

5. A band sold 677 concert tickets. There are 800 seats. How many more tickets are available?

6. Natalie took 67 pictures of animals. Of the pictures, 19 were of birds. How many were not birds?

7. The table shows how many stickers each student earned. How many stickers did Corina and Suzanne earn in all?

8. The table shows how many waste baskets are emptied each week. How many baskets are there in all?

Stickers Earned

Name	Number of Stickers
Corina	44
Kat	37
Suzanne	57

Waste Baskets Emptied Each Week

Day	Number of Waste Baskets
Monday	14
Wednesday	35
Friday	22

H.O.T. Problems

9. **CHALLENGE** Mac's garden has 45 tulips, 32 carnations, and 18 lilies. If he cuts 8 of each to put in a vase, how many flowers would be left?

10. **WRITING IN ► MATH** Write a real-world word problem that uses addition. Solve.

FOLDABLES™ Study Organizer GET READY to Study

Be sure the following Key Vocabulary words and Key Concepts are written in your Foldable.

Subtraction
Two-Digit Subtraction
Estimate Differences
Subtract Money
Regrouping in Subtraction
Subtract Across Zeros
Subtract 3-digit and 4-digit Numbers
Addition or Subtraction

Key Concepts

- You may need to **regroup** to subtract.
  (p. 112)

$$
\begin{array}{r}
3\ 11 \\
4\cancel{1} \\
-27 \\
\hline
14
\end{array}
$$

  You cannot take 7 ones from 1 one.
  Regroup 1 ten as 10 ones.
  1 one + 10 ones = 11 ones

- Sometimes, you have to regroup more than one time. (p. 128)

$$
\begin{array}{r}
6\ 1215 \\
\cancel{7}\cancel{3}\cancel{5} \\
-247 \\
\hline
488
\end{array}
$$

  Regroup 1 ten as 10 ones.
  Regroup 1 hundred as 10 tens.

- Sometimes, before subtracting, you have to regroup more than one time. (p. 138)

$$
\begin{array}{r}
9\ \ 9 \\
4\ 10\ 10\ 10 \\
5,\cancel{0}\cancel{0}\cancel{0} \\
-3,283 \\
\hline
1,717
\end{array}
$$

Key Vocabulary

difference (p. 111)

estimate (p. 114)

regroup (p. 112)

Vocabulary Check

Choose the vocabulary word that completes each sentence.

1. If there are not enough ones to subtract from, you will need to ____?____ one ten.

2. When you find an answer that is close to the exact answer, you ____?____ .

3. The answer to a subtraction problem is the ____?____ .

4. In the number sentence 12 − 8 = 4, the 4 is the ____?____ .

5. A number close to the exact value is a(n) ____?____ .

Lesson-by-Lesson Review

3-1 Two-Digit Subtraction (pp. 111–113)

Example 1
Find 23 − 15.

Step 1 Regroup.

$$
\begin{array}{r}
\overset{1\ 13}{\cancel{23}} \\
-15 \\
\end{array}
$$
Regroup 1 ten as 10 ones.
3 ones + 10 ones = 13 ones

Step 2 Subtract ones.

$$
\begin{array}{r}
\overset{1\ 13}{\cancel{23}} \\
-15 \\
\hline
8 \\
\end{array}
$$
13 ones − 5 ones = 8 ones

Step 3 Subtract tens.

$$
\begin{array}{r}
\overset{1\ 13}{\cancel{23}} \\
-15 \\
\hline
08 \\
\end{array}
$$
1 ten − 1 ten = 0 tens

So, 23 − 15 = 8.

Subtract. Check your answer.

6. $\begin{array}{r} 17 \\ -4 \\ \end{array}$ 7. $\begin{array}{r} 38 \\ -6 \\ \end{array}$

8. 83 − 49   9. 62 − 28

10. Jackie has 37 pieces of paper. She gives 14 to Sharon. How many are left for Jackie?

11. There are 18 days left in the month. If there are 31 days total, how many days have passed?

12. Mario wrote a report for extra credit. There were 25 points possible. Mario earned 19 points. How many points did he miss?

3-2 Estimate Differences (pp. 114–117)

Example 2
Estimate 679 − 325. Round to the nearest hundred.

Round each number then subtract.

$$
\begin{array}{l}
679 \longrightarrow 700 \\
325 \longrightarrow -300 \\
\hline
400 \\
\end{array}
$$

So, 679 − 325 is about 400.

Estimate. Round to the nearest ten.

13. 94 − 55   14. 43 − 29

Estimate. Round to the nearest hundred.

15. 732 − 280   16. 668 − 325

17. A striped kite costs $108. A blue kite costs $91. About how much more does the striped kite cost?

Chapter 3 Study Guide and Review **145**

3-3 **Subtract Money** (pp. 118–120)

Example 3
Find 74¢ − 58¢.

```
  6 14     Regroup.
  7̶4̶¢     Subtract the ones.
−58¢      Subtract the tens.
  16¢      Place the cents sign after the answer.
```

Example 4
Find $74 − $58.

```
  6 14
  $7̶4̶     Regroup.
−$58      Subtract the ones.
  $16     Subtract the tens.
```

Subtract. Check your answer.

18.
```
  73¢
−  6¢
```

19.
```
  $92
− $48
```

20. $77 − $38

21. 63¢ − 58¢

22. Denny has 50¢. If he buys a goldfish for 38¢, how much change will he receive?

23. One bag of oranges costs $6. Jaleesa buys two bags of oranges. She gives the cashier $20. How much change will Jaleesa receive?

3-4 **Problem-Solving Skill:** Reasonable Answers (pp. 124–125)

Example 5
Reuben's book has 96 pages. He read 47 pages today and wants to finish his book tomorrow. Is his goal reasonable? Explain.

Use estimation to check for reasonableness.

```
  96  ⟶   100   total pages
− 47  ⟶  − 50   pages read today
          50   more pages to read
```

Reuben has about 50 pages to read.

Reuben can read 47 pages in one day. 47 is close to 50. So, his goal is reasonable.

Solve.

24. All 30 seats on a bus are taken. After the first stop, only 18 seats are taken. Is it reasonable to say that about 10 passengers got off of the bus? Explain.

25. Marra is saving her money to buy a bike. The bike costs $90. She saved $23 this week and $19 last week. Is it reasonable to say Marra will have enough to buy a bike after 1 more week? Explain.

Subtract Three-Digit Numbers with Regrouping (pp. 128–131)

Example 6

A group of friends made bracelets to sell at a craft fair. How many bracelets did they have left?

Craft Fair Bracelets	
Bracelets Made	Bracelets Sold
133	98

To find how many bracelets are left, you need to find 133 − 98.

Step 1 Subtract the ones.

$$\begin{array}{r} {\scriptstyle 2\ 13} \\ 1\cancel{3}\cancel{3} \\ -\ 98 \\ \hline 5 \end{array}$$

Regroup 1 ten as 10 ones.

3 ones + 10 ones = 13 ones

13 ones − 8 ones = 5 ones

Step 2 Subtract the tens.

$$\begin{array}{r} {\scriptstyle 0\,12\,13} \\ 1\cancel{3}\cancel{3} \\ -\ 98 \\ \hline 35 \end{array}$$

Regroup 1 hundred as 10 tens.

2 tens + 10 tens = 12 tens

12 tens − 9 tens = 3 tens

There are no hundreds to subtract. So, the friends have 35 bracelets left.

Check Check the answer by adding.

$$\begin{array}{r} 133 \\ -\ 98 \\ \hline 35 \end{array} \qquad \begin{array}{r} 35 \\ +\ 98 \\ \hline 133 \end{array}$$

So, the answer is correct. ✓

Subtract. Check your answer.

26. $\begin{array}{r} 213 \\ -155 \\ \hline \end{array}$

27. $\begin{array}{r} \$633 \\ -\$486 \\ \hline \end{array}$

28. $\begin{array}{r} 577 \\ -\ 98 \\ \hline \end{array}$

29. $\begin{array}{r} \$431 \\ -\$252 \\ \hline \end{array}$

30. 767 − 78

31. $333 − $265

32. 538 − 329

33. 875 − 677

34. 728 − 527

35. 492 − 235

36. **Measurement** There are 365 days in one year. There was sunshine 173 days this year. How many days did the sun not shine?

37. Students are having a car wash to raise money for the drama club. Their goal is to raise $150. How much more money do they need to meet their goal?

Drama Club Car Wash	
1st hour	$23
2nd hour	$29
3rd hour	$25

38. Della needs to jump rope 433 times in a row to beat the school jump rope record. So far, Della has jumped 284 times. How many more jumps will beat the record?

3-6 Problem-Solving Investigation: Choose a Strategy (pp. 132–133)

Example 7
Students need to make 425 cards for hospital patients. The second graders made 75 cards. The third graders made 90 cards. How many cards still need to be made?

Find the number of cards made.

```
   75    second grade made
 + 90    third grade made
  165    total made so far
```

Subtract to find how many cards are still needed.

```
   3 12
   4̷2̷5
 −  165
    260
```

So, 260 cards still need to be made.

Solve. Tell what strategy you used.

39. Ellis has a collection of 711 leaves. Of those leaves, he collected 126 of them this year. How many leaves were in his collection before this year?

40. The table below shows how long it takes each child to walk to school. How much longer does Salma walk than Ernie and Jesse combined?

The Walk To School	
Ernie	14 minutes
Jesse	18 minutes
Salma	36 minutes

3-7 Subtract Greater Numbers (pp. 134–136)

Example 8
Find 5,236 − 2,477.

Subtract as you would with smaller numbers. Remember to regroup when needed.

```
  11 12
 4 1̷ 2̷ 16
  5̷,2̷3̷6̷
 − 2,477
   2,759
```

Subtract. Check your answer.

41. $4,246
      − $1,781

42. 7,624
      − 5,937

43. A theater compared weekend ticket sales. In which month were fewer tickets sold? Explain.

Theater Ticket Sales		
Days	January	February
Saturday	789	829
Sunday	677	617

3-8 Subtract Across Zeros (pp. 138–141)

Example 9
A baseball stadium holds 3,000 people. If there are 588 empty seats, how many people are at the game? Find 3,000 − 588.

Step 1 Regroup the thousands.

```
 2 10
 3̶,0̶00   You cannot take 8 ones from 0 ones.
−  588   There are no tens or hundreds to
         regroup. Regroup 1 thousand as
         10 hundreds.
```

Step 2 Regroup the hundreds.

```
   9
 2 1̶0 10
 3,0̶0̶0̶    Regroup 1 hundred as 10 tens.
− 588
```

Step 3 Regroup the tens.

```
   9  9
 2 1̶0 1̶0 10
 3, 0̶ 0̶ 0̶    Regroup 1 ten as 10 ones.
− 5 8 8
 2, 4 1 2    Subtract.
```

Check Add up to check.

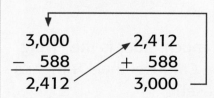

```
 3,000      2,412
−  588     + 588
 2,412      3,000
```

The answer is correct. ✓

So, there are 2,412 people at the game.

Subtract. Check your answer.

44.
```
  400
−  39
```

45.
```
  $600
 −$322
```

46.
```
  202
 −174
```

47.
```
  $800
 −$712
```

48. 8,000 − 456 49. 4,000 − 823

50. 5,007 − 669 51. 3,002 − 518

52. Norma bought a doll house for $189. She paid for the doll house with two $100-bills. How much change should Norma receive?

53. The school sold 1,677 tickets to the choir concert. There are 3,000 seats in the auditorium. How many more tickets are available?

54. Students in Ms. Turner's class earn points for reading books. Students can use their points to buy rewards. Kimberly has 74 reading points. She wants to earn an extra recess. How many more points does Kimberly need?

Ms. Turner's Reading Rewards	
Sticker	100 points
Homework pass	200 points
Extra recess	300 points

3-9 **Select Addition or Subtraction** (pp. 142–143)

Example 10
A farmer sold 354 melons on Saturday and 466 on Sunday. How many melons did he sell in all?

The words *in all* imply that you will need to add.

$$\begin{array}{r} 1\ 1 \\ 354 \\ +\ 466 \\ \hline 820 \end{array}$$

So, the farmer sold 820 melons in all.

Example 11
A school plants $1,109 worth of flowers and trees. The trees cost $768. How much did the flowers cost?

To find the difference in the cost of the trees and flowers you need to subtract.

$$\begin{array}{r} 10 \\ 0\ \backslash\ 10 \\ \$1,1\cancel{0}9 \\ -\ \$768 \\ \hline \$341 \end{array}$$

So, the flowers cost $341.

Check Add up to check

```
   ┌─── same ───┐
   ↓            │
$1,109      $341
– $768    + $768
 $341      $1,109 ←
```

So, the answer is correct. ✓

Select addition or subtraction to solve.

55. The school band practices 62 times in the winter and 48 times in the spring. How many more times do they practice in the winter than in the spring?

56. Bethany bought the bike shown. She used a $25-off coupon. How much did Bethany have to give the clerk for her bike?

$179

57. A store sold 346 clocks last year. This year they sold 251 clocks. How many clocks did they sell for the two years?

58. An apartment building has 32 apartments on the first floor and 25 on the second floor. How many apartments are there in all?

59. How much older is the Flying Horse Carousel than Trimper's Carousel?

Old Carousels	
Date built	**Name**
1867	Flying Horse Carousel
1902	Trimper's Carousel

For Exercises 1–2, decide whether each statement is *true* or *false*.

1. Always begin with the tens place when subtracting.

2. Sometimes before you can begin subtracting, you have to regroup more than one time.

Select addition or subtraction to solve.

3. Vianca ate 2 pieces of fruit today and 3 pieces yesterday. How many pieces of fruit did she eat altogether?

4. The movie store had 8 copies of my favorite movie. Then they sold 3. How many copies do they have left?

Estimate. Round to the nearest hundred.

5.   632
    − 151

6.   862
    − 305

7. **MULTIPLE CHOICE** How much more does the black pair of ballet shoes cost than the pink pair?

Price of Ballet Shoes	
Black pair	$108
Pink pair	$91

A $9   C $27

B $17   D $117

8. A cake was cut into 40 pieces. Is it reasonable to say that there was enough cake for 32 people? Explain.

Subtract. Check your answer.

9.   394
    − 271

10. $927
     − $439

11.   3,079
     − 674

12. $8,000
     − $2,174

Select addition or subtraction to solve.

13. Each can below has a different type of nut in it. How many servings of pecans and almonds are there altogether?

Almonds — 23 servings   Peanuts — 58 servings   Pecans — 39 servings

14. **MULTIPLE CHOICE** Rusty's book has 285 pages. He read 24 pages on Monday, 37 pages on Tuesday, and 41 pages on Wednesday. How many pages does he have left?

F 102   H 187

G 183   J 309

15. **WRITING IN ►MATH** Explain why you should always check over your work.

PART 1   Multiple Choice

Read each question. Then fill in the correct answer on the answer sheet provided by your teacher or on a sheet of paper.

1. What is $9,000 + 400 + 50 + 2$ in standard form?

A 2,549

C 9,452

B 4,925

D 9,542

2. Each year the Garden Club collects $1,200. So far, the club has $958. How much more does the club need to collect?

F $242

H $348

G $252

J $358

3. Which of the following shows the numbers in order from least to greatest?

A 115, 119, 122, 127

B 115, 122, 119, 127

C 119, 115, 122, 127

D 127, 122, 119, 115

4. What is the best estimate of the difference rounded to the nearest hundred?

$$721 - 293$$

F 300

H 500

G 400

J 350

5. Alina has 145 stickers. Which of these equals 145?

A $1 + 4 + 5$

B $1 + 40 + 500$

C $100 + 50 + 4$

D $100 + 40 + 5$

6. How can you find the total quiz score?

Quiz #1	Score
Part 1	18
Part 2	16
Part 3	19

F 18×3

H $18 + 16 + 19$

G $18 + 16$

J $18 + 16 - 19$

7. What number is missing in the pattern 12, 18, 24, 30, ___?

A 34

C 38

B 36

D 40

8. The table shows the number of students in each grade at Glenview Elementary School. How many more third graders are there than first graders?

Glenview Elementary School	
1st graders	216
2nd graders	194
3rd graders	233
4th graders	205

F 17   **H** 194

G 39   **J** 233

9. Last year, the theater spent $7,625. This year the theater will spend $9,910. How much more is the theater spending this year?

A $2,285   **C** $2,325

B $2,315   **D** $2,395

10. On a car trip, Jerry counts 125 white cars. Marla counts 67 sports cars. How many more cars did Jerry count?

F 58   **H** 68

G 62   **J** 192

PART 2   Short Response

Record your answers on the answer sheet provided by your teacher or on a sheet of paper.

11. Estimate the difference of 376 − 269. Show your work.

12. Draw models of base-ten blocks to show 137 − 25.

13. A grocery store clerk put 12 more cans on each shelf. What is the total number of cans now? What is the difference in the number of cans on shelf 2 and shelf 3 now?

Shelf	Cans
1	16
2	48
3	61

PART 3   Extended Response

Record your answers on the answer sheet provided by your teacher or on a sheet of paper.

14. The students sold muffins and juice at the bake sale. They earned $25 from selling muffins. They earned more than $40 in all. How much could they have earned from selling juice? Explain.

NEED EXTRA HELP?														
If You Missed Question...	1	2	3	4	5	6	7	8	9	10	11	12	13	14
Go to Lesson...	1-3	3-3	1-7	3-2	1-3	2-4	1-1	3-5	3-3	3-5	3-2	3-5	3-6	3-6

Develop Multiplication Concepts and Facts

BIG Idea What is Multiplication?

Multiplication is an operation on two numbers that can be thought of as repeated *addition*.

Example Tarantula spiders like this one can be seen at the Louisville Zoo in Kentucky. Suppose there are 4 spiders. Spiders have 8 legs. So, there would be 4×8 or 32 legs in all.

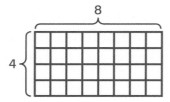

What will I learn in this chapter?

- Explore the meaning of multiplication.
- Use models and patterns to multiply.
- Multiply by 2, 4, 5, 10, 0, and 1.
- Use multiplication properties and rules.
- Solve a problem by identifying extra or missing information.

Key Vocabulary

multiply

multiplication

Zero Property of Multiplication

array

Commutative Property of Multiplication

Math Online > **Student Study Tools**
at **macmillanmh.com**

Make this Foldable to help organize information about multiplication concepts and facts. Begin with four sheets of $8\frac{1}{2}'' \times 11''$ paper.

1 **Fold** one piece of paper in half.

2 **Fold** one side up 5 inches. Glue the outer edges.

3 **Label** with the lesson titles. Record what you learn.

Meaning of Multiplication   Arrays and Multiplication

4 **Repeat** Steps 1–3 with 3 more pieces of paper.

Chapter 4 Develop Multiplication Concepts and Facts   **155**

You have two ways to check prerequisite skills for this chapter.

Option 2

Math Online ▷ Take the Chapter Readiness Quiz at macmillanmh.com.

Option 1

Complete the Quick Check below.

QUICK Check

Find each sum. (Prior grade)

1. $2 + 2 + 2 + 2$ **2.** $4 + 4$ **3.** $5 + 5 + 5$

4. $10 + 10 + 10 + 10$ **5.** $0 + 0 + 0$ **6.** $1 + 1 + 1 + 1 + 1$

Identify a pattern. Then find the missing numbers. (Lesson 1-1)

7. 5, 10, 15, ■, ■, 30   **8.** 2, ■, 6, 8, ■, 12   **9.** 3, 6, 9, ■, 15, ■

10. ■, 8, 12, 16, ■   **11.** ■, 20, 30, ■, 50   **12.** 6, 12, ■, 24, ■

Write an addition sentence for each picture. (Prior grade)

13.
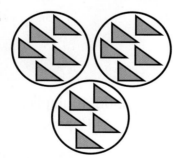

14.

15.

Solve. Use repeated addition. (Prior grade)

16. Larisa has 2 cups with 4 crackers in each cup. How many crackers does she have in all?

17. On Monday and Tuesday, Lance rode his bike around the block 3 times each day. How many times in all did he ride his bike around the block?

Multiplication is an operation on two numbers that can be thought of as repeated *addition*. The sign (×) means to **multiply**. You can use models to explore multiplication.

MAIN IDEA

I will use models to explore multiplication.

You Will Need
connecting cubes

New Vocabulary

multiplication
multiply

Math Online

macmillanmh.com

• Concepts in Motion

ACTIVITY **Find how many are in 5 groups of 4.**

Step 1 **Model 5 groups of 4.**

Use connecting cubes to show 5 groups of 4 cubes.

There are 5 groups. There are 4 cubes in each group.

Step 2 **Find 5 groups of 4.**

Model the groups of cubes with numbers. Use repeated addition.

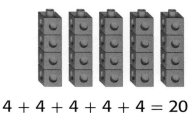

$$4 + 4 + 4 + 4 + 4 = 20$$

Step 3 Record the results.

Copy the table. Record the number of groups, the number in each group, and the total.

Use connecting cubes to explore other ways to group the 20 cubes equally.

Number of Groups	Number in Each Group	Total
5	4	20

Think About It

1. How can addition help you find the total number when multiplying?

2. How did you find the total number of cubes in Step 2?

3. What do the numbers stand for in the number sentence in Step 2?

4. Explain another way to group 20 cubes equally.

CHECK What You Know

Use models to find the total number.

5. 2 groups of 3

6. 3 groups of 4

7. 1 group of 5

8. 8 groups of 2

9. 5 groups of 5

10. 6 groups of 4

11. 6 groups of 2

12. 4 groups of 5

13. 7 groups of 2

14. WRITING IN ►MATH Explain how addition and multiplication are similar.

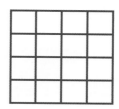

EXTRA PRACTICE
See page R11.

Write two multiplication sentences for each array.

See Examples 1 and 2 (pp. 159–160)

5.

6.

7.

Algebra Use the Commutative Property of Multiplication to find each missing number. See Example 2 (p. 160)

8. $5 \times 2 = 10$
$2 \times \blacksquare = 10$

9. $3 \times 5 = 15$
$\blacksquare \times 3 = 15$

10. $3 \times 9 = 27$
$9 \times 3 = \blacksquare$

11. Geometry Hope drew the area model at the right. Write a multiplication sentence to represent her model.

See Example 2 (p. 160)

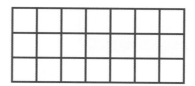

Multiply. Use an array if needed. See Examples 1 and 2 (pp. 159–160)

12. Adult tickets to the talent show cost $8. How much will 4 adult tickets cost?

13. Tamika gives her dog 2 treats every day. How many treats does Tamika's dog get in one week?

H.O.T. Problems

14. FIND THE ERROR Marita and Tyrone are using the numbers 3, 4, and 12 to show the Commutative Property of Multiplication. Who is correct? Explain.

Marita
$3 \times 4 = 12$
$12 \div 3 = 4$

Tyrone
$4 \times 3 = 12$
$3 \times 4 = 12$

15. **WRITING IN** ►**MATH** Describe how an array can help you find the answer to a multiplication problem.

MAIN IDEA

I will multiply by 2.

Math Online

macmillanmh.com
• Extra Examples
• Personal Tutor
• Self-Check Quiz

GET READY to Learn

The students in an art class are working on an art project. They are told to work in 8 groups of 2. How many students are there in all?

There are many different ways to multiply by 2. One way is to model an array. Another way is to draw a picture.

Real-World EXAMPLE   Multiply by 2

1 SCHOOL How many students are there in the art class if there are 8 groups of 2?

You need to model 8 groups of 2 or 8 × 2.

One Way: Model an Array	Another Way: Draw a picture
Model 8 rows and 2 columns.	Draw 8 groups of 2.
⚬⚬ ⚬⚬ ⚬⚬ ⚬⚬ ⚬⚬ ⚬⚬ ⚬⚬ ⚬⚬	Ⓧ Ⓧ Ⓧ Ⓧ Ⓧ Ⓧ Ⓧ Ⓧ
2 + 2 + 2 + 2 + 2 + 2 + 2 + 2 or 16	2 + 2 + 2 + 2 + 2 + 2 + 2 + 2 or 16

So, there are 8 × 2 or 16 students in the art class.

To multiply by 2, you can use skip counting.

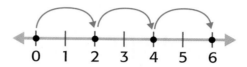 **Real-World EXAMPLE** Use Skip Counting

② **FRIENDS** Sybil rides her bike to the park Monday, Wednesday, and Friday. It is a 2-mile round trip. How many miles does she ride for the three days?

There are 3 days. Each day she rides 2 miles.

To find how many miles she rides for the three days, find 3 × 2.

$$\overset{0\quad 1\quad 2\quad 3\quad 4\quad 5\quad 6}{\longleftrightarrow}$$

Count equal jumps of 2 three times.
Read *2, 4, 6*

So, Sybil rides 3 × 2, or 6 miles for the three days.

CHECK What You Know

Multiply. See Examples 1 and 2 (pp. 162–163)

1.

4 groups of 2

2.

3 groups of 2

3.

5 rows of 2

Multiply. Use an array or draw a picture if needed. See Example 1 (p. 162)

4. $\begin{array}{r} 6 \\ \times 2 \\ \hline \end{array}$

5. $\begin{array}{r} 2 \\ \times 2 \\ \hline \end{array}$

6. $\begin{array}{r} 9 \\ \times 2 \\ \hline \end{array}$

7. $\begin{array}{r} 8 \\ \times 2 \\ \hline \end{array}$

8. Ten students each have 2 pieces of chalk. How many pieces of chalk are there?

9. **Talk About It** Explain the different strategies you can use to remember the multiplication facts for 2.

Multiply. See Examples 1 and 2 (pp. 162–163)

10.

2 groups of 2

11.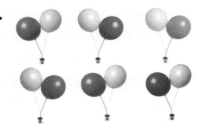

6 groups of 2

12.

7 groups of 2

13.

3 rows of 2

14.

4 rows of 2

15.

5 rows of 2

Multiply. Use an array or draw a picture if needed. See Example 1 (p. 162)

16. $\begin{array}{r} 2 \\ \times\, 5 \\ \hline \end{array}$

17. $\begin{array}{r} 2 \\ \times\, 3 \\ \hline \end{array}$

18. $\begin{array}{r} 5 \\ \times\, 2 \\ \hline \end{array}$

19. $\begin{array}{r} 4 \\ \times\, 2 \\ \hline \end{array}$

20. 2×7

21. 2×9

22. 6×2

23. 10×2

24. 2×4

25. 2×8

26. 2×2

27. 3×2

Solve. Use models if needed. See Examples 1 and 2 (pp. 162–163)

28. There are 3 students. Each has 2 pencils. How many pencils are there in all?

29. There are 2 dogs. How many eyes are there in all?

30. There are 2 spiders. Each has 8 legs. How many legs are there in all?

31. There are 2 squares. How many sides are there in all?

H.O.T. Problems

32. OPEN ENDED Write a real-world multiplication word problem whose answer is between 11 and 19.

33. WRITING IN MATH Write a problem about a real-world situation in which a number is multiplied by 2.

You can use the counters from the *Math Tool Chest* to help you model a multiplication problem.

EXAMPLE

1 **TOYS** A pet store sells dog toys. Three toys come in a package. How many toys are in 4 packages?

There are 3 toys in each package. To find how many toys are in 4 packages, find 4 groups of 3.

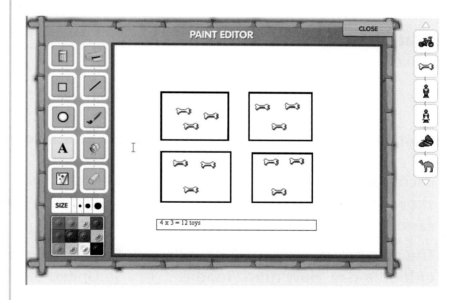

- Choose the rectangle tool, to make 4 rectangles. These are the groups.
- Choose the dog bone stamp tool to the right of the screen to stamp 3 dog bones in each rectangle. These are the toys.
- Choose the letter tool and type a multiplication sentence.

So, 4 equal groups of 3 equals 12. There are 12 toys altogether.

Tech Link

② SPORTS A football player made 3 touchdowns. Each touchdowns earned 6 points. How many points did the football player earn?

- Make 3 circles.
- Stamp 6 footballs in each circle.
- Write a multiplication sentence to represent the situation.

$$3 \times 6 = 18$$

So, the player earned 18 points.

CHECK What You Know

Use Math Tool Chest to model each multiplication sentence.

1. 5×3 **2.** 8×2 **3.** 7×6 **4.** 5×9

Use Math Tool Chest to model each situation. Then solve.

5. Each hour there are 2 cartoons shown on television. How many cartoons are shown in 6 hours?

6. A car has 4 tires. How many tires are on 7 cars?

7. The pigs below each eat 4 pounds of food a day. How much food is needed for one day?

8. Eight friends are each making a United States flag from craft paper. How many red strips of paper will they need for 8 flags?

9. Talk About It How do models help you solve multiplication problems?

Write two multiplication sentences for each array. Then multiply. (Lesson 4-1)

1.

2.

Multiply. (Lesson 4-1)

3. 3 × 2   **4.** 4 × 2

5. Mallory has 3 bags of shells. Each bag has 4 shells. How many shells are there in all? (Lesson 4-1)

6. MULTIPLE CHOICE Which multiplication sentence is modeled below? (Lesson 4-1)

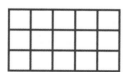

A 4 × 5 = 20   **C** 4 × 5 = 9

B 3 × 5 = 15   **D** 5 × 3 = 8

Algebra Use the Commutative Property of Multiplication to find each missing number.

(Lesson 4-1)

7. 9 × 2 = 18   **8.** 3 × 7 = 21
2 × ▇ = 18   7 × ▇ = 21

Multiply. (Lesson 4-2)

9.

10.

11. MULTIPLE CHOICE If 9 × 4 = 36, then what is 4 × 9? (Lesson 4-1)

F 28   **H** 36

G 32   **J** 40

Multiply. (Lesson 4-2)

12.   7
        × 2

13.   2
        × 3

Solve. Use models if needed. (Lesson 4-2)

14. There are 2 elephants. How many legs in all?

15. There are 4 dogs. How many tails in all?

16. Ebony is practicing her multiplication facts for 2. She has counted to 18. How many 2s has she counted?

17. WRITING IN ▶MATH Explain how multiplication and addition are related. (Lesson 4-1)

4-3 Multiply by 4

MAIN IDEA

I will multiply by 4.

Math Online

macmillanmh.com
• Extra Examples
• Personal Tutor
• Self-Check Quiz

A car transport has 5 new cars. Each car has 4 wheels. How many wheels are there in all on the cars?

To multiply by 4, you can use the same strategies you used to multiply by 2.

Real-World EXAMPLE Multiply by 4

① **WHEELS Each car on the car transport has 4 wheels. How many wheels are there in all on the 5 new cars?**

You need to find 5 groups of 4 or 5 × 4.

One Way: Model Using Counters

Model 5 groups of 4.

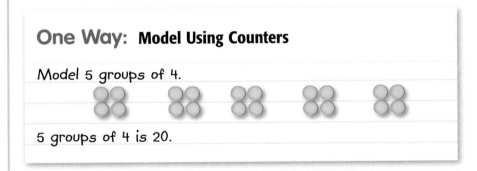

5 groups of 4 is 20.

Another Way: Draw a Picture

Use repeated addition to find 5 × 4.

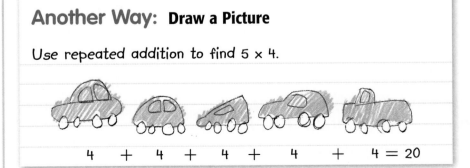

4 + 4 + 4 + 4 + 4 = 20

So, there are 5 × 4 or 20 wheels in all.

Multiply. Use models or draw a picture if needed. See Example 1 (p. 168)

1. 4
    × 4

2. 4
    × 2

3. 4
    × 8

4. 6 × 4

5. 5 × 4

6. 4 × 10

7. Tomas read 8 books. Each book had 4 chapters. How many chapters did he read in all?

8. (Talk About It) Explain how knowing 7 × 2 can help you find 7 × 4.

Practice and Problem Solving

EXTRA PRACTICE
See page R11.

Multiply. Use models or draw a picture if needed. See Example 1 (p. 168)

9. 3
    × 4

10. 4
     × 2

11. 5
     × 4

12. 4
     × 6

13. 4
     × 7

14. 4
     × 5

15. 9 × 4

16. 8 × 4

17. 10 × 4

**Write a multiplication sentence for each situation.
Use models if needed.**

18. Ruben and Tonisha each have an umbrella. How many umbrellas are there after 2 friends join them with their umbrellas?

19. There are 9 rows of seats on a bus. Four children can sit in each row. If there are 48 children, how many will not be able to ride on the bus?

20. Write a multiplication sentence to show that 4 dimes equal 40 cents.

21. A factory packs 4 science kits in each box. If they packed 28 kits, how many boxes did they pack?

H.O.T. Problems

22. OPEN ENDED Explain the strategy you would use to find 4×6. Why do you prefer this strategy?

23. FIND THE ERROR Anica and Roberta are finding 8×4. Who is correct? Explain your reasoning.

Anica

8×4 is the same as $4 + 4 + 4 + 4 + 4 + 4 + 4 + 4$. The answer is 32.

Roberta

8×4 is the same as $8 + 4$. The answer is 12.

24. WRITING IN ►MATH Write and solve a real-world problem that involves multiplying by 4.

TEST Practice

25. Which multiplication sentence is modeled below? (Lesson 4-1)

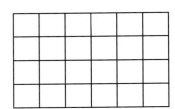

A $5 \times 7 = 35$ **C** $8 \times 3 = 24$

B $6 \times 6 = 36$ **D** $4 \times 6 = 24$

26. If $7 \times 5 = 35$, then what is 5×7? (Lesson 4-1)

F 30

G 35

H 40

J 45

Spiral Review

Model with an array or draw a picture to multiply. (Lesson 4-2)

27. $\begin{array}{r} 2 \\ \times 6 \\ \hline \end{array}$

28. $\begin{array}{r} 7 \\ \times 2 \\ \hline \end{array}$

29. $\begin{array}{r} 2 \\ \times 9 \\ \hline \end{array}$

Write a multiplication sentence for each array. Then multiply. (Lesson 4-1)

30.

31.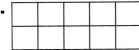

Factor Power

Factors and Products

Get Ready!

Players: 2 players

Get Set!

Write a product for the multiplication facts you have learned so far on each card. Then shuffle.

Go!

- Player 1 turns over the top card. He or she names two factors of the product on the card.

- Player 1 then colors an array anywhere on the graph paper to match the factors.

You will need: 20 index cards, crayons, 1-inch graph paper for each player

- Player 2 checks to see if the product matches the number of squares colored in. If it is correct, Player 1 gets 1 point.

- The game continues, taking turns, until neither player can make any more arrays on their graph paper or until one player reaches 10 points after a complete round.

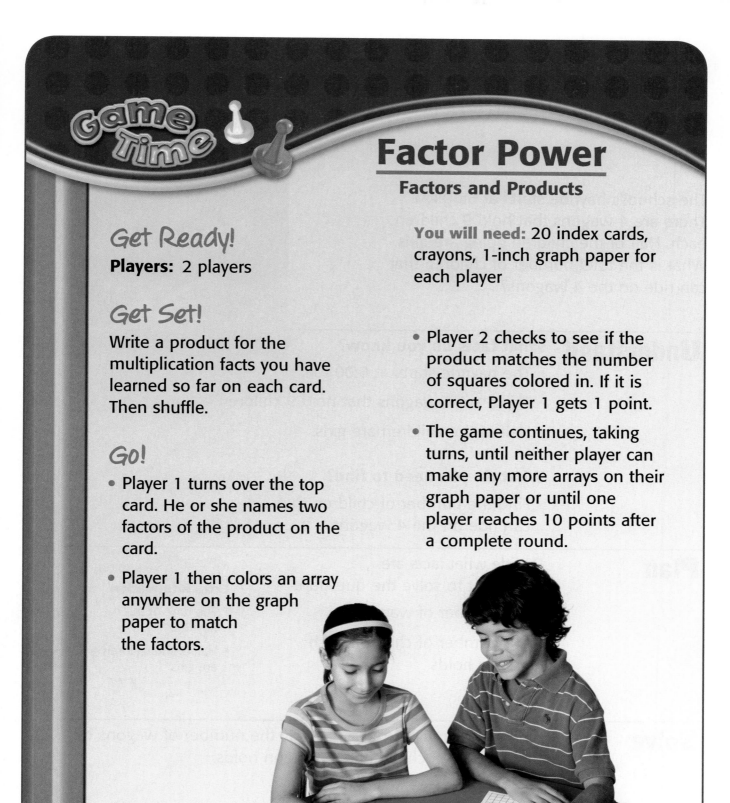

Problem-Solving Skill

<u>**MAIN IDEA**</u> I will solve a problem by identifying extra or missing information.

The school's hayride starts at 6:00 P.M. There are 4 wagons that hold 9 children each. Half of the children going are girls. What is the total number of children that can ride on the 4 wagons?

Understand	**What facts do you know?** • The hayride starts at 6:00 P.M. • There are 4 wagons that hold 9 children. • Half of the children are girls. **What do you need to find?** • Find the number of children that can ride on the 4 wagons.
Plan	Decide what facts are important to solve the question. • the number of wagons • the number of children each wagon holds **Extra Information** • The time of the hayride. • Half of the children are girls.
Solve	To find the total number, multiply the number of wagons by the number of children each wagon holds. $4 \times 9 = \blacksquare$ $4 \times 9 = 36$ So, 36 children can ride on the hay wagons.
Check	Look back. Since $9 + 9 + 9 + 9 = 36$, you know the answer is correct.

ANALYZE the Skill

Refer to the problem on the previous page.

1. How did you know what information was important and what was not?

2. Suppose there are 36 children but only 3 wagons. How many children will ride on each wagon?

3. Look back to your answer for Exercise 2. How do you know that the answer is correct?

4. Draw an array to verify that your answer to Exercise 3 is correct.

PRACTICE the Skill

EXTRA PRACTICE
See page R12.

Solve. If there is missing information, tell what facts you need to solve the problem.

5. Below is a list of the items that Bert bought at a store. How much change did he get back?

Item	Cost
Pencils	$2
Paper	$1
Binder	$3

6. **Measurement** Nina is 58 inches tall. Her sister is in the first grade and is 48 inches tall. How much taller is Nina than her sister?

7. Mrs. Friedman bought 2 boxes of chalk. She bought 4 more boxes with 10 pieces each. She paid $2 per box. How much did she spend in all?

8. Ten of Eduardo's baseball cards are All Star cards. His friend has twice as many cards. How many cards does Eduardo's friend have?

9. Every day for 5 days, the third grade class had 4 chicks hatch. Nine of the chicks were yellow, and the rest were brown. How many chicks hatched in all?

10. The graph below shows the number of cats and dogs at an animal shelter. How much would it cost to adopt 1 cat and 1 dog if a cat costs $35 and a dog costs $40?

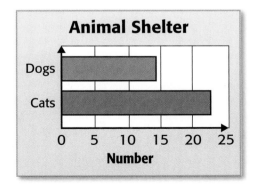

11. **WRITING IN MATH** Rewrite Exercise 5 so it has enough information to solve. Then solve.

Multiply by 5

MAIN IDEA

I will multiply by 5.

Math Online

macmillanmh.com
• Extra Examples
• Personal Tutor
• Self-Check Quiz

GET READY to Learn

A watermelon patch has 6 rows of watermelons. Each row has 5 watermelons. How many watermelons are in the patch?

There is more than one way to multiply by 5.

Real-World EXAMPLE   Multiply by 5

① **WATERMELONS** There are 6 rows and each row has 5 watermelons. How many watermelons are in the farmer's watermelon patch?

You need to find 6×5.

One Way: Use Counters to Model	Another Way: Draw a Picture
○○○ ○○○ ○○ ○○ ○○○ ○○○ ○○ ○○ ○○○ ○○○ ○○ ○○ So, $6 \times 5 = 30$.	🍉🍉 🍉🍉 🍉 🍉 🍉🍉 🍉 🍉 🍉🍉 🍉🍉 🍉🍉 🍉 🍉🍉 🍉🍉 🍉🍉 🍉 🍉🍉 🍉 🍉🍉 🍉🍉 Use repeated addition. $5 + 5 + 5 + 5 + 5 + 5 = 30$ So, $6 \times 5 = 30$ watermelons.

You can use patterns to multiply by 5.

Real-World EXAMPLE   **Multiply Using Patterns**

Remember

Multiplying by a number is the same as skip-counting by that number.

2 **NICKELS** **Jorge has 7 nickels. How much money does he have?**

You know that a nickel is 5¢. Count by fives for each nickel to find 7 × 5¢.

Read   5   10   15   20   25   30   35

Notice the patterns in the answers.

0 × 5 = 0   ◄─── | All of the answers
1 × 5 = 5   ◄─── | end in 0 or 5.
2 × 5 = 10
3 × 5 = 15

Extend the pattern.
4 × 5 = 20
5 × 5 = 25
6 × 5 = 30
7 × 5 = 35

So, Jorge has 7 × 5¢ or 35¢.

✓ CHECK What You Know

Multiply. Use counters to model or draw a picture if needed. See Example 1 and 2 (pp. 174–175)

1.   5
   × 4

2.   5
   × 3

3.   6
   × 5

4.   5
   × 8

5.   5
   × 7

6.   5
   × 5

7. Kai, Lakita, and Maxwell have a box of pretzels. If each gets 5 pretzels, how many pretzels are in the box? Explain.

8. **Talk About It** Explain why the 5s facts might be easier to remember than most sets of facts.

Multiply. Use counters to model or draw a picture if needed. See Example 1 and 2 (pp. 174–175)

9. 5
    × 2

10. 3
    × 5

11. 5
    × 6

12. 7 × 5

13. 8 × 5

14. 5 × 10

15. 5 × 5

16. 5 × 3

17. 4 × 5

18. A pan of corn bread is cut into 5 rows with 4 pieces in each row. How many pieces are there in all?

19. A sunflower costs $6. Evelyn wants to buy 4. Does she have enough if she has four $5-bills? Explain.

20. Bernardo's dad paid for his new roller blades with seven $5-bills. If his dad's change was $2, how much did his roller blades cost?

21. There are 82 members in a marching band. Part of the band divides into 5 groups of 9. How many members are not divided into a group?

Data File

The rose is one of the world's most popular flowers. It is the state flower of New York.

22. A flower shop sells each rose for $3. How much would it cost to buy 5 roses?

23. If you buy a dozen roses, you can save $1 on the dozen. Write a number sentence to describe how much youwould save on 5 dozen roses.

H.O.T. Problems

24. WHICH ONE DOESN'T BELONG? Identify the strategy that will not help you find 5 × 6.

| skip counting | rounding | make an array | draw a picture |

25. WRITING IN ►MATH Can the ones-digit in the product ever end in 2 when you are multiplying by 5? Explain.

Facts Practice

Multiply.

1. $\begin{array}{r} 2 \\ \times\ 3 \\ \hline \end{array}$
2. $\begin{array}{r} 2 \\ \times\ 10 \\ \hline \end{array}$
3. $\begin{array}{r} 4 \\ \times\ 4 \\ \hline \end{array}$
4. $\begin{array}{r} 2 \\ \times\ 9 \\ \hline \end{array}$

5. $\begin{array}{r} 5 \\ \times\ 8 \\ \hline \end{array}$
6. $\begin{array}{r} 4 \\ \times\ 3 \\ \hline \end{array}$
7. $\begin{array}{r} 5 \\ \times\ 2 \\ \hline \end{array}$
8. $\begin{array}{r} 5 \\ \times\ 5 \\ \hline \end{array}$

9. $\begin{array}{r} 2 \\ \times\ 6 \\ \hline \end{array}$
10. $\begin{array}{r} 4 \\ \times\ 8 \\ \hline \end{array}$
11. $\begin{array}{r} 4 \\ \times\ 2 \\ \hline \end{array}$
12. $\begin{array}{r} 2 \\ \times\ 2 \\ \hline \end{array}$

13. $\begin{array}{r} 5 \\ \times\ 7 \\ \hline \end{array}$
14. $\begin{array}{r} 8 \\ \times\ 4 \\ \hline \end{array}$
15. $\begin{array}{r} 2 \\ \times\ 7 \\ \hline \end{array}$
16. $\begin{array}{r} 2 \\ \times\ 3 \\ \hline \end{array}$

17. $\begin{array}{r} 5 \\ \times\ 6 \\ \hline \end{array}$
18. $\begin{array}{r} 2 \\ \times\ 4 \\ \hline \end{array}$
19. $\begin{array}{r} 5 \\ \times\ 4 \\ \hline \end{array}$
20. $\begin{array}{r} 4 \\ \times\ 7 \\ \hline \end{array}$

21. 2×5
22. 5×3
23. 5×10
24. 4×9

25. 2×8
26. 4×6
27. 2×5
28. 4×7

29. 5×6
30. 5×9
31. 4×5
32. 5×4

33. 7×5
34. 4×2
35. 9×2
36. 6×5

4-6 Multiply by 10

GET READY to Learn

Walking on the beach, Oliver saw footprints. He counted 10 toes on each of the 3 sets of footprints. How many toes did he count in all?

MAIN IDEA

I will multiply by 10.

Math Online

macmillanmh.com
• Extra Examples
• Personal Tutor
• Self-Check Quiz

Patterns can help you multiply by 10 to solve the problem.

Real-World EXAMPLE   Use Patterns to Multiply

1 **TOES** **How many toes did Oliver count in all?**

Find 10×3.

Notice the pattern when multiplying by 10.

$10 \times 1 = 10$ ← The ones digit of the answer is zero.
$10 \times 2 = 20$
$10 \times 3 = 30$
$10 \times 4 = 40$
$10 \times 5 = 50$

same

The pattern can also be seen when skip counting on a number line. Count equal jumps of 10 three times.

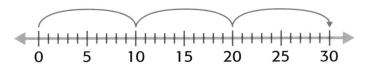

Read *10, 20, 30*

The pattern shows that $10 \times 3 = 30$.
So, Oliver counted 30 toes.

Dimes can be used to model multiplying by 10.

Remember

When you *multiply*, you add the same number *multiple* times.

Real-World EXAMPLE   Use Models

② **MONEY** Orlando found 8 dimes under his bed while cleaning. How much money did Orlando find?

You need to find 8 × 10¢.
Dimes can be used as models to count by 10.

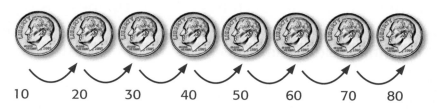

10   20   30   40   50   60   70   80

Read *10, 20, 30, 40, 50, 60, 70, 80*

8 dimes shows 80¢.

8 × 10¢ = 80¢.

So, Orlando found 80¢.

Check Use repeated addition to check.

10 + 10 + 10 + 10 + 10 + 10 + 10 + 10 = 80

So, the answer is correct.

CHECK What You Know

Multiply. Use patterns or models if needed. See Examples 1 and 2 (pp. 178–179)

1. 10
  × 2

2. 10
  × 4

3. 10
  × 7

4. 5 × 10

5. 3 × 10

6. 10 × 10

7. Mina bought a dress for $50. How many $10-bills will she need to pay for the dress?

8. **Talk About It** How can knowing the 5s facts help you with your 10s facts?

Multiply. Use patterns or models if needed. See Examples 1 and 2 (pp. 178–179)

9. 10
$\times 2$

10. 10
$\times 6$

11. 10
$\times 5$

12. 10×3

13. 10×9

14. 10×10

15. 4×10

16. 10×5

17. 10×6

18. There are 10 cars. Each has 4 wheels. How many wheels are there altogether?

19. Ines has 6 packs of whistles. There are 10 whistles in each pack. How many whistles does she have altogether?

20. Measurement There are 3 feet in one yard. How many feet are in 10 yards?

21. There are 5 giraffes and 10 monkeys. How many legs are there altogether?

For Exercises 22–24, use the bar graph.

22. How much money do the children have altogether?

23. Algebra Write a multiplication sentence comparing the amount of money that Robin has with the amount Hakeem has.

24. What is the difference in the least amount of money and the most?

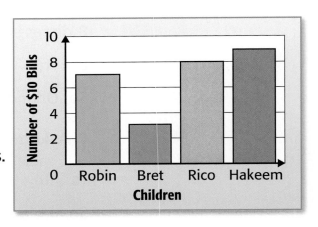

Real-World PROBLEM SOLVING

Art Some of the world's largest glass sculptures are found in the United States. Find the length of each sculpture.

25. Fiori di Como: 5 less than 7×10.

26. Chihuly Tower: 5 more than 10×5.

27. Cobalt Blue Chandelier: 9 more than 2×10.

28. River Blue: 4 more than 10×1.

World's Largest Glass Sculptures	
Sculpture Name	**Length (feet)**
Fiori di Como, NV	
Chihuly Tower, OK	
Cobalt Blue Chandelier, WA	
River Blue, CT	

Source: Book of World Records

H.O.T. Problems

29. WHICH ONE DOESN'T BELONG? Identify the pair of multiplication sentences that is false.

| $2 \times 5 = 10 \times 1$ | $4 \times 3 = 6 \times 2$ | $5 \times 4 = 2 \times 10$ | $10 \times 0 = 5 \times 1$ |

30. **WRITING IN** ►**MATH** Explain how you know that a multiplication fact with an answer of 25 cannot be a 10s fact.

TEST Practice

31. Which of the following is used to find out how many legs are on 6 chairs? (Lesson 4-3)

A 4×6 **C** $4 + 6$

B $4 \div 6$ **D** $6 - 4$

32. What number makes this number sentence true? (Lesson 4-6)

$$12 + 8 = \blacksquare \times 2$$

F 5 **H** 9

G 8 **J** 10

Spiral Review

Multiply. (Lesson 4-5)

33. 9×5 **34.** 7×5 **35.** 4×5

36. An adult ticket to the zoo is $6. A child's ticket is $4. How much would tickets cost for 2 adults and one child? (Lesson 4-3)

Write a multiplication sentence for each array. (Lesson 4-1)

37.
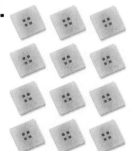

38.

Subtract. Check your answer. (Lesson 3-8)

39.
$$\begin{array}{r} 200 \\ -\ 199 \\ \hline \end{array}$$

40.
$$\begin{array}{r} 500 \\ -\ 208 \\ \hline \end{array}$$

41.
$$\begin{array}{r} 2,300 \\ -\ 576 \\ \hline \end{array}$$

Problem Solving in Science

LOTS OF ARMS AND LEGS

Have you ever wondered why a cheetah has 4 legs instead of 3? Or why an octopus has 8 arms instead of 4? The number of arms or legs an animal has helps it hunt for food and escape from predators.

A cheetah has 4 legs that balance its body. Its legs help it run as fast as 70 miles per hour. An octopus has an unprotected body and no claws or teeth. So, 8 arms are more helpful to an octopus than only 4 or 6 arms.

ANIMAL	NUMBER OF LEGS OR ARMS
Sugar stars (Sea star)	5
Ant	6
Ostrich	2
Hermit crab	10
Sea turtle	4

Real-World Math

Use the information on page 182 to solve each problem. Write a multiplication sentence to solve. Then write an addition sentence to check.

 1. Three ants are on a park bench. How many legs are there in all?

 2. You see 7 ostriches. How many legs do you see altogether?

 3. If you see a pack of 3 cheetahs, how many legs are there in all?

 4. If there are 4 octopuses, how many octopus arms are there total?

 5. You count 30 sugar star arms in the aquarium. How many sugar star are there? Explain.

 6. There are 3 sea turtles and 2 sugar stars in another aquarium. How many arms and legs are there altogether?

 7. How many legs in all do 6 hermit crabs have?

Did You Know?

An octopus has 240 suction cups on each of its 8 arms.

Problem-Solving Investigation

MAIN IDEA I will choose the best strategy to solve a problem.

P.S.I. TEAM +

DENZELL: Our third grade class will make 6 holiday baskets to give away. We will fill each basket with 7 food items.

YOUR MISSION: Find how many items are needed to fill the baskets.

Understand	You know the class will make 6 baskets with 7 items each. Find out the total number of food items needed.
Plan	You can use the *draw a picture* strategy to solve the math problem.
Solve	Draw a picture to represents the situation. 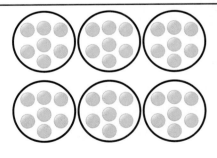 The picture shows that 6 × 7 = 42. So, the third grade class needs 42 food items to fill the baskets.
Check	Look back. Check by using repeated addition 7 + 7 + 7 + 7 + 7 + 7. The answer is 42, so you know your answer is correct and reasonable.

Use any strategy shown below to solve. Tell what strategy you used.

PROBLEM-SOLVING STRATEGIES
• Act it out.
• Draw a picture.
• Look for a pattern.

1. At a space museum, there are 15 large rockets, 8 space capsules, and 12 small rockets. How many rockets are there altogether?

2. George paid $5 for a movie. Is it reasonable to say he spent more money on the food than he did on the movie? Explain.

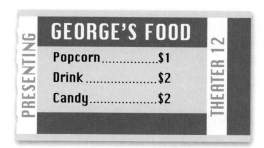

GEORGE'S FOOD

Popcorn...............$1
Drink..................$2
Candy.................$2

PRESENTING   THEATER 12

3. Jonas sells angelfish. He has 6 tanks, and each tank has 5 fish. After he sold some, he had 22 fish left. How many did he sell? How much did he make if he sold each fish for $5?

4. Mr. Trevino spent $63 on 7 berry bushes. Each bush will give 10 pints of berries. He sells each pint for $2. Find the difference in the amount of money he spent and what he makes from selling the berries.

5. **Measurement** Paula put 6 books on one side of a balance scale. On the other side, she put 2 books and her baseball glove. The sides balanced. Each book weighs 3 ounces. How much does her glove weigh?

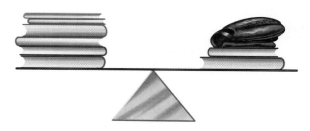

6. Grandmother picked 8 pears. Then, she picked 4 times as many apples. What is the difference in the number of apples and pears picked?

7. Suki made a picture graph to show which insects she collected on a nature hike. Hannah collected two times as many insects. What is the total number of insects Hannah collected?

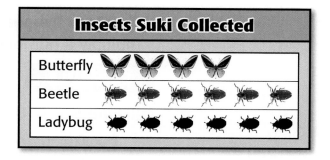

Insects Suki Collected

Butterfly	
Beetle	
Ladybug	

8. **WRITING IN ►MATH** Refer to Exercise 7. Explain how multiplication is used to find the answer.

4-8 Multiply by 0 and 1

MAIN IDEA

I will multiply by 0 and 1.

New Vocabulary

Identity Property of Multiplication

Zero Property of Multiplication

Math Online

macmillanmh.com
• Extra Examples
• Personal Tutor
• Self-Check Quiz

GET READY to Learn

There are 4 flowerpots. Each has 1 daisy. How many daisies are there in all?

There are special properties for multiplying by 1 and 0.

The **Identity Property of Multiplication** says that when any number is multiplied by 1, the product is that number.

Real-World EXAMPLE   Multiply by 1

1 **FLOWERS Find 4 × 1 to find how many daisies there are in all.**

Model 4 groups of 1.

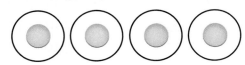

The model shows 4 groups of 1.
So, $4 \times 1 = 4$.

The **Zero Property of Multiplication** says that when you multiply a number by 0, the product is zero.

EXAMPLE   Multiply by 0

2 **Find 6 × 0. Use the pattern shown.**

$1 \times 0 = 0$ ◄── Any number times zero equals 0.
$2 \times 0 = 0$
$3 \times 0 = 0$
$4 \times 0 = 0$

So, $6 \times 0 = 0$.

✓ CHECK What You Know

Multiply. See Examples 1 and 2 (p. 186)

1.   6
   × 0

2.   1
   × 7

3.   5
   × 0

4.   8
   × 1

5. There is 1 student sitting at each of the 9 tables in the cafeteria. How many students are there altogether?

6. **Talk About It** If 100 is multiplied by 0, what will be the answer? Explain your reasoning.

▶ Practice and Problem Solving

EXTRA PRACTICE
See page R13.

Multiply. See Example 1 and 2 (p. 186)

7.   7
   × 1

8.   5
   × 0

9.   10
   × 1

10.   10
   × 0

11.   3
   × 1

12.   1
   × 0

13.   4
   × 0

14.   1
   × 1

15. 8×0

16. 1×2

17. 0×1

18. 4×1

19. 9×0

20. 9×1

21. 0×2

22. 1×5

Solve. Use models if needed.

23. How many pouches does 1 kangaroo have?

24. How many legs do 8 snakes have?

25. In a fantasy story, a pirate found 3 empty treasure chests with no jewels. How many jewels were there?

26. There is only one book on the shelf. It has 90 pages. How many pages are there altogether?

27. Thomas saw a group of 8 lizards. Each lizard had one spot on its back. How many spots were there in all?

28. How many legs do 15 fish have?

Algebra Find each missing number.

29. ■ $\times 7 = 7$

30. ■ $\times 8 = 0$

31. $6 \times$ ■ $= 0$

32. $1 \times$ ■ $= 0$

33. ■ $\times 5 = 5$

34. $10 \times 0 =$ ■

35. $9 \times$ ■ $= 9$

36. ■ $\times 2 = 2$

H.O.T. Problems

37. OPEN ENDED Write a problem using one of the multiplication properties that you have just learned. Explain how to find the answer.

CHALLENGE Find the missing number.

38. 2,684 × ▇ = 2,684　　**39.** 1,039 × 1 = ▇　　**40.** 27 × ▇ = 0

41. **WRITING IN ►MATH** Explain the Zero Property of Multiplication.

TEST Practice

42. Mrs. Smyth reads aloud to her class for 10 minutes each day. Which number sentence tells how to find the number of minutes she reads in a 5-day week? *(Lesson 4-7)*

A 10 + 5

B 10 × 5

C 10 − 5

D 10 ÷ 5

43. What number can be multiplied by 3,859 to give the product 3,859? *(Lesson 4-8)*

F 0

G 1

H 2

J 10

Spiral Review

44. Harold collected at least 9 shells while at the beach every day. How many shells did he collect over his 10-day vacation? *(Lesson 4-7)*

A survey was taken of people's favorite water activity. Use the data to answer the questions. Write a multiplication sentence. *(Lesson 4-6)*

45. How many people enjoy surfing?

46. How many people prefer swimming?

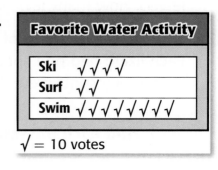

Favorite Water Activity
Ski √√√√
Surf √√
Swim √√√√√√√

√ = 10 votes

Algebra Compare. Use >, <, or =. *(Lessons 4-2 and 4-5)*

47. 2 × 7 ● 8　　　　**48.** 8 × 5 ● 18　　　　**49.** 10 × 2 ● 20

Facts Practice

Multiply.

1. $\begin{array}{r} 4 \\ \times\,6 \\ \hline \end{array}$

2. $\begin{array}{r} 10 \\ \times\,5 \\ \hline \end{array}$

3. $\begin{array}{r} 0 \\ \times\,9 \\ \hline \end{array}$

4. $\begin{array}{r} 2 \\ \times\,9 \\ \hline \end{array}$

5. $\begin{array}{r} 4 \\ \times\,8 \\ \hline \end{array}$

6. $\begin{array}{r} 2 \\ \times\,3 \\ \hline \end{array}$

7. $\begin{array}{r} 10 \\ \times\,8 \\ \hline \end{array}$

8. $\begin{array}{r} 0 \\ \times\,6 \\ \hline \end{array}$

9. $\begin{array}{r} 1 \\ \times\,9 \\ \hline \end{array}$

10. $\begin{array}{r} 5 \\ \times\,5 \\ \hline \end{array}$

11. $\begin{array}{r} 4 \\ \times\,0 \\ \hline \end{array}$

12. $\begin{array}{r} 2 \\ \times\,7 \\ \hline \end{array}$

13. $\begin{array}{r} 10 \\ \times\,0 \\ \hline \end{array}$

14. $\begin{array}{r} 5 \\ \times\,3 \\ \hline \end{array}$

15. $\begin{array}{r} 1 \\ \times\,6 \\ \hline \end{array}$

16. $\begin{array}{r} 2 \\ \times\,10 \\ \hline \end{array}$

17. $\begin{array}{r} 4 \\ \times\,5 \\ \hline \end{array}$

18. $\begin{array}{r} 2 \\ \times\,2 \\ \hline \end{array}$

19. $\begin{array}{r} 1 \\ \times\,1 \\ \hline \end{array}$

20. $\begin{array}{r} 5 \\ \times\,8 \\ \hline \end{array}$

21. 4×3

22. 10×1

23. 0×3

24. 4×9

25. 0×8

26. 10×7

27. 1×4

28. 2×6

29. 5×10

30. 0×7

31. 1×0

32. 10×6

33. 4×7

34. 5×6

35. 10×3

36. 2×0

37. 10×10

38. 0×10

39. 1×5

40. 0×4

FOLDABLES Study Organizer GET READY to Study

Be sure the following Key Vocabulary words and Key Concepts are written in your Foldable.

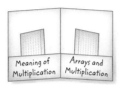

Meaning of Multiplication   Arrays and Multiplication

Key Concepts

• Use an **array** to multiply 2 rows of 5. (p. 159)

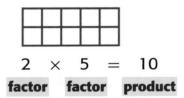

$2 \times 5 = 10$

factor   factor   product

• The **Commutative Property of Multiplication** states that the order in which numbers are multiplied does not change the product. (p. 160)

$3 \times 2 = 6$ $2 \times 3 = 6$

• **Zero Property of Multiplication** states that any number multiplied by 0 equals zero. (p. 186)

$2 \times 0 = 0$

• **Identity Property of Multiplication** states that any number multiplied by 1 equals that number. (p. 186)

$1 \times 2 = 2$

Key Vocabulary

array (p. 159)

Commutative Property of Multiplication (p. 160)

factor (p. 159)

multiply (p. 159)

product (p. 159)

Zero Property of Multiplication (p. 186)

Vocabulary Check

Choose the vocabulary word that completes each sentence.

1. The ____?____ says that a number multiplied by zero has the product of 0.

2. The ____?____ states that the order in which numbers are multiplied does not change the product.

3. In $3 \times 5 = 15$, 3 and 5 are ____?____.

4. To ____?____ is to put equal groups together.

5. A(n) ____?____ is an arrangement of equal rows and columns.

Lesson-by-Lesson Review

4-1 **Arrays and Multiplication** (pp. 159–161)

Example 1
There are 3 rows of 4 muffins.
How many muffins altogether?

You can use addition or multiplication.

Add: $4 + 4 + 4 = 12$
Multiply: $3 \times 4 = 12$

So, 3 equal groups of 4 is 12.

Write two multiplication sentences for each array.

6.   7.

Algebra Use the Commutative Property of Multiplication to find each missing number.

8. $6 \times 4 = 24$ 9. $8 \times 2 = 16$
$4 \times \blacksquare = 24$ $\blacksquare \times 8 = 16$

4-2 **Multiply by 2** (pp. 162–164)

Example 2
How many wings are there if there are 5 butterflies?

There are 5 butterflies. Each has 2 wings. Find 5 groups of 2 or 5×2.

Count 5 jumps of 2.

$2 + 2 + 2 + 2 + 2 = 10$

So, 5 groups of $2 = 10$.

Multiply.

10. $\begin{array}{r} 2 \\ \times 3 \\ \hline \end{array}$ 11. $\begin{array}{r} 7 \\ \times 2 \\ \hline \end{array}$ 12. $\begin{array}{r} 2 \\ \times 4 \\ \hline \end{array}$

Multiply. Use an array or draw a picture if needed.

13. There are 4 birds and each has 2 legs. How many legs are there in all?

14. There are 7 dogs. How many ears are there in all?

15. There are 6 bicycles. Each has 2 wheels. How many wheels are there in all?

4-3 **Multiply by 4** (pp. 168–170)

Example 3
How many legs are there altogether on 6 cats?

There are 6 cats. Each has 4 legs. Find 6 groups of 4 or 6 × 4.

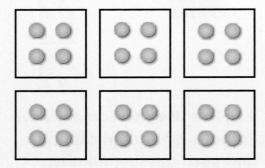

4 + 4 + 4 + 4 + 4 + 4 = 24
So, there are 6 × 4 or 24 legs.

Multiply. Use models or draw a picture if needed.

16. 3 × 4   **17.** 6 × 4

18. 4 × 4   **19.** 7 × 4

20.   4   **21.**   8
       × 5   × 4

22. Algebra Write a multiplication sentence to show that 4 nickels equal 20 cents.

23. The bread factory packs 4 buns in each bag. If they packed 4 bags, how many buns did they pack?

4-4 **Problem-Solving Skill: Extra or Missing Information** (pp. 172–173)

Example 4
A troop leader drives 5 miles to and from the troop meeting. He leaves at 4 P.M. and gets to the meeting at 4:30 P.M. How many miles does he drive there and back?

Decide what facts are important.

• He drives 5 miles to the meeting.
• He drives 5 miles from the meeting.

Multiply to solve.

2 × 5 = 10

So, he travels a total of 10 miles.

Solve. If there is missing information, tell what facts you need to solve the problem.

24. The troop ordered pizza. Each pizza was cut into 8 slices. They ate a total of 32 slices. How many pizzas did they order?

25. The troop left the pizza parlor at 6 P.M. Their van has 4 rows of seats, and each row holds 3 people. How many people can the van hold altogether?

4-5 Multiply by 5 (pp. 174–176)

Example 5
There are 3 groups of toy cars.
5 cars are in each group. How many cars are there in all?

You need to find 3 × 5.

Use counters to model.

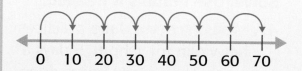

So, there are 3 groups of 5 = 3 × 5 or 15 cars.

Multiply. Use counters if needed.

26. 5
   × 3

27. 7
   × 5

28. 2
   × 5

29. 8
   × 5

30. 4
   × 5

31. 9
   × 5

32. Marlena's mom paid for her new school uniform with ten $5-bills. If her mom's change was $3, how much did her uniform cost?

4-6 Multiply by 10 (pp. 178–181)

Example 6
Pedro practices his trumpet 10 minutes each day for 7 days. How much time has Pedro spent practicing after 7 days?

You need to find 7 × 10.

Use the pattern of counting by 10s.

0 10 20 30 40 50 60 70

Read 10, 20, 30, 40, 50, 60, 70

So, 7 × 10 = 70.

Multiply.

33. 10
   × 3

34. 10
   × 6

35. 10 × 5

36. 4 × 10

37. Each book costs $10. Harvey wants to buy 2, and Bree wants to buy 4. Will they have enough with five $10-bills? Explain.

38. A Cub Scout group sold boxes of popcorn for $10. How much money did Javier raise?

Scout	Money Raised							Total
Jarred	✪	✪	✪	✪	✪	✪	✪	$70
Bartolo	✪	✪	✪	✪				$40
Javier	✪	✪	✪	✪	✪	✪		■

✪ = $10

4-7 Problem-Solving Strategies: Choose a Strategy (pp. 184–185)

Example 7

Samir has $25. Does he have enough money to purchase 6 toys for $5 each.

Find the total cost of the toys.

6 × $5 = $30

Since $25 < $30, Samir does not have enough money to buy all the toys.

Solve.

39. Marlon buys a backpack for $18, a skateboard for $37, and a visor for $13. Estimate the cost of his purchases.

40. School lunches cost $3. If Vera and her sister had to share $5, would they have enough for lunch? Explain.

4-8 Multiply by 0 and 1 (pp. 186–188)

Example 8

There are 3 chairs. Each chair has 0 students. How many students are there in all?

You need to find 3 × 0.

3 × 0 = 0

So, there are 0 students.

Example 9

Find 1 × 5.

When any number is multiplied by 1, the answer is that number.

$$1 × 5 = 5$$

Multiply.

41. 8
     × 1

42. 0
     × 5

43. 12
     × 1

44. 0
     × 6

Algebra Find each missing number.

45. ▦ × 5 = 0   **46.** 4 × ▦ = 4

Solve. Use models if needed.

47. Measurement Jade practices her violin for 1 hour, 5 days a week. How many hours does she practice?

48. A new restaurant has been open for 11 days. Each day 0 people order a cheese sandwich. How many cheese sandwiches have been sold in all?

For Exercises 1 and 2, tell whether each statement is *true* or *false*.

1. The Commutative Property of Multiplication says that the order in which numbers are multiplied changes the product.

2. When you multiply by a 5, you will always have either 5 or 0 in the ones place.

Multiply.

3. 5×3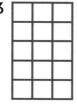
4. 4×1

5. 3×2
6. 5×4

7. 2×6
8. 4×8

9. The movie theater had 6 rows of seats. Each row had 10 people sitting in it. How many people were in the theater?

Algebra Find each missing number.

10. $7 \times \blacksquare = 35$
11. $\blacksquare \times 5 = 40$

12. MULTIPLE CHOICE Which of the following is used to find how many toes are on 7 people?

  A 7×10 **C** $7 + 10$

  B $10 \div 7$ **D** $10 - 7$

Multiply.

13. $\begin{array}{r} 6 \\ \times\ 5 \\ \hline \end{array}$
14. $\begin{array}{r} 3 \\ \times\ 10 \\ \hline \end{array}$

15. $\begin{array}{r} 7 \\ \times\ 5 \\ \hline \end{array}$
16. $\begin{array}{r} 10 \\ \times\ 9 \\ \hline \end{array}$

17. $\begin{array}{r} 9 \\ \times\ 1 \\ \hline \end{array}$
18. $\begin{array}{r} 6 \\ \times\ 0 \\ \hline \end{array}$

Solve. If there is missing information, tell what facts you need to solve the problem.

19. Morgan buys packages of bookmarks. Each package has 30 bookmarks and cost $2. How many bookmarks did she get?

20. Each playground slide has 7 steps. If the playground has 3 slides, how many steps is that altogether?

21. MULTIPLE CHOICE What number can be multiplied by 9,250 to give the answer 9,250?

  F 0   **H** 2

  G 1   **J** 10

22. **WRITING IN** ▶**MATH** Can the answer of a multiplication problem ever end in 2 when you are multiplying by 10? Explain.

Read each question. Then fill in the correct answer on the answer sheet provided by your teacher or on a sheet of paper.

1. Parker swims 5 times a week for 2 hours. How many hours does he swim in a week?

A 7   **C** 15

B 10   **D** 25

2. What number would make the multiplication sentence true?

$$\blacksquare \times 4 = 0$$

F 0   **H** 4

G 1   **J** 8

3. Ben bought an eraser for $2, a box of pencils for $6, and a notebook for $3. He gave the clerk $20. How much change should he get back?

A $5   **C** $8

B $7   **D** $9

4. What is another way to write 4×3?

F $3+3+3$ **H** $3+4+3+4$

G $3+3+3+3$ **J** $4+4+4+4$

5. Which number sentence is modeled by the figure below?

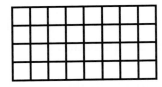

A $5 \times 8 = 40$

B $4 \times 8 = 32$

C $8 + 8 + 8 = 24$

D $3 \times 8 = 24$

6. Which set of numbers is in order from least to greatest?

F 645, 449, 437, 345

G 449, 345, 645, 437

H 437, 449, 645, 345

J 345, 437, 449, 645

7. Write the number the model represents in standard form.

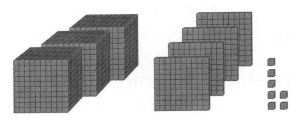

A 4,370   **C** 3,407

B 3,470   **D** 3,047

8. Gabe works at a car wash 6 hours a day. It takes Gabe 1 hour to wash a car. Which number sentence shows how many cars he washes in a day?

F $6 - 6 = 0$ **H** $6 \times 1 = 6$

G $6 \times 0 = 0$ **J** $6 + 1 = 7$

9. Tenell collected 54 shells. Janet collected 82 shells. How many more shells did Janet collect?

A 28   **C** 32

B 30   **D** 38

10. The figure below is a model of a number sentence.

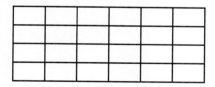

Which number sentence is modeled by the figure?

F $3 \times 6 = 18$ **H** $4 \times 6 = 24$

G $6 + 6 + 6 = 18$ **J** $5 \times 6 = 30$

PART 2   **Short Response**

Record your answer on the answer sheet provided by your teacher or on a sheet of paper.

11. Draw an array to model the multiplication sentence $6 \times 2 = \blacksquare$. Multiply to solve.

12. Tickets to a high school football game costs $5 for adults and $4 for students. How much would 3 adult and 4 student tickets cost?

PART 3   **Extended Response**

Record your answers on the answer sheet provided by your teacher or on a sheet of paper.

13. Ginny joined a soccer team in April. Two weeks later she played her first game. What day of the week was her first game? If there is not enough information, tell what facts you need to solve the problem. Then solve.

NEED EXTRA HELP?													
If You Missed Question...	1	2	3	4	5	6	7	8	9	10	11	12	13
Go to Lesson...	4-2	4-8	3-3	4-1	4-1	1-7	1-3	4-8	3-1	4-1	4-1	4-3	4-4

CHAPTER 5

Develop More Multiplication Facts

BIG Idea When will I use multiplication?

When you combine equal amounts, you can use multiplication. It is useful when you buy items in a store, keep score in a game, or plant a garden.

Example Benny planted a garden. It has 3 rows with 7 vegetable plants in each row. The model shows that Benny planted 3 × 7 or 21 plants.

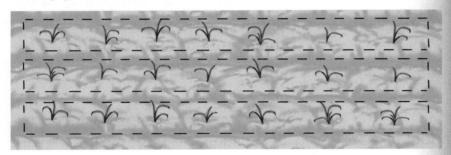

What will I learn in this chapter?

- Explore using the multiplication table.
- Multiply by 3, 6, 7, 8, 9, 11, and 12.
- Use properties of multiplication.
- Solve problems by looking for a pattern.

Key Vocabulary

Associative Property of Multiplication

Commutative Property of Multiplication

factor

product

Math Online > **Student Study Tools**
at macmillanmh.com

FOLDABLES®
Study Organizer

Make this Foldable to help you organize information about multiplication facts. Begin with three sheets of $8\frac{1}{2}$″ × 11″ paper.

1 **Fold** one sheet of paper in half as shown.

2 **Open** and fold a 2″ pocket. Glue the outside edges.

3 **Label** with the lesson titles. Write what you learn.

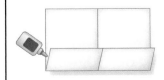

Multiply by 3   Multiply by 6

4 **Repeat** steps 1–3 with two more pieces of paper.

Chapter 5 Develop More Multiplication Facts   **199**

You have two ways to check prerequisite skills for this chapter.

Option 2

Math Online Take the Chapter Readiness Quiz at macmillanmh.com.

Option 1

Complete the Quick Check below.

QUICK Check

Multiply. (Chapter 4)

1. 6×4

2. 1×5

3. 5×2

4. 7×2

Draw an array for each fact. Multiply.

(Chapter 4)

5. 5×4 **6.** 1×6 **7.** 4×7 **8.** 2×9

Solve. (Lessons 3-3 and 4-2)

9. Louis has 2 quarters. Yellow whistles cost 5¢ each. Louis wants to buy 8 whistles. Does he have enough money to buy 8 whistles? Explain.

10. There were 9 oak trees on each side of a street. After some trees were cut down, there were a total of 7 left. How many trees were cut down?

Algebra   **Identify a pattern. Then find the missing numbers.**

(Lesson 1-1)

11. 15, 20, 25, 30, ■, ■

12. 9, 12, 15, 18, ■, ■

13. 11, 21, 31, 41, ■, ■

14. 60, 50, 40, 30, ■, ■

Explore

Math Activity for 5-1
Multiplication Table

In Chapter 4, you learned different strategies for finding products. Recall that a product is the answer to a multiplication problem. Patterns you find in the multiplication table can help you remember products.

MAIN IDEA

I will explore the multiplication table.

Math Online

macmillanmh.com
• Concepts in Motion

ACTIVITY   **Make a Multiplication Table**

Step 1   **Find the Factors.**

To find the product of two factors, find the first factor in the left column and the second factor across the top row.

$2 \times 3 = 6$

factors   product

factors

×	0	1	2	3	4	5	6	7	8	9	10
0											
1											
2				6							
3											
4											
5											
6											
7											
8											
9											
10											

factors

Write the product of 2 × 3 where row 2 and column 3 meet.

Step 2   **Fill in the Grid.**

Write the products of the multiplication facts you know. Remember you can use the Commutative Property of Multiplication, a known fact, or patterns.

Step 3 Use Models.

For the products you do not know, you can use a model. For example, the array shows 3 × 4.

So, 3 × 4 = 12. Fill in the square where 3 and 4 meet with the product 12.

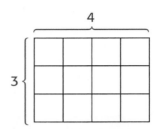

THINK ABOUT IT

1. What is the product when you multiply a number by 1? Explain.

2. What pattern do you see for row 10?

3. What do you notice about row 6 and column 6? Does it apply to all rows and columns of the same number?

CHECK What You Know

Multiply. Use your multiplication table.

4. 2
  × 5

5. 4
  × 0

6. 10
  × 3

7. 5
  × 6

Identify the factor row or column where each part of the multiplication table is found.

8.

	3	6
	4	8
	5	1
	6	1

9.

		15	18
12	16	20	24
15	20	25	30

10.

	6	8
	9	1
	12	1
	15	2

11. **WRITING IN ►MATH** Write two other patterns you found in the multiplication table.

Multiply by 3

GET READY to Learn

In the previous activity, you used a multiplication table to explore multiplication.

×	0	1	2	3	4	5	6	7	8	9	10
0	0	0	0	0	0	0	0	0	0	0	0
1	0	1	2	3	4	5	6	7	8	9	10
2	0	2	4	6	8	10	12	14	16	18	20
3	0	3	6	9	12	15	18	21	24	27	30
4	0	4	8	12	16	20	24	28	32	36	40
5	0	5	10	15	20	25	30	35	40	45	50
6	0	6	12	18	24	30	36	42	48	54	60
7	0	7	14	21	28	35	42	49	56	63	70
8	0	8	16	24	32	40	48	56	64	72	80
9	0	9	18	27	36	45	54	63	72	81	90
10	0	10	20	30	40	50	60	70	80	90	100

There are different ways you can find products.

Real-World EXAMPLE   Use Models

① **PETS There are 4 dogs. Each dog buried 3 bones in a yard. How many bones are buried in the yard?**

Use counters to model 4 groups of 3 bones or 4 × 3.

So, there are 12 bones buried in the yard.

You can draw a picture to help you solve a problem.

Remember

Multiplication can be thought of as repeated addition. So, add three 8 times.

Real-World EXAMPLE   Draw a Picture

2 **GAMES Eight friends have 3 marbles each. How many marbles are there in all?**

Each friend has a group of 3 marbles. There are 8 friends. Draw a picture to find 8 × 3.

$3 + 3 + 3 + 3 + 3 + 3 + 3 + 3 = 24$

So, there are 24 marbles in all.

Multiplication Strategies   **Key Concept**

There are different ways to find answers for multiplication problems.

- Use models or draw a picture.
- Use repeated addition or skip count.
- Draw an array or an area model.
- Use a related multiplication fact.
- Use patterns.

CHECK What You Know

Multiply. Use models or draw a picture if needed. See Examples 1 and 2 (pp. 203–204)

1.   4
     × 3

2.   3
     × 5

3. 3 × 8

4. 3 × 9

5. The branches on a tree have leaves that grow in groups of 3. How many leaves are on 9 branches?

6. **Talk About It** Explain two ways to find the product 3 × 7.

Multiply. Use models or draw a picture if needed. See Examples 1 and 2 (pp. 203–204)

7. 1
× 3

8. 10
× 3

9. 3
× 4

10. 5
× 3

11. 3
× 6

12. 3
× 9

13. 3
× 0

14. 8
× 3

15. 9 × 3

16. 3 × 7

17. 3 × 3

18. 6 × 3

19. 7 × 3

20. 3 × 4

21. 3 × 5

22. 3 × 10

Algebra Copy and complete each table.

23.

Rule: Multiply by 5.	
Input	Output
3	■
7	■
■	40
■	0
1	■

24.

Rule: Multiply by 4.	
Input	Output
5	■
■	28
■	40
9	■
0	■

25.

Rule: Multiply by 3.	
Input	Output
9	■
■	18
4	■
■	24
7	■

26. There are 9 students. They each put 3 books on a shelf. How many books did they place on the shelf?

27. There are 9 daisies and 9 tulips. Every flower has 3 petals. How many petals are there in all?

28. Hoshi, Joan, and Kita each had 3 snacks packed in their lunch boxes. They each ate one snack in the morning. How many snacks are left in all?

29. Thom is buying 4 packages of seeds. Each package costs $3 and contains 5 envelopes of seeds. What will be the total cost? How many envelopes will he have?

H.O.T. Problems

30. OPEN ENDED Look at the 3s row in a multiplication table. Describe the pattern.

31. **WRITING IN MATH** Write a real-world problem that contains groups of 3. Ask a classmate to solve. Check the answer.

5-2 Multiply by 6

MAIN IDEA

I will multiply by 6.

Math Online

macmillanmh.com
• Extra Examples
• Personal Tutor
• Self-Check Quiz

GET READY to Learn

There are 4 frogs sitting on a log. Each frog eats 6 flies. How many flies were eaten altogether?

In this lesson you will learn to multiply by 6.

Real-World EXAMPLE Use a Model

1 FROGS If each frog ate 6 flies, how many flies did they eat in all? Write a multiplication sentence.

There are 4 frogs and each frog ate 6 flies. Use counters to model an array showing 4 rows with 6 in each row.

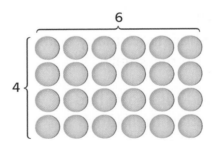

$6 + 6 + 6 + 6 = 24$

4 groups of 6 is 24.

The multiplication sentence is $4 \times 6 = 24$.
So, the frogs ate 24 flies.

Check

Use the multiplication table that you made in the Explore 5–1 Lesson. It shows that $4 \times 6 = 24$. So, the answer is correct. ✔

Real-World EXAMPLE Model a Missing Factor

2 ALGEBRA Clara's jewelry box can hold 48 pairs of earrings. The box has 8 rows. Each row has the same number of spaces. How many spaces are in each row?

To solve this problem, you can use counters to model and solve a multiplication sentence.

Model 8 rows. Place one counter in each row until there are 48 counters total.

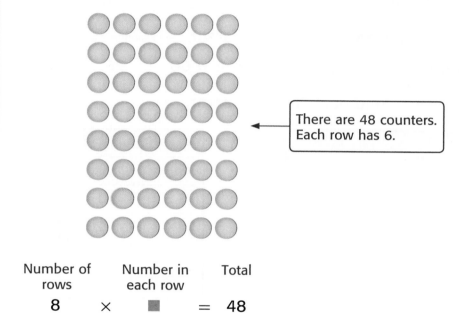

There are 48 counters. Each row has 6.

Number of rows		Number in each row		Total
8	×	■	=	48

Since 8 × 6 = 48, there are 6 spaces in each row.

Remember

There are many different ways to multiply.

CHECK What You Know

Multiply. Use models or draw a picture if needed. See Example 1 (p. 206)

1. 2
     × 6

2. 0
     × 6

3. 6
     × 4

4. 6
     × 6

Algebra Find each missing factor. See Example 2 (p. 207)

5. 5 × ■ = 30

6. ■ × 6 = 6

7. 7 × ■ = 42

8. ■ × 6 = 54

9. Gil has 5 friends. He and each friend have 5 video games. How many video games do they have in all?

10. **Talk About It** Explain why the product of 8 and 3 is double the product of 4 and 3.

Lesson 5-2 Multiply by 6   **207**

Multiply. Use models or draw a picture if needed. See Example 1 (p. 206)

11. 6
    × 2

12. 5
    × 6

13. 4
    × 6

14. 3
    × 6

15. 6
    × 6

16. 10
    × 6

17. 6
    × 9

18. 7
    × 6

19. 0 × 6

20. 6 × 3

21. 8 × 6

22. 6 × 5

Algebra Find each missing factor. See Example 2 (p. 207)

23. 4 × ■ = 24

24. ■ × 6 = 60

25. 6 × ■ = 36

26. 6 × ■ = 18

Algebra Find each missing factor.

27.

Multiply by ■.	
Input	Output
2	6
3	9
4	12
5	15

28.

Multiply by ■.	
Input	Output
3	15
4	20
5	25
6	30

29.

Multiply by ■.	
Input	Output
5	20
6	24
7	28
8	32

Solve. Use models if needed.

30. Six students bought 5 pretzels each. If they gave away 6 of the pretzels, how many pretzels do they have left?

31. If Ida has 6 dimes, does she have enough money for 8 pieces of bubble gum that costs 6¢ each? Explain.

32. In the morning, 6 eggs hatched. By the evening, it was 9 times that. How many total eggs hatched?

33. There are 7 vans. If each holds 6 students, is there enough room for 45 students? Explain.

H.O.T. Problems

34. **OPEN ENDED** Use one of the multiplication strategies to explain how you would find the product 6 × 6.

35. **WRITING IN ►MATH** Write a real-world problem that can be solved by multiplying by 6.

36. Mr. Lobo buys 3 of the same item at a store. The total is $27. What did he buy? (Lesson 5-1)

A  $8

C  $4

B  $9

D  $3

37. Which sign goes in the box to make the sentence true? (Lesson 5-1)

$$3 \blacksquare 10 = 30$$

F +   **H** ×

G −   **J** ÷

38. Mr. Baxter bought 6 boxes of light bulbs. Each box has 4 bulbs. Which number sentence shows how to find the total number of bulbs? (Lesson 5-2)

A $6 - 4 = 2$

B $24 \div 6 = 4$

C $6 + 4 = 10$

D $6 \times 4 = 24$

Spiral Review

39. Henry put all the shells he collected into 7 equal groups for his friends. Each group had 3 shells. How many shells did Henry collect? (Lesson 5-1)

Multiply. (Lesson 4-8)

40. 0×9 **41.** 6×0 **42.** 8×1

43. Jena and Crystal each have $2. Is it reasonable to say that they have enough money to buy a box of crayons for $5? Explain your reasoning. (Lesson 3-4)

Round to the nearest hundred. (Lesson 1-8)

44. 555   **45.** 209   **46.** 499

Add. Check for reasonableness. (Lesson 2-7)

47.   748
       + 212

48.   136
       + 299

49.   374
       + 158

Facts Practice

Multiply.

1. 4
    × 9

2. 5
    × 3

3. 6
    × 4

4. 3
    × 6

5. 3
    × 2

6. 4
    × 4

7. 2
    × 2

8. 4
    × 5

9. 4
    × 6

10. 2
     × 5

11. 2
     × 7

12. 8
     × 2

13. 8
     × 3

14. 2
     × 3

15. 6
     × 3

16. 4
     × 3

17. 6
     × 5

18. 3
     × 9

19. 6
     × 2

20. 4
     × 7

21. 7 × 2

22. 5 × 2

23. 6 × 6

24. 2 × 4

25. 6 × 7

26. 3 × 3

27. 5 × 6

28. 7 × 4

29. 3 × 4

30. 4 × 4

31. 7 × 3

32. 9 × 2

33. 5 × 5

34. 9 × 4

35. 2 × 6

36. 5 × 7

Three in a Row
Multiplication Facts

Get Ready!

Players: 2 players

Get Set!

- Each player chooses a color for his or her counter.
- Make a game board like the one shown.

Go!

- Player 1 places two pennies on any two factors. Then Player 1 places a counter on the product.
- Player 2 moves only one penny to a new factor. Then Player 2 places a counter on the product.
- Players take turns moving only one penny each turn and placing their counter on the corresponding product.
- The first player to get three of their counters in a row wins the game.

You will need: 2 pennies
2-color counters

Factors

2	3	4	5	6	7	8	9

Products

20	36	12	14	30
54	45	8	24	40
28	16	27	20	32
42	15	10	21	18
6	24	12	48	35

5-3 Problem-Solving Strategy

MAIN IDEA I will solve a problem by looking for a pattern.

Christina is making a pattern with colored tiles. In the first row, she uses 2 tiles. She uses 4 tiles in the second row and 8 tiles in the third row. If she continues the pattern, how many tiles will be in the sixth row?

Understand	**What facts do you know?** • There will be 2 tiles in the first row, 4 tiles in the second row, and 8 tiles in the third row. **What do you need to find?** • How many tiles will be in row six?
Plan	You can first make a table of the information. Then look for a pattern.
Solve	• First, put the information in a table. • Look for a pattern. The numbers double. • Once you know the pattern, you can continue it. So, there will be 64 tiles in the sixth row.
Check	Look back. Complete the table using the pattern. There are 64 tiles in the sixth row. ✔

Solve table:

1st	2nd	3rd	4th	5th	6th
2	4	8			

+2 +4 +8

$8 + 8 = 16$
$16 + 16 = 32$
$32 + 32 = 64$

Check table:

1st	2nd	3rd	4th	5th	6th
2	4	8	16	32	64

+2 +4 +8 +16 +32

Refer to the problem on the previous page.

1. Look back to the Example. Check your answer. How do you know that it is correct? Show your work.

2. Explain how you identified the pattern for this problem.

3. Suppose there are 4 tiles in the first row, 8 in the second, and 16 in the third. How many tiles are in row 6?

4. Why is it a good idea to put the information in a table first?

PRACTICE the Strategy

EXTRA PRACTICE
See page R14.

Solve. Use the *look for a pattern* strategy.

5. Algebra A set of bowling pins is shown. If there are 3 more rows, how many pins are there in all?

6. Yutaka is planting 24 flowers. He uses a pattern of 1 daisy and then 2 tulips. If the pattern continues, how many tulips will he use?

7. Measurement Adelina makes 3 hops forward and then 1 hop back. Each hop is 1 foot. How many hops does she make before she has gone 2 yards? (*Hint:* There are 3 feet in 1 yard.)

8. Jacy mows lawns every other day. She earns $5 the first day. After that, she earns $1 more than the time before. If she starts mowing on the first day of the month, how much money will she earn on day 19?

Day of month	1	3	5	7	9
Earned	$5	$6	$7		

9. Algebra Shandra is collecting cans for a recycling drive. If the pattern continues, how many cans will she collect in week 5?

Week	1	2	3	4	5
Cans	6	12	24		

10. **WRITING IN ►MATH** Explain how the *look for a pattern* strategy helps you solve problems.

Multiply by 7

MAIN IDEA

I will multiply by 7.

Math Online

macmillanmh.com

• Extra Examples
• Personal Tutor
• Self-Check Quiz

GET READY to Learn

A ride at an amusement park has 5 cars. Each car has 7 seats. How many people can go on the ride at the same time?

You can use models to multiply by 7. Use your multiplication table to help you learn to multiply by 7.

Real-World EXAMPLE Use Models

① **RIDES If there are 5 cars with 7 seats in each car, how many can ride at the same time?**

Find 5 × 7. Use counters to model 5 groups of 7.

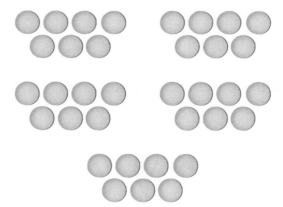

5 × 7 = 35
So, 35 people can ride at the same time.

Check

Use the Commutative Property of Multiplication. Since 7 × 5 = 35, the answer is correct. ✔

2 **ALGEBRA** **A bug box has a total of 28 beetles. There are an equal number of different sizes of beetles. There are 7 of each size. How many sizes are there?**

To solve the problem, you can draw pictures to model and solve a multiplication sentence.

Different Sizes		Number of each size		Total
■	×	7	=	28

> THINK What times 7 equals 28?

Draw beetles in groups of 7 until you have 28 beetles.

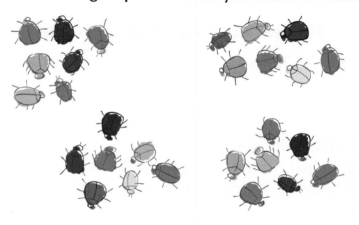

Four groups of 7 is 28. The missing factor in the multiplication sentence is 4.

So, 7 × 4 = 28. There are 4 sizes of beetles.

CHECK What You Know

Multiply. Use models or draw a picture if needed. See Example 1 (p. 214)

1.   2
    × 7

2.   7
    × 8

3. 9 × 7

4. 7 × 10

Algebra **Find each missing factor.** See Example 2 (p. 215)

5. 7 × ■ = 0

6. ■ × 7 = 49

7. 7 × ■ = 70

8. Brianna gave 7 friends 4 pencils each. How many pencils did she give them in all?

9. **Talk About It** Describe two different strategies for multiplying a number by 7.

Multiply. Use models or draw a picture if needed. See Example 1 (p. 214)

10. $\begin{array}{r} 3 \\ \times 7 \\ \hline \end{array}$

11. $\begin{array}{r} 1 \\ \times 7 \\ \hline \end{array}$

12. $\begin{array}{r} 4 \\ \times 7 \\ \hline \end{array}$

13. $\begin{array}{r} 2 \\ \times 7 \\ \hline \end{array}$

14. $\begin{array}{r} 0 \\ \times 7 \\ \hline \end{array}$

15. $\begin{array}{r} 7 \\ \times 7 \\ \hline \end{array}$

16. $\begin{array}{r} 9 \\ \times 7 \\ \hline \end{array}$

17. $\begin{array}{r} 7 \\ \times 6 \\ \hline \end{array}$

18. 7×4

19. 5×7

20. 7×8

21. 7×10

22. 7×2

23. 10×7

24. 7×9

25. 7×5

Algebra Find each missing factor. See Example 2 (p. 215)

26. $4 \times \blacksquare = 28$

27. $7 \times \blacksquare = 49$

28. $8 \times \blacksquare = 56$

29. $\blacksquare \times 7 = 63$

30. $\blacksquare \times 7 = 21$

31. $7 \times \blacksquare = 42$

32. Ryan and 6 friends played basketball. They made a total of 35 baskets. If each made the same number of baskets, how many baskets did each person make?

33. During 9 weeks of summer vacation, Bradley spent 2 weeks at soccer camp. How many days did he not spend at camp?

34. Elian has 5 packs of rubber spiders. If each pack has 7 spiders, how many does he have in all?

35. Inez has 8 CDs. How many songs are there if each CD has 7 songs?

H.O.T. Problems

36. NUMBER SENSE Is 3×7 greater than 3×8? How do you know without multiplying? Explain.

37. WHICH ONE DOESN'T BELONG? Identify which multiplication sentence is incorrect. Explain.

| $7 \times 9 = 63$ | $7 \times 7 = 48$ | $5 \times 7 = 35$ | $7 \times 0 = 0$ |

38. WRITING IN ►MATH Explain why using repeated addition is not the best strategy for finding a product like 7×9.

Multiply. Use models or draw a picture if needed. (Lesson 5-1)

1. 3
$\times\,8$

2. 3
$\times\,4$

3. 3×7

4. 3×9

5. MULTIPLE CHOICE Three times as many students are having a hot lunch than a packed lunch. If there are 8 students with packed lunches, how many are having hot lunches? (Lesson 5-1)

A 5

C 24

B 11

D 32

Multiply. Use models or draw a picture if needed. (Lesson 5-2)

6. 6×0

7. 8×6

Algebra Find each missing factor. (Lesson 5-2)

8. $6 \times \blacksquare = 42$

9. $\blacksquare \times 6 = 36$

10. Algebra Gretchen made a wall for her sand castle. First, she made 4 sand bricks and 1 sand tower. Next, she made 8 bricks and 1 tower. Twelve bricks and 1 tower followed. If she continues in this pattern, how many bricks will come next? (Lesson 5-3)

11. Algebra Find a pattern. Complete the table. (Lesson 5-3)

1st	2nd	3rd	4th	5th	6th
2	7	12	■	■	■

12. Chloe's mom told her to place biscuit dough in equal rows. She fit 3 balls of dough across the top of the baking sheet and 7 balls going down. How many biscuits can Chloe put on the sheet? (Lesson 5-4)

Multiply. Use models or draw a picture if needed. (Lesson 5-4)

13. 7
$\times\,4$

14. 5
$\times\,7$

15. MULTIPLE CHOICE The school district has 6 elementary schools with 7 third grade classes in each school. How many third grade classes are there in all? (Lesson 5-4)

F 13

H 42

G 36

J 49

16. WRITING IN ►MATH Explain how patterns in a multiplication table can help you find the product of 6×9. (Lesson 5-1)

Multiply by 8

GET READY to Learn

There are 6 trees on the side of a street. In each tree, there are 8 birds. How many birds are there in all?

There are many ways to multiply by 8. You can use your multiplication table to help you learn to multiply by 8.

Real-World EXAMPLE   Model an Array

1 BIRDS Find the number of birds in all if there are 8 birds in each of the 6 trees. Use a multiplication sentence to solve.

Think of each tree as a group of 8 birds.

You need to find 6 × 8. Use an array to show 6 rows of 8.

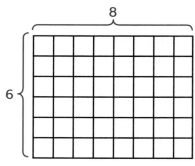

The array shows that 6 × 8 = 48.
So, there are 48 birds in all.

Check

You can use the Commutative Property of Multiplication to check.
8 × 6 = 48 ✓

You can change the order of the factors to find a related fact.

Real-World EXAMPLE   Use a Known Fact

2 **BUTTONS** **Suzy has 4 shirts. There are 8 buttons on each shirt. How many buttons are there altogether?**

Think of each shirt as a group with 8 buttons in each group. You need to find 4 × 8.

You know that 8 × 4 = 32.

So, 4 × 8 = 32.   Commutative Property

There are 32 buttons on the shirts.

Check

You can use a model to check.

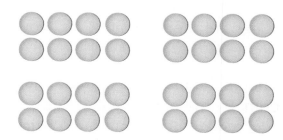

Four groups of 8 is 32. So, 4 × 8 = 32. ✔

> **Remember**
>
> Use the Commutative Property of Multiplication to see a known fact.

✓ CHECK What You Know

Multiply. Use models or a known fact if needed. See Examples 1 and 2 (pp. 218–219)

1. 8
   × 2

2. 0
   × 8

3. 4
   × 8

4. 8
   × 5

5. 8 × 1

6. 6 × 8

7. 8 × 3

8. 8 × 7

9. Nate buys 8 cans of dog food for $4 every week. How much does he spend in 4 weeks?

10. **Talk About It** If there are 4 groups of 8 students and 8 groups of 8 students, how many students are there in all? Explain.

Multiply. Use models or a known fact if needed. See Example 1 and 2 (pp. 218–219)

11. 2
× 8

12. 1
× 8

13. 7
× 8

14. 8
× 8

15. 0
× 8

16. 8
× 9

17. 10
× 8

18. 8
× 3

19. 6×8

20. 5×8

21. 8×4

22. 9×8

Algebra Find each missing factor.

23. $8 \times \blacksquare = 64$

24. $\blacksquare \times 8 = 40$

25. $8 \times \blacksquare = 56$

26. $8 \times \blacksquare = 80$

27. There are 3 large and 4 small spiders in a web. Each has 8 legs. How many legs are there altogether?

28. Admission for one person to the Science Center is $8. How much would it cost a family of 5?

29. Jolon worked 5 hours the first week of the month. By the end of the month, he had worked 8 times as many hours as the first week. How many hours had he worked by the end of the month?

30. There are 9 crates, each with 8 cases of oranges, on a delivery truck. How many cases of oranges will be left if 2 crates are delivered at the first stop?

Real-World PROBLEM SOLVING

Food A recipe for banana bread is shown. Marlo will make 8 times as much for a party.

31. How many bananas will she need?

32. Will 15 cups of flour be enough? Explain.

33. The first four times Marlo makes the recipe, she will make large loaves. The other four times she will make small loaves. How many loaves will she have in all?

34. If there are 8 teaspoons of vanilla in a bottle, how many bottles of vanilla will she need?

Banana Bread

3 bananas, mashed

$\frac{3}{4}$ cup oil

$\frac{3}{4}$ cup sugar

2 teaspoons vanilla

2 cups flour

1 teaspoon baking soda

$\frac{1}{2}$ teaspoon baking powder

$\frac{1}{2}$ cup walnuts

Stir the first four ingredients well. Mix all dry ingredients alone. Then add to liquid, stirring well. Add walnuts last. Pour into prepared pans. Bake 45 minutes at 350°.

Makes 2 large or 6 small loaves.

H.O.T. Problems

35. OPEN ENDED Explain a strategy that you would use to find 9×8. Why do you prefer this strategy?

36. NUMBER SENSE Explain how you can use the Commutative Property of Multiplication to find 8×7.

37.  **MATH** Write a real-world problem that involves multiplying by 8.

TEST Practice

38. What number makes this number sentence true? (Lesson 5-4)

$$7 \times 5 < 4 \times \blacksquare$$

A 3   **C** 7

B 5   **D** 10

39. Which multiplication sentence is modeled below? (Lesson 5-5)

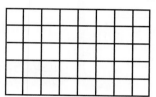

F $5 \times 8 = 40$ **H** $40 \times 8 = 5$

G $5 \times 9 = 40$ **J** $40 \times 5 = 8$

Spiral Review

Multiply. Use models or draw a picture if needed. (Lesson 5-4)

40. 8×7 **41.** 7×7 **42.** 9×7

43. Algebra Martha is building shapes with straws. She uses 3 straws for a triangle and 4 straws for a square. Next, she makes a 5-sided shape. She continues this pattern. How many straws will she have used by the time she makes a 6-sided shape? (Lessons 5-3)

Write a multiplication sentence for each array. (Lesson 4-1)

44.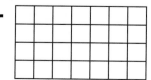

45.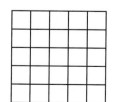

46.

Write each number in word form. (Lesson 1-4)

47. 12,021   **48.** 4,910   **49.** 90,009

Multiply by 9

A grocery store sold 8 boxes of jam. Each box holds 9 jars. How many jars of jam were sold?

MAIN IDEA

I will multiply by 9.

Math Online

macmillanmh.com

• Extra Examples
• Personal Tutor
• Self-Check Quiz

To multiply by 9, you can use models.

Real-World EXAMPLE   Use Models

1 **JAM If the grocery store sold 8 boxes of jam and each box holds 9 jars, how many jars of jam were sold?**

To solve the problem you can use counters to model.

Model 8 groups of 9.

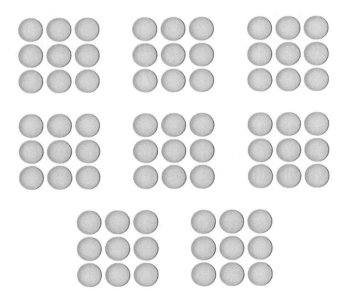

8 groups of 9 is 72.

So, $8 \times 9 = 72$.

72 jars of jam were sold.

You can use patterns to help remember the 9s facts. The second factor and the product in the 9s table create a pattern.

- The tens digit of the product is always 1 less than the factor that is multiplied by 9.

- The sum of the digits of the product equals 9.

$$1 \times 9 = 9$$
$$2 \times 9 = 18$$
$$3 \times 9 = 27$$
$$4 \times 9 = 36$$ ← 3 is one less than 4.
$$5 \times 9 = 45$$
$$6 \times 9 = 54$$
$$7 \times 9 = 63$$
$$8 \times 9 = 72$$ ← In 72, the sum of 7 and 2 is 9.
$$9 \times 9 = 81$$

Real-World EXAMPLE Use Patterns

2 MONEY Mr. Clancy bought 6 cans of paint. Each can costs $9. How much did he spend?

Since the total cost is needed, multiply. Find $6 \times \$9$.

$6 \times \$9 \rightarrow \$5$ ← THINK $6 - 1 = 5$

$6 \times \$9 = \$54$ ← THINK $5 + ? = 9$
$5 + 4 = 9$

So, $6 \times \$9 = \$54$. Mr. Clancy spent $54.

CHECK What You Know

Multiply. Use models or patterns if needed. See Examples 1 and 2 (pp. 222–223)

1. $\begin{array}{r} 9 \\ \times 1 \\ \hline \end{array}$

2. $\begin{array}{r} 4 \\ \times 9 \\ \hline \end{array}$

3. $\begin{array}{r} 9 \\ \times 2 \\ \hline \end{array}$

4. $\begin{array}{r} 6 \\ \times 9 \\ \hline \end{array}$

5. 0×9

6. 9×3

7. 10×9

8. 7×9

9. Lyle has 63 rocks in his collection. He places them into bags. Each bag holds 9 rocks. How many bags are there?

10.  How can patterns help you when multiplying by 9?

Multiply. Use models or patterns if needed. See Examples 1 and 2 (pp. 222–223)

11. 3
       × 9

12. 9
       × 6

13. 4
       × 9

14. 2
       × 9

15. 5
       × 9

16. 8
       × 9

17. 9
       × 10

18. 9
       × 9

19. 1×9

20. 7×9

21. 9×5

22. 10×9

23. 9×0

24. 9×3

25. 6×9

26. 9×7

Algebra Find each missing factor.

27. $\blacksquare \times 9 = 18$

28. $3 \times \blacksquare = 27$

29. $5 \times \blacksquare = 45$

30. $9 \times \blacksquare = 54$

31. $6 \times \blacksquare = 54$

32. $9 \times \blacksquare = 72$

Solve. Use models if needed.

33. Opal and Ela have 9 marbles each. How many marbles are there in all?

34. Cecilia sold 5 books for $9 each. How much money did she get?

35. There were 4 car races on Saturday and 3 on Sunday. If there were 9 cars racing in each race, how many cars raced over the two days?

36. **Measurement** Phil uses 9 yards of rope for each rope ladder he makes. He makes 4 rope ladders. How many yards of rope will he use?

H.O.T. Problems

37. **NUMBER SENSE** Is 9×2 the same as $3 \times 3 \times 2$? Explain.

38. **FIND THE ERROR** Jacinda and Zachary are finding 9×9. Who is correct? Explain.

Jacinda

If $9 \times 8 = 72$, then 9×9 must be 9 more, so $9 \times 9 = 81$.

Zachary

If $9 \times 8 = 72$, then 9×9 must be 8 more, so $9 \times 9 = 80$.

39. **WRITING IN ▸MATH** Describe how the number 10 can help you to solve multiplication problems with 9 as a factor.

Facts Practice

Multiply.

1. $\begin{array}{r} 4 \\ \times\ 6 \\ \hline \end{array}$

2. $\begin{array}{r} 6 \\ \times\ 7 \\ \hline \end{array}$

3. $\begin{array}{r} 3 \\ \times\ 9 \\ \hline \end{array}$

4. $\begin{array}{r} 5 \\ \times\ 9 \\ \hline \end{array}$

5. $\begin{array}{r} 4 \\ \times\ 2 \\ \hline \end{array}$

6. $\begin{array}{r} 9 \\ \times\ 5 \\ \hline \end{array}$

7. $\begin{array}{r} 2 \\ \times\ 8 \\ \hline \end{array}$

8. $\begin{array}{r} 9 \\ \times\ 6 \\ \hline \end{array}$

9. $\begin{array}{r} 8 \\ \times\ 9 \\ \hline \end{array}$

10. $\begin{array}{r} 7 \\ \times\ 4 \\ \hline \end{array}$

11. $\begin{array}{r} 8 \\ \times\ 3 \\ \hline \end{array}$

12. $\begin{array}{r} 4 \\ \times\ 8 \\ \hline \end{array}$

13. $\begin{array}{r} 5 \\ \times\ 8 \\ \hline \end{array}$

14. $\begin{array}{r} 5 \\ \times\ 3 \\ \hline \end{array}$

15. $\begin{array}{r} 8 \\ \times\ 6 \\ \hline \end{array}$

16. $\begin{array}{r} 3 \\ \times\ 3 \\ \hline \end{array}$

17. $\begin{array}{r} 4 \\ \times\ 5 \\ \hline \end{array}$

18. $\begin{array}{r} 7 \\ \times\ 3 \\ \hline \end{array}$

19. $\begin{array}{r} 2 \\ \times\ 7 \\ \hline \end{array}$

20. $\begin{array}{r} 5 \\ \times\ 8 \\ \hline \end{array}$

21. 6×5

22. 8×10

23. 9×8

24. 7×6

25. 6×6

26. 4×8

27. 8×5

28. 9×4

29. 6×2

30. 9×2

31. 3×7

32. 9×9

33. 1×1

34. 7×7

35. 5×5

36. 6×9

Not Just a Blanket

People have been making quilts for 2,000 years. The oldest existing quilt is between 1,000 and 1,500 years old.

Quilts are made of two layers of fabric, with padding in between. Different shapes of cloth are sewn together in detailed patterns. Some quilts are very small, but others are very large. The largest quilt in the world weighs 800 pounds. It is 85 feet wide and 134 feet long. Quilts are much more than blankets. They are pieces of art!

Did You Know?

Some quilts are so small they can almost fit in your hand.

Real-World Math

Use the information on page 226 and the picture of the quilt above to solve each problem.

1 How many feet longer is the length than the width of the largest quilt in the world?

2 How can you use repeated addition to find how many squares are in the quilt pictured?

3 Suppose you need to make a quilt that uses twice as many squares as the quilt shown. How many squares do you need to make your quilt?

4 How many squares do you need if you make 3 quilts with 9 squares each?

5 If you need to make 6 quilts, how many squares do you need?

6 Each quilt square is 7 inches wide and 7 inches long. How long is the quilt?

7 A quilt is 9 squares wide and 7 squares long. How many squares are there in all?

8 You have 7 quilts. Each quilt is 3 squares long and 3 squares wide. Do you have 63 squares? Explain.

Problem-Solving Investigation

MAIN IDEA I will choose the best strategy to solve problems.

P.S.I. TEAM +

ALEC: I have a goal to ride my bike 20 miles this week. Last night, I rode my bike 2 miles each way going to and from softball practice. I will ride this distance for 6 more days.

YOUR MISSION: Find if Alec will meet his goal and ride his bike 20 miles this week.

Understand	Alec wants to ride 20 miles each week. He will ride 2 miles each way to and from practice for 7 days. Find if he will meet his goal.
Plan	Find the total miles he will ride each day and for the week. Multiply to find a total.
Solve	Alec will ride 2 miles to practice and 2 miles home, or $2 \times 2 = 4$ miles each day. 4 miles each day $\times$ 7 days = 28 miles So, Alec rides his bike 28 miles this week. Since 28 miles > 20 miles, Alec will meet his goal.
Check	Look back. Use an array to check. $4 \times 7 = 28$ So, the answer is correct. ✔

Use any strategy shown below to solve. Tell what strategy you used.

PROBLEM-SOLVING STRATEGIES
• Act it out.
• Draw a picture.
• Look for a pattern.

1. There are 2 ladybugs. Together they have 12 spots. If one has 4 more spots than the other, how many spots do they each have?

2. **Measurement** A train travels the distances shown.

Day	Distance (miles)
Monday	75
Tuesday	■
Wednesday	200

If the train travels a total of 500 miles, how far did it travel on Tuesday?

3. **Algebra** What are the next three numbers in the pattern?

5, 8, 11, 14, 17, ■, ■, ■

4. **Geometry** What will be the measure of the two labeled sides of the next smallest triangle if the pattern continues?

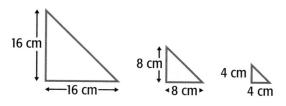

5. **Algebra** The picture shows the cost of sodas at a restaurant. Identify and use a pattern to find the cost of a large soda.

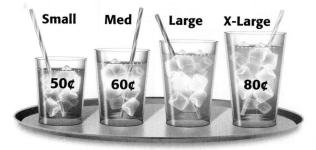

Small	Med	Large	X-Large
50¢	60¢		80¢

6. Carmen has $5. Darby has two times as much money as Carmen. Frank has $3 more than Carmen. What is the total in all?

7. **Algebra** A weaver is creating a design for a scarf. The first row has 3 hearts. The second row has 7 hearts, and the third row has 11. As she continues the pattern, how many hearts will be in the seventh row?

8. Diego, Marco, and Andrea earn money raking leaves after school. If they evenly share the money they make, how much will each get?

Money Earned Raking
Friday $6
Saturday $10
Sunday $8

9. **WRITING IN ►MATH** Write a real-world problem that could be solved in more than one way. Explain.

GET READY to Learn

A hardware store sells wrench sets. Each set holds 11 wrenches. A customer orders 7 sets. How many wrenches are there in 7 sets?

MAIN IDEA

I will multiply by 11 and 12.

Math Online

macmillanmh.com
• Extra Examples
• Personal Tutor
• Self-Check Quiz

When multiplying by 11, you can use patterns and models.

Real-World EXAMPLE   Use Patterns or Models

1. **TOOLS How many wrenches are there in 7 sets?**

   Find 7 groups of 11 or 7 × 11. Look at the pattern.

Multiply by 11.		
1	× 11 =	11
2	× 11 =	22
3	× 11 =	33
4	× 11 =	44
5	× 11 =	55
6	× 11 =	66
7	× 11 =	77

The pattern shows that when a single digit number is multiplied by 11, the product is the digit repeated.

The model also shows that 7 × 11 = 77.

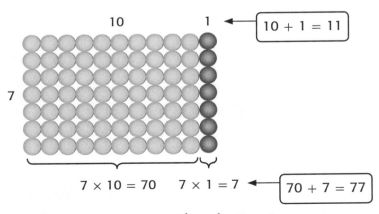

10 + 1 = 11

7 × 10 = 70   7 × 1 = 7   70 + 7 = 77

So, there are 77 wrenches in 7 sets.

EXAMPLE Use Models

2 MEASUREMENT There are 12 inches in one foot. How many inches are in 8 feet?

You need to find 8×12.

Model 8×12. That is, 8 groups of 12.

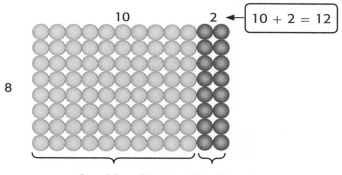

10 + 2 = 12

$8 \times 10 = 80$ $8 \times 2 = 16$

$80 + 16 = 96$

There are a total of 96 counters.
So, $8 \times 12 = 96$.
There are 96 inches in 8 feet.

 CHECK What You Know

Multiply. Use models or patterns if needed. See Examples 1 and 2 (pp. 230–231)

1. 11
$\times 3$

2. 4
$\times 11$

3. 11
$\times 11$

4. 11
$\times 5$

5. 8
$\times 11$

6. 10
$\times 11$

Multiply. Use models if needed. See Example 2 (p. 231)

7. 12×2

8. 12×6

9. 3×12

10. 12×5

11. 9×12

12. 12×12

13. Measurement Each bag at the right has 12 bagels. How many bagels are there altogether?

14. **Talk About It** Why does the pattern shown in Example 1, only work on single digit numbers?

Multiply. Use models or patterns if needed. See Examples 1 and 2 (pp. 230–231)

15. 11×2

16. 11×1

17. 11×7

18. 11×6

19. 8×11

20. 9×11

Multiply. Use models if needed. See Example 2 (p. 231)

21. 12×1

22. 5×12

23. 12×4

24. 12×8

25. 11×12

26. 7×12

Solve.

27. The crackers shown are one serving. There are 10 servings in the box. How many crackers are in the box?

28. A restaurant orders 12 dozen eggs. The picture shows how many eggs broke during shipping. How many eggs were left?

29. The soccer team runs 11 laps every day. How many laps will they run from Monday to Friday?

30. Each aquarium has 10 goldfish and 6 puffer fish. There are 11 aquariums. How many fish in all?

Data File

In South Carolina, sweetgrass basket weaving is a tradition that has been passed from one family to another for over 300 years. How many yards of grass are needed for each decoration?

31. 2 baskets with 1 braid and 1 flower

32. 3 baskets with fringe

33. 2 baskets with 2 flowers and 2 braids

BASKET WEAVING

Decoration	Needed Grass (yard)
Braid	12
Fringe	5
Flower	11

H.O.T. Problems

34. OPEN ENDED Draw a model to show a multiplication sentence that has 11 as a factor. Write the multiplication sentence.

35. CHALLENGE There are 12 inches in a foot and 3 feet in a yard. How many inches are in 2 yards?

36. WRITING IN ►MATH Write a real-world problem that can be represented by 5 × 12. Solve.

TEST Practice

37. If 11 horses each ate 4 apples, how many apples did they eat?
(Lesson 5-8)

A 4   C 15

B 11   D 44

38. Which point best represents 245 on the number line? (Lesson 1-8)

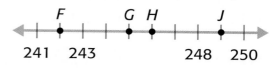

241   243   248   250

F *F*   H *H*

G *G*   J *J*

Spiral Review

39. The table shows how much money Sofia has saved each day. If she continues this pattern, how many more days will she need to put money in the jar so she can buy her mom a $15 gift? Extend the table to solve the problem. (Lesson 5-7)

Multiply. Use models or patterns if needed. (Lesson 5-6)

40.   9
   × 6

41.   9
   × 7

42.   9
   × 9

43. Brock earned an 83 on his first test and a 91 on his second test. How can you find how many more points he scored on the second test? Explain. (Lesson 3-9)

 **5-9**

Algebra: Associative Property

MAIN IDEA

I will identify and use the Associative Property of Multiplication.

New Vocabulary

Associative Property of Multiplication

Math Online

macmillanmh.com

• Extra Examples
• Personal Tutor
• Self-Check Quiz

 **GET READY to Learn**

Write a multiplication sentence with three numbers and two multiplication signs to find the total number of smiley faces.

To find a product, like $2 \times 3 \times 4$, you can use properties of multiplication. The properties make multiplying easier.

Associative Property	Key Concept
Words	The **Associative Property of Multiplication** states that the grouping of factors does not change the product.
Examples	$(2 \times 3) \times 4$ $2 \times (3 \times 4)$
	$6 \times 4 = 24$ $2 \times 12 = 24$

The parentheses tell you which factors to multiply first.

EXAMPLE

① Find $5 \times 2 \times 3$.

One Way
Multiply 5 and 2 first.
$(5 \times 2) \times 3$
$10 \times 3 = 30$

Another Way
Multiply 2 and 3 first.
$5 \times (2 \times 3)$
$5 \times 6 = 30$

So, $5 \times 2 \times 3 = 30$.

You can use the Associative Property of Multiplication to solve real-world problems.

Real-World EXAMPLE

2 BOOKS Amada reads 3 books. Each book has 6 pages. There are 2 pictures on each page. How many pictures are there altogether?

To find the total number of pictures, you can write a multiplication sentence. You can group the easier factors.

$(3 \times 2) \times 6$

$6 \qquad \times 6 = 36$

THINK It is easier to multiply 3×2.

So, $3 \times 2 \times 6 = 36$. There were 36 pictures.

Remember

No matter how you group the factors, the product will be the same. This is the Associative Property.

To find missing factors when multiplying more than two numbers, use the Associative Property of Multiplication.

Real-World EXAMPLE Find Missing Factors

3 ALGEBRA Cheryl has 2 photos. Each photo shows 5 friends holding the same number of flowers. There are 30 flowers altogether. How many flowers is each person holding?

You can write a multiplication sentence to help you find the missing factor.

Number of photos		Number of friends		Flowers each is holding		Total
2	$\times$	5	$\times$	■	=	30

Use the Associative Property to find 2×5 first.

$(2 \times 5) \times$ ■ $= 30$

$10 \qquad \times$ ■ $= 30$

$10 \qquad \times 3 \ = 30$

THINK 10 times what number equals 30?

So, $2 \times 5 \times 3 = 30$. Each person is holding 3 flowers.

CHECK What You Know

Find each product. See Examples 1 and 2 (pp. 234–235)

1. $2 \times 4 \times 6$

2. $5 \times 2 \times 8$

3. $4 \times 1 \times 3$

Algebra Find each missing factor. See Example 3 (p. 235)

4. $\blacksquare \times 2 \times 3 = 30$

5. $\blacksquare \times 8 \times 1 = 72$

6. $4 \times 2 \times \blacksquare = 40$

7. There are 3 tables with 4 books. Each book has 2 bookmarks. Find the number of bookmarks.

8. **Talk About It** Explain how the Associative Property of Multiplication can help you find missing numbers.

Practice and Problem Solving

EXTRA PRACTICE See page R16.

Find each product. See Examples 1 and 2 (pp. 234–235)

9. $5 \times 2 \times 3$

10. $6 \times 2 \times 2$

11. $4 \times 5 \times 1$

12. $2 \times 4 \times 9$

13. $3 \times 2 \times 8$

14. $2 \times 7 \times 2$

Algebra Find each missing factor. See Example 3 (p. 235)

15. $3 \times \blacksquare \times 4 = 24$

16. $6 \times \blacksquare \times 5 = 30$

17. $\blacksquare \times 3 \times 3 = 27$

18. $2 \times 5 \times \blacksquare = 20$

19. $3 \times 3 \times \blacksquare = 63$

20. $6 \times \blacksquare \times 3 = 36$

Write a multiplication sentence for each situation. Then solve. See Example 3 (p. 235)

21. Mrs. Flanagan has 2 new book stands with 3 shelves each. She put 10 books on each shelf. How many books has she set up?

22. A grocer unpacked 2 boxes of soup. Each box held 4 cartons with 10 cans of soup in each. How many cans did the grocer unpack?

Algebra Find the value of each number sentence if ☺ = 2, ♡ = 3, and ☆ = 4.

23. $6 \times 1 \times$ ☺

24. $4 \times$ ♡ $\times 2$

25. ☆ $\times$ ☺ $\times 5$

26. ♡ $\times$ ♡ $\times 4$

27. There are 5 apples. Britt cuts each apple into 2 pieces. Then Emma cuts each piece into 4 slices. Write a multiplication sentence to show the total number of apple slices.

28. Britt and Emma each cut up 2 bananas into 4 pieces. Write a multiplication sentence to show the total number of banana pieces.

H.O.T. Problems

29. OPEN ENDED Write three factors that have a product of 24.

30. WHICH ONE DOESN'T BELONG? Identify which of the following is not true. Explain.

$(2 \times 3) \times 3 = 2 \times (3 \times 3)$

$(4 \times 4) \times 2 = (4 \times 4) \times 4$

$3 \times (1 \times 5) = (3 \times 1) \times 5$

$6 \times (4 \times 2) = (6 \times 4) \times 2$

31. WRITING IN ►MATH Explain why order does not matter when finding $(3 \times 4) \times 2$.

TEST Practice

32. What number makes this number sentence true? (Lesson 5-9)

$$(6 \times 3) \times 7 = 6 \times (\blacksquare \times 7)$$

A 3   **C** 6

B 4   **D** 7

33. There were 9 horses. Each horse ate 4 apples. Which of the following shows how to find the number of apples in all? (Lesson 5-6)

F $9 + 4$ **H** 9×4

G $9 - 4$ **J** $9 \div 4$

Spiral Review

Use any strategy to solve. Tell what strategy you used. (Lesson 5-7)

34. Measurement A saltwater crocodile weighs up to 1,150 pounds. An anaconda weighs 500 pounds. Together, how much can these two animals weigh?

35. Last summer, Terrence made $34 selling lemonade at his stand. This summer he made $56. How much more money did he make this summer?

Multiply. Use models or patterns if needed. (Lesson 5-6)

36. 9×6 **37.** 9×7 **38.** 9×9 **39.** 9×0

Compare. Use >, <, or =. (Lesson 1-3)

40. 3,839 ● 3,973   **41.** 2,371 ● 237   **42.** 209 ● 290

FOLDABLES Study Organizer **GET READY** to Study

Be sure the following Key Concepts and Key Vocabulary words are written in your Foldable.

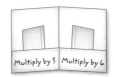

Multiply by 3 | Multiply by 6

Key Concepts

• When **factors** are multiplied, the result is a **product**. (p. 201)

$$8 \times 4 = 32$$

factor   factor   product

• The **Commutative Property of Multiplication** states that the order in which the factors are multiplied does not change the product. (p. 214)

$$6 \times 7 = 42 \qquad 7 \times 6 = 42$$

• The **Associative Property of Multiplication** states that the grouping of factors does not change the product. (p. 234)

One Way
$$(3 \times 2) \times 4$$
$$6 \quad \times 4 = 24$$

Another Way
$$3 \times (2 \times 4)$$
$$3 \times 8 = 24$$

Key Vocabulary

array (p. 206)

Associative Property of Multiplication (p. 234)

Commutative Property of Multiplication (p. 214)

factor (p. 201)

product (p. 201)

Vocabulary Check

Choose the vocabulary word that completes each sentence.

1. The ____?____ Property of Multiplication states that the grouping of factors does not change the product.

2. Two ____?____ are multiplied together to get a product.

3. The ____?____ of 8 and 7 is 56.

4. A(n) ____?____ is an arrangement of equal rows and columns.

Lesson-by-Lesson Review

5-1 **Multiply by 3** (pp. 203–205)

Example 1
There are 8 parrots. Each has 3 red feathers on its head. How many feathers are there in all?

There are 8 groups of 3 feathers. Model 8 groups of 3, or 8 × 3.

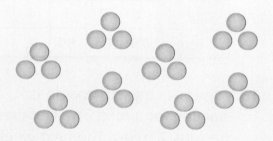

8 × 3 = 24
So, there are 24 feathers.

Multiply. Use models or draw a picture if needed.

5. $\begin{array}{r} 3 \\ \times 7 \\ \hline \end{array}$
6. $\begin{array}{r} 4 \\ \times 3 \\ \hline \end{array}$

7. 6 × 3
8. 8 × 3

9. **Algebra** Copy and complete.

Rule: Multiply by 3.				
Input	7	■	2	■
Output	■	18	■	24

10. There are 4 trees with 3 rabbits sitting under each tree. How many rabbits are there after 2 rabbits hop away?

5-2 **Multiply by 6** (pp. 206–209)

Example 2
Each package of yogurt has 6 cups. If Sue Ellen buys 3 packages, how many cups will she have?

Use counters to model 3 rows of 6.

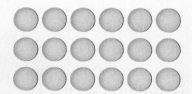

6 + 6 + 6 = 18

So, 3 × 6 = 18.

She will have 18 cups of yogurt.

Multiply. Use models or draw a picture if needed.

11. 6 × 5
12. 7 × 6

13. $\begin{array}{r} 6 \\ \times 8 \\ \hline \end{array}$
14. $\begin{array}{r} 6 \\ \times 6 \\ \hline \end{array}$

Algebra Find each missing factor.

15. 6 × ■ = 18
16. ■ × 6 = 6

17. Chantal invited 3 friends over for pizza. Each person gets 6 slices. How many slices of pizza will she need for her friends and herself?

5-3 **Problem-Solving Strategy:** Look for a Pattern (pp. 212–213)

Example 3
Tate made rows of toys. He lined up 3 toys in the first row, 5 in the second, and 7 in the third. How many toys will be in the seventh row if he continues his pattern?

Understand

You know the number of toys in the first three rows. Find how many toys will be in the seventh row.

Plan

Organize the data in a table. Then look for and extend the pattern.

Solve

Row	1	2	3	4	5	6	7
Toys	3	5	7	9	11	13	15

+2 +2 +2 +2 +2 +2

$7 + 2 = 9$

$9 + 2 = 11$

$11 + 2 = 13$

$13 + 2 = 15$

So, there will be 15 toys in the seventh row.

Check

The answer makes sense for the problem. The answer is correct. ✔

Solve. Use the *look for a pattern* strategy.

18. **Algebra** The following table shows Rich's test scores. All of the scores follow a pattern. What are the last two scores he received?

Test	1	2	3	4	5
Score	79	84	89	■	■

19. **Measurement** Arnaldo is training for a race. The first week, he runs 2 miles. The next week he runs 5 miles, and the third week he runs 8 miles. How many weeks will it take him to reach his goal of 20 miles?

20. Kashi earned a $4 allowance this week. If he continues to do his chores every day, he will earn $2 more than the week before. What will his allowance be for week 5?

21. Lee takes care of the two family dogs. He feeds each dog 3 treats each day. How many treats does Lee give the dogs in one week?

Multiply by 7 (pp. 214–216)

Example 4
Jermaine split 28 football cards equally among his 7 friends. How many cards did each friend receive?

You can use a number sentence.

■ × 7 = 28

THINK What times 7 equals 28?

Since 4 × 7 = 28, he gave each friend 4 cards.

Multiply. Use models or draw a picture if needed.

22. 7 × 3   **23.** 5 × 7

24. 7   **25.** 7
      × 7   × 9

Algebra **Find each missing factor.**

26. 6 × ■ = 42

27. 7 × ■ = 28

28. There are 9 flights of stairs. Each flight has 7 steps. What is the total number of steps?

Multiply by 8 (pp. 218–221)

Example 5
Jamal has 6 bags. Each bag contains 8 coins. How many coins does Jamal have in all?

You need to find 6 groups of 8 or 6 × 8.

Use the Commutative Property of Multiplication.

You know that 8 × 6 = 48.

So, 6 × 8 = 48.

Jamal has 48 coins.

Multiply. Use models or known facts if needed.

29. 8 × 3   **30.** 8 × 7

31. 8   **32.** 8
      × 4   × 5

Algebra **Find each missing factor.**

33. ■ × 8 = 64

34. 8 × ■ = 40

35. Each box contains 9 bunches of bananas. If the manager orders 8 boxes, how many bunches of bananas will arrive?

5-6
Multiply by 9 (pp. 222–224)

Example 6
Karen has 3 picture frames. There are 9 photos in each frame. How many photos are there?

To find 3 × 9, you can use a pattern.

$3 \times 9 = 2$ THINK 3 − 1 = 2

The sum of the digits of the product is 9.

$3 \times 9 = 27$ THINK 2 + ? = 9
$2 + 7 = 9$

So, 3 × 9 = 27 photos.

Multiply. Use models or patterns if needed.

36. 5
     ×9

37. 9
     ×2

38. 8
     ×9

39. 7
     ×9

Algebra Find each missing factor.

40. ■ × 9 = 36   41. ■ × 9 = 45

42. Each of Judie's dolls is worth $9. If she sold some and made $81, how many dolls did she sell?

5-7
Problem-Solving Investigation: Choose a Strategy (pp. 228–229)

Example 7
Gonzalo sold 9 raffle tickets today and 7 yesterday. Each costs $5. How many more tickets does he need to sell to raise $100?

You need to find how many more tickets he must sell to raise $100.

$5 × 9 = $45 today
$5 × 7 = $35 yesterday

He raised $45 + $35 or $80.

He still needs to raise $100 − $80 or $20.

Since 4 × $5 is $20, he must sell 4 more tickets.

Solve. Tell what strategy you used.

43. To win a contest, students must read 24 books in 3 months. How many books does a student need to read each month?

44. Mrs. Larkin's students earned 4 points for their contest on Monday, 6 on Tuesday and 8 on Wednesday. With this pattern, how many points will be added on Friday, the fifth day?

45. **Measurement** It takes 3 minutes for each child to get through the lunch line. How long will it take for 7 students?

Multiply by 11 and 12 (pp. 230–233)

Example 8
Four actors in a play each received 12 roses. How many roses were received in all?

Use counters to find 4 × 12.

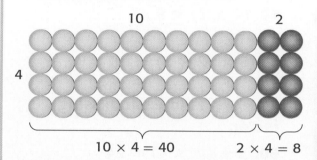

10 × 4 = 40   2 × 4 = 8

40 + 8 = 48

So, 48 roses were received in all.

Example 9
Martin has 6 packages of rubber insects. Each package has 11 insects. How many insects does Martin have in all?

To find how many insects Martin has, you can use a pattern to multiply 6 by 11.

1 × 11 = 11

2 × 11 = 22

3 × 11 = 33

4 × 11 = 44

5 × 11 = 55

6 × 11 = 66 ◄───

> The pattern shows that when a single digit is multiplied by 11, the product is the digit repeated.

So, Martin had 66 insects altogether.

Multiply. Use models or patterns if needed.

46. 11 × 5   **47.** 8 × 11

48. 3 × 12   **49.** 12 × 6

Algebra Find each missing factor.

50. 11 × ■ = 44

51. ■ × 6 = 66

52. Gabriel earned $11 each week for helping his neighbor rake leaves. How much did he earn in 4 weeks?

53. Measurement Annie is celebrating her ninth birthday. How many months will have passed by the time she turns 14 years old? (*Hint:* There are 12 months in one year.)

54. Rayna's father bought 3 dozen eggs. When he got home he found 4 eggs were broken. How many eggs does Rayna's Dad have now?

55. The school store is selling packages of pencils. There are 4 packages with 12 pencils in each. Seven of the packages have 11 pencils. What is the total number of pencils the school store has to sell?

5-9 **Algebra: Associative Property** (pp. 234–237)

Example 10
There are 5 farmers. Each farmer has 4 sheep. Each sheep has 2 ears. How many ears are there in all?

You can use the Associative Property of Multiplication to find the total number of ears.

Find the product of $4 \times 2 \times 5$.

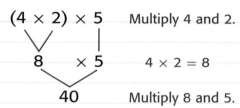

One Way First find 4×2.

$(4 \times 2) \times 5$ Multiply 4 and 2.

$8 \quad \times 5$ $4 \times 2 = 8$

40 Multiply 8 and 5.

Another Way First find 2×5.

$4 \times (2 \times 5)$ Multiply 2 and 5.

$4 \times \quad 10$ $2 \times 5 = 10$

40 Multiply 4 and 10.

So, $4 \times 2 \times 5 = 40$. There are a total of 40 ears.

Find each product.

56. $7 \times (2 \times 2)$

57. $(2 \times 5) \times 3$

58. $(4 \times 2) \times 9$

59. $6 \times (2 \times 3)$

Algebra Find each missing factor.

60. $3 \times (\blacksquare \times 4) = 24$

61. $(6 \times \blacksquare) \times 5 = 30$

62. Jerrica has 3 bookcases. Each one has 4 shelves. She will put 3 stuffed animals on each shelf. How many stuffed animals does Jerrica have on the three bookcases?

63. A third grade classroom has 3 rows. Each row has 10 desks. Upon each desk are 3 pencils. How many pencils are there in all?

64. There are 4 families at the fair. Each of these families has 3 children. If each of the children buys 5 ride tickets, how many tickets will the children have?

Multiply.

1. 3
×6

2. 3
×9

3. 9
×4

4. 6
×4

5. A 60-member marching band stood in 12 equal rows. How many band members are in each row?

Algebra Find each missing factor.

6. 8 × ■ = 32

7. ■ × 9 = 54

8. 7 × ■ = 35

9. ■ × 8 = 56

10. 3 × ■ = 24

11. ■ × 5 = 20

12. MULTIPLE CHOICE Four students are in this year's spelling bee. Each student had to pass 5 tests to be in the spelling bee. How many tests is that in all?

A 7

C 12

B 9

D 20

13. There are 7 gardeners. Each picked the number of tomatoes shown below. How many tomatoes were picked in all?

14. During gym class, the teacher gave the students a number as they stood in line. She counted 1, 2, 3, 1, 2, 3, . . . What number did the 22nd student get?

15. Algebra Find a pattern for 2, 6, 5, 9, 8, . . . Then give the next three numbers.

Multiply.

16. 8
×8

17. 7
×7

18. 9
×7

19. 11
×5

20. 11
×8

21. 12
×7

22. MULTIPLE CHOICE Mr. Thompson bought 7 of the same item at the store. He paid a total of $42. What did he buy?

F $6 shirts

G $7 pants

H $35 shoes

J $49 jacket

23. WRITING IN ►MATH If 2 × 7 × 4 = 56, then what is 7 × 4 × 2? Explain.

PART 1 Multiple Choice

Read each question. Then fill in the correct answer on the answer sheet provided by your teacher or on a sheet of paper.

1. For the school play, chairs were set in 8 rows of 8 seats. How many seats are there in all?

A 40 **C** 63

B 56 **D** 64

2. Tavis bought 4 packages of juice boxes. Each package has 4 juice boxes. Which number sentence shows how to find the number of juice boxes in all?

F $4 + 4 = \blacksquare$ **H** $4 \times 4 = \blacksquare$

G $4 - 4 = \blacksquare$ **J** $16 - 4 = \blacksquare$

3. Each package of markers has 6 markers. How many markers are in 7 packages?

A 35 **C** 42

B 36 **D** 48

4. In the pattern below, each number is 2 times greater than the number before it. What is the next number in the pattern?

1, 2, 4, 8, _____

F 10 **H** 16

G 12 **J** 18

5. What is the best estimate of the difference rounded to the nearest hundred?

809 − 327

A 300 **C** 500

B 400 **D** 600

6. Which number sentence is modeled by the figure below?

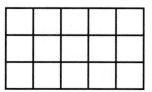

F $3 + 3 + 3 = 9$ **H** $3 \times 6 = 18$

G $4 \times 5 = 20$ **J** $3 \times 5 = 15$

7. A paperback book costs $7. How much will 4 of these books cost?

A $21　　　　**C** $28

B $25　　　　**D** $35

8. The Main Street School bought 1,250 erasers. They gave out 867 to students. How many are left?

F 383　　　　**H** 417

G 393　　　　**J** 483

9. What does 2×5 mean?

A $5 + 5$

B $2 + 5 + 2 + 5 + 2$

C $5 + 5 + 5 + 5 + 5$

D $2 + 2$

10. Tony has 8 packs of gum. Each pack contains 5 pieces of gum. If Tony chews 3 pieces, how many pieces does he have left?

F 37　　　　**H** 13

G 32　　　　**J** 8

PART 2　　Short Response

Record your answer on the answer sheet provided by your teacher or on a sheet of paper.

11. Which factor will make the multiplication sentence true?

$$\blacksquare \times 3 = 0$$

12. Write two multiplication sentences that model the Commutative Property of Multiplication.

PART 3　　Extended Response

Record your answer on the answer sheet provided by your teacher or on a sheet of paper.

13. Enrique has 2 sets of paints that are arranged in 5 rows each. There are 8 paints in each row. How many paints does Enrique have altogether? Show your work. Explain how the properties of multiplication can make solving this problem easier.

NEED EXTRA HELP?													
If You Missed Question...	1	2	3	4	5	6	7	8	9	10	11	12	13
Go to Lesson...	5-5	4-3	4-2	5-2	3-2	4-1	5-4	3-7	4-2	5-7	4-8	4-1	5-9

CHAPTER 6 Develop Division Concepts and Facts

BIG Idea   What is division?

Division is an operation with two numbers. One number tells you how many things you have. The other tells you how many equal groups to form.

Example Lonnie has 15 pennies to share among 5 friends. If Lonnie gives each friend the same number of pennies, each friend will get $15 \div 5$, or 3 pennies.

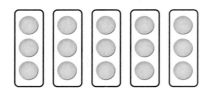

What will I learn in this chapter?

- Explore the meaning of division.
- Relate division to subtraction and multiplication.
- Divide by 0, 1, 2, 5, and 10.
- Write number sentences to show division problems.
- Choose an operation to solve problems.

Key Vocabulary

divide

dividend

divisor

quotient

fact family

Math Online > **Student Study Tools**
at macmillanmh.com

FOLDABLES®
Study Organizer

Make this foldable to help you organize information about division. Begin with one sheet of 11″ × 17″ paper.

1 **Fold** the shorter edges so they meet in the middle.

2 **Fold** in half as shown.

3 **Unfold** and cut along the two outside folds.

4 **Label** as shown. Record what you learn.

| Divide by 2 | Divide by 10 |
| Divide by 5 | Divide by 0 and 1 |

You have two ways to check prerequisite skills for this chapter.

Option 2

Math Online Take the Chapter Readiness Quiz at macmillanmh.com.

Option 1

Complete the Quick Check below.

QUICK Check

Subtract. (Lesson 3-1)

1. $14 - 7$ **2.** $36 - 6$ **3.** $45 - 9$ **4.** $56 - 8$

5. There are 18 children reading books. If 6 of them are reading mystery books, how many are reading other kinds of books?

Tell whether each pair of groups are equal. (Prior grade)

6.

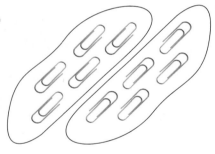

7.

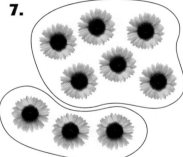

8. Camille, Delia, and Emily are sharing a full box of crackers. They each get 7 crackers. How many crackers are in the box?

Multiply. (Lessons 4-3, 5-2, and 5-5)

9. 2×4 **10.** 3×6 **11.** 5×4 **12.** 7×8

13. Write the multiplication sentences for the two arrays shown.

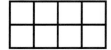

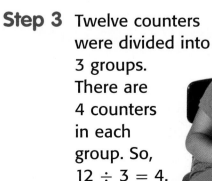

Division is an operation with two numbers. One number tells you how many things you have. The other tells you how many equal groups to form.

$$10 \div 5 = 2$$

Read ÷ as *divided by*.
10 divided by 5 = 2.

To **divide** means to separate a number into equal groups, to find the number of groups, or the number in each group.

MAIN IDEA

I will explore the meaning of division.

You Will Need
counters
paper plates

Math Online

macmillanmh.com
• Concepts in Motion

ACTIVITY

① **Divide 12 counters into 3 equal groups.**

Step 1 Count out 12 counters. Using paper plates, show 3 groups.

Step 2 Place counters equally among the 3 groups until all of the counters are gone.

Step 3 Twelve counters were divided into 3 groups. There are 4 counters in each group. So, $12 \div 3 = 4$.

Hands-On Activity

2 **There are 12 counters. Group the counters 3 at a time.**

Step 1 Count out 12 counters.

Step 2 Make equal groups of 3 until all the counters are gone.

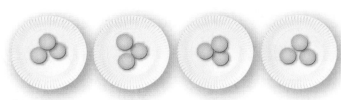

There are 4 equal groups of 3.
So, $12 \div 4 = 3$.

Think About It

1. Explain how you divided 12 counters into equal groups.

2. When you divided the counters into groups of 3, how did you find the number of equal groups?

CHECK What You Know

3. Make equal groups to find the number of counters in each group.

4. Find the number of equal groups of 5.

5. Copy the chart. Then use counters to help complete.

Number of Counters	Number of Equal Groups	Number in Each Group	Division Sentence
9	3	3	$9 \div 3 = 3$
14	2	▨	▨
15	▨	5	▨
6	▨	3	▨

6. **WRITING IN ►MATH** Can 13 counters be divided equally into groups of 3? Explain.

6-1

Relate Division to Subtraction

MAIN IDEA

I will use models to relate division to subtraction.

New Vocabulary

division

divide

Math Online

macmillanmh.com

• Extra Examples
• Personal Tutor
• Self-Check Quiz

GET READY to Learn

There are 15 pencils in a box. Each pencil is either red, blue, or yellow. There is the same number of each color. How many pencils of each color are there?

You have used counters to model **division**. Recall that to **divide** means to separate a number into equal groups, to find the number of groups, or the number in each group.

Real-World EXAMPLE   Use Models to Divide

1 PENCILS How many pencils of each color are there? Use a number sentence to record the solution.
Using counters, divide 15 counters equally into 3 groups until all the counters are gone.

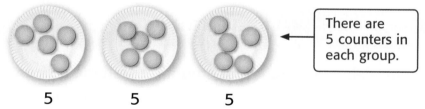

There are 5 counters in each group.

5   5   5

The number sentence that describes the model is
$15 \div 3 = 5$. So, there are 5 pencils of each color.

You can also divide using *repeated subtraction*. Subtract equal groups repeatedly until you get to zero.

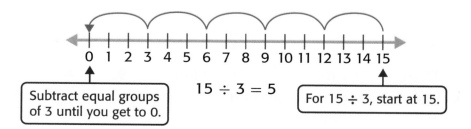

Subtract equal groups of 3 until you get to 0.

$15 \div 3 = 5$

For $15 \div 3$, start at 15.

Lesson 6-1 Relate Division to Subtraction   **253**

Real-World EXAMPLE Repeated Subtraction

② **SPORTS** Nathan wants to put his 10 baseball cards in equal groups of 2. How many groups can he make?

Use repeated subtraction to find 10 ÷ 2. Write a number sentence.

One Way: Number Line	Another Way: Paper and Pencil
⑤ ④ ③ ② ① 0 1 2 3 4 5 6 7 8 9 10	① ② ③ ④ ⑤ $\begin{array}{r}10\\-2\\\hline 8\end{array}$ $\begin{array}{r}8\\-2\\\hline 6\end{array}$ $\begin{array}{r}6\\-2\\\hline 4\end{array}$ $\begin{array}{r}4\\-2\\\hline 2\end{array}$ $\begin{array}{r}2\\-2\\\hline 0\end{array}$
Start at 10. Count back by 2s until you reach 0. How many times did you subtract?	Subtract groups of 2 until you reach 0. How many groups did you subtract?

So, the number sentence 10 ÷ 2 = 5 shows that Nathan will have 5 groups of cards.

✓ CHECK What You Know

Use models to divide. Write a number sentence. See Example 1 (p. 253)

1. There are 16 flowers. Each vase has 4 flowers. How many vases are there?

2. There are 14 ears. Each dog has 2 ears. How many dogs are there?

Use repeated subtraction to divide. See Example 2 (p. 254)

3.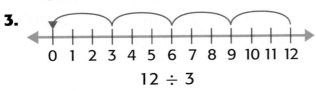

0 1 2 3 4 5 6 7 8 9 10 11 12

12 ÷ 3

4.

0 1 2 3 4 5 6 7 8

8 ÷ 2

5. 6 ÷ 2

6. 12 ÷ 6

7. 25 ÷ 5

8. There are 14 mittens. Each student wears 2 mittens. How many students are there?

9. 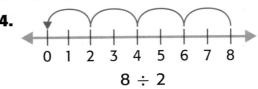 **Talk About It** Explain how to use a number line to find 18 ÷ 9.

Use models to divide. Write a number sentence. See Example 1 (p. 253)

10. There are 16 orange slices. Each orange has 8 slices. How many oranges are there?

11. Measurement There are 16 miles. Each trip is 2 miles. How many trips are there?

12. There are 25 marbles, with 5 marbles in each bag. How many bags are there?

13. There are 12 muffins and 4 friends. How many muffins will each friend get if they get the same number of muffins?

Use repeated subtraction to divide. See Example 2 (p. 254)

14.

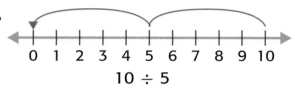

$10 \div 5$

15.

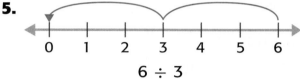

$6 \div 3$

16.

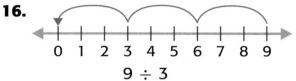

$9 \div 3$

17.

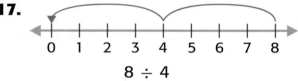

$8 \div 4$

18. $18 \div 3$

19. $12 \div 2$

20. $24 \div 6$

21. $12 \div 3$

22. $27 \div 3$

23. $28 \div 7$

24. A house has 16 windows. Each room has 4 windows. How many rooms does the house have?

25. Felecia wants to read 9 stories. If each magazine has 3 stories, how many magazines should she read?

26. There are 12 erasers in a bag. Rodrigo wants to share them equally among himself and his 2 friends. How many erasers will each person get?

27. Chester has 24 pencils. He kept 4 and shared the others equally among his 4 brothers. How many pencils did each brother get?

H.O.T. Problems

28. OPEN ENDED Write a real-world problem that could be represented by $18 \div 6$.

29. **WRITING IN** ►**MATH** How is division related to subtraction?

Explore

Relate Division to Multiplication

You can relate division to multiplication.

MAIN IDEA

I will relate division to multiplication.

You Will Need
counters

Math Online

macmillanmh.com
• Concepts in Motion

ACTIVITY Relate division to multiplication.

Step 1 **Find 21 ÷ 3.**

Model 21 counters divided into 3 equal groups.

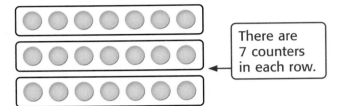

There are
7 counters
in each row.

Step 2 **Write a division sentence.**

number in all   number of groups   number in each group

$$21 \div 3 = 7$$

The **dividend** is the number to be divided.

The **divisor** is the number the dividend is divided by.

The answer is the **quotient**.

Step 3 **Write a multiplication sentence.**

number of groups   number in each group   number in all

$$3 \times 7 = 21$$

Think About It

1. Explain how you used models to show $21 \div 3$.

2. Explain how the array shows that $21 \div 3 = 7$ is related to $3 \times 7 = 21$.

3. What pattern do you notice in the number sentences?

4. How can multiplication facts be used to divide?

CHECK What You Know

Use counters to model each problem. Then write related division and multiplication sentences to help find the answer.

5. $12 \div 6$ **6.** $18 \div 3$ **7.** $25 \div 5$

8. $15 \div 3$ **9.** $16 \div 2$ **10.** $24 \div 8$

Write a related division and multiplication sentence for each picture.

11.   **12.**

13. 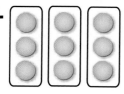   **14.**

15. WRITING IN ►MATH How do you know what multiplication sentence to use to find $28 \div 4$?

Relate Division to Multiplication

GET READY to Learn

A pan of blueberry muffins is shown. The pan represents an array. The array shows 3 rows of muffins with 4 muffins in each row.

MAIN IDEA

I will divide using related multiplication facts.

New Vocabulary

dividend
divisor
quotient
fact family

Math Online

macmillanmh.com
• Extra Examples
• Personal Tutor
• Self-Check Quiz

In the Explore Activity, you used arrays to help you understand how division and multiplication are related.

Real-World EXAMPLE   Relate Division to Multiplication

1. **MUFFINS** Use the array of muffins to write related multiplication and division sentences.

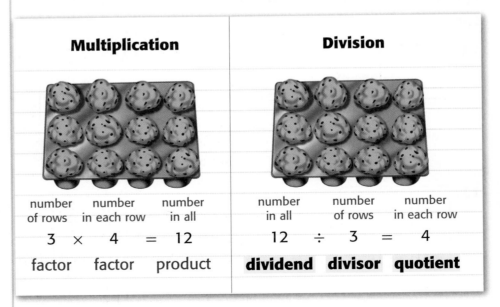

Multiplication	Division
number of rows number in each row number in all	number in all number of rows number in each row
3 × 4 = 12	12 ÷ 3 = 4
factor factor product	**dividend** **divisor** **quotient**

The related multiplication and division sentences are
3 × 4 = 12 and 12 ÷ 3 = 4.

A group of related facts using the same numbers is a **fact family**. Each fact family follows a pattern by using the same numbers.

Fact Family 3, 4, and 12	Fact Family 7 and 49
$3 \times 4 = 12$	$7 \times 7 = 49$
$4 \times 3 = 12$	$49 \div 7 = 7$
$12 \div 3 = 4$	
$12 \div 4 = 3$	

Remember

Thinking about numbers in a fact family can help you remember related facts.

EXAMPLE Write a Fact Family

2 **Use the fact family 3, 6, and 18 to write four related multiplication and division sentences.**

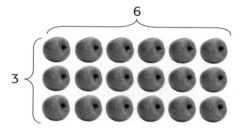

$3 \times 6 = 18$

$6 \times 3 = 18$

$18 \div 3 = 6$

$18 \div 6 = 3$

The pattern shows that 3, 6, and 18 are used in each number sentence.

CHECK What You Know

Use the array to complete each pair of number sentences. See Example 1 (p. 258)

1. $\blacksquare \times 5 = 15$

   $\blacksquare \div 3 = 5$

2. $4 \times \blacksquare = 24$

   $24 \div \blacksquare = 6$

Write the fact family for each set of numbers. See Example 2 (p. 259)

3. 2, 6, 12

4. 4, 5, 20

5. 3, 9, 27

6. Gwen will divide 20 marbles equally into 5 bags. Show this with a number sentence.

7.  Why are the product and the dividend the same in $3 \times 7 = 21$ and $21 \div 3 = 7$?

Use the array to complete each pair of number sentences. See Example 1 (p. 258)

8. ▦ × 2 = 8
▦ ÷ 4 = 2

9. 2 × ▦ = 4
4 ÷ ▦ = 2

10. ▦ × 2 = 14
▦ ÷ 2 = 7

11. 4 × ▦ = 20
20 ÷ ▦ = 4

Write the fact family for each set of numbers. See Example 2 (p. 259)

12. 2, 3, 6

13. 2, 7, 14

14. 4, 16

15. 4, 8, 32

16. 4, 3, 12

17. 4, 7, 28

Identify the pattern by writing the set of numbers for each fact family.

18. 5 × 9 = 45
9 × 5 = 45
45 ÷ 5 = 9
45 ÷ 9 = 5

19. 7 × 2 = 14
2 × 7 = 14
14 ÷ 2 = 7
14 ÷ 7 = 2

20. 3 × 3 = 9
9 ÷ 3 = 3

Solve. Write a number sentence.

21. All 5 members of the Malone family went to the movies. Their tickets cost a total of $30. How much was each ticket?

22. The petting zoo has 21 animals. There are 7 types of animals, each with an equal number. How many animals does each type have?

23. Measurement Mr. Thomas travels 20 miles each week to and from work. If he works 5 days a week, how many miles does Mr. Thomas travel each day to go to work?

24. Stacia and her friend are each making a bracelet. They have 18 beads to share. If they use the same number of beads, how many beads will each bracelet have?

H.O.T. Problems

25. NUMBER SENSE What multiplication fact will help you find $27 \div 9$?

26. WHICH ONE DOESN'T BELONG? Identify the number sentence that does not belong. Explain.

| $3 \times 6 = 18$ | $18 \div 2 = 9$ | $18 \div 6 = 3$ | $6 \times 3 = 18$ |

27. WRITING IN ▸MATH Explain how multiplication facts can help you with division facts. Give an example.

TEST Practice

28. The figure below is a model for $4 \times 6 = 24$.

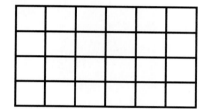

Which number sentence is in the same fact family? (Lesson 6-2)

A $6 \div 4 = 24$ **C** $24 \div 4 = 6$

B $24 \div 3 = 8$ **D** $24 \div 6 = 6$

29. Which number sentence is modeled by repeated subtraction on the number line? (Lesson 6-1)

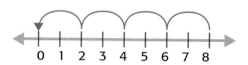

F $4 \div 2 = 8$

G $16 \div 2 = 8$

H $8 \div 2 = 4$

J $24 \div 8 = 3$

Spiral Review

Use repeated subtraction to divide. (Lesson 6-1)

30. $12 \div 4$ **31.** $18 \div 3$ **32.** $28 \div 7$ **33.** $25 \div 5$

Multiply. Use models or patterns if needed. (Lesson 5-8)

34. 8×11 **35.** 11×9 **36.** 3×12 **37.** 12×5

38. One frog sat on a log for 29 minutes. A second frog sat for 16 minutes longer. How long did the second frog sit on the log? (Lesson 2-2)

6-3 Problem-Solving Skill

MAIN IDEA I will choose an operation to solve a problem.

Latesha's doctor saw 20 patients in 5 hours today. The doctor saw the same number each hour. How many patients did the doctor see each hour?

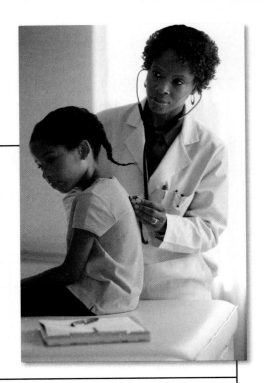

Understand	**What facts do you know?** • The doctor saw patients for 5 hours. • She saw 20 patients in all. **What do you need to find?** • The number of patients the doctor saw each hour.
Plan	There were 20 patients. You need to find how many patients the doctor saw each hour in the 5 hours. Use division.
Solve	Find $20 \div 5$. total number of patients number of hours number of patients $20 \quad \div \quad 5 \quad = \quad 4$ So, the doctor saw 4 patients each hour.
Check	You can use multiplication to check division. $5 \times 4 = 20$ So, it makes sense that 4 patients would have been seen each hour.

Refer to the problem on the previous page.

1. Explain why division was used to solve the problem. What other operation might you use to solve this problem?

2. Explain how the four-step plan helped you solve this problem.

3. Suppose the doctor has seen the same number of patients, but in 4 hours. How many patients would have been seen each hour then?

4. Check your answer to Exercise 3. How do you know it is correct?

> PRACTICE the Skill

EXTRA PRACTICE
See page R17.

Choose an operation to solve.

5. Write a multiplication sentence to find how many plants are in the garden. What other operation can help you solve this problem?

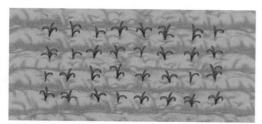

6. **Measurement** Melody has 14 feet of string. She wants to make necklaces that are 2 feet long. How many necklaces can she make?

7. Shaun and Dean went to the park. The leaves were brown, red, and orange. They picked 7 of each color. How many leaves did they pick in all?

8. **Measurement** It rained 6 inches each month for the last 5 months. If it rains 6 more inches this month, what will be the total rainfall?

9. Sondra and Wanda were making jewelry for the school fund raiser. They each made the amount listed. How many items were made in all?

Jewelry Made

Item	Number
Earrings	5
Pins	4
Bracelets	6

10. **Measurement** The Prudential Tower in Boston is 750 feet tall. A TV tower in North Dakota is 2,063 feet tall. How much taller is the TV tower than the Prudential Tower? Explain.

11. **Geometry** Jerome has a square garden. Each side is 5 yards. How many yards of fence does he need to border the garden? Explain.

12. **WRITING IN MATH** Explain how to read a problem and decide what operation to use.

6-4 Divide by 2

MAIN IDEA

I will use models to divide by 2.

Math Online

macmillanmh.com

- Extra Examples
- Personal Tutor
- Self-Check Quiz

GET READY to Learn

Jose and Jenna are sharing an apple equally. If there are 8 apple slices on the plate, how many slices will each of them get?

In Lesson 6-1, you learned about the division symbol ÷ . Another symbol for division is $\overline{)}$.

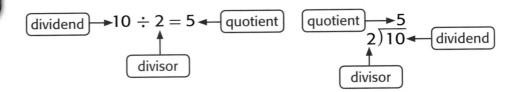

Real-World EXAMPLE   Make Equal Groups

1 **FRUIT Jose and Jenna share an apple equally. If there are 8 slices, how many slices will each of them get?**

To share equally between 2 people means to divide by 2. So, find $8 \div 2$ or $2\overline{)8}$.

Use counters to model 8 divided into 2 groups.

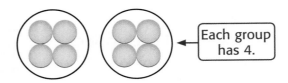

The model shows $8 \div 2 = 4$ or $2\overline{)8}^{\,4}$.
So, if they share the apple equally, each person will get 4 apple slices.

Divide. Write a related multiplication fact. See Example 1 (p. 264)

1.

$2\overline{)4}$

2.

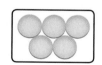

$12 \div 2$

3. $6 \div 2$

4. $14 \div 2$

5. $2\overline{)8}$

6. Victor and his sister each read an equal number of books. Together they read 16 books. Write the number sentence to show how many books each read.

7. **Talk About It** What are two different ways to find $10 \div 2$?

Practice and Problem Solving

EXTRA PRACTICE
See page R17.

Divide. Write a related multiplication fact. See Example 1 (p. 264)

8.

$10 \div 2$

9.

$2\overline{)18}$

10. $20 \div 2$

11. $16 \div 2$

12. $18 \div 2$

13. $2\overline{)2}$

14. $2\overline{)10}$

15. $2\overline{)6}$

Solve. Write a number sentence.

16. Damian will plant 12 seeds in groups of 2. How many groups of 2 will he have once they are all planted?

17. Kyle and Alan equally divide a package of 14 erasers. How many erasers will each person get?

18. Lydia shared her 16 pattern tiles equally with Pilar. Pilar then shared her tiles equally with Toby. How many tiles does each student have?

19. Each car on the Supersonic Speed ride can hold 18 people. If the seats are in groups of 2, how many groups of two are there on 3 cars?

Algebra Copy and complete each table.

20.

Rule: Divide by 2.			
In	10	■	18	14
Out	■	4	■	7

21.

Rule: Multiply by 5.			
In	7	■	6	■
Out	■	25	■	15

Data File

Kentucky is a state with many weather extremes.

22. Which city has half the yearly snowfall of Ashland?

23. Which city's annual snowfall is 18 ÷ 2?

24. Which two cities' yearly snowfall, when added together, is the same as 14 ÷ 2?

25. Write a number sentence for the amount of snow Lexington would get if its snowfall is half of its usual amount.

Average Yearly Snowfall	
City	Snowfall (in.)
Ashland	10
Hopkinsville	9
Somerset	2
Bowling Green	5
Lexington	18

Source: Kentucky Almanac

H.O.T. Problems

26. OPEN ENDED Write a number that when divided by 2 is more than 8.

CHALLENGE Divide.

27. 36 ÷ 2　　　**28.** 50 ÷ 2　　　**29.** 80 ÷ 2　　　**30.** 42 ÷ 2

31. FIND THE ERROR Andres and Muna are finding 8 ÷ 2. Who is correct? Explain your reasoning.

Andres
8 ÷ 2 = 16
because
2 × 8 = 16

Muna
8 ÷ 2 = 4
because
2 × 4 = 8

32. **WRITING IN** ►**MATH** Can you divide 9 into equal groups of 2? Explain.

Extend

Technology Activity for 6-4
Model Division

You can use *Math Tool Chest* to model division problems.

Activity

① **PARTY** **Eight friends went to a party. They formed 2 teams for a game. How many friends were on each team?**

You want to find 8 ÷ 2.

Choose the counters icon from the *Math Tool Chest*.

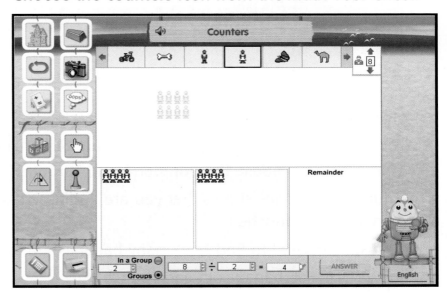

- Choose division for the mat type.
- Choose a boy or girl stamp.
- Stamp out 8 boys or girls.
- At the bottom of the screen, choose 2 and "Groups."

The number sentence shows that you are finding 8 ÷ 2, 8 children in 2 groups.

- Choose "Answer" to find how many friends were on each team.

  So, 8 ÷ 2 = 4 teams.

Activity

2 **FOOD** The grocer is selling 24 bananas. There are 4 bunches of bananas. How many bananas are in each bunch?

To find 24 ÷ 4, choose the Counters icon from *Math Tool Chest.*

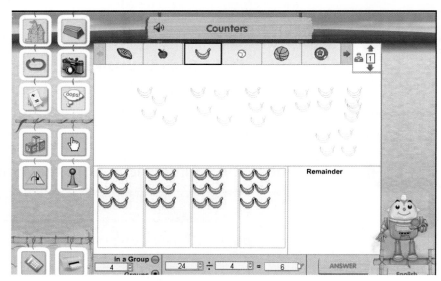

- Choose division for the mat type.
- Choose the banana stamp, and stamp out 24 bananas.
- At the bottom, choose the number 4 and "Groups."

The number sentence shows that you are finding 24 ÷ 4, 24 bananas in 4 bunches.

- Choose "Answer" to find how many bananas are in each bunch.

  So, 24 ÷ 4 = 6 bananas in each bunch.

CHECK What You Know

Use *Math Tool Chest* to solve.

1. Sixteen people are camping in 4 tents. Each tent holds the same number of people. How many people are in each tent?

2. The zoo has 18 monkeys. The monkeys will only share their tree with one other monkey. How many trees are needed?

3. Each circus car holds 7 clowns. There are a total of 21 clowns. How many cars will be needed?

4. 32 people are on floats at the parade. Each float holds 8 people. How many floats are there?

Use repeated subtraction to divide.
(Lesson 6-1)

1.

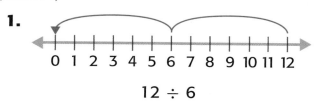

$$12 \div 6$$

2. $8 \div 2$

3. $16 \div 4$

4. $15 \div 3$

5. $10 \div 2$

6. Tory wants to read 3 chapters a day. His book has 18 chapters. How many days will it take for him to finish the book? (Lesson 6-1)

7. MULTIPLE CHOICE The figure shows $3 \times 6 = 18$.

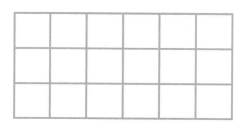

Which division sentence is in the same fact family? (Lesson 6-2)

A $6 \div 3 = 2$ **C** $18 \div 6 = 3$

B $24 \div 8 = 3$ **D** $36 \div 6 = 6$

Algebra Find each missing number.
(Lesson 6-2)

8. $16 \div \blacksquare = 2$ **9.** $14 \div 2 = \blacksquare$

Complete each pair of sentences.
(Lesson 6-2)

10. $\blacksquare \times 2 = 12$ **11.** $8 \times \blacksquare = 24$
$\blacksquare \div 6 = 2$ $24 \div \blacksquare = 3$

Write the fact family for each set of numbers. (Lesson 6-2)

12. 5, 2, 10   **13.** 9, 3, 27

Choose an operation to solve. (Lesson 6-3)

14. All the students in the Art Club must pay $2 each for supplies. If $20 was collected, how many students are in the club?

15. Twelve students are going on a field trip. There are 2 vans that hold the same number of students. How many students will go in each van?

16. MULTIPLE CHOICE Sophie divided $16 \div 2 = 8$. Which problem could she use to check her answer?
(Lesson 6-4)

F $8 - 2 = \blacksquare$ **H** $8 + 2 = \blacksquare$

G $8 \times 2 = \blacksquare$ **J** $8 \div 2 = \blacksquare$

17. **WRITING IN ►MATH** Can 6 roses be divided equally between 2 vases? Explain. (Lesson 6-4)

6-5 Divide by 5

MAIN IDEA

I will use models to divide by 5.

Math Online

macmillanmh.com

• Extra Examples
• Personal Tutor
• Self-Check Quiz

> **GET READY to Learn**
>
> A group of friends have a lemonade stand. The price of one glass of lemonade is 5¢. They earned 30¢ selling lemonade. How many glasses of lemonade did they sell?

There are different ways to divide by 5.

Real-World EXAMPLE Use Models

① **MONEY** How many glasses of lemonade did they sell? Write a number sentence to show the solution.

You need to find 30¢ ÷ 5¢. Use counters to model 30 ÷ 5.

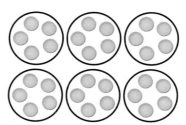

There are 30 counters and 5 counters are in each group. There are 6 equal groups.

The model shows 30¢ ÷ 5¢ = 6 or $5¢\overline{)30¢}$ with quotient 6.

The number sentence is 30¢ ÷ 5¢ = 6.

So, the friends sold 6 glasses of lemonade.

Check You can use multiplication to check.

Since 6 × 5¢ = 30¢, the answer is correct. ✓

In addition to using models, you can use related multiplication facts to divide.

Real-World EXAMPLE Use Related Facts

2 **MONEY** The school store is selling pencils for 5¢ each. If Percy has 45¢, how many pencils can he buy with all his money?

Write a related multiplication fact to find 45¢ ÷ 5¢.

$45¢ ÷ 5¢ = \blacksquare$

$5¢ × \blacksquare = 45¢$ ← What number times 5 is 45¢?

$5¢ × 9 = 45¢$

So, $45¢ ÷ 5¢ = 9$ or $5¢\overline{)45¢}^{\,9}$. Percy can buy 9 pencils.

Pencils
5¢ each

Remember

Nickels can be used to represent 5.

Check
The picture shows the number sentence $45¢ ÷ 5¢ = 9$.

45¢ divided into groups of 5¢ forms 9 groups.
9 groups of 5¢ = 45¢. ✓

CHECK What You Know

Divide. Use models or related facts. See Examples 1 and 2 (pp. 270–271)

1. 35 ÷ 5

2. 5 ÷ 5

3. $5\overline{)20}$

4. $5\overline{)40}$

5. Measurement Lucia's classroom has rows of tables that are 25 feet wide. If 5 tables are in each row, how wide is each table? Write a number sentence to show the solution.

6. **Talk About It** How can you tell if a number is divisible by 5?

Divide. Use models or related facts. See Examples 1 and 2 (pp. 270–271)

7. $20 \div 5$

8. $40 \div 5$

9. $45 \div 5$

10. $50 \div 5$

11. $5\overline{)5}$

12. $5\overline{)15}$

13. $5\overline{)10}$

14. $5\overline{)45}$

For Exercises 15–18, use the recipe for corn bread. Find how much of each is needed to make 1 loaf of corn bread.

15. cornmeal

16. flour

17. eggs

18. vanilla extract

Buttermilk Corn Bread	
10 cups cornmeal	3 cups butter
5 cups flour	8 cups buttermilk
1 cup sugar	5 tsp vanilla extract
5 Tbsp baking powder	15 eggs
4 tsp salt	2 tsp baking soda
Makes: 5 loaves	

Solve. Write a number sentence.

19. Measurement Rose has a 35-inch piece of ribbon. If she divides the ribbon into 5 equal pieces, how many inches long will each piece be?

20. Helen is reading a book with 50 pages. If she reads 5 pages every day, how many days will it take her to finish the book?

21. Measurement Garrison has 45 minutes to do his homework. He has 9 problems left. How long can he spend on each problem if each one takes an equal amount of time?

22. Addison got 40 points on yesterday's 10-question math quiz. If each question is worth 5 points and there is no partial credit, how many questions did he miss?

Real-World PROBLEM SOLVING

Science The grizzly bear is one of the largest and most powerful animals.

23. About how long is a grizzly bear's foot?

24. What is the length of one grizzly bear?

25. The grizzly runs at about a speed of 35 miles per hour. What is that divided by 5?

40 inches

14 feet

H.O.T. Problems

26. OPEN ENDED Write a division sentence with a quotient of 9.

27. WHICH ONE DOESN'T BELONG? Identify the division sentence that does not belong. Explain your reasoning.

| $30 \div 5 = 6$ | $20 \div 2 = 10$ | $30 \div 6 = 5$ | $35 \div 5 = 7$ |

28. WRITING IN ▸MATH Explain the method you would use to find $45 \div 5$ and why you prefer that method.

TEST Practice

29. This is a model for which number sentence? (Lesson 6-5)

A $15 \div 3 = 5$

B $3 + 5 = 8$

C $3 + 3 + 3 + 3 = 12$

D $5 \times 5 = 25$

30. Robert solved this division problem.

$$20 \div 2 = 10$$

Which problem could he do to check his answer? (Lesson 6-4)

F $10 + 2 = \blacksquare$

G $10 - 2 = \blacksquare$

H $10 \times 2 = \blacksquare$

J $10 \div 2 = \blacksquare$

Spiral Review

Divide. Write a related multiplication fact. (Lesson 6-4)

31. $18 \div 2$

32. $16 \div 2$

33. $2\overline{)12}$

34. Angelica had $40 to buy her mother a birthday present. She bought flowers for $16 and a new pen set for $8. How much money does she have left? (Lesson 6-3)

Add. Check for reasonableness. (Lesson 2-4)

35. 48
  $+ 24$

36. 83
  $+ 29$

37. 54
  $+ 94$

Problem Solving in Community

Communities Within Communities

A community is a group of people who work, live, or play together. There are over 1,000 public schools in Texas, and each school is a community. Your classroom is also a community.

Often, a class works together on an art project. A mural is an art project that many people can work on together. It is a large painting that sometimes covers an entire wall.

There are many other examples of communities. In Texas, there are over 200 cities. Each city is a community. There are about 1 million businesses in Texas that also form communities. Each of the 7 million families in Texas is a community, too.

Did You Know?

There are more than 33 million elementary students in the United States.

Real-World Math

Use the information on page 274 to solve each problem.

1 A group is painting a mural of their community. They have 14 pictures of places around their community. The mural is 7 feet wide. If they want to place the pictures evenly, how many will go on each foot?

2 A community has a Clean Up Litter Day. Thirty-five community members come to help. They divide into groups of five. How many groups are there?

3 There are 20 students in a class. The teacher splits the class in 5 equal groups. How many students are in each group?

4 Suppose a community has 45 stores. There are 5 stores on each street. How many streets are in the community?

5 The schools in your community donate 30 boxes of clothes to the local charity. There are 10 schools. If each school donates the same number of boxes, how many boxes of clothes does each school donate?

6 There are 3 million more families in Texas than in Florida. Texas has about 7 million families. About how many families are in Florida?

Problem-Solving Investigation

MAIN IDEA I will choose the best strategy to solve a problem.

P.S.I. TEAM ✛

KINAH: I want to plant a vegetable garden. I have 6 tomato plants, 5 pepper plants, and 5 zucchini plants. I want to put the plants into 4 equal rows.

YOUR MISSION: Find how many plants should be planted in each row.

Understand	Kinah has 6 tomato, 5 pepper, and 5 zucchini plants. She wants to plant them in 4 equal rows. Find how many plants to put in each row.
Plan	You need to look at how to arrange items. So, the *draw a picture* strategy is a good choice.
Solve	The picture shows 6 + 5 + 5, or 16 plants. There are 4 equal groups of 4 plants. 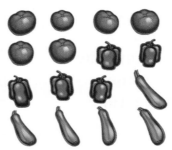 Since 16 ÷ 4 = 4, Kinah needs to plant 4 plants in each row.
Check	Look back at the problem. Since 4 × 4 = 16, you know that the answer is correct.

Use any strategy shown below to solve. Tell what strategy you used.

> PROBLEM-SOLVING STRATEGIES
> • Act it out.
> • Draw a picture.
> • Look for a pattern.

1. Mrs. Hunt bought 18 cans of cat food. The cats eat the same amount each day. This amount lasts for 6 days. How many cans of food do her cats eat each day?

2. Bill catches 3 more fish than Anton, and Sarita catches 3 more fish than Bill. Bill catches 5 fish. How many fish did each person catch?

3. How much will lunch cost for 5 people if each person buys all of the items on the menu?

Lunch	
Item	**Cost**
Chicken	$2
Apple	$1
Milk	$1

4. One day, 3 children played together. The next day, 5 children played together. On the third day, 7 children played together. If this pattern continues, how many children could be playing together on the sixth day?

5. Ivan bought 6 gifts at a store. His sister bought 5 gifts and their mother bought 7 gifts. How many gifts did they buy in all?

6. **Algebra** Latoya and Latisha use red, blue, and yellow rubber bands to make bracelets. Each bracelet's rubber bands make the pattern red, blue, yellow, red, blue, yellow. If they make 10 bracelets, how many yellow rubber bands will they use?

7. **Algebra** If Rita draws 15 more shapes, how many of those shapes could be triangles?

8. Marvina and Gustavo went to the grocery store. Marvina bought 5 items for $6 each. Gustavo bought 7 items for $8 each. How much did they spend in all?

9. Selena, Ronnie, Jared, and Patty each have 5 books. Selena and Jared each read 4 of their books. Ronnie did not read 2 of his books. Patty read all of her books. How many books did they read in all?

10. **WRITING IN ▶MATH** Look back at Exercise 9. Which strategy did you use to solve it? Explain your reason for using that strategy.

> **GET READY to Learn**
>
> Juice bars come 10 in a box. The third grade class needs 50 juice bars for a party treat. How many boxes will they need?

Base-ten blocks can be used to model dividing by 10.

Real-World EXAMPLE Use Models

1 **SCHOOL** How many boxes of juice bars will the third grade class need? Write a number sentence.

You need to find 50 ÷ 10.

One Way: Use Repeated Subtraction

$$
\begin{array}{ccccc}
① & ② & ③ & ④ & ⑤ \\
50 & 40 & 30 & 20 & 10 \\
-10 & -10 & -10 & -10 & -10 \\
\hline
40 & 30 & 20 & 10 & 0
\end{array}
$$

Subtract groups of 10 until you reach 0. Count the number of groups you subtracted. You subtracted groups of 10 five times.

Another Way: Use a Related Fact

You know that $10 \times 5 = 50$.

So, $50 \div 10 = 5$ or $10\overline{)50}$.

So, $50 \div 10 = 5$. They will need 5 boxes of juice bars.

Divide. See Example 1 (p. 278)

1. $20 \div 10$

2. $40 \div 10$

3. $10\overline{)60}$

4. $10\overline{)10}$

5. There are 40 chairs at 10 tables. Each table has an equal number of chairs. How many chairs are at each table? Write a number sentence.

6. **Talk About It** When you divide by 10, what do you notice about the quotient and dividend?

Practice and Problem Solving

EXTRA PRACTICE See page R18.

Divide. See Example 1 (p. 278)

7. $30 \div 10$

8. $50 \div 10$

9. $80 \div 10$

10. $90 \div 10$

11. $10\overline{)20}$

12. $10\overline{)70}$

13. $10\overline{)30}$

14. $10\overline{)40}$

Solve. Write a number sentence.

15. A vase holds 40 flowers. There are an equal number of daisies, roses, tulips, and lilies. How many of each kind of flower are there in the vase?

16. Rona went to the car show and saw 60 cars. If he saw 10 of each kind of car, how many different kinds of cars were there?

For Exercises 17–19, use the sign shown.

17. Julius spent 40¢ on sunflower seeds. How many packages did he buy?

18. How much did Beth pay for 1 yogurt?

19. How much would it cost to buy 1 of everything, including 1 piece of dried fruit?

HEALTH SHACK'S SNACKS

Sunflower seeds ......10¢ per package
Dried fruit...............10 pieces for 50¢
Juice ....20¢ each
Yogurt........................2 for 80¢

H.O.T. Problems

20. OPEN ENDED Use the numerals 7, 0, 8, 5 to write two 2-digit numbers that can be divided by 10.

21. **WRITING IN ▸ MATH** Explain how counting by 10s can help you find $80 \div 10$.

22. Mr. Gomez bought 30 frozen pizzas, with 5 in each box. Which number sentence shows how to find the number of boxes of pizza he bought? (Lesson 6-5)

A $30 - 5 = $ ■   **C** $30 \times 5 = $ ■

B $30 + 5 = $ ■   **D** $30 \div 5 = $ ■

23. Look at the number sentence below.

$$90 \div \blacksquare = 9$$

Which number will make the number sentence true? (Lesson 6-7)

F 1   **H** 81

G 10   **J** 100

Spiral Review

24. Measurement On Monday, Nelson rode a horse 12 miles and Ramon rode a horse 14 miles. If they ride the same amount 4 more days this week, how many more miles will Ramon have ridden than Nelson? (Lesson 6-6)

Divide. (Lesson 6-5)

25. $25 \div 5$ **26.** $45 \div 5$ **27.** $50 \div 5$

28. There are 40 soccer players. Each team will have the same number of players and 1 coach. If there are 5 coaches, how many players will be on each team? Write a number sentence to record the solution. (Lesson 6-5)

Write a multiplication sentence for each array. (Lesson 4-1)

29.    **30.**    **31.**

32. Algebra The table below shows Dion's reading pattern. If the pattern continues, how many pages will he read the rest of the week?

(Lesson 5-3)

Mon.	Tues.	Wed.	Thur.	Fri.	Sat.
4	8	16	■	■	■

Number Cubes
Multiply and Divide Numbers

Get Ready!

Players: 2 players

Get Set!

Make a chart to record each roll.

Go!

- Player 1 rolls the number cubes.

- Each player writes down the numbers.

- Each player uses the two numbers in number sentences that use multiplication and division facts.

- Each player gets 1 point for each correctly written number sentence.

- The game ends when 1 player earns 50 points.

You will need: one 0–5 number cube, one 5–10 number cube

Cube 1	Cube 2	Multiplication sentence	Division sentence
5	4	$5 \times 4 = 20$	$20 \div 5 = 4$

MAIN IDEA

I will use division rules to divide with 0 and 1.

Math Online

macmillanmh.com

• Extra Examples
• Personal Tutor
• Self-Check Quiz

GET READY to Learn

Suppose that you have 3 toys. One storage box will hold 3 toys. How many boxes will you need?

There are rules you can use when 0 and 1 are divisors.

Real-World EXAMPLE

① **TOYS** **How many boxes will you need for 3 toys?**

You need to find $3 \div 1$ or $1\overline{)3}$. Since 3 toys fit in 1 box, make groups of 3, using counters.

There is 1 group of 3.

So, $3 \div 1 = 3$ or $1\overline{)3}^{\,3}$.

Division Rules   Key Concepts

Words	When you divide any number (except 0) by itself, the quotient is 1.
Example	$4 \div 4 = 1$ $4\overline{)4}^{\,1}$ 

Words	When you divide any number by 1, the quotient is the same as the dividend.
Example	$4 \div 1 = 4$ $1\overline{)4}^{\,4}$

Words	When you divide 0 by any number (except 0), the quotient is 0.
Example	$0 \div 4 = 0$ $4\overline{)0}^{\,0}$

Words	You cannot divide by 0.

Divide. See Example 1 (p. 282)

1. $5 \div 1$ **2.** $0 \div 1$ **3.** $1 \div 1$ **4.** $1\overline{)9}$

5. $0 \div 7$ **6.** $10 \div 1$ **7.** $6\overline{)0}$ **8.** $7\overline{)7}$

9. If 6 people show up at the theater and there are 6 seats left, how many seats will each person get?

10. **Talk About It** Can you divide a number by 0? Can you divide 0 by a number other than 0? Explain.

Practice and Problem Solving

EXTRA PRACTICE
See page R19.

Divide. See Example 1 (p. 282)

11. $2 \div 1$ **12.** $10 \div 10$ **13.** $0 \div 5$ **14.** $6 \div 1$

15. $0 \div 3$ **16.** $0 \div 9$ **17.** $1\overline{)4}$ **18.** $5\overline{)5}$

19. $1\overline{)7}$ **20.** $2\overline{)2}$ **21.** $1\overline{)10}$ **22.** $10\overline{)0}$

Solve. Write a number sentence.

23. There are 35 students in Mr. Macy's class. To play a game, each person needs 1 playing piece. How many playing pieces are needed for the class to play the game?

24. Mr. Carrington has a pack of paper with 5 different colors. If he gives 1 of each color to his students, how many pieces of paper will they each have?

25. Kari wants to give 5 friends an apple. She finds that she has no apples. How many apples can she give to her friends?

26. Marcy and her 4 friends have 5 glasses of juice. How many glasses of juice will each person get?

H.O.T. Problems

27. **OPEN ENDED** Write a real-world division problem in which a number is divided by itself. Ask a classmate to answer it.

28. **WRITING IN MATH** Explain how you could divide any number someone gives you by 1 or itself.

FOLDABLES
Study Organizer **GET READY to Study**

Be sure the following Key Vocabulary words and Key Concepts are written in your Foldable.

Divide by 2	Divide by 10
Divide by 5	Divide by 0 and 1

Key Concepts

- **Division** is an operation with two numbers. One number tells you how many things you have. The other tells you how many equal groups to form. (p. 253)

$$8 \div 2 = 4 \qquad 2\overline{)8}^{\,4}$$

dividend   **divisor**   **quotient**

- You can relate division to multiplication. (p. 258)

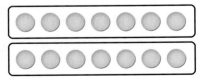

$$2 \times 7 = 14$$
number of groups   number in each   number in all

$$14 \div 2 = 7$$
number in all   number of groups   number in each group

Key Vocabulary

divide (p. 253)
dividend (p. 258)
divisor (p. 258)
quotient (p. 258)
fact family (p. 259)

Vocabulary Check

Choose the vocabulary word that completes each sentence.

1. The answer to a division problem is called the ____?____ .

2. The number to be divided is the ____?____ .

3. A ____?____ is a group of related facts using the same numbers.

4. In the sentence $36 \div 4 = 9$, 9 is the ____?____ .

5. ____?____ means to separate a number into equal groups, to find the number of groups, or the number in each group.

6. The ____?____ is the number by which the dividend is divided.

Lesson-by-Lesson Review

6-1 **Relate Division to Subtraction** (pp. 253–255)

Example 1

One Way: **Number Line**

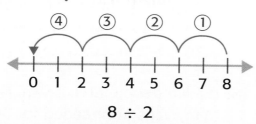

$8 \div 2$

Start at 8. Count back by 2s until you reach 0. Count how many times you subtracted. So, $8 \div 2 = 4$.

Another Way: **Repeated Subtraction**

①	②	③	④
8	6	4	2
-2	-2	-2	-2
6	4	2	0

So, $8 \div 2 = 4$.

Use repeated subtraction to divide.

7.

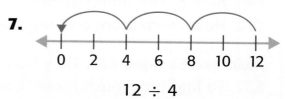

$12 \div 4$

8.

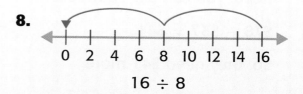

$16 \div 8$

9. $6 \div 2$ **10.** $27 \div 3$

11. Chang has 15 frogs in his pond. If he catches 3 each day, how many days will it take him to catch all of the frogs?

6-2 **Relate Division to Multiplication** (pp. 258–261)

Example 2
Write the fact family for 4, 2, and 8.

$4 \times 2 = 8$

$8 \div 2 = 4$

$2 \times 4 = 8$

$8 \div 4 = 2$

Write the fact family for each set of numbers.

12. 6, 7, 42   **13.** 8, 4, 2

14. 5, 4, 20   **15.** 4, 9, 36

16. Measurement Last week, Jacqui rode her bike 4 days in a row for a total of 20 miles. She rode the same number of miles each day. How many miles did she ride her bike each day?

6-3 **Problem-Solving Skill:** **Choose an Operation** (pp. 262–263)

Example 3
Alvin and Eli have $37 to build a tree house. The materials will cost $78. How much more do they need?

Alvin and Eli need $78. They have $37. To find how much more they need, you can use subtraction.

$78 − $37 = $41

So, they need $41 more.

Use addition to check subtraction.

41 + 37 = 78

So, the answer is correct.

Choose an operation to solve.

17. Measurement Caleb ran a mile in 18 minutes. He wants to be able to finish the mile in 15 minutes. How much faster does Caleb need to run?

18. Twelve people fit in a van. How many vans are needed for 36 people?

19. The coach has assigned 3 people to compete in each event. There are a total of 27 people. In how many events did each team member compete?

6-4 **Divide by 2** (pp. 264–266)

Example 4
Maros wants to share 6 dog biscuits with his 2 dogs. How many biscuits will each dog get? Write the number sentence.

To find 6 ÷ 2 use counters to model 6 divided into 2 groups.

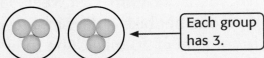

Each group has 3.

The model shows 6 ÷ 2 = 3. Each dog will get 3 biscuits.

Divide. Write a related multiplication fact.

20. 12 ÷ 2   **21.** 14 ÷ 2

22. 16 ÷ 2   **23.** 20 ÷ 2

24. Measurement Veronica and Koko want to equally share a piece of paper that is 14 inches long. How long will be each of their pieces? Write a number sentence.

6-5 Divide by 5 (pp. 270–273)

Example 5
Marion has 20 tadpoles. She will divide them equally between 5 fishbowls. How many tadpoles will be in each bowl? Write a number sentence.

You can use models to find 20 ÷ 5.

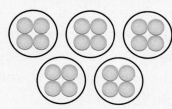

The model shows 20 ÷ 5 = 4 tadpoles in each.

Divide. Use models or related facts.

25. 20 ÷ 5   **26.** 35 ÷ 5

27. 45 ÷ 5   **28.** 15 ÷ 5

29. Lalo has 45 books to put in her bookcase. The bookcase has 5 shelves. If she wants to put the same number of books on each shelf, how many will there be on each shelf? Write a number sentence to record the solution.

6-6 Problem-Solving Investigation: Choose an Strategy (pp. 276–277)

Example 6
Mora bought 3 toys. William bought 2 more toys than Mora. How many toys did they buy?

Mora bought 3 toys. William bought 2 more than Mora. Find how many they bought together. You can model the problem with counters.

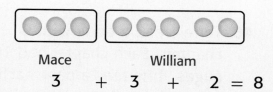

Mace   William
   3   +   3   +   2   =   8

So, Mace and William bought 8 toys.

30. Algebra One day, Juana received 2 gifts. The next day she received 4 gifts. The third day she received 6 gifts. If the pattern continues, how many gifts will she receive on the 6th day? How many gifts did she receive altogether?

31. You need to read 5 books a month during the school year. The school year is from August to May. How many books will you read in a year?

6-7 **Divide by 10** (pp. 278–280)

Example 7

Find 40 ÷ 10.
Use repeated subtraction.

①   ②   ③   ④

    40   30   20   10
  – 10   – 10   – 10   – 10
  ────   ────   ────   ────
    30   20   10   0

Subract groups of 10 until you reach 0. You subtracted 10 four times.

So, 40 ÷ 10 = 4 or $10\overline{)40}$ with quotient 4.

Divide.

32. 90 ÷ 10   **33.** 80 ÷ 10

34. 70 ÷ 10   **35.** 50 ÷ 10

36. 30 ÷ 10   **37.** 100 ÷ 10

38. There are 40 baskets of grapes on store shelves. If there are 10 baskets on each shelf, how many shelves are there? Write a number sentence to record the solution.

6-8 **Divide with 0 and 1** (pp. 282–283)

Example 8

Ginger wants to give 8 gifts to her friends. If she gives each friend one gift, how many friends does Ginger have?

Use the rules of division to solve 8 ÷ 1.

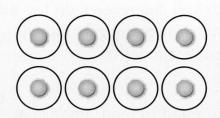

There are 8 groups of 1.

So, 8 ÷ 1 = 8.
Ginger has 8 friends.

Divide.

39. 5 ÷ 1   **40.** 0 ÷ 5

41. 0 ÷ 2   **42.** 10 ÷ 1

43. 10 ÷ 10   **44.** 0 ÷ 10

Solve. Write a number sentence.

45. Five boys want to go fishing. They find 5 fishing poles in the garage. Did they have enough poles to go fishing? Explain.

46. Terrell read a book with 9 chapters. Each chapter had 10 pages. If he read a page each day for 90 days, did he finish the book? Explain.

For Exercises 1-3, tell whether each statement is *true* or *false*.

1. When you divide any number by 1, the quotient is that number.

2. In $32 \div 4 = 8$, the 4 is the dividend.

3. Repeated subtraction can help you solve a division problem.

Divide. Write a related multiplication fact.

4. $30 \div 5$

5. $25 \div 5$

6. $0 \div 7$

7. $10 \div 2$

8. There were 28 students at the beginning of the year. Since then 4 have moved away and 3 have joined the class. How many students are there now?

9. **MULTIPLE CHOICE** During gym class 16 students were equally divided into 8 different teams. How many were on each team?

   A 2   C 24

   B 3   D 128

10. There are 48 students on the debate team. If 8 students from the debate team fit in each row in the room, how many rows will they need? Write a number sentence to record the solution.

Divide.

11. $12 \div 2$

12. $35 \div 5$

13. $0 \div 8$

14. $2 \div 2$

15. **MULTIPLE CHOICE** Benita did this division problem.

$$15 \div 5 = 3$$

Which problem could she do to check her answer?

F $5 + 3$

G $3 - 5$

H 5×3

J $3 \div 5$

Write the fact family for each set of numbers.

16. 3, 7, 21

17. 8, 4, 32

18. **Algebra** Copy and complete.

Rule: Divide by 7.	
Input	Output
■	6
56	■
■	10
63	■

19. **WRITING IN ►MATH** Neil does not understand why any number divided by 1 is that number. Explain it to Neil.

PART 1 — Multiple Choice

Read each question. Then fill in the correct answer on the answer sheet provided by your teacher or on a sheet of paper.

1. Marquez has 16 baseball cards. He puts the cards in piles of 8. How many piles does he make?

   A 2   **C** 6

   B 4   **D** 8

2. Pam arranged a group of buttons in the pattern shown.

   What operation best shows how she arranged them?

   F 6 + 4   **H** 4 − 6

   G 6 ÷ 4   **J** 4 × 6

3. Which problem could Leo do to check 60 ÷ 10 = 6?

   A 10 + 6 = ▣

   B 10 − 6 = ▣

   C 10 × 6 = ▣

   D 10 ÷ 6 = ▣

4. Which set of numbers is in order from least to greatest?

   F 537, 453, 387, 345

   G 345, 387, 453, 537

   H 387, 537, 345, 453

   J 453, 345, 537, 387

5. What number can be divided into 8 to give the answer 8?

   A 0   **C** 8

   B 1   **D** 16

6. Hayden's groceries cost $72. His 5 coupons saved him $8. Which of the following shows how to find the amount Hayden paid?

   F 72 + 8   **H** 72 − 5

   G 72 + 5   **J** 72 − 8

7. 28 students are divided into 7 equal groups. Which of the following shows the number of students in each group?

   A 28 × 7   **C** 28 + 7

   B 28 ÷ 7   **D** 28 − 7

Preparing for Standardized Tests
For test-taking strategies and practice,
see pages R50–R63.

8. Which number sentence belongs to the family of facts?

$3 \times 5 = 15, 5 \times 3 = 15, 15 \div 3 = 5$

F $3 \times 15 = 45$

H $15 \div 1 = 15$

G $15 \div 15 = 1$

J $15 \div 5 = 3$

9. Ming-Su has 6 fish. He puts them into 3 fish tanks. Each tank has the same number of fish. Which picture shows Ming-Su's fish?

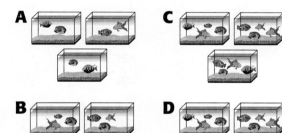

10. Adrian wants to place 15 apples into 3 baskets. Each basket will have the same number of apples. How many apples will be in each basket?

F 6   **H** 4

G 5   **J** 3

Record your answers on the answer sheet provided by your teacher or on a sheet of paper.

11. Jenny has 20 beads to make bracelets. Each bracelet has 10 beads. Write a number sentence to show how many bracelets Jenny can make?

12. The model below shows $16 \div 2 = 8$.

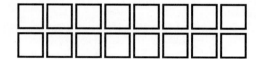

Write the remaining number sentences in this family of facts.

Record your answers on the answer sheet provided by your teacher or on a sheet of paper.

13. Draw a picture and explain in words how you would use a number line to solve this problem. Then solve by writing a number sentence.

$$20 \div 5 = \blacksquare$$

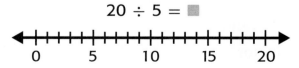

NEED EXTRA HELP?													
If You Missed Question...	1	2	3	4	5	6	7	8	9	10	11	12	13
Go to Lesson...	6-4	4-1	6-7	1-7	6-8	3-3	6-2	6-2	6-4	4-1	6-5	6-7	6-2

BIG Idea What are division facts and strategies?

Division facts and strategies will help you divide.

Example One of the greatest places to fish is off the coast of North Carolina. Suppose there are 16 boats in a marina. Each dock holds 8 boats. So, 16 ÷ 8 or 2 docks are needed to hold the boats.

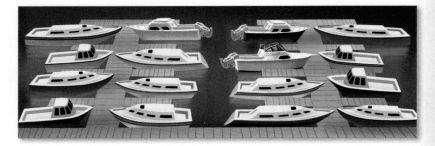

What will I learn in this chapter?

- Use models, arrays, repeated subtraction, and related facts to divide.
- Divide by 3, 4, 6, 7, 8, 9, 11, and 12.
- Write and solve number sentences.
- Solve problems by making a table.

Key Vocabulary

dividend

divisor

quotient

| Math Online > | Student Study Tools at macmillanmh.com |

FOLDABLES®
Study Organizer

Make this Foldable to help you organize information about division facts. Begin with one sheet of 11″ × 17″ paper.

① **Fold** the sheet of paper in half.

② **Fold** the paper in half again as shown.

③ **Unfold** the paper and label.

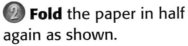

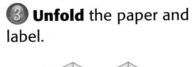

Divide by 3. Divide by 4. Divide by 6 & 7. Divide by 8 & 9.

You have two ways to check prerequisite skills for this chapter.

Option 2

Math Online > Take the Chapter Readiness Quiz at macmillanmh.com.

Option 1

Complete the Quick Check below.

QUICK Check

Algebra Use the array to complete each pair of number sentences.
(Lesson 6-2)

1. $2 \times \blacksquare = 8$
$8 \div \blacksquare = 4$

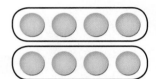

2. $1 \times 4 = \blacksquare$
$4 \div \blacksquare = 4$

Divide. (Chapter 6)

3. $25 \div 5$

4. $18 \div 2$

5. $10\overline{)20}$

6. Luther and Sheila have 49 marbles. They are playing with 5 friends. Will there be enough marbles for each player to have an equal number of marbles? Explain.

Subtract. (Chapter 3)

7. $8 - 2$

8. $10 - 5$

9. $12 - 4$

Algebra Find each missing factor. (Lesson 5-4)

10. $4 \times \blacksquare = 20$

11. $3 \times \blacksquare = 30$

12. $5 \times \blacksquare = 45$

13. Fidaa and Joseph each caught 8 grasshoppers. How many did they catch in all?

Math Activity for 7-1
Model Division

You can use counters to model division.

ACTIVITY

1 **Find 20 ÷ 5.**

Step 1 Use 20 counters. Make a column with 5 of the counters. Continue making columns of 5 with the rest of the counters.

Step 2 Line up the columns next to each other.

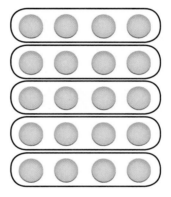

Step 3 Count the counters in each row. There are 4 counters in each row.

So, $20 \div 5 = 4$ or $5\overline{)20}^{\,4}$.

ACTIVITY

2 **Write number sentences using division with a dividend of 12.**

Step 1 Using 12 counters, make an array. Write a number sentence using division that describes the array.

 $12 \div 2 = 6$

Step 2 Model other arrays using 12 counters. Write the number sentences using division.

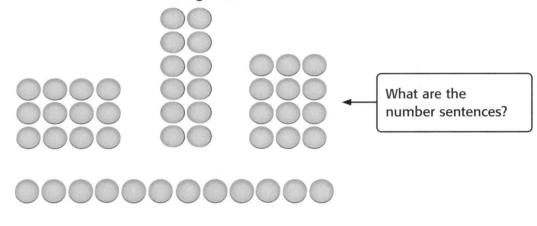

What are the number sentences?

Think About It

1. Why can arrays be used to divide?

2. Refer to Activity 2. Which pairs of number sentences shown in Step 2 are related?

CHECK What You Know

Use models to divide.

3. $21 \div 3$ **4.** $49 \div 7$ **5.** $36 \div 9$ **6.** $72 \div 8$

Write a number sentence using division that has the dividend shown.

7. 10 **8.** 9 **9.** 15 **10.** 16

11. **WRITING IN** ►**MATH** Explain how arrays are used for division.

7-1 Divide by 3

MAIN IDEA

I will use models and related multiplication facts to divide by 3.

Math Online

macmillanmh.com

• Extra Examples
• Personal Tutor
• Self-Check Quiz

GET READY to Learn

Martino, Maria, and Weston have 24 jacks in all. If each person has the same number of jacks, how many jacks does each person have?

In the Explore Activity, you used counters to model division. To find the quotient, you divide the dividend by the divisor.

Real-World EXAMPLE Make Equal Groups

1 **GAMES** There are 24 jacks in all. Martino, Maria, and Weston each have the same number of jacks. Write a number sentence to show how many jacks each has.

Divide 24 counters into 3 equal groups.

Find $24 \div 3$ or $3\overline{)24}$.

$$\underset{\text{dividend}}{24} \div \underset{\text{divisor}}{3} \quad \text{or} \quad 3\overline{)24}$$

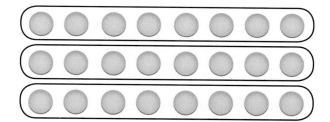

$$24 \div 3 = 8 \quad \text{or} \quad 3\overline{)24}^{\,8} \leftarrow \text{quotient}$$

The number sentence $24 \div 3 = 8$ shows that each person has 8 jacks.

You can use related facts to help you divide.

 Real-World EXAMPLE Use Related Facts

② **TRAVEL To travel to the beach, Angela and her 14 friends will divide up equally into 3 cars. Write a number sentence to show how many friends will be in each car.**

You need to find $15 \div 3$ or $3\overline{)15}$.

$15 \div 3 = \blacksquare$

$3 \times \blacksquare = 15$ ← THINK 3 times what number equals 15?

$3 \times 5 = 15$

$15 \div 3 = 5$ or $3\overline{)15}^{5}$

So, $15 \div 3 = 5$ shows that 5 friends will be in each car.

You can use repeated subtraction on a number line to divide.

EXAMPLE Use Repeated Subtraction

③ **Find $6 \div 3$ or $3\overline{)6}$.**

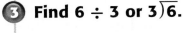
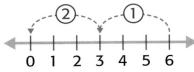

- Start at 6 and count back by 3s to 0.

- 3 was subtracted two times.

So, $6 \div 3 = 2$ or $3\overline{)6}^{2}$.

Remember

A division number sentence like $3\overline{)6}$ is read *six divided by three*. Always read the dividend under the symbol first.

✓ **CHECK What You Know**

Divide. Use models or related facts. See Examples 1–3 (pp. 297–298)

1. $12 \div 3$ **2.** $18 \div 3$ **3.** $3\overline{)9}$ **4.** $3\overline{)27}$

5. Rosa spent $30 on 2 skirts and a purse. Each item cost the same. How much did each item cost?

6. **Talk About It** How can you use 8×3 to find $24 \div 3$?

Divide. Use models or related facts. See Examples 1–3 (pp. 297–298)

7. 15 ÷ 3

8. 9 ÷ 3

9. 6 ÷ 3

10. 0 ÷ 3

11. 16 ÷ 2

12. 20 ÷ 10

13. 3)‾12‾

14. 3)‾3‾

15. 3)‾30‾

16. 3)‾27‾

17. 5)‾25‾

18. 10)‾100‾

Algebra **Copy and complete each table.**

19.

Rule: Divide by 3.				
Input	24	▪	30	▪
Output	▪	4	▪	6

20.

Rule: Subtract 3.				
Input	28	▪	33	▪
Output	▪	15	▪	16

Solve. Write a number sentence.

21. A soccer coach buys 3 new soccer balls for $21. What is the price for each ball?

22. There are 27 bananas on a counter. They will be divided equally into 3 piles. How many will be in each pile?

23. **Measurement** Kevin is on a 3-day hike. He will hike a total of 18 miles. If he hikes the same number of miles each day, how many miles will he hike the first day?

24. Makenna placed 20 stickers in equal rows of 5. She gave away 2 stickers. Now she wants to make equal rows of 3. How many stickers will be in each row?

H.O.T. Problems

25. **NUMBER SENSE** Mr. Marcos buys 4 bottles of glue, 1 stapler, and 2 notebooks. Can the total amount spent be divided equally by 3? Explain why or why not.

Item	Cost
Glue	$2
Stapler	$5
Notebook	$3

26. **WHICH ONE DOESN'T BELONG?** Which fact does not belong with the others? Explain your reasoning.

| 18 ÷ 3 | 3)‾18‾ | 6 ÷ 3 | 6)‾18‾ |

27. **WRITING IN ►MATH** Explain how to find 18 ÷ 3 in two different ways.

MAIN IDEA

I will use models and related multiplication facts to divide by 4.

Math Online

macmillanmh.com

• Extra Examples
• Personal Tutor
• Self-Check Quiz

> **GET READY to Learn**

The distance around a window in Peter's house is 12 feet. Each side has the same length. What is the length of each side?

You can use models and related multiplication facts to divide by 4. Use number sentences to show division problems.

Real-World EXAMPLE   Make Equal Groups

① **MEASUREMENT Write a number sentence to show the length of each side of the window.**

Divide 12 feet by the number of sides, 4.

There are 12 counters divided into 4 equal groups. There are 3 counters in each group.

So, $12 \div 4 = 3$. Each side is 3 feet long.

Real-World EXAMPLE   Use Related Facts

② **BIRDS An ostrich egg weighs 4 pounds. The total weight of the eggs in a nest is 28 pounds. Write a number sentence to show how many ostrich eggs are in the nest.**

Use a related multiplication fact to find $28 \div 4$ or $4\overline{)28}$.

$28 \div 4 = \blacksquare$

$4 \times \blacksquare = 28$ ← THINK What number times 4 equals 28?

$4 \times 7 = 28$

So, $28 \div 4 = 7$ or $4\overline{)28}^{\,7}$. There are 7 eggs in the nest.

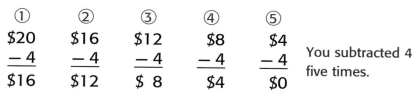
3 **MONEY** Trevor has $20 to divide equally among 4 people. Write a number sentence to show how much each person will get.

You need to find $20 ÷ 4 or 4$\overline{)\$20}$.

①	②	③	④	⑤	
$20	$16	$12	$8	$4	You subtracted 4
− 4	− 4	− 4	− 4	− 4	five times.
$16	$12	$ 8	$4	$0	

So, $20 ÷ 4 = $5 or 4$\overset{5}{\overline{)20}}$. Each person will get $5.

Check The number line shows there are 5 groups of 4 in 20. ✓

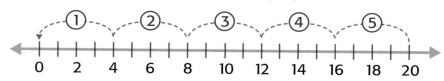

Remember

A number line can also be used for repeated subtraction.

Division Strategies   Key Concepts

There are several methods you can use to divide.

- Use models or draw an array.
- Use repeated subtraction.
- Use a related multiplication fact.

✓ CHECK What You Know

Divide. Use models or related facts. See Examples 1–3 (pp. 300–301)

1. 16 ÷ 4

2. 4 ÷ 4

3. 32 ÷ 4

4. 4$\overline{)8}$

5. 7$\overline{)28}$

6. 4$\overline{)36}$

7. Amel has 36 quarters. If each video game machine takes 4 quarters, how many games can he play?

8. **Talk About It** Without dividing, how do you know that 12 ÷ 3 is greater than 12 ÷ 4?

Divide. Use models or related facts. See Examples 1–3 (pp. 300–301)

9. $0 \div 4$ **10.** $4 \div 4$ **11.** $24 \div 4$ **12.** $36 \div 9$

13. $5\overline{)20}$ **14.** $3\overline{)12}$ **15.** $4\overline{)12}$ **16.** $4\overline{)40}$

Algebra Find each missing number.

17. $36 \div \blacksquare = 4$ **18.** $\blacksquare \div 4 = 6$ **19.** $4 \times \blacksquare = 40$ **20.** $\blacksquare \times 4 = 28$

Measurement Find the measure of the shaded part.

21.

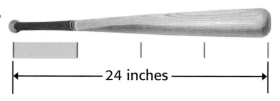

— 24 inches —

22.

— 40 miles —

Solve. Write a number sentence.

23. Grace, Clark, Elvio, and Trent will be traveling for 20 days. They are dividing the planning equally. How many days will Clark have to plan?

24. There are 36 pieces of luggage on a bus. If each person brought 4 pieces of luggage, how many people are on the trip?

25. It costs $40 for a family of 4 to ride go-carts for 1 hour. How much does it cost 1 person to ride for 2 hours?

26. There are 4 bananas, 3 apples, and 5 pears. If an equal number of fruit is placed in 4 baskets, how many pieces will be in each basket?

Berto wants to make a pictograph from the data he collected. He will use a key where each 🎺 **= 4.**

27. Each symbol equals 4 friends. How many symbols should he use to show the number of friends that marched in the parade? Explain.

28. If the number of friends who watched the parade are put into groups of 4, how many groups are there?

Did You Go to Saturday's Parade?

Responses	Number
Marched	20
Watched	16
Did not go	4

H.O.T. Problems

29. FIND THE ERROR Noelle and Brady are finding $12 \div 4$. Who is correct? Explain your reasoning.

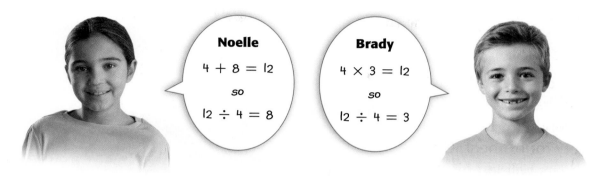

Noelle

$4 + 8 = 12$

so

$12 \div 4 = 8$

Brady

$4 \times 3 = 12$

so

$12 \div 4 = 3$

30. **WRITING IN ►MATH** Write a real-world problem that uses the division fact $36 \div 9$.

TEST Practice

31. 3 students each bought the same number of computer games. They bought a total of 21 games. Which number sentence shows how many games each student bought? **(Lesson 7-1)**

A $21 \times 3 = 63$

B $21 \div 3 = 7$

C $21 + 3 = 24$

D $21 - 3 = 18$

32. Which symbol goes in the box to make the number sentence true?
(Lesson 7-2)

$$28 \ \blacksquare \ 4 = 7$$

F $+$

G $-$

H $\times$

J $\div$

Spiral Review

Divide. Use models or related facts. **(Lesson 7-1)**

33. $30 \div 3$ **34.** $15 \div 3$ **35.** $9 \div 3$ **36.** $12 \div 3$

Divide. **(Lesson 6-8)**

37. $9 \div 9$ **38.** $8 \div 1$ **39.** $6 \div 6$ **40.** $0 \div 4$

41. There are 7 trucks at a rest stop. Each truck has 8 wheels. How many truck wheels are there in all? **(Lesson 5-5)**

MAIN IDEA I will solve a problem by making a table.

Ian plays the drum and triangle in the school band. He has to hit the drum every third beat and the triangle every fourth beat. On what two beats will Ian hit the drum and the triangle together?

Understand	**What facts do you know?** • Ian hits the drum on every third beat. • He hits the triangle on every fourth beat. **What do you need to find?** • When Ian will hit the drum and triangle together.
Plan	Organize the information in a table. Then use the table to solve.
Solve	The table shows the beats Ian hits the drum and the beats he hits the triangle. Circle the numbers that are the same in both rows of the table. + 3 + 3 + 3 + 3 + 3 + 3 + 3 | **Drum**	3	6	9	(12)	15	18	21	(24) | | **Triangle** | 4 | 8 | (12) | 16 | 20 | (24) | 28 | 32 | + 4 + 4 + 4 + 4 + 4 + 4 + 4 So, Ian will hit both the drum and the triangle on beats 12 and 24.
Check	Look back. Since 12 and 24 can both be evenly divided by 3 and 4, you know the answer is correct. ✓

Refer to the problem on the previous page.

1. Describe a problem in which you might use the *make a table* strategy to solve.

2. Explain how you use information in a table to help solve a problem.

3. Continue the table. What will be the next beat when Ian hits the drum and the triangle together?

4. Suppose Ian hits the drum every third beat and the triangle every fifth beat. On what two beats will he hit the drum and triangle together?

PRACTICE the Strategy

EXTRA PRACTICE
See page R20.

Solve. Use the *make a table* strategy.

5. **Algebra** Vicky is training for a 20-lap swimming race. The table shows the laps she swims each week. If the pattern continues, how many weeks will it take to reach 20 laps a day?

Training Record	
Weeks	**Laps**
1	2
2	5
3	8

6. Julian bought 32 books in one year. How many free books did he get?

7. Lucas is saving to buy a new watch that costs $45. He has $27 saved. If he saves $3 a week, how long will it take until he has enough money?

8. A group of 16 people want to go to the zoo. Use the sign below to find how they can get the lowest cost for admission.

Zoo Admission Prices

Per person............... $6

Group rate......... $30 for 6

9. Dennis and his friends played a water balloon game. At station 1, each person picks up 1 water balloon. At each of the next four stations, they pick up 2 more than the time before. What is the greatest number of balloons one person can collect if none break?

10. **WRITING IN ▶MATH** Write a problem that you could use the *make a table* strategy to solve.

MAIN IDEA

I will use an array and repeated subtraction to divide by 6 and 7.

Math Online

macmillanmh.com

• Extra Examples
• Personal Tutor
• Self-Check Quiz

GET READY to Learn

Paco set each picnic table with 6 dinner plates. He used 24 plates to set the tables. How many tables did he set?

You learned that arrays can help you understand how division and multiplication are related.

Real-World EXAMPLE   Model an Array

① **PICNIC** Use a number sentence to find how many tables Paco set.

Use an array to find $24 \div 6$ or $6\overline{)24}$. It will help you relate division and multiplication.

In the array, each table is represented by one column, which contains 6 plates. There are 4 columns. So, there will be 4 tables.

So, $24 \div 6 = 4$ or $6\overline{)24}^{\,4}$.

Paco will set 4 tables.

Check

The number line below shows that $24 \div 6$ is 4. ✔

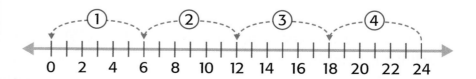

You can use many different methods to divide by 6 and 7.

Real-World EXAMPLES

Remember

Some division strategies are more useful than others when working with large numbers.

2 READING Markel read 28 books in 7 months. She read the same number each month. How many books did she read each month?

You want to find how many groups of 7 are in 28. Use repeated subtraction to find $28 \div 7$ or $7\overline{)28}$.

①	②	③	④	
28	21	14	7	The number 7 is
− 7	− 7	− 7	− 7	subtracted four times
21	14	7	0	to reach 0.

So, $28 \div 7 = 4$ or $7\overline{)28}^{\,4}$. Markel read 4 books each month.

3 TEACHING Mr. Jeremiah has 21 papers to grade. He wants to grade the same number of papers each day for 7 days. How many papers will he grade each day?

Use a related multiplication fact to find $21 \div 7$ or $7\overline{)21}$.

$21 \div 7 = \blacksquare$

$7 \times \blacksquare = 21$ ◄── THINK 7 times what equals 21?
$\qquad\qquad\qquad\qquad 7 \times 3 = 21$

$7 \times 3 = 21$

So, $21 \div 7 = 3$ or $7\overline{)21}^{\,3}$. He will grade 3 papers each day.

CHECK What You Know

Divide. Use models or repeated subtraction. See Examples 1–3 (pp. 306–307)

1. $24 \div 6$

2. $18 \div 6$

3. $7\overline{)35}$

4. $21 \div 7$

5. $14 \div 7$

6. $6\overline{)30}$

7. Measurement One kite tail measures 7 feet long. Elena has 56 feet of tail fabric. How many kite tails can she make?

8. **Talk About It** Are using related multiplication and division facts the same as using fact families? Explain.

Lesson 7-4 Divide by 6 and 7 **307**

Practice and Problem Solving

Divide. Use models or repeated subtraction. See Examples 1–3 (pp. 306–307)

9. 6 ÷ 6

10. 42 ÷ 6

11. 28 ÷ 7

12. 70 ÷ 7

13. 6)‾36‾

14. 6)‾60‾

15. 7)‾0‾

16. 7)‾42‾

Algebra Find each missing number.

17. 7 × ▇ = 63

63 ÷ 7 = ▇

18. 7 × ▇ = 35

35 ÷ 7 = ▇

19. 7 × ▇ = 70

70 ÷ 7 = ▇

Algebra Copy and complete each table.

20.

Rule: Divide by 6.			
Input	36	12	48	▇
Output	▇	▇	▇	10

21.

Rule: Divide by 7.			
Input	28	42	▇	70
Output	▇	▇	7	▇

Solve. Write a number sentence.

22. A rose bush has 42 rose buds. The 7 stems of the rose bush have an equal number of buds. How many buds are on each stem?

23. The sewing club is making a quilt with 63 squares. The squares are quilted in 7 equal rows. How many quilt squares are there in each row?

24. For every tree that is cut down, 7 new trees are planted. If 56 new trees have been planted, how many trees were cut down?

25. There are 7 groups of 5 students and 5 groups of 7 students at the tables in the cafeteria. What is the total number of students?

H.O.T. Problems

26. OPEN ENDED Write two numbers that cannot be divided by 7.

27. WHICH ONE DOESN'T BELONG? Identify the division sentence that does not belong with the others. Explain.

56 ÷ 7   7)‾48‾   49 ÷ 7   7)‾63‾

28. WRITING IN ►MATH When you know that 42 ÷ 6 = 7, you also know that 42 ÷ 7 = 6. Explain why.

Divide. Use models or related facts.
(Lessons 7-1 and 7-2)

1. $27 \div 3$

2. $18 \div 3$

3. $3\overline{)12}$

4. $3\overline{)9}$

5. $12 \div 4$

6. $36 \div 4$

Algebra Find each missing number.
(Lessons 7-1 and 7-2)

7. $\blacksquare \div 3 = 7$

8. $15 \div \blacksquare = 5$

9. $24 \div \blacksquare = 6$

10. $\blacksquare \div 4 = 2$

11. MULTIPLE CHOICE Which number will make the number sentence true? (Lesson 7-2)

$$40 \div \blacksquare = 4$$

A 10

C 14

B 11

D 100

12. Measurement On Monday, Wednesday, and Friday, Kimi runs 3 miles. On the other weekdays, she runs 2 miles. She does not run on Saturday. Sunday she runs twice as much as on Monday. How many miles does Kimi run in a week? Use the *make a table* strategy.

(Lesson 7-3)

13. Samuel earns $20 for each lawn that he mows. He can mow 2 lawns in one day. How long will it take Samuel to earn $200? (Lesson 7-3)

Algebra Find each missing number.
(Lesson 7-4)

14. $6 \times \blacksquare = 48$
$48 \div 6 = \blacksquare$

15. $7 \times \blacksquare = 70$
$70 \div 7 = \blacksquare$

Algebra Copy and complete the table. (Lesson 7-4)

C07-66A-105709

16.

Rule: Divide by 7.				
Input	28	35	42	49
Output	$\blacksquare$	$\blacksquare$	$\blacksquare$	$\blacksquare$

17. MULTIPLE CHOICE Aisha picked 42 apples. She placed an equal number in 6 bags. How many apples were in each bag?

(Lesson 7-4)

F 6

H 8

G 7

J 9

18. Horatio is placing an equal number of raisins on each muffin. He has 49 raisins, and there are 7 muffins. How many raisins will he place on each muffin?

(Lesson 7-4)

19. WRITING IN MATH Sophia said that if she knows $36 \div 4 = 9$, then she can find $36 \div 9$. What is the answer? Explain her reasoning.

(Lesson 7-2)

Stars and Stripes

For more than 200 years, the American flag has been a symbol of the United States. However, it has not always looked like it does now. The first United States flag had 13 stripes and only 13 stars. George Washington wanted the stars in the flag to have 6 points. However, Betsy Ross, the maker of the first flag, chose a star with 5 points.

The American flag has changed 27 times. Today, the flag has 7 red stripes and 6 white stripes. There are 50 stars on the flag, one for each state.

Did You Know?

The world's largest flag is 550 feet long and 225 feet wide.

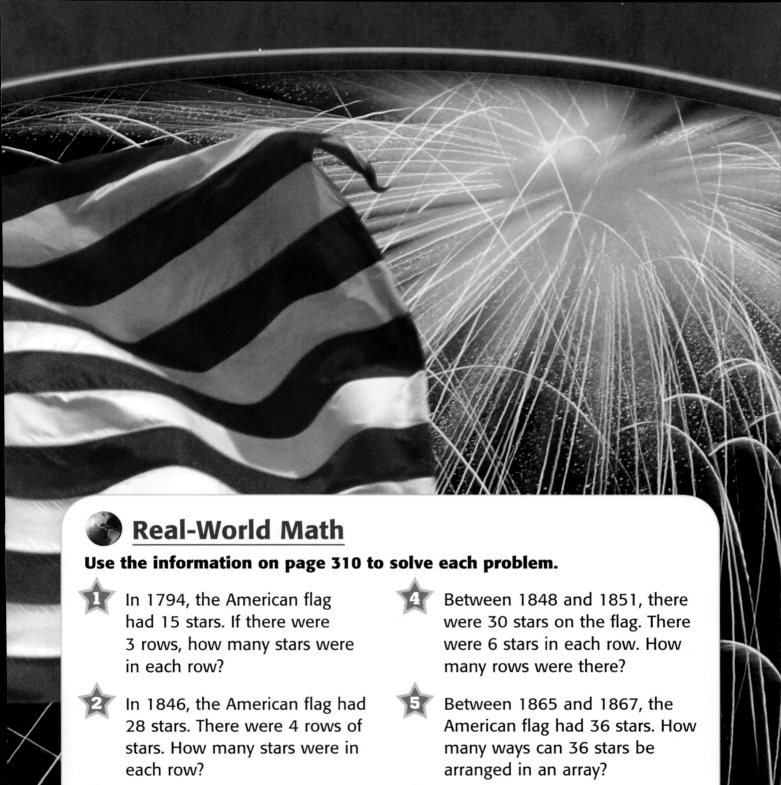

Real-World Math

Use the information on page 310 to solve each problem.

1 In 1794, the American flag had 15 stars. If there were 3 rows, how many stars were in each row?

2 In 1846, the American flag had 28 stars. There were 4 rows of stars. How many stars were in each row?

3 Suppose a flag has 7 rows of stars with 4 stars in each row. Does this flag have the same number of stars as the flag in Exercise 2? Explain.

4 Between 1848 and 1851, there were 30 stars on the flag. There were 6 stars in each row. How many rows were there?

5 Between 1865 and 1867, the American flag had 36 stars. How many ways can 36 stars be arranged in an array?

6 A flag has 42 stars. There are 7 stars in each row. How many rows does this flag have?

Divide by 8 and 9

The pictograph shows the number of times each student visited the Navy Pier in Chicago, Illinois. If 32 students visited 2 or more times, how many symbols should be drawn in that row?

MAIN IDEA

I will learn to divide by 8 and 9.

Math Online

macmillanmh.com

• Extra Examples
• Personal Tutor
• Self-Check Quiz

Navy Pier Visits

Number of Visits	Number of Students
Never	🖐
1	🖐 🖐
2 or more	

Each 🖐 = 8 Students

You can use either a related multiplication or division fact within a family of facts to find a quotient.

Real-World EXAMPLE Use Related Facts

1 GRAPHS How many symbols should be drawn in the row for 2 or more visits?

There are 32 students being divided into groups of 8.

One Way: Multiplication	**Another Way:** Division
$32 \div 8 = \blacksquare$	$32 \div 8 = \blacksquare$
$8 \times \blacksquare = 32$	$32 \div \blacksquare = 8$
$8 \times 4 = 32$	$32 \div 4 = 8$
So, $32 \div 8 = 4$.	So, $32 \div 8 = 4$.

So, there should be 4 symbols in the row.

Real-World EXAMPLE Make Equal Groups

2 **ART** Kyra and 8 of her friends made 63 paper stars. They will each take home an equal number. How many paper stars will each take home?

Find $63 \div 9$ or $9\overline{)63}$.
Model 63 counters in 9 groups.

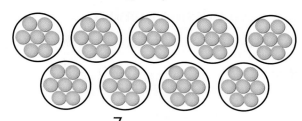

So, $63 \div 9 = 7$ or $9\overline{)63}^{\,7}$. Kyra and her friends will each have 7 stars.

Real-World EXAMPLE Repeated Subtraction

3 **QUARTERS** John collected 27 state quarters over the last 9 years. Each year he added the same number. How many quarters did he add each year?

Use repeated subtraction to find $27 \div 9$ or $9\overline{)27}$.

①	②	③	The number 9 is subtracted three times to reach zero.
27	18	9	
$-\ 9$	$-\ 9$	$-\ 9$	
18	9	0	

So, $27 \div 9 = 3$ or $9\overline{)27}^{\,3}$. He added 3 quarters each year.

CHECK What You Know

Divide. Use related facts or repeated subtraction. See Examples 1–3 (pp. 312–313)

1. $8 \div 8$

2. $18 \div 9$

3. $8\overline{)48}$

4. Each art project uses 9 tiles. If there are 36 tiles, how many projects can be made?

5. **Talk About It** How do multiplication facts help you check to see if your division is correct?

Divide. Use related facts or repeated subtraction. See Examples 1–3 (pp. 312–313)

6. $16 \div 8$

7. $72 \div 8$

8. $63 \div 9$

9. $27 \div 9$

10. $8\overline{)80}$

11. $8\overline{)32}$

12. $9\overline{)90}$

13. $9\overline{)54}$

Algebra Find each missing number.

14. $9 \times \blacksquare = 36$

$36 \div 9 = \blacksquare$

15. $8 \times \blacksquare = 40$

$40 \div 8 = \blacksquare$

16. $8 \times \blacksquare = 48$

$48 \div 8 = \blacksquare$

Solve. Write a number sentence.

17. Tionna has 24 party favors for each of the 8 guests coming to her party. How many favors will each get?

18. One baseball game has 9 innings. If 36 innings out of 54 have been played, how many games remain?

Data File

Mrs. Benson's class of 9 students decided to adopt an animal at New Jersey's Cohanzick Zoo.

19. Which animal could the students adopt if they each paid $3?

20. If each student paid $8, would they be able to adopt the pheasant and falcon? Explain.

21. Which animals would cost each student more than $10 each to adopt? Explain.

Cohanzick Zoo Adopt An Animal

Animal	Price
pheasant	$25
falcon	$50
alligator	$100
reindeer	$250

Source: Cohanzick Zoo

H.O.T. Problems

22. OPEN ENDED Choose two facts from Exercises 6–13. Explain a strategy for remembering them.

23. WRITING IN ►MATH Write a real-world multiplication problem in which you would divide by 8 or 9.

Facts Roll
Division Facts

Get Ready!
Players: 3 or more

Get Set!
- Choose 1 player to be the timer.

Go!
- The timer sets the stopwatch for 45 seconds.

- The timer rolls the number cube and starts the timer. The number rolled is the quotient.

- The other players write as many division facts as they can that have the same quotient as the number on the cube.

You will need: stopwatch, number cube, paper

- After 45 seconds, the timer calls, "Time!"

- The player with the most correct facts wins.

$$20 \div 5$$
$$16 \div 4$$
$$12 \div 3$$

GET READY to Learn

For a field trip, 33 students went to the science museum. There were 11 microscopes. Each was used by an equal number of students in a group. How many students were in each group?

You can use models, repeated subtraction, or related facts to divide by 11 and 12.

Real-World EXAMPLE Divide

1 **SCIENCE** Write a number sentence that shows the number of students using each microscope.

Find $33 \div 11$ or $11\overline{)33}$.

One Way: Use Models	Another Way: Use Repeated subtraction
Place 33 counters into 11 equal groups.	Count how many times 11 is subtracted from 33 until the difference is 0.

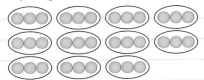

One Way: There are 3 counters in each group. So, $33 \div 11 = 3$.

Another Way:

  ① ② ③

33 22 11
−11 −11 −11
22 11 0

11 was subtracted from 33, 3 times. So, $33 \div 11 = 3$.

$33 \div 11 = 3$ or $11\overline{)33}^{3}$.

So, 3 students used each microscope.

Check You can use multiplication to check.

$3 \times 11 = 33$ ✔

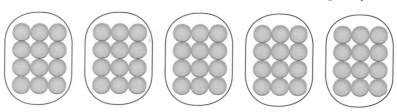

Real-World EXAMPLE Make Equal Groups

2 **EGGS** Pierce places 60 eggs into cartons. Each carton holds 12 eggs. How many cartons did he fill?

Find $60 \div 12$ or $12\overline{)60}$. Put 60 counters in groups of 12.

There are 5 equal groups of 12. $60 \div 12 = 5$ or $12\overline{)60}^{5}$.

So, Pierce filled 5 cartons.

Multiplication facts can be used to solve division problems.

EXAMPLE Use Related Facts

3 **Find $48 \div 12$ or $12\overline{)48}$.**

$12 \times \blacksquare = 48$

↑

The missing factor is 4.

↓

$12 \times 4 = 48$

So, $48 \div 12 = 4$ or $12\overline{)48}^{4}$.

Remember

In Lesson 5-8, you learned multiplication facts for 12.

CHECK What You Know

Divide. See Examples 1–3 (pp. 316–317)

1. $22 \div 11$

2. $77 \div 11$

3. $11\overline{)33}$

4. $36 \div 12$

5. $72 \div 12$

6. $12\overline{)48}$

7. Tonya shares 44 carnival tickets with 11 friends. Write a number sentence to show how many tickets Tonya and her friends will each get.

8. **Talk About It** Describe two-digit quotients that are divided by 11.

Divide. See Examples 1–3 (pp. 316–317)

9. $11 \div 11$

10. $55 \div 11$

11. $44 \div 11$

12. $66 \div 11$

13. $11 \overline{)99}$

14. $11 \overline{)77}$

15. $24 \div 12$

16. $12 \div 12$

17. $36 \div 12$

18. $84 \div 12$

19. $12 \overline{)96}$

20. $12 \overline{)120}$

Algebra **Find each missing number.**

21. $22 \div \blacksquare = 2$

22. $48 \div \blacksquare = 4$

23. $\blacksquare \div 11 = 3$

24. $\blacksquare \div 12 = 5$

25. $\blacksquare \div 9 = 11$

26. $\blacksquare \div 11 = 12$

Solve. Write a number sentence.

27. The books below have a total of 66 chapters. Suppose each book has an equal number of chapters. How many chapters does each of the books have?

28. Darla made a gift bag for each guest at her party. Each bag had 12 gifts. There were a total of 84 gifts. Some of the bags are shown. How many bags are not shown?

29. Allen has a total of 60 sports cards. He divided them equally among his friends and himself. Each person got 12 cards. How many friends got sports cards?

30. A restaurant served 144 onion rings to 12 customers. Each customer got an equal number of onion rings. How many onion rings did each customer receive?

H.O.T. Problems

31. OPEN ENDED Tell about a real-world situation in which you could divide by 12.

32. FIND THE ERROR Jeff and Jamie each solved a division problem. Who is correct? Explain your reasoning.

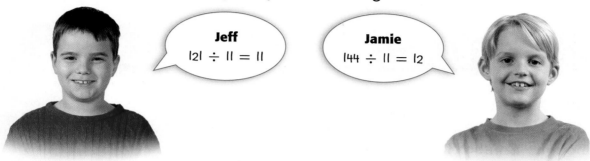

Jeff
$121 \div 11 = 11$

Jamie
$144 \div 11 = 12$

33. WRITING IN ▶MATH Write a division problem in which the quotient is 11.

TEST Practice

34. Sandra planted 18 seeds. She put an equal number of seeds in 9 pots. Which number sentence shows how many seeds Sandra put in each pot? (Lesson 7-5)

A $18 \div 9 = 2$

B $18 \times 9 = 162$

C $18 + 9 = 27$

D $18 - 9 = 9$

35. There were 77 flowers placed equally in 11 vases. Which number sentence shows how many flowers are in each vase? (Lesson 7-6)

F $77 \times 11 = 847$

G $77 \div 11 = 7$

H $77 + 11 = 88$

J $77 - 11 = 66$

Spiral Review

Algebra Find each missing number. (Lesson 7-5)

36. $56 \div 8 = \blacksquare$
$8 \times \blacksquare = 56$

37. $32 \div 8 = \blacksquare$
$8 \times \blacksquare = 32$

38. $81 \div 9 = \blacksquare$
$9 \times \blacksquare = 81$

39. There are 42 windows on a street. Each house has 2 windows on the front, 3 on the back, and 1 on each side. How many houses are there on the street? (Lesson 7-4)

Problem-Solving Investigation

MAIN IDEA I will choose the best strategy to solve a problem.

P.S.I. TEAM ✚

SELMA: I bought 3 shorts and 2 shirts. My sister Lexie bought 4 shorts and 2 shirts.

YOUR MISSION: Find out how many different shirt and shorts combinations each girl can make.

Understand	You know what each girl bought. Find out how many different shirt and shorts combinations they can each make.
Plan	Organize the information in a table.
Solve	Set up a table for each girl. Make a row for each pair of shorts and a column for each shirt. List the possible shirt and shorts combinations.

Selma	Shirt 1	Shirt 2
Shorts A	A1	A2
Shorts B	B1	B2
Shorts C	C1	C2

Lexie	Shirt 1	Shirt 2
Shorts A	A1	A2
Shorts B	B1	B2
Shorts C	C1	C2
Shorts D	D1	D2

Selma: 3 × 2 = 6
shorts   shirts   combinations

Lexie: 4 × 2 = 8

So, Lexie can make more combinations.

Check	Look back. Since 3 × 2 = 6 and 4 × 2 = 8, you know that the number of clothing combinations is correct. ✓

Use any strategy shown below to solve. Tell what strategy you used.

PROBLEM-SOLVING STRATEGIES
• Act it out.
• Draw a picture.
• Look for a pattern.
• Make a table.

1. Claudia and Danielle went to the store to buy paint for an art project. They chose 5 colors. Each bottle of paint costs $3. Find the total cost.

2. **Algebra** What is the next number in the pattern?

    25, 26, 29, 30, 33, 34, ▦

3. **Measurement** Devon and his sister have 42 bottles of water. Devon drinks the amount shown each day. His sister drinks 4 bottles each day. How many days will the water last?

4. Dasan planted 30 tomato seeds in his garden. Three out of every 5 seeds grew into plants. How many tomato plants did he have?

5. Carlita spent $1 on popcorn and $3 on a soda. She had $1 left. How much money did she start with?

6. Would it cost more to send 2 letters or 3 postcards? Explain.

    letter

    postcard

7. Nela's class has 6 more students than Meg's class. Meg's class last year had 5 more than it does this year. Meg has 24 students in her class this year. How many students are in Nela's class this year?

8. **WRITING IN ►MATH** There are 42 students going on a picnic. Each car can take 6 students. Each van can take 7 students. Would it be less expensive to take cars or vans to the picnic? Explain.

Picnic Transportation	
Vehicle	**Cost**
Car	$10
Van	$11

FOLDABLES
Study Organizer GET READY to Study

Be sure the following Key Vocabulary words and Key Concepts are written in your Foldable.

Key Concepts

Division (p. 297)

Division separates amounts into equal groups.

$$20 \div 5 = 4 \leftarrow \text{quotient}$$

↑ dividend ↑ divisor

Model Division (p. 297)

Counters can be put in equal groups to model division.

$$18 \div 3 = 6$$

Division Strategies (p. 301)

There are several division strategies.
- models
- arrays
- related facts
- repeated subtraction

Key Vocabulary

dividend (p. 297)
divisor (p. 297)
quotient (p. 297)

Vocabulary Check

Choose the vocabulary word that completes each sentence.

1. The ____?____ is the answer of a division problem.

2. The ____?____ is the number that gets divided in a division problem.

3. A division problem divides the ____?____ by the ____?____.

4. In the division problem $15 \div 3 = 5$, 3 is the ____?____.

5. The ____?____ of $20 \div 4$, is 5.

6. The 24 in the division problem $24 \div 6 = 4$ is the ____?____.

Lesson-by-Lesson Review

7-1 **Divide by 3** (pp. 298–299)

Example 1
There are 21 wheels. Each tricycle has 3 wheels. Find how many tricycles there are.

To find $21 \div 3$ or $3\overline{)21}$, you can use a related multiplication fact.

$21 \div 3 = \blacksquare$

$3 \times \blacksquare = 21$ ← THINK 3 times what equals 21?

$3 \times 7 = 21$

Since $21 \div 3 = 7$ or $3\overline{)21}^{\,7}$, there are 7 tricycles.

Divide. Use models or related facts.

7. $18 \div 3$ **8.** $24 \div 3$

9. $3\overline{)27}$ **10.** $3\overline{)9}$

Algebra Compare. Use >, <, or =.

11. $18 \div 3 \;\bullet\; 18 + 3$

12. $3 \times 10 \;\bullet\; 27 \div 3$

13. Herman spent $24 on 2 CDs and a poster. Each item cost the same. How much was each item?

7-2 **Divide by 4** (pp. 300–303)

Example 2
Measurement **A square playground measures 40 feet around its outside edge. How long is one side of the playground?**

Find $40 \div 4$ or $4\overline{)40}$.

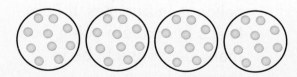

There are 10 objects in each group.

$40 \div 4 = 10$ or $4\overline{)40}^{\,10}$

So, each side is 10 feet.

Check
Since $4 \times 10 = 40$, you know $40 \div 4$ is 10. ✓

Divide. Use models or related facts.

14. $32 \div 4$ **15.** $16 \div 4$

16. $36 \div 4$ **17.** $28 \div 4$

18. There are 4 soccer teams and 24 children who want to play soccer. How many children will be on each team if each team has the same number of players?

19. There are 36 buttons to be sewn on 9 jackets. How many buttons will go on each jacket if they are divided equally?

Algebra Find each missing number.

20. $\blacksquare \div 4 = 5$ **21.** $\blacksquare \div 4 = 3$

7-3 **Problem-Solving Strategy:** **Make a Table** (pp. 304–305)

Example 3

At 8:00 A.M., Graham saw 24 birds sitting on a wire. At 10:00 A.M., he saw 21 birds on the wire. There were 18 birds on the wire at noon. If the pattern continues, how many birds would Graham see at 4:00 P.M.?

Understand You know the number of birds Graham saw. You need to find how many birds will be on the wire at 4:00 P.M.

Plan Make a table.

Solve The table shows a pattern. Continue the pattern.

Bird Watching	
Time	**Birds**
8:00 A.M.	24
10:00 A.M.	21
12:00 A.M.	18
2:00 P.M.	15
4:00 P.M.	12

+2 +2 +2 +2   −3 −3 −3 −3

So, Graham will see 12 birds at 4:00 P.M.

Check Graham sees 12 birds at 4:00 P.M. Add 3 birds for each of the previous 4 hours.

$$12 + 3 = 15$$
$$15 + 3 = 18$$
$$18 + 3 = 21$$
$$21 + 3 = 24 \checkmark$$

Solve. Use the *make a table* strategy.

22. A toy store is having a sale. How many games will you have if you buy 6 at regular price?

23. Measurement Jamil rides his bike 3 days a week. He rides for 10 minutes on Mondays and twice that long on Wednesdays. On Fridays he rides three times longer than he did on Wednesdays. How many minutes does he ride in two weeks?

24. Algebra Polly is placing balloons in bunches. If Polly keeps her pattern going, how many balloons will be in the sixth bunch?

Balloon Bunches	
Bunch	**Balloons**
First	3
Second	5
Third	7

7-4 Divide by 6 and 7 (pp. 306–308)

Example 4

There are 28 students. The desks are in 7 equal rows. How many desks are in each row?

To find $28 \div 7$ or $7\overline{)28}$, you can use a related multiplication fact.

$28 \div 7 = \blacksquare$

$7 \times \blacksquare = 28$ ← THINK 7 times what equals 28?

$7 \times 4 = 28$

So, $28 \div 7 = 4$ or $7\overline{)28}^{\,4}$.

There are 4 desks in each row.

Divide. Use models or repeated subtraction.

25. $54 \div 6$ **26.** $63 \div 7$

27. $14 \div 7$ **28.** $36 \div 6$

Algebra Find each missing number.

29. $7 \times \blacksquare = 35$ **30.** $6 \times \blacksquare = 30$

    $35 \div 7 = \blacksquare$ $30 \div 6 = \blacksquare$

31. Maggie went to dance class for 42 days without missing a day. How many 7-day weeks is that?

7-5 Divide by 8 and 9 (pp. 312–314)

Example 5

Hugo passed out 36 paper clips for an experiment. Nine people were given the same number of clips. How many paper clips did each person get?

You can find $36 \div 9$ or $9\overline{)36}$. Use a related division fact.

$36 \div 9 = \blacksquare$

$36 \div \blacksquare = 9$ ← THINK 36 divided by what number equals 9?

$36 \div 4 = 9$

Since $36 \div 9 = 4$ or $9\overline{)36}^{\,4}$, each person got 4 clips.

Divide. Use related facts or repeated subtraction.

32. $81 \div 9$ **33.** $64 \div 8$

34. $45 \div 9$ **35.** $48 \div 8$

Algebra Find each missing number.

36. $9 \times \blacksquare = 36$ **37.** $8 \times \blacksquare = 80$

    $36 \div 9 = \blacksquare$ $80 \div 8 = \blacksquare$

38. There are 80 marshmallows for 8 campers. If each camper uses two marshmallows to make a s'more, how many s'mores can each camper have?

7-6 **Divide by 11 and 12** (pp. 316–319)

Example 6
Kent used 22 slices of cheese to make 11 sandwiches. Each sandwich has the same number of cheese slices. How many slices of cheese are on each sandwich?

Put 22 counters into 11 equal groups.

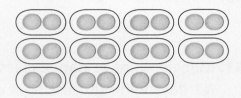

Since 22 ÷ 11 = 2, each sandwich has 2 slices of cheese.

Divide.

39. 24 ÷ 12   **40.** 66 ÷ 11

41. 60 ÷ 12   **42.** 99 ÷ 11

Algebra Find each missing number.

43. 11 × ■ = 33   **44.** ■ × 11 = 77
33 ÷ 11 = ■   77 ÷ 11 = ■

45. A chef cooks 72 eggs at the restaurant. Each carton holds 12 eggs. How many cartons does the chef need?

7-7 **Problem-Solving Investigation: Choose a Strategy** (pp. 320–321)

Example 7
There are 146 students. Of the students, 32 take a bus home, 13 students ride home in a car, and the rest walk. How many students walk home?

Find the number of students that take a bus or ride in a car. Then subtract this amount from 146.

$$\begin{array}{r} 32 \\ +13 \\ \hline 45 \end{array} \qquad \begin{array}{r} 146 \\ -45 \\ \hline 101 \end{array}$$

So, 101 students walk to school.

Solve. Use any strategy.

46. Joyce read 5 more books than Jeremy. Yoko read twice as many books as Joyce. Yoko read 16 books. How many books did Jeremy read?

47. Of 48 balls sold, 3 times as many tennis balls were sold as soccer balls. How many soccer balls were sold?

48. Benito had $20. He took out $8 and later put back $6. How much money does he have?

For Exercises 1–3, tell whether each statement is *true* or *false*.

1. An example of a number sentence is $15 \div 5 = 3$.

2. The dividend is the answer to a division problem.

3. In $16 \div 2 = 8$, the divisor is 2, and the quotient is 8.

Divide. Use models or related facts.

4. $28 \div 4$

5. $21 \div 3$

6. $36 \div 6$

7. $42 \div 7$

8. $72 \div 8$

9. $81 \div 9$

10. $48 \div 12$

11. $55 \div 11$

12. **MULTIPLE CHOICE** Ryder did this division problem.

$$56 \div 7 = 8$$

Which problem could he do to check his answer?

A $56 + 7$

B 7×8

C $8 + 7$

D $7 \div 56$

13. Chandra has 64 autographed baseballs in her collection. If 8 baseballs fit on each display shelf, how many shelves will she need?

14. **MULTIPLE CHOICE** A cook boils 16 potatoes in 2 pots. Each pot has the same number of potatoes. Which number sentence shows how many potatoes are in each pot?

F $16 + 2 = 18$

G $16 - 2 = 18$

H $16 \times 2 = 32$

J $16 \div 2 = 8$

Solve. Use the *make a table* strategy.

15. On Monday, Oleta swims 5 laps. She swims 5 laps more each day, from the day before. What is the total number of laps she swims from Monday to Friday?

Solve. Write a number sentence.

16. There are 9 students. Each student has a sweater with 3 buttons. How many buttons are there?

17. A baker made 48 muffins for a birthday party. Each muffin pan held 8 muffins. How many muffin pans did he use?

18. **WRITING IN ►MATH** Write an equation that uses $18 - \blacksquare$ and $10 + 2$. Explain how to decide which numbers would make it true.

PART 1 **Multiple Choice**

Read each question. Then fill in the correct answer on the answer sheet provided by your teacher or on a sheet of paper.

1. Hamburger buns are sold in packages of 8. How many packages are needed for 48 buns in all?

A 5   **C** 7

B 6   **D** 8

2. The figure below is a model for $5 \times 9 = 45$.

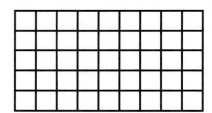

Which division sentence is modeled by the same figure?

F $36 \div 9 = 4$ **H** $45 \div 5 = 9$

G $36 \div 4 = 9$ **J** $50 \div 5 = 10$

3. Chandler worked a total of 36 hours the last 4 weeks. He worked the same number of hours each week. How many hours did he work each week?

A 3   **C** 8

B 6   **D** 9

4. Ella placed apples in a bowl. Of the 15 apples, 5 are green. The others are red. What number makes the number sentence true?

$$15 - 5 = \blacksquare$$

F 3   **H** 10

G 5   **J** 20

5. Seven footballs are shared equally by 49 students. Which sign shows the operation you would use to find the number of students that will share each football?

A $+$ **C** $\times$

B $-$ **D** $\div$

6. Mateo recorded 2,348 minutes of reading this year. Bianca recorded 2,483. If Ross recorded a number between Mateo and Bianca, how many minutes did he record?

F 2,446   **H** 2,834

G 2,487   **J** 3,446

7. If $3 \times 5 \times 2 = 30$, then what is $2 \times 3 \times 5$?

A 10   **C** 30

B 25   **D** 60

8. Which number sentence is modeled by the array?

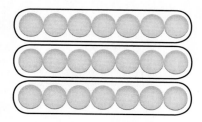

F $24 \div 8 = 3$ **H** $21 \div 3 = 7$

G $18 \div 3 = 6$ **J** $21 \div 4 = 5$

9. Which number sentence below is in the same family as $18 \div 6 = 3$?

A $18 \div 2 = 9$ **C** $18 \div 3 = 6$

B $6 \div 3 = 2$ **D** $6 \times 6 = 36$

10. Bella collected 24 shells. She arranged them in 6 equal-size groups. How many were in each group?

F 3 **H** 6

G 4 **J** 8

PART 2   Short Response

Record your answers on the answer sheet provided by your teacher or on a sheet of paper.

11. Mrs. Williams divided her class of 28 into 4 groups. Write a number sentence to describe the number of students in each group?

12. Jeslin bought 3 pairs of socks. Each pair of socks cost the same price. The total cost was $18. How much money did each pair of socks cost?

PART 3   Extended Response

Record your answers on the answer sheet provided by your teacher or on a sheet of paper.

13. An art teacher has planned a project that uses 2 wiggle eyes for each student. Wiggle eyes come in packages of 12. Explain how you would use the *make a table* strategy to find the number of packages needed so she has 84 wiggle eyes. Show your work.

NEED EXTRA HELP?													
If You Missed Question...	1	2	3	4	5	6	7	8	9	10	11	12	13
Go to Lesson...	7-5	7-5	7-2	3-1	7-4	1-7	5-9	7-1	6-2	7-4	7-2	7-1	7-3

CHAPTER 8 Use Patterns and Algebraic Thinking

BIG Idea What are patterns and functions?

Patterns are seen everywhere. **Functions** help us use information from patterns to solve problems. One way to show this information is a table.

Example Each bee has 6 legs. The function table shows how many total legs there are on several bees.

Number of Legs on Bees		
Number of Bees	Each has 6 legs	Total Number of Legs
2	2 × 6	12
3	3 × 6	18
4	4 × 6	24
5	5 × 6	30

What will I learn in this chapter?

- Model and use number sentences and expressions.
- Make a table to show functions.
- Identify and describe patterns in a table.
- Solve a problem by acting it out.

Key Vocabulary

number sentence

expression

functions

rule

Math Online > **Student Study Tools**
at macmillanmh.com

FOLDABLES®
Study Organizer

Make this Foldable to help you understand patterns and algebraic thinking. Begin with one sheet of $8\frac{1}{2}'' \times 11''$ paper.

1 **Fold** the sheet of paper as shown.

2 **Fold** again in half as shown.

3 **Unfold** and cut along the two inside valley folds.

4 **Label** as shown. Record what you learn.

Number Sentences	Expressions
Find a Rule	Function Tables

You have two ways to check prerequisite skills for this chapter.

Option 2

Math Online Take the Chapter Readiness Quiz at macmillanmh.com.

Option 1

Complete the Quick Check below.

QUICK Check

Compare. Use <, >, =. (Lesson 1-6)

1. 5 ● 8

2. 62 ● 26

3. 298 ● 199

4. 824 ● 842

5. 3 + 7 ● 10

6. 2 + 9 ● 10

7. 17 − 9 ● 8

8. 14 − 2 ● 16

Add or subtract. (Lesson 2-4 and 3-1)

9. 9 + 3

10. 12 + 7

11. 16 + 5

12. 32 + 43

13. 11 − 4

14. 20 − 6

15. 25 − 8

16. 38 − 22

Multiply or divide. (Chapters 5, 6, and 7)

17. 5 × 6

18. 3 × 8

19. 18 ÷ 2

20. 28 ÷ 4

21. Lin sold 1 more candle for the fundraiser than Joan. Together, they sold 15 candles. Draw a picture that shows how many candles they each sold.

22. Daniela spent $20 at the grocery store and $15 at the toy store. How much did she spend in all? Show how you can solve this problem using numbers.

23. Each toy shown costs $5. Show how you can find the total cost using an addition sentence.

$5 $5

8-1 Model and Write Number Sentences

MAIN IDEA

I will model and write addition and subtracton number sentences.

New Vocabulary

number sentence

Math Online

macmillanmh.com
- Extra Examples
- Personal Tutor
- Self-Check Quiz

GET READY to Learn

The Nine-Banded Armadillo can weigh about 15 pounds. The Giant Armadillo can weigh up to 100 pounds! Write a number sentence that shows the difference in their weights.

A **number sentence** is an expression using numbers and the equal sign (=) or the < or > sign. Number sentences can be *modeled* or shown using pictures and words.

Real-World EXAMPLE Model and Write Number Sentences

1 **ANIMALS** Model and write a number sentence that shows the difference in the weights.

To find the difference, you can subtract.

Models

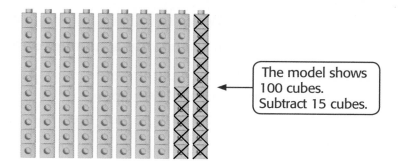

The model shows 100 cubes. Subtract 15 cubes.

Words After subtracting 15 cubes from 100, there are 85 cubes left. 100 minus 15 equals 85.

Number Sentence 100 − 15 = 85

So, 100 − 15 = 85 shows the difference in the weights.

EXAMPLE Model and Write Number Sentences

2 Use pictures and words to model 8 + 13 = 21.

Pictures

8 {😊😊😊😊😊😊😊😊

13 {😊😊😊😊😊😊😊😊😊😊😊😊😊

21 total

Words Thirteen added to eight equals twenty-one.

Number Sentence 8 + 13 = 21

Real-World EXAMPLE Model and Write Number Sentences

3 There are 3 students on the swings, 4 playing ball, and 2 jumping rope. Model and write a number sentence to show how many students there are in all.

Pictures

Words Three students plus four students plus two students equals nine students.

Number Sentence 3 + 4 + 2 = 9

CHECK What You Know

Model each problem. Then write a number sentence. See Examples 1–3 (pp. 333–334)

1. On Monday 212 newspapers are delivered. On Tuesday, 189 papers are delivered. How many papers are delivered altogether?

2. Felisha had $20. She spent $3 on her lunch, $5 at the movies, and $9 at the toy store. How much money was left?

Model each number sentence. Use pictures and words. See Examples 1–3 (pp. 333–334)

3. 14 + 7 = ▦

4. 30 − ▦ = 18

5. 12 + 3 + 4 = ▦

6. Mark has 25 toys. Draw a picture and write a number sentence to show how many toys Mark will give away if he keeps 4.

7. **Talk About It** Describe a real-world problem that uses a number sentence with several numbers.

Model each problem. Then write a number sentence. See Examples 1 and 3 (pp. 333–334)

8. Jessie picked 85 strawberries and 72 blueberries. How many berries did Jessie pick altogether?

9. A truck driver drove 548 miles one day and 163 miles during the next. How much farther did the truck driver travel on the first day?

10. Al fed his iguana 15 beans. The iguana ate 4 beans by lunch, 7 more by dinner, and 3 more by bedtime. How many beans were left at the end of the day?

11. Twenty customers ordered a turkey sandwich. Three ordered a ham sandwich. Thirteen ordered a chicken sandwich. How many customers were there in all?

Model each number sentence. Use pictures and words. See Examples 1–3 (pp. 333–334)

12. $14 - 8 = \blacksquare$

13. $24 + 9 = \blacksquare$

14. $32 + \blacksquare = 36$

15. $6 + 4 + 11 = \blacksquare$

16. $12 + 3 + \blacksquare = 17$

17. $\blacksquare - 7 - 6 = 22$

For Exercises 18–20, use the table.

18. Write a number sentence using subtraction.

19. Write a number sentence using addition.

20. Write a problem using the number sentence $258 - 75 = 183$.

Miles Between Cities in New York		
From	**To**	**Miles**
Albany	Buffalo	258
Rochester	Ithaca	75
Buffalo	New York City	297
New York City	Syracuse	199

H.O.T. Problems

21. OPEN ENDED Complete the number sentence with two different numbers to make it true. $874 - \blacksquare = 444 - \blacksquare$

22. FIND THE ERROR Kaylee and Lu each wrote a number sentence. Who is correct? Explain your reasoning.

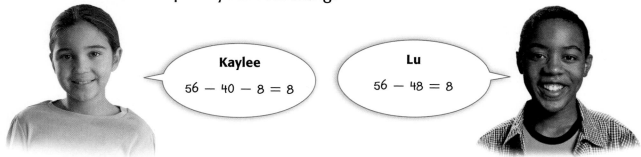

Kaylee
$56 - 40 - 8 = 8$

Lu
$56 - 48 = 8$

23. **WRITING IN ▶MATH** Write a problem that uses the number sentence $48 + \blacksquare = 55$. Solve.

An **expression** is a combination of numbers and operations that represents a mathematical quantity.

ACTIVITY **Model Addition Expressions**

MAIN IDEA

I will model addition and subtraction expressions using pictures, words, and numbers.

You Will Need
counters

① **Alice invited three friends to play in her backyard. Model this expression with pictures, numbers, and words.**

Step 1 Use Pictures.

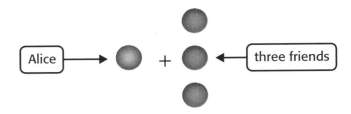

Step 2 Use Numbers.

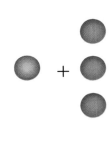

1 + 3

Step 3 Use Words.

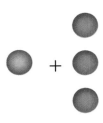

one plus three, or three more than one

ACTIVITY   Model Subtraction Expressions

2 There were seven cartons of milk. Josh drank one.
Model this expression with pictures, numbers, and words.

Step 1 Use pictures.

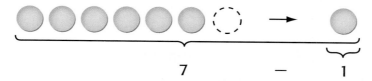

Seven cartons of milk. — Josh drank one.

Step 2 Use numbers.

7 — 1

Step 3 Use words.

seven minus one, or one less than seven

Think About It

1. In Activity 2, why is one of the counters separate from the others?

2. Which operation do the words *gained* and *bought more* show?

3. Which operation do the words *lost* and *went away* show?

CHECK What You Know

Model each expression with pictures, numbers and words.

4. Jim had 12 crayons. He lost 5 of them.

5. The soccer team scored 1 point. They gained 4 more points.

6. A carpenter had 6 nails. She went to the store to buy 8 more.

7. 10 sandwiches were served 6 sandwiches were eaten

8. **WRITING IN ►MATH** Write an expression using numbers. Model the expression using words and pictures.

8-2 Expressions and Number Sentences

▶ GET READY to Learn

There are 16 apples in a basket. Lenora buys 4 apples. The expression 16 − 4 tells how many apples were left.

total apples ➔ 16 − 4 ◀ apples sold

MAIN IDEA

I will write and simplify expressions.

New Vocabulary

expression

Math Online

macmillanmh.com
- Extra Examples
- Personal Tutor
- Self-Check Quiz

An **expression** is a combination of numbers and operations. A few examples of expressions are shown.

$$5 + 7 \qquad 3 + 2 + 5 \qquad 12 - 8$$

A number sentence is an expression using numbers and the equal sign (=), or the < or > sign. A few examples are shown.

$$5 + 7 = 12 \qquad 3 + 2 + 5 = 10 \qquad 12 - 8 = 4$$

Real-World EXAMPLE

① **APPLES** Use the information shown. Write an expression about the red and green apples and a number sentence to show how many apples there are in all.

Apples
Red........5
Yellow....3
Green......4

Use counters to model the expression.

red apples   green apples

$$5 \qquad + \qquad 4$$

So, the expression is 5 + 4. The number sentence is 5 + 4 + 3 = 12.

Write a Number Sentence

2 **Tell whether + or − makes the number sentence
4 ⬤ 3 = 7 true.**

Use counters to model.

4 ⬤ 3 = 7	4 ⬤ 3 = 7
4 − 3 = 7	4 + 3 = 7
1 = 7	7 = 7
false	true

4 + 3 = 7

So, the symbol + makes the number sentence
4 ⬤ 3 = 7 true.

CHECK What You Know

**Write an expression and a number sentence for each
problem. Use models if needed.** See Example 1 (p. 338)

1. Jin wrote 3 letters today and
2 letters yesterday. How many
letters did she write in all?

2. The animal shelter had 6 puppies.
They sold 3 of them. How many
puppies are left?

**Tell whether + or − makes each number sentence true.
Use models if needed.** See Example 2 (p. 339)

3. 9 ⬤ 2 = 11

4. 18 = 28 ⬤ 10

5. 14 ⬤ 7 = 10 + 11

6. 18 ⬤ 9 = 9

7. 18 ⬤ 20 = 38

8. 45 ⬤ 40 = 5

9. Talk About It   What is the difference between an expression and a
number sentence?

Write an expression and a number sentence for each problem. Use models if needed. See Example 1 (p. 338)

10. A basketball team won 11 games. A soccer team won 14 games. How many games were won in all?

11. Of the girls in a group, 14 have long hair and 9 have short hair. How many more have long hair?

12. Monisha scored 15 points Monday and 13 today. How many fewer points were made today?

13. Mick needs 4 yellow beads, 16 red, 2 white, and 14 green. How many beads are needed?

14. Cara caught 37 fish and threw 9 back. How many fish were left?

15. There are 143 goats and 291 cows. How many animals are there?

Tell whether + or − makes each number sentence true. Use models if needed. See Example 2 (p. 339)

16. $444 \bullet 6 = 460 - 10$

17. $74 \bullet 47 = 17 + 10$

18. $125 - 27 = 23 \bullet 75$

19. $345 - 126 > 217 \bullet 4$

20. $520 \bullet 317 < 400 + 150$

21. $715 - 617 < 25 \bullet 75$

Real-World PROBLEM SOLVING

Ice Cream Use the data to write a number sentence for each phrase.

22. difference of votes for the two most favorite flavors

23. sum of votes for vanilla and cookie dough flavors

24. difference of votes for vanilla and strawberry flavors

25. sum of all the votes

Favorite Ice Cream Flavors

Strawberry 51
Moose Tracks 65
Cookie Dough 133
Vanilla 88
Chocolate 97

Flavors / Votes
0 20 40 60 80 100 120 140

26. CHALLENGE Use the numbers 13, 16, and 29 to write two expressions. Then compare. Use <, >, or =.

27. WHICH ONE DOESN'T BELONG? Identify the example that is not an expression. Explain.

| 41 + 66 | 17 + 3 | 28 − 9 = 19 | 12 + 2 + 6 |

28. **WRITING IN ►MATH** Write a real-world problem that can be solved using a subtraction number sentence.

TEST Practice

29. Identify the answer to this number sentence. (Lesson 8-1)

$$352 - 199 = \blacksquare$$

A 1,153   **C** 153

B 157   **D** 147

30. Which sign makes the number sentence true? (Lesson 8-2)

$$79 \;\blacksquare\; 26 = 105$$

F +   **H** ×

G −   **J** ÷

Spiral Review

31. There are 20 drummers in the parade. 11 of them are men. How many are women? Model the problem using a number sentence. (Lesson 8-1)

32. Measurement In a race, Josh ran 206 feet and Dwayne ran 181 feet. How much farther did Josh run than Dwayne? (Lesson 7-7)

33. Use the picture shown. Donna has $10. She bought a pencil and a book. She wants to buy one more thing. What is reasonable? (Lesson 3-4)

Round to the nearest hundred. (Lesson 1-8)

34. 729

35. 750

36. 542

37. 903

Problem-Solving Strategy

MAIN IDEA I will solve a problem by acting it out.

Five students are throwing paper balls into a basket. They line up in order from tallest to shortest. Adriana is taller than Bryce, but she is shorter than Jenelle. Delmar is shorter than Evan, but taller than Jenelle. In what order are they lined up?

Understand	**What facts do you know?** • Adriana is taller than Bryce. • Adriana is shorter than Jenelle. • Delmar is shorter than Evan. • Delmar is taller than Jenelle. • The students line up from tallest to shortest. **What do you need to find?** • The order in which they line up.
Plan	Act out the problem using 5 students.
Solve	Use the facts in the problem to arrange the students. Work with the facts that make sense first.

	tallest				shortest
Adriana is taller than Bryce.	A	B			
Adriana is shorter than Jenelle.	J	A	B		
Delmar is taller than Jenelle.	D	J	A	B	
Delmar is shorter than Evan.	E	D	J	A	B

So, the order is Evan, Delmar, Jenelle, Adrianna, and Bryce.

Check	Look back. The answer makes sense for the facts given in the problem.

Refer to the problem on the previous page.

1. Would the results have been the same if any of the facts were missing? Explain.

2. Explain why this strategy is a good choice for this kind of problem.

3. Could you use this strategy if 5 students are not available to act it out? Explain.

4. Could you use another strategy to solve this problem? Explain.

PRACTICE the Strategy

EXTRA PRACTICE
See page R22.

Solve. Use the *act it out* strategy.

5. Skylar has 8 baseball cards now. Her sister gave her 4 cards and took 2 from her. How many baseball cards did Skylar have to begin with?

6. **Measurement** The length of a football field is 100 yards. Alonso ran 20 yards in one play and another 40 yards for the next play. Suppose he started at the 5 yard line. How many more yards does he need to run to make a touchdown?

7. An empty bus picks up 5 people at the first stop. At the second stop, 4 people get on and 2 people get off. At the third stop, 5 people get on. At the last stop 1 person gets on and 4 people get off. How many passengers are now on the bus?

8. **Geometry** One face of a geometry cube is shown. All six faces look the same. How many squares of the cube are blue?

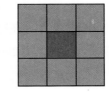

9. Kamal and Alisa go to a park to play basketball, feed the ducks, and have a picnic. In how many different orders can they do these activities at the park?

10. There was a total of 10 fish in three different tanks. Four of the fish were sold by noon. Each tank had a different number of fish. How many fish were left in each tank?

11. **WRITING IN ▶MATH** Look back at Exercise 10. Explain how you solved the problem by using the *act it out* strategy.

Make a Table to Find a Rule

 **GET READY to Learn**

Katie is building 5 separate triangles with straws. The first triangle used 3 straws, and the second triangle used another 3 straws. How many straws does she need to make 5 triangles?

The number of straws Katie uses follows a pattern. The pattern is called a rule. A **rule** tells you what to do to the first number (input) to get the second (output).

Real-World EXAMPLE Find and Extend a Rule

1 **GEOMETRY** Find the total number of straws Katie needs to make 5 triangles.

Make a table to find and extend the rule.

Step 1 Find the rule.
You know that 1 triangle = 3 straws.

$1 \times 3 = 3$

2 triangles = 6 straws

$2 \times 3 = 6$

3 triangles = 9 straws

$3 \times 3 = 9$

The rule is to multiply the number of triangles by 3.

Rule: Multiply by 3.	
Number of Triangles	**Number of Straws**
1	3
2	6
3	9
4	■
5	■

Step 2 Extend the rule.
4 triangles = 4 × 3 or 12 straws

5 triangles = 5 × 3 or 15 straws

So, Katie will need 15 straws to make 5 triangles.

Remember

You can make a table to help you see a pattern.

2 **PLANTS** Mitch found one 4-leaf clover. He then found another. Now, he has 2 clovers and there are a total of 8 leaves. How many leaves will there be if he finds 5 clovers that have 4 leaves each?

Step 1 Find the rule.

Rule: Multiply by 4.	
Clover	**Leaves**
1	4
2	8
3	▣
4	▣
5	▣

Step 2 Extend the rule.

$3 \times 4 = 12$
$4 \times 4 = 16$
$5 \times 4 = 20$

So, there are 20 leaves on 5 clovers.

✓ CHECK What You Know

Find and extend the rule for the table. Then copy and complete.

See Examples 1 and 2 (pp. 344–345)

1. The table shows how many shoes are needed for different numbers of people.

Rule: ▣				
Input	1	2	3	4
Output	2	4	6	▣

2. There are 2 books about Italy, 4 books about Japan, 6 books about China, and 8 books about Russia. If the pattern continues and the next set of books is about England, how many books will there be about England? Make a table to solve the problem.

3. **Talk About It** Explain how multiplication can be used to help you extend a pattern.

Find and extend the rule for each table. Then copy and complete.

See Examples 1 and 2 (pp. 344–345)

4. The table shows the number of sails needed for several boats. Each boat has the same number of sails.

Rule: ■	
Boats	Sails
7	63
■	36
3	■
■	18

5. The table shows how much movie tickets cost for different numbers of people.

Rule: ■				
Input	6	4	9	■
Output	30	20	■	35

For Exercises 6 and 7, make a table to find a rule. Then extend the rule to solve. See Examples 1 and 2 (pp. 344–345)

6. The amusement park sold ride tickets in packs of 5, 10, 15, and 20 tickets. What would a pack of 5 tickets cost if 20 tickets cost $4?

7. Mrs. Glenn planted 5 flowers in the front row of her garden. The second row had 10 flowers and the third row had 15. How many flowers will be in the 5th row?

Data File

The folk dance of Illinois is the square dance. The table shows how many dancers complete each square.

8. Find and extend the rule for the table.

Rule: ■				
Input	8	24	32	48
Output	1	■	■	■

H.O.T. Problems

9. **CHALLENGE** Create a table that uses a multiplication rule. Write input and output pairs.

10. **WHICH ONE DOESN'T BELONG?** Identify the number pair that would not be found in a table with a rule of $\times$ 6. Explain.

| 5 and 30 | 8 and 24 | 10 and 60 | 7 and 42 |

11. **WRITING IN MATH** Explain how to find a rule when given a pattern.

TEST Practice

12. The table shows the number of crayons needed. (Lesson 8-4)

Crayons Needed	
Students	**Crayons**
3	15
4	20
6	30

Each student gets the same number of crayons. How many are needed for 8 students?

A 20   C 35

B 30   D 40

13. If $3 \times 7 \times 8 = 168$, then what is $7 \times 8 \times 3$? (Lesson 5-9)

F 75   H 158

G 97   J 168

14. One pencil costs $2. Two pencils cost $4. Three pencils cost $6. How much will 4 pencils cost? (Lesson 8-4)

A $7   C $10

B $8   D $12

Spiral Review

15. The barber had appointments at 1:00, 2:00, 3:00, and 4:00. Sam could not arrive until after 2:30. James went right after Roger. Bryan did not go first or last. What order did each person have their appointment? Use the *act it out* strategy to solve. (Lesson 8-3)

Tell whether + or − makes each number sentence true. (Lesson 8-2)

16. 14 ● 8 = 22   17. 36 ● 6 = 30   18. 28 ● 5 = 23

Make Function Tables (+, −)

GET READY to Learn

The table shows the amount of money four children each have saved. If each child receives $5 to add to their savings, how much money will each child have?

Savings Accounts	
Name	**Amount ($)**
Lorena	25
Nina	23
Shelly	22
Trey	21

MAIN IDEA

I will use addition and subtraction to complete function tables.

New Vocabulary

function

Math Online

macmillanmh.com
• Extra Examples
• Personal Tutor
• Self-Check Quiz

The amount each child will have depends on the amount they each will receive. A relationship where one quantity depends upon another quantity is a **function**.

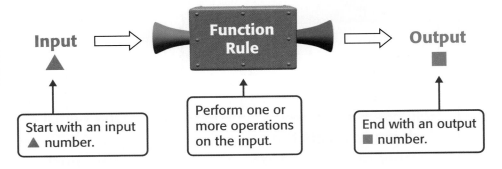

Input ⟹ **Function Rule** ⟹ **Output**
▲ ■

Start with an input ▲ number.

Perform one or more operations on the input.

End with an output ■ number.

A symbol such as ▲ or ■ represents an unknown number. The input symbol ▲, output symbol ■, and function rule can be shown in a table.

Real-World EXAMPLE Make a Function Table

1 MONEY Make a function table to find how much money each child will have in savings after receiving $5.

Rule: Add $5.		
Input (△)	△ + $5	Output (□)
$25	$25 + $5	$30
$23	$23 + $5	$28
$22	$22 + $5	$27
$21	$21 + $5	$26

Real-World EXAMPLE Complete a Function Table (+)

2 **AGE** Eric is 3 years older than his brother. Find Eric's age when his brother is 2, 3, 4, and 5 years old. Use the rule to extend the function table.

Rule: △ + 3	
Input △	Output □
2	▦
3	▦
4	▦
5	▦

The rule is brother's age + 3 or add 3.

Start with each input (△) number. Apply the rule to find the output (□) number.

Rule: △ + 3		
Input (△)	△ + 3	Output (□)
2	2 + 3	5
3	3 + 3	6
4	4 + 3	7
5	5 + 3	8

Remember

To check your answer, reverse the operation of your function and see if it works out.

$$5 - 3 = 2$$
$$6 - 3 = 3$$
$$7 - 3 = 4$$
$$8 - 3 = 5$$

You can use subtraction to complete a function table.

Real-World EXAMPLE Complete a Function Table (−)

3 **CHAIRS** Each third grade class always has 2 extra chairs in the room. Find how many students there are based on the number of chairs. Use the rule to extend the function table.

Rule: △ − 2	
Input (△)	Output (□)
20	▦
21	▦
22	▦
23	▦

The rule is △ − 2 or subtract 2.

Start with each input (△) number. Apply the rule to find the output (□) number.

Rule: △ − 2		
Input (△)	△ − 2	Output (□)
20	20 − 2	18
21	21 − 2	19
22	22 − 2	20
23	23 − 2	21

Copy the function table and extend the pattern. See Examples 1–3 (pp. 348–349)

1. Kia is 5 years older than her sister. Use the function table to find Kia's age when her sister is 1, 2, 3, or 4 years old.

Rule: △ + 5	
Input (△)	Output (□)
1	■
2	■
3	■
4	■

For Exercises 2 and 3, use the following information.

Alicia is four years older than her pet turtle.

2. Make a function table to find how old her pet turtle will be when she is 13, 14, 15, or 16 years old.

3. Write the function rule.

4. **Talk About It** How do function tables show patterns?

Practice and Problem Solving

EXTRA PRACTICE
See page R23.

Copy each function table and extend the pattern. See Examples 1–3 (pp. 348–349)

5. Jaime always rides his bike 6 miles farther than Sean. Use the function table to find how many miles Jaime rides when Sean rides 1, 3, 5, or 7 miles.

Rule: △ + 6	
Input (△)	Output (□)
1	■
3	■
5	■
7	■

6.

Rule: △ − 9	
Input (△)	Output (□)
17	■
18	■
19	■
20	■

7.

Rule: △ − 4	
Input (△)	Output (□)
15	■
12	■
9	■
6	■

Find the rule for the function table.

8. A book has 44 pages. Ely read the same number of pages each day until he finished. The table shows how many pages were left to read before and after he read each day.

Rule: ■	
Input (△)	Output (□)
44	33
33	22
22	11
11	0

Make a function table for each situation. Write the function rule.

9. Pasqual and his friends will each get $7 for allowance. How much money will each child have if they already have $1, $2, $3, or $4?

10. A store orders 3 more boxes of strawberries than oranges. How many boxes of oranges will the store order if they order 8, 9, 10, or 11 boxes of strawberries?

11. Caroline is reading a book that has 122 pages. If she reads 25 pages every day, how many pages will she have left to read after 1, 2, 3, or 4 days?

12. Every week, Mr. Montoya pays $3 to send a package. He began with $30. How much money will he have after 4 weeks?

H.O.T. Problems

13. OPEN ENDED Make a function table for the rule add 5.

14. FIND THE ERROR Dai and Lonzo are making a function table for $\square = \triangle + 9$. Who is correct? Explain your reasoning.

Dai

△	7	5	6
□	16	13	15

Lonzo

△	8	10	15
□	17	19	24

15. **WRITING IN ►MATH** Write a real-world problem that would result in the function table to the right. What is the function rule?

Rule: ■			
Input (△)	250	251	252
Output (□)	260	261	262

Rolling Digits

Expressions and Number Sentences

Get Ready!
Players: 2 players

Get Set!
• Find a partner.

Go!
• Player 1 rolls the number cube twice to create a 2-digit number. The first roll is the tens digit. The second roll is the ones digit.

• Player 2 does the same.

• Player 1 writes an expression with the numbers from both players' rolls.

• Player 2 writes the number sentence. If the answer is correct, Player 2 gets a point.

• Switch and play again.

• The player who has more points after 10 rounds wins.

You will need: paper, pencil and number cube

Player 1		Player 2	

Expression: 24 + 35
Number Sentence:
24 + 35 = 59

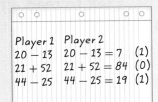

Player 1   Player 2
20 − 13   20 − 13 = 7   (1)
21 + 52   21 + 52 = 84 (0)
44 − 25   44 − 25 = 19 (1)

Model each problem. Then write a number sentence. (Lesson 8-1)

1. Katrina took 47 pictures yesterday and 32 pictures today. How many pictures did she take in all?

2. Anthony had 97 baseball cards. He gave 31 to Derek. How many cards does Anthony have left?

Model each number sentence. Use pictures and words. (Lesson 8-1)

3. $32 - 14 = $ ■   4. $60 - 26 = $ ■

Tell whether + or − makes each number sentence true. (Lesson 8-2)

5. $538 + 112 = 569$ ● 81

6. $824 - 719 > 261$ ● 158

7. **MULTIPLE CHOICE** The table below shows how much water to use when making different amounts of rice. How much water is need if 4 cups of rice are used? (Lesson 8-2)

Rice	2	4	6	8
Water	4	■	12	16

A 2   C 6
B 4   D 8

8. Marcel blew up 12 balloons. Seven of them did not pop. Of the balloons that popped, 1 is red. The rest are blue. How many are blue? (Lesson 8-3)

9. **Algebra** Find and extend the rule for the table. Then copy and complete. (Lesson 8-4)

Rule: ■				
Input	3	6	■	25
Output	7	10	16	■

Make a table to find a rule. Then extend the rule to solve. (Lesson 8-4)

10. Ana bought movie tickets for 20 of her friends. Five tickets cost $10. How much money did she spend?

11. Copy and complete the function table. (Lesson 8-5)

Rule: $\triangle + 5$	
Input ($\triangle$)	Output ($\square$)
4	■
6	■
8	■
10	■

12. **MULTIPLE CHOICE** Erasers are sold in packages of 3. Which number below could not be a total of erasers bought? (Lesson 8-5)

F 6   H 13
G 9   J 15

13. **WRITING IN ▶MATH** How do you find the rule in a table? Explain. (Lesson 8-5)

Problem-Solving Investigation

MAIN IDEA I will choose the best strategy to solve a problem.

P.S.I. TEAM ✚

JANE: I planted 30 tomato seeds in my garden. Three out of every 5 seeds grew into tomato plants.

YOUR MISSION: Find how many seeds grew into tomato plants.

Understand	You know 30 seeds were planted and that 3 out of every 5 seeds grew into tomato plants. Find out how many seeds grew into tomato plants.
Plan	You can *draw a picture* to help solve the problem. Use tallies to represent the seeds.
Solve	Put the tallies in groups of 5 until there are 30 tallies. ‖‖‖ ‖‖‖ ‖‖‖ ‖‖‖ ‖‖‖ ‖‖‖ Three of each group became plants. ‖‖‖ ‖‖‖ ‖‖‖ ‖‖‖ ‖‖‖ ‖‖‖  3 + 3 + 3 + 3 + 3 + 3 So, $6 \times 3 = 18$ seeds grew into tomato plants.
Check	Look back. You can add to check. $3 + 3 + 3 + 3 + 3 + 3 = 18$ So, the answer is correct. ✔

Use any strategy shown below to solve. Tell what strategy you used.

PROBLEM-SOLVING STRATEGIES
- Guess and check.
- Work a simpler problem.
- Make an organized list.
- Draw a picture.
- Act it out.

1. Darin has 17 apples in a basket. He wants to share them with 3 of his friends. How many will each friend get if each one has the same amount? How many will be left?

2. Two boys and a girl share $80. The girl gets twice as much money as each of the boys. How much money do each of the children get?

3. How many more sticks of butter need to be added to the right side to balance the scale?

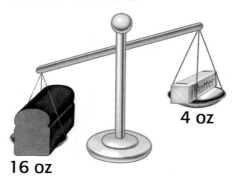

16 oz   4 oz

4. Josie has one folder of each color red, green, and blue. How many different ways can she order them?

5. Sei played marbles with her friends and lost some. She now has 25 marbles. She lost 3 to Hakeem, 6 to Mavis, 1 to Mikayla and 4 to Ramous. How many marbles did she have to start with?

6. Suni earned an allowance of $2. She also takes care of her baby sister for $1. She spent $1 on a snack and drink. How much money does she have left?

7. Measurement Jocelyn wants to install a fence around a garden. How many feet of fence will be needed? Use 3 ft = 1 yd.

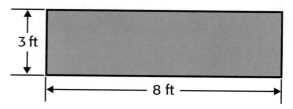

3 ft

8 ft

8. Measurement Mr. Glover walks 200 meters to a store. He walks 30 meters while in the store and then walks back home. How many meters does he walk in all?

9. **WRITING IN ▸MATH** When you add 8 to a number, subtract 10 from the sum, and double the difference, you get 44. What is the number? Explain.

8-7 Make Function Tables ($\times$, $\div$)

GET READY to Learn

MAIN IDEA

I will use multiplication and division to complete function tables.

Math Online

macmillanmh.com

- Extra Examples
- Personal Tutor
- Self-Check Quiz

Christine's neighbor owns a farm. On the farm, one of the pigs is 5 feet long. Christine made a table to convert 5 feet to inches. What pattern do you see in the input and output numbers?

Change Feet to Inches

Feet Input ($\triangle$)	Inches Output ($\square$)
1	12
2	24
3	36
4	48
5	■

Function rules can also involve multiplication or division.

Real-World EXAMPLE   Make a Function Table

1 MEASUREMENT Make a function table to find the length of the pig if it is 5 feet long.

There are 12 inches in one foot. To find the output values ($\square$), multiply each input value ($\triangle$) by 12.

Change Feet to Inches

Input ($\triangle$)	Rule: $\triangle \times 12$	Output ($\square$)
1	1×12	12
2	2×12	24
3	3×12	36
4	4×12	48
5	5×12	60

There are 60 inches in 5 feet.
So, the length of the pig in inches is 60 inches.

You can identify or describe the rule, or pattern, in a function table.

Real-World EXAMPLE

2 **QUARTERS** The table shows how many quarters (□) are in different numbers of dollar bills (△). Use the function table to identify the rule.

Start with each input (△) number. Identify the rule that gives the output (□) number.

So, the rule is △ × 4.

Identify the Rule In a Function Table (×)

Rule: ■	
Input (△)	Output (□)
1	4
2	8
3	12
4	16

Rule: △ × 4		
Input (△)	△ × 4	Output (□)
1	1 × 4	4
2	2 × 4	8
3	3 × 4	12
4	4 × 4	16

Real-World EXAMPLE

3 **TRICYCLES** The table shows how many tricycles (□) can be made with different numbers of wheels (△). Describe the rule.

Start with each input (△) number. Apply the rule to find the output (□) number.

The pattern shows that as △ decreases by 3, □ decreases by 1.

Describe the Rule In a Function Table (÷)

Rule: △ ÷ 3	
Input (△)	Output (□)
27	■
24	■
21	■
18	■

Rule: △ ÷ 3		
Input (△)	△ ÷ 3	Output (□)
27	27 ÷ 3	9
24	24 ÷ 3	8
21	21 ÷ 3	7
18	18 ÷ 3	6

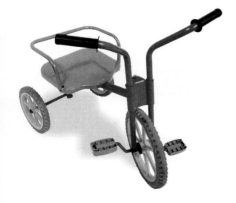

Lesson 8-7 Make Function Tables (×, ÷) **357**

Copy the function table and extend the pattern. See Example 1 (p. 356)

1. The table shows how many pairs (□) of socks can be matched up after different numbers of socks (△) are taken out of the clothes dryer.

Rule: △ ÷ 2	
Input (△)	Output (□)
8	■
10	■
12	■
14	■

2. A butterfly has 2 wings. Make a function table to show the total number of wings for 4, 5, 6, or 7 butterflies. Then write the rule and describe the pattern.

See Examples 2–3 (p. 357)

3. (Talk About It) Can you look only at the input numbers in a function table to determine the function rule? Explain.

Practice and Problem Solving

EXTRA PRACTICE
See page R23.

Copy each function table and extend the pattern. See Example 1 (p. 356)

4. Each ladybug that Karley sees has 6 spots on it. Use the table to find the total number of spots (□) for different numbers of ladybugs (△).

Rule: △ × 6	
Input (△)	Output (□)
5	■
6	■
7	■
8	■

5. Each week the total number of snacks (△) are divided evenly amoung 9 students in the travel club. Use the table to find how many snacks each student gets when different numbers of snacks are served.

Rule: △ ÷ 9	
Input (△)	Output (□)
18	■
27	■
36	■
45	■

Make a function table for each situation. Write the rule. See Example 2 (p. 357)

6. The price for admission to a zoo is $7. How many tickets can you buy for $63, $56, $49, or $42?

7. Each box holds 12 water bottles. How many boxes are there if there are 60, 48, 36, or 24 bottles?

8. Rama bought 6 bags of chips. He spent $12. How many bags of chips would he have bought if he spent $14, $16, $18, or $20?

9. Dorian and her friends went to the movies. Each ticket cost $5. How much would they spend if there were 2, 3, 4, or 5 friends?

Describe the pattern of each function table. See Example 3 (p. 357)

10.

Rule: $\triangle \div 3$	
Input ($\triangle$)	Output ($\square$)
27	9
21	7
15	5
9	3

11.

Rule: $\triangle \div 6$	
Input ($\triangle$)	Output ($\square$)
72	12
54	9
36	6
18	3

12.

Rule: $\triangle \times 2$	
Input ($\triangle$)	Output ($\square$)
12	24
13	26
14	28
15	30

13.

Rule: $\triangle \times 4$	
Input ($\triangle$)	Output ($\square$)
6	24
7	28
8	32
9	36

H.O.T. Problems

14. OPEN ENDED Name two pairs of input and output values for the function rule $2 \times \triangle = \square$.

15. CHALLENGE Look at the function table shown. What is the function rule?

Input ($\triangle$)	15	25	40	50
Output ($\square$)	4	6	9	11

16. NUMBER SENSE In the function rule $\triangle + 3$, the output value is 8. How can you determine the value of $\triangle$?

17. WRITING IN ▶MATH Write a real-world math problem where using a function table for multiplication or division will help you solve the problem.

Problem Solving in Science

A Visit to the Supermarket

A supermarket is a busy place. Some people shop at a supermarket every day of the week. Today, some supermarkets are as big as 4 football fields!

Supermarkets sell many kinds of food, including healthy foods. Some of the healthiest foods you can buy are yogurt, broccoli, citrus fruits, nuts, oatmeal, and orange juice.

These foods come in different containers and sizes. The prices of some foods are based on size. For example, 1 kilogram of pears costs around $2. The price of a liter of soda is around $3.

Did You Know?

Tomatoes are the most widely grown fruit in the world.

Aisle 3
Healthy Goods

Old Fashioned Oats

Tomato Sauce

Water 500 ml

Orange Juice 1 liter

Yogurt

Spaghetti 250 grams

 Real-World Health

Use the items on the shelf above to solve each problem.

1 Suppose a customer buys 1 item from the shelf. The next customer buys 2 items. The next buys 3 items. If this pattern continues, how many items would be bought in all after the fifth customer?

2 Write a number sentence to show how many items are on the bottom shelf.

3 A function table shows that 4 more containers of yogurt are sold than apples. What is the rule?

4 Write a number sentence to show how many more apples are shown than boxes of spaghetti.

5 A store receives twice as many tomato sauce cans as orange juice bottles. Write a number sentence to show how many juice bottles are received if 12 cans of tomato sauce are received.

6 A bottle of water costs $1 and a package of 6 bottles costs $5. How much money is saved if you buy 4 packages instead of 24 bottles?

FOLDABLES
Study Organizer

GET READY to Study

Be sure the following Key Vocabulary words and Key Concepts are written in your Foldable.

Number Sentences | Expressions

Find a Rule | Function Tables

Key Concepts

- A **number sentence** is an **expression** using numbers and the equal sign (=) or the > or < sign. The number sentence 20 − 10 = 10 is modeled below. (p. 333)

Models

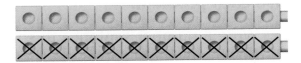

Words Twenty cubes minus ten cubes equals ten cubes

Number Sentence 20 − 10 = 10

- A **function** table uses a **rule** to show the relationship between an input and output value. (p. 344)

Rule: △ + 2	
Input (△)	Output (□)
1	3
2	4
3	■

Key Vocabulary

number sentence (p. 333)

expression (p. 338)

function (p. 348)

rule (p. 344)

Vocabulary Check

Decide which vocabulary word best completes each sentence.

1. Omar wrote 3 + 53 + 26 = 82 on the chalkboard. This is an example of a(n) _____?_____.

2. An example of a(n) _____?_____ is 7 + 4.

3. In a function, the _____?_____ tells you what to do to the input number to get the output number.

4. A(n) _____?_____ is a relationship between input and output values. An example would be △ + 5 = ■.

Lesson-by-Lesson Review

8-1 Model and Write Number Sentences (pp. 333–335)

Example 1
There are 2 ladybugs on the window sill. On a flower there are 3 butterflies. How many insects are there in all?

Pictures

Words Two insects plus three insects equals five insects.

Number Sentence $2 + 3 = 5$

Model each problem. Then write a number sentence.

5. TJ picked 67 raspberries. He gave his sister 29 of them. How many raspberries does TJ have left?

6. Rashid hit 27 baseballs at practice yesterday. Today he hit 41. How many baseballs did he hit in all?

Model each number sentence. Use pictures and words.

7. $4 + 16 = \blacksquare$

8. $30 - \blacksquare = 12$

8-2 Expressions and Number Sentences (pp. 338–341)

Example 2
Write an expression and number sentence for the total number of points scored.

Lions 21
Panthers 17

The expression is $21 + 17$. The number sentence is $21 + 17 = 38$.

Write an expression and a number sentence for the problem.

9. Julian's basketball team won 10 games and his brother's team won 12 games. How many games were won in all?

Tell whether + or − makes each number sentence true.

10. $65 \bullet 13 > 599 - 534$

11. $147 + 32 = 106 \bullet 73$

8-3 **Problem-Solving Strategy:** Act it Out (pp. 342–343)

Example 3
Alvar, Brian, Charlotte, Diana, and Ethan line up in the order of their ages. Diana was born after Alvar but before Brian. Charlotte is the oldest. Ethan is younger than Brian. In what order are they lined up?

Act out the problem using 5 students.

D born after A	A	D		
C born first (oldest)	C	A	D	
D born before E	C	A	D	E
E born after Brian (younger)	C	A	D	B	E

So, the order is Charlotte, Alvar, Diana, Brian, and Ethan.

Solve. Use the *act it out* strategy.

12. In the first quarter of the football game, the Eagles scored 7 points. In the second quarter, 14 more points were scored. No points were scored in the third quarter. They finished the game by scoring 3 more points. How many points in all?

13. Marlin had a party. Twelve people came on time. Three left. Six more came later. Two people left and everyone else spent the night. How many people spent the night?

8-4 **Make a Table to Find a Rule** (pp. 344–347)

Example 4
Alek is making cubes. Find how many sides he needs to make 4 cubes.

Rule: ■	
Number of Cubes	**Number of Sides**
1	6
2	12
3	18
4	■

The rule is multiply by 6. So, Alek needs 24 sides.

14. Find and extend the rule for the table. Then copy and complete.

Rule: ■	
Input	**Output**
3	9
5	15
■	21
9	■

Make Function Tables (+, −) (pp. 348–351)

Example 5
Use the rule to complete the function table.

Rule: △ − 3	
Input (△)	Output (□)
16	▩
15	▩
14	▩
13	▩

Start with each input (△) number. Apply the rule to find the output (□) number.

Rule: △ − 3		
Input (△)	△ − 3	Output (□)
16	16 − 3	13
15	15 − 3	12
14	14 − 3	11
13	13 − 3	10

15. Copy and complete.

Rule: △ + 5	
Input (△)	Output (□)
5	▩
7	▩
9	▩
11	▩

Make a function table for each situation. Write the function rule.

16. Miguel is 10 years old. His uncle is 14 years older than him. How old will Miguel be when his uncle is 28, 30, 32, or 34 years old?

17. For every bead project, Zita will use 5 less red than blue beads. How many blue beads will she use if she uses 10, 20, 30, or 40 red beads?

Problem-Solving Investigation: Choose a Strategy (pp. 354–355)

Example 6
Connie gave 6 baseball cards to Ellie, 4 to Emmanuel, and 7 to Augustine. She has 7 left. How many cards did she have to start?
Choose an operation.

$6 + 4 + 7 + 7 = 24$

So, Connie had 24 cards to start.

Solve.

18. A girl walks 20 meters north, 30 meters east, 10 meters south, and 20 meters west. Is she where she started? Explain.

8-7 **Make Function Tables (×, ÷)** (pp. 336–339)

Example 7

There are 3 sticks of clay in 1 package. How many sticks of clay would there be in 3, 5, and 7 packages?

Use the rule to complete the function table.

Rule: △ × 3	
Input (△)	Output (□)
1	▩
3	▩
5	▩
7	▩

Start with each input (△) number. Apply the rule to find the output (□) number.

Rule: △ × 3		
Input (△)	△ × 3	Output (□)
1	1 × 3	3
3	3 × 3	9
5	5 × 3	15
7	7 × 3	21

So, there are 9 sticks of clay in 3 packages, 15 sticks in 5 packages, and 21 sticks in 7 packages of clay.

19. Copy and complete the function table.

Rule: △ ÷ 5	
Input (△)	Output (□)
25	▩
20	▩
15	▩
10	▩

Make a function table for each situation. Write the function rule.

20. There are 6 eggs in my recipe for a cake. If I used 48, 42, 36, or 30 eggs, how many cakes would I have made?

21. Erin blows up 2 balloons every minute. How many balloons will she have after 18, 20, 22, and 24 minutes?

22. Complete the function table.

Rule: △ ÷ 4	
Input (△)	Output (□)
36	▩
32	▩
28	▩
24	▩

Chapter Test

Lessons 8-1 through 8-7

Decide whether each statement is true or false.

1. $4 + 8 + 9 = 21$ is an expression.

2. Solve for the output by doing the opposite of the rule listed.

Model the problem. Then write a number sentence.

3. Alexa picked 20 daffodils and 16 daisies for a bouquet. How many flowers did she pick in all?

Tell whether + or − makes the number sentence true.

4. $36 + 114 = 156 \bullet 6$

5. **Algebra** Copy and complete.

Rule: ■				
Input	7	■	11	13
Output	16	18	20	■

6. **MULTIPLE CHOICE** While he was in the hospital for 3 days, Emil received 23 get well cards. He received 12 on the first day and 6 on the second. Which number sentence can be used to find the number of cards he received on the third day?

   A $23 - 12 - 6 = \blacksquare$

   B $23 \times 12 \div 3 = \blacksquare$

   C $23 - 6 + 3 = \blacksquare$

   D $23 + 6 + 12 = \blacksquare$

7. **Measurement** Allie's family is putting a fence around their swimming pool. How many yards of fence will they need?

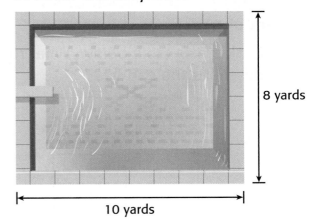

8 yards

10 yards

Make a function table for the situation. Write the function rule.

8. Alejandro runs 3 miles in 21 minutes. If he continues to run at the same speed, how many minutes will it take him to run 6, 9, or 12 miles?

9. **MULTIPLE CHOICE** Ray gives each of his 2 dogs 3 biscuits in each day. If he counted the biscuits in groups of 6, which list shows numbers Ray could have named?

   F 12, 18, 24   H 6, 12, 18, 28

   G 6, 12, 16, 26   J 12, 24, 46

10. **WRITING IN ► MATH** Can $\triangle = 2$ and $\triangle = 5$ be in the same problem? Why or why not?

Read each question. Then fill in the correct answer on the answer sheet provided by your teacher or on a sheet of paper.

1. Look at the table below. What rule describes the pattern?

Rule: ■	
Input (△)	Output (□)
5	1
10	6
15	11
20	16

   A Add 3.   **C** Add 4.

   B Subtract 3.   **D** Subtract 4.

2. Each student is given 5 crayons. The crayons are sold in boxes of 12. How many boxes are needed for 60 crayons in all?

   F 4   **H** 6

   G 5   **J** 7

3. Gail has 22 marbles, Nikki has 29 marbles, and Ann has 34 marbles. How can you find the total number of marbles?

   A 34 + 29   **C** 22 + 29 − 34

   B 22 + 29 + 34   **D** 22 − 29 − 34

4. Martina writes these five numbers on a chalkboard. Which rule describes the numbers?

   15,   12,   9,   6,   3

   F Add 3.

   G Subtract 3.

   H Add 2.

   J Subtract 2.

5. Estimate 532 + 493 to the nearest thousand.

   A 900   **C** 1,000

   B 925   **D** 1,030

6. Coach Jarvis divided 25 third graders into 5 equal teams. Which expression describes the number of students on each team?

   F 25 + 5   **H** 25 × 5

   G 25 − 5   **J** 25 ÷ 5

7. Dominic has 21 stamps. He puts the stamps in 3 piles. How many stamps are in each pile?

   A 4   **C** 6

   B 5   **D** 7

8. Which number sentence below is in the same family as 20 ÷ 5 = 4?

F 20 ÷ 4 = 5   **H** 4 ÷ 20 = 5

G 20 ÷ 2 = 10   **J** 20 × 1 = 20

---

9. Jimmy is buying bottled water. Based on the table, how many bottles come in one case?

Bottles of Water	
Number of Cases	**Number of Bottles**
2	20
4	40
6	60
8	80

A 10   **C** 15

B 20   **D** 25

---

10. Gina babysat 8 hours each week for 5 weeks. How many hours did she babysit in all?

F 13 hours   **H** 32 hours

G 20 hours   **J** 40 hours

PART 2   Short Response

Record your answer on the answer sheet provided by your teacher or on a sheet of paper.

11. There are 5 ladybugs with the same number of spots. There are 40 spots total. How many spots are on each bug? Tell what strategy you used.

---

12. Write a number sentence that has a product of 24.

PART 3   Extended Response

Record your answer on the answer sheet provided by your teacher or on a sheet of paper.

13. Explain how a number sentence and expression are different. Give an example of each.

---

14. Horatio is buying ping-pong balls for a party. Each package contains 4 balls. How many balls will he have if he buys 7, 8, 9, or 10 packages? Make a function table for the situation. Write the rule.

NEED EXTRA HELP?														
If You Missed Question...	1	2	3	4	5	6	7	8	9	10	11	12	13	14
Go to Lesson...	8-5	7-6	8-2	8-5	1-9	8-2	7-1	6-2	8-4	4-5	8-6	8-2	8-2	8-4

CHAPTER 9

Measure Length, Area, and Temperature

BIG Idea **What are customary units of measurement?**

Customary units for **length** are inch, foot, yard, and mile.

Example Monarch butterflies can be seen in great numbers in New Jersey during their migration south. The table lists the wingspan of a Monarch butterfly and two others in inches.

Wingspan of Butterflies	
Butterfly	**Wingspan (in.)**
Zebra Longwing	2
Monarch	4
Painted Lady	3

What will I learn in this chapter?

- Choose appropriate measurement tools and units.
- Estimate and measure length.
- Solve problems by working backward.

Key Vocabulary

length	perimeter
inch (in.)	area
meter	thermometer

Math Online > **Student Study Tools** at macmillanmh.com

FOLDABLES®
Study Organizer

Make this foldable to help you organize information about measurement. Begin with one sheet of 11″ × 17″ paper.

① Fold paper vertically in 5 equal sections. Unfold.

② Fold paper horizontally in 5 equal sections. Unfold.

③ Label as shown.

Measurement	Tool	Unit	Estimate	Measure
Length				
Perimeter				
Area				
Temperature				

You have two ways to check prerequisite skills for this chapter.

Option 2

Math Online ▷ Take the Chapter Readiness Quiz at macmillanmh.com.

Option 1

Complete the Quick Check below.

QUICK Check

Identify which figure or object is longer.

(Prior grade)

1.

Figure A

Figure B

2.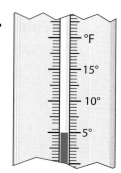

3. Sancho walked 15 miles, and Alberto walked 15 yards. Who walked farther? Explain.

Find the area of each figure. (Prior grade)

4.

5.

6. Can two figures have a different shape, but have the same area? Explain.

Write the temperature. (Prior grade)

7.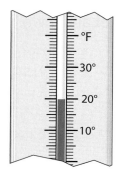

8.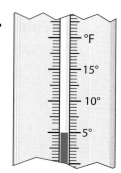

9. Which temperature is warmer, 60°F or 35°F?

Length to the Nearest Inch

Length is the measure of the distance between two points. You can estimate and measure length with a nonstandard unit.

ACTIVITY

1 **Estimate and measure length using a paper clip.**

Step 1 **Estimate**

Estimate the length of a pencil in paper clips.

Step 2 **Measure**

Arrange paper clips end to end as shown. Count the paper clips.

How close was your estimate to the actual length in paper clips?

An **inch (in.)** is a standard unit of measure in the customary measurement system.

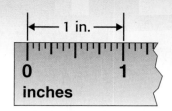

ACTIVITY

2 **Estimate and measure length using a ruler.**

> **Step 1** **Estimate**
>
> Estimate the length of the comb in inches.
>
> **Step 2** **Measure**
>
> Line up the comb as shown. Find the closest inch mark at the other end of the comb. How close was your estimate to the actual length?

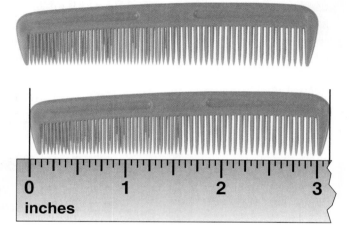

Think About It

1. In both activities, how did you estimate length?

2. Is a paper clip or a ruler more accurate for measuring? Explain.

CHECK What You Know

Estimate. Then measure each length to the nearest inch.

3.

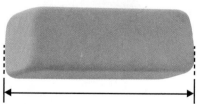

4.

5. Find objects in your classroom that measure about 1 inch, 4 inches, and 6 inches. Then copy and complete the table shown.

Measure	Object
1 inch	?
4 inches	?
6 inches	?

6. **WRITING IN** ►**MATH** How did you decide which objects to choose?

Length to the Nearest Half Inch

GET READY to Learn

Jamari has a rubber bug collection. About how long is the rubber bug shown?

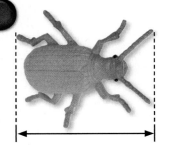

In the Explore Activity for Lesson 9-1, you measured **length** to the nearest **inch (in.)**. You can also measure length to the nearest half inch.

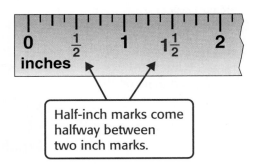

Half-inch marks come halfway between two inch marks.

Real-World EXAMPLE   Nearest Half Inch

1   TOYS **What is the length of Jamari's rubber bug to the nearest half inch?**

Estimate The bug is a little longer than the length of a paper clip. So, the bug is a little longer than 1 inch.

Line up one end of the bug with the 0 mark. Find the half inch mark that is closest to the other end of the bug.

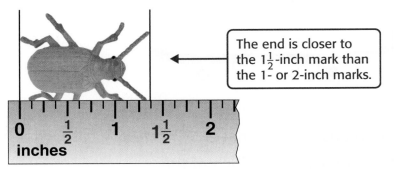

The end is closer to the $1\frac{1}{2}$-inch mark than the 1- or 2-inch marks.

To the nearest half inch, the bug is $1\frac{1}{2}$ inches long.

Sometimes when measuring to the nearest half inch, the length is a whole number.

Real-World EXAMPLE Nearest Half Inch

2 **SCIENCE** **What is the length of the red feather to the nearest half inch?**

Estimate The feather is a little shorter than the length of 3 paper clips. So, the feather is about 3 inches long.

Even though you are measuring to the nearest half inch, the end of the feather is closer to a whole number, 3.

Remember

To measure length, line up one end of the object to be measured with the 0 mark on the ruler.

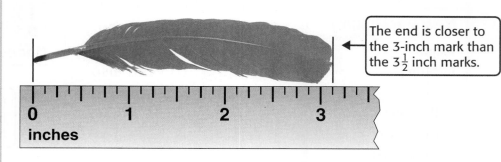

The end is closer to the 3-inch mark than the $3\frac{1}{2}$ inch marks.

So, to the nearest half inch, the red feather is 3 inches long.

✓ CHECK What You Know

Estimate each length. Then measure each to the nearest half inch.

See Examples 1 and 2 (pp. 375–376)

1.

2.

3. Alton has a carrot that is $5\frac{1}{2}$ inches long. He cuts off $\frac{1}{2}$ inch of the carrot. How long is the carrot now?

4. **Talk About It** How do you know where to find a half-inch mark on the ruler?

Estimate each length. Then measure each to the nearest half inch.

See Examples 1 and 2 (pp. 375–376)

5.

6.

7.

8.

9.

10. Ramiro needs two pieces of string that are $2\frac{1}{2}$ inches long and 3 inches long. What is the total length of string needed?

11. Tia's hair ribbon is 10 inches long, and Jan's is $8\frac{1}{2}$ inches long. What is the difference in the lengths of the two ribbons?

12. Kerri's comb is $4\frac{1}{2}$ inches long. What are the inch marks on either side of this measurement?

13. Tamara's ruler is starts at $2\frac{1}{2}$ inches. Can she use her ruler to draw a line 5 inches long? Explain.

H.O.T. Problems

14. CHALLENGE Explain how you could use a piece of string to measure this line.

15. OPEN ENDED Draw a design using straight lines. Measure the total length of all the lines.

16. **WRITING IN ►MATH** Explain how you would use a ruler to measure the length of the stick bug.

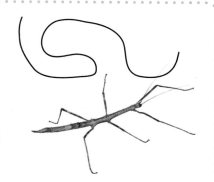

Customary Units of Length

GET READY to Learn

Patterson stood with both of his feet together and then jumped as far as he could. How far do you think he jumped?

To measure longer lengths and distances, you need to use other customary units of measure.

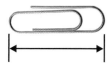

The length of a paper clip is about 1 inch long.

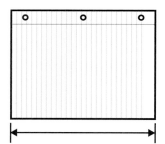

A sheet of notebook paper is about a **foot (ft)** long.

A baseball bat is about 1 **yard (yd)** long.

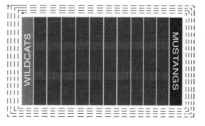

A **mile (mi)** equals 4 times around a football field.

Real-World EXAMPLE   Choose Units of Length

① **MEASUREMENT What unit should Patterson use to measure the distance he jumped?**

An inch is too short. A mile is too long. Even a yard is too long. The most appropriate unit to use is the foot.

You can estimate length.

> **Real-World EXAMPLE**   Estimate Length

2 **BOOKS** **Choose the better estimate for the width of your math book, 10 inches or 10 feet.**

Think about an inch and a foot.

A paper clip is 1 inch.   A ruler is 1 foot.

The width of your math book is about 10 paper clips not 10 rulers. So, the better estimate is 10 inches.

You can use estimation to check if your actual measurement is reasonable.

> **Real-World EXAMPLE**   Estimate and Measure Length

3 **MONEY** **Measure the width of a quarter.**

Estimate A quarter is about the same width as a paper clip, so it is probably 1 inch wide.

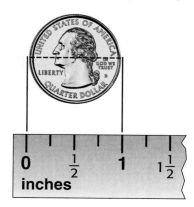

Remember

Estimating helps you check for reasonableness.

So, the width of a quarter is 1 inch.

Check Based on the estimate, the answer is reasonable. ✔

Choose the most appropriate unit to measure each length.
Write *inch*, *foot*, *yard*, or *mile*. See Example 1 (p. 378)

1. height of a calculator

2. width of your desk

Choose the better estimate. See Example 2 (p. 379)

3. length of a cricket
1 inch or 1 foot

4. the height of a chair
3 miles or 3 feet

Estimate then measure each length. Use *inch*, *foot*, or *yard*. See Example 3 (p. 379)

5. your shoe length

6. length of your smile

7. width of a door

8. Mrs. Frisk is measuring the length of her cat. What unit should she use to measure the cat? Why?

9. **Talk About It** Describe a situation when it may be helpful to know how to estimate units of length.

Practice and Problem Solving

EXTRA PRACTICE
See page R24.

Choose the most appropriate unit to measure each length.
Write *inch*, *foot*, *yard*, or *mile*. See Example 1 (p. 378)

10. the length of a pencil

11. the height of a boy

12. the distance between two cities

13. the length of a swing set

Choose the better estimate. See Example 2 (p. 379)

14. the length of your foot
8 inches or 8 feet

15. the width of a computer screen
16 yards or 16 inches

16. the height of a wall
11 inches or 11 feet

17. the width of a window
3 miles or 3 feet

Estimate then measure each length. Use *inch*, *foot*, or *yard*. See Example 3 (p. 379)

18. your pencil

19. your height

20. arm length

21. A piece of rope can stretch from the front of a classroom to the back. Are there 25 miles of rope or 25 feet of rope? Explain.

22. The distance from Oscar's home to school is 100 units. What is the most appropriate unit to measure the length? Explain.

H.O.T. Problems

23. OPEN ENDED Give an example of an object that is a little longer than 1 yard.

24. WHICH ONE DOESN'T BELONG? Identify the length that does not belong. Explain your reasoning.

36 inches   3 feet   48 inches   1 yard

25. WRITING IN ▶MATH Explain why a yardstick is a better choice than a ruler for measuring the distance around a playground.

TEST Practice

26. Melissa's ruler is broken and starts at the 2-inch mark. If she wants to draw a line that is 4 inches long, at what inch mark will she stop drawing?
(Explore 9-1)

A 2 inch

B $4\frac{1}{2}$ inch

C 6 inch

D 7 inch

27. Which real-life object is longer than 1 foot? (Lesson 9-2)

F

G

H

J

Spiral Review

28. Measure the brush to the nearest half inch. (Lesson 9-1)

29. Write the expression that can be used to represent the phrase *5 girls have 2 braids each.* (Lesson 8-2)

Find each sum. Use estimation to check for reasonableness.
(Lesson 2-7)

30. $227 + $96

31. $384 + $807

9-3 Problem-Solving Strategy

MAIN IDEA I will solve a problem by working backward.

Clarissa is at a fair. She wants to play three games. The line for the second game is twice as long as the first line. The third line has 5 fewer people than the second line. There are 17 people in line for her last game. How many people are in line for the first game?

Understand	**What facts do you know?**
	• The second line is twice as long of the first one.
	• The third line has 5 fewer people than the second one.
	• The third line has 17 people.
	What do you need to find?
	• The number of people in the first line.
Plan	Work backward to find the number of people in the first line.
Solve	17 People in the third line. +5 Second line has 5 more people than third line. 22 People in the second line. $\begin{array}{r} 11 \\ 2\overline{)22} \end{array}$ First line is half as long as the second one. So, there are 11 people in the first line.
Check	Look back. Does 11 people make sense for the situation? $11 + 11 = 22$ $22 - 5 = 17$ So, the answer is correct. ✔

ANALYZE the Strategy

Refer to the problem on the previous page.

1. Why was the *work backward* strategy the best way to solve the example?

2. If you knew how long the first line was, would you use the *work backward* strategy to solve? Explain.

3. How can you tell when to use the *work backward* strategy?

4. What would you need to do if the answer was incorrect when you checked it?

PRACTICE the Strategy

EXTRA PRACTICE
See page R24.

Solve. Use the *work backward* strategy.

5. Mrs. Keys has 36 pens. Yesterday she had half that amount plus 2. How many pens did she have yesterday?

6. There are three lines. The first line is 3 times as long as the second. The second line is 4 inches longer than the third. The third line is 2 inches long. How long is the first line?

7. The table below shows Adelais' schedule. She finished at 5:00 P.M. At what time did she start her lunch?

Adelais' Schedule	
Activity	**Time**
Lunch	1 hour
Scrapbooking	3 hours

8. Dirk walked 15 miles this week. He walked twice as much Tuesday as he did on Monday. He walked 5 miles on Wednesday and 2 miles on both Thursday and Friday. How many miles did he walk on Monday?

9. **Algebra** The table shows the number of shapes after a pattern repeats itself five times. How many of each shape are in the original pattern?

Pattern			
Shape	Circles	Squares	Hearts
Number	15	5	10

10. In the morning, Mr. Lawrence gave 9 pencils to students. Later, 5 of the students returned the pencils. After lunch, he handed out 5 more pencils. Now he has 15 pencils. How many did he start with?

11. Leslie played with some friends on Monday. She played with 2 times as many friends on Wednesday. This was 4 more than on Friday. On Friday there were 4. How many did she play with on Monday?

12. **WRITING IN MATH** Look back at Exercise 11. Explain another strategy you could use to solve the problem.

You can use metric units to measure length. One metric unit is the **centimeter (cm)**. A centimeter is about the width of your index finger.

MAIN IDEA

I will measure length to the nearest centimeter and millimeter.

You Will Need
centimeter ruler, crayon, small paper clip

New Vocabulary

centimeter (cm)

millimeter (mm)

Math Online >
macmillanmh.com
• Concept in Motion

ACTIVITY

1 **Estimate and Measure in Centimeters**

Step 1 Estimate

About how many finger-widths long would you estimate the crayon to be?

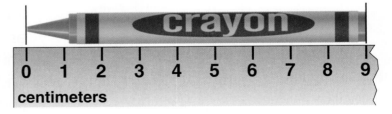

Step 2 Measure

Use the width of your finger to measure the length of the crayon.

• How close was your estimate to the actual finger-width measure?

Step 3 Measure

Align the left end of the crayon with the 0 at the end of the centimeter ruler. Find the tick mark closest to the other end of the crayon.

• What is the length in centimeters?

• How close was your finger-width measure to the actual number of centimeters?

Step 4 A **millimeter (mm)** is smaller than a centimeter. It is used to measure very small lengths. 1 centimeter = 10 millimeters.

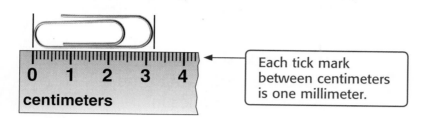

Each tick mark between centimeters is one millimeter.

- How many millimeters long is the paper clip?

Think About It

1. When you compare the millimeter and centimeter measurements, which is more accurate? Explain.

CHECK What You Know

Estimate. Then measure each length to the nearest millimeter and centimeter.

2.

3.

4.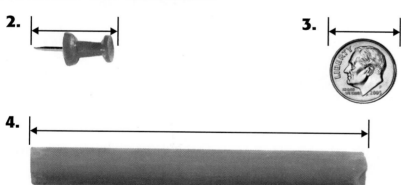

Choose the better estimate.

5. length of a marker
10 mm or 10 cm

6. width of your thumb
2 mm or 2 cm

7. length of your arm
30 mm or 30 cm

8. **WRITING IN MATH** Are there more centimeters or more millimeters in something 5 centimeters long? Explain.

Metric Units of Length

Math Online

macmillanmh.com
• Extra Examples
• Personal Tutor
• Self-Check Quiz

MAIN IDEA

I will choose the appropriate tools and units to estimate and measure metric units of length.

New Vocabulary

centimeter (cm)
millimeter (mm)
meter (m)
kilometer (km)

> **GET READY to Learn**

Owen's big sister runs high school track. She runs one time around the track. Would you measure the distance in millimeters, centimeters, meters, or kilometers?

To measure short lengths, you can use the **centimeter (cm)** and **millimeter (mm)**. To measure longer lengths, use the **meter (m)** and the **kilometer (km)**.

A millimeter is about the width of a dime.

A centimeter is about the width across your index finger.

 1 meter

The width of a door.

It takes about 10 minutes to walk 1 kilometer.

> **Real-World EXAMPLE**   Choose Metric Units

① **SPORTS Choose the unit that should be used to find how far Owen's big sister runs around the track one time.**

You need to determine whether to use *millimeter, centimeter, meter,* or *kilometer.*

A millimeter and centimeter are too short. A kilometer is too long to measure one time around the track. Use a meter.

Real-World EXAMPLE   Choose Metric Units

② **INSECTS** **Carl has a bumblebee in his insect collection. Choose the unit that he should use to measure the length of the bee.**

You need to determine whether it is best to use *millimeter, centimeter, meter,* or *kilometer.*

A meter and kilometer are too long. A centimeter is used to measure short lengths, but it is best to use a millimeter to measure very short lengths.

So, use a millimeter.

Use what you know about meter and kilometer to estimate.

Real-World EXAMPLE   Estimate Length

③ **LONG JUMP** **The world record for jumping the greatest distance in the long jump has not been broken in many years. Choose the better estimate for the distance jumped, 9 m or 9 km.**

Think about a meter and a kilometer. A meter is about the width of a door. It makes more sense that someone can jump 9 door widths.

Thus 9 m is reasonable and 9 km is not reasonable.

So, 9 meters is the better estimate.

Choose the most appropriate unit to measure each length. Write
millimeter, centimeter, meter, **or** *kilometer.* See Examples 1 and 2 (pp. 386–387)

1. length of an ant

2. length of a car

3. length of a hiking trail

4. length of a pencil

Estimate then measure each length. Use millimeter, centimeter, or meter.

5. width of book

6. width of your classroom

7. Choose the best estimate for the length of a chalkboard, 5 m or 5 km. See Example 3 (p. 387)

8. **Talk About It** Why is it important to know both the customary and metric systems of measurement?

Practice and Problem Solving

EXTRA **PRACTICE**
See page R25.

Choose the most appropriate unit to measure each length.
Write *millimeter, centimeter, meter,* **or** *kilometer.* See Examples 1 and 2 (pp. 386–387)

9. height of a flagpole

10. distance a plane travels

11. length of an insect

12. length of a crayon

Choose the better estimate. See Example 3 (p. 387)

13. distance you could travel on a train
500 km or 5,000 cm

14. length of a sofa
2 m or 20 cm

Estimate then measure each length. Use millimeters, centimeter, or meter.

15. width of your hand

16. height of a door

For Exercises 17 and 18, use the chart that shows the wood needed to build a fort.

17. The wood for the floor needs to be cut into 4 pieces. Two of the pieces will be 1 meter long. If the remaining wood is cut equally in half, how long will the remaining boards be?

18. The wood for the walls will be cut into 5 equal pieces. How long will each piece be in centimeters?

Build a Fort
Amount of wood needed
Floor	6m
Walls	5m
Roof	6m

H.O.T. Problems

Algebra Compare. Use >, <, or =.

19. 30 cm ⬤ 30 m   **20.** 4 m ⬤ 400 cm   **21.** 2 m ⬤ 3 mm

22. Measurement Reggie is in a race that is 5 kilometers long. Jim is in a race that is 500 meters long. Whose race is longer? Explain.

23. **WRITING IN ►MATH** Suppose you are measuring the length of an object in centimeters. What should you do if the object does not line up exactly with the centimeter mark on the ruler?

TEST Practice

24. Mr. Rockwell gave 9 students one pencil each. That afternoon, he gave 5 more students one pencil each. He has 15 pencils left. How many pencils did he start with? (Lesson 9-3)

   A 14   **C** 20

   B 15   **D** 29

25. Choose the best unit for measuring distance across the United States. (Lesson 9-4)

   F liter   **H** centimeter

   G meter   **J** kilometer

Spiral Review

26. Kareem gave 25 football cards to Mali, 13 cards to Millie, and 14 cards to Justin. He has half of the cards he started with. How many cards did he start with? (Lesson 9-3)

Choose the most appropriate unit to measure each length. Write *inch, foot, yard,* or *mile*. (Lesson 9-2)

27. distance between two countries   **28.** distance from floor to ceiling

Algebra Write an expression and a number sentence. (Lesson 8-2)

29. 4 groups with 3 bananas in each   **30.** 81 campers in 9 equal groups

Hit the Target

Metric Measurement

Get Ready!

Players: 2 players

Get Set!

Copy the game card shown.

Go!

- Player 1 tosses the pattern block at the target.

- Player 2 measures the distance between the target and the pattern block to the nearest centimeter.

- Player 1 records the distance on the game card.

- Player 2 stands the same distance from the target as Player 1 and tosses the pattern block. Player 1 measures the distance.

You will need: game card, metric ruler, a target (i.e., an eraser), and 1 square pattern block

Player's Name	Toss #1	Toss #2	Toss #3	Toss #4	Total Score

- Move the target and toss again.

- Both players will take four turns at tossing the pattern block at the target. Add the distances of the four tosses together. The player with the lowest total wins.

Estimate each length. Then measure each to the nearest half inch. (Lesson 9-1)

1.

2.

Choose the most appropriate unit to measure each length. Write *inch, foot, yard,* or *mile*. (Lesson 9-2)

3. height of a DVD container

4. height of a basketball player

5. **MULTIPLE CHOICE** Which of the following is longer than 1 yard? (Lesson 9-2)

   A your bed   **C** a toy car

   B a crayon   **D** a notebook

Choose the better estimate. (Lesson 9-2)

6. height of a ladder
   4 miles or 4 yards

7. length of a spoon
   6 feet or 6 inches

Solve.

8. Some children offered to clean Ms. Dawson's garage. It took 2 days. She paid each child $4 per day. If she paid them $24 total, how many children were there? (Lesson 9-3)

9. Daphne took her younger cousins to the water park. It cost her $5 to get in, and $2 less for her cousins. She spent a total of $26. How many cousins were with Daphne?
   (Lesson 9-3)

Choose the most appropriate unit to measure each length. Write *millimeter, centimeter, meter,* or *kilometer*. (Lesson 9-4)

10. distance driven in a car

11. length of a school bus

Choose the better estimate. (Lesson 9-4)

12. length of a window
    2 m or 2 km

13. length of a classroom
    18 km or 18 m

14. **MULTIPLE CHOICE** Choose the best unit for measuring the distance across the state of South Carolina.
    (Lesson 9-4)

    F liter

    G centimeter

    H meter

    J kilometer

15. **WRITING IN ►MATH** Name something that you would measure in millimeters. Explain why you would not use centimeters to measure it. (Lesson 9-4)

Perimeter

MAIN IDEA

I will find the perimeter of a figure.

New Vocabulary

perimeter

Math Online

macmillanmh.com
- Extra Examples
- Personal Tutor
- Self-Check Quiz

GET READY to Learn

Hands-On Mini Activity

Perimeter is the distance around the outside of a *figure,* or shape. You can estimate and measure perimeter.

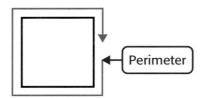

Step 1 Copy the table shown below.

Object	Estimate (cm)	Exact Measure (cm)
Math book		
Desk top		
Chalkboard eraser		

Step 2 Estimate the perimeter of your math book.

Step 3 Use a centimeter ruler to find the exact perimeter.

Step 4 Record the results. Repeat the steps for each object listed.

1. Write the number sentence for the perimeter of your math book.

2. What operation did you use to find perimeter?

Find Perimeter
Key Concept

Words   The perimeter of a figure is the sum of the side lengths.

Model

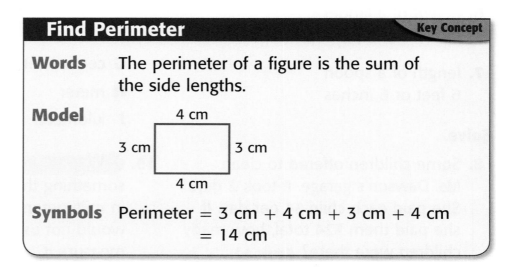

Symbols   Perimeter = 3 cm + 4 cm + 3 cm + 4 cm
                                      = 14 cm

1 **Find the perimeter of the triangle.**

To find the perimeter, add the lengths of the sides.

5 in. + 3 in. + 5 in. = 13 in.

So, the perimeter is 13 inches.

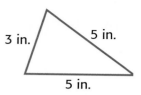

3 in.   5 in.

5 in.

Remember

On grid paper, think of each square as one unit.

2 **Find the perimeter of the shaded rectangle.**

To find the perimeter add the lengths of the sides.

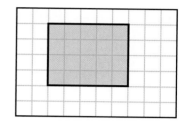

4 units + 5 units + 4 units + 5 units = 18 units

So, the perimeter is 18 units.

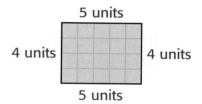

5 units

4 units   4 units

5 units

CHECK What You Know

Find the perimeter of each figure. See Examples 1 and 2 (p. 393)

1.

3 ft

10 ft

2.

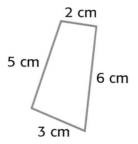

2 cm

5 cm

6 cm

3 cm

3.

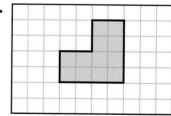

4. Geometry The front of the birdhouse at the right is shaped like a pentagon. The lengths of all of the sides are equal. Find the perimeter of the birdhouse.

8 in.

5. **Talk About It** A triangle has three equal sides. Its perimeter is 15 units. How would you find the length of each side?

Find the perimeter of each figure. See Examples 1 and 2 (p. 393)

6.
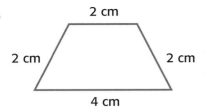
2 cm
2 cm
2 cm
4 cm

7.
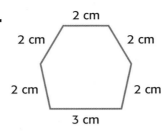
2 cm
2 cm
2 cm
2 cm
2 cm
3 cm

8.
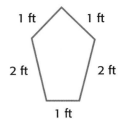
1 ft
1 ft
2 ft
2 ft
1 ft

9.
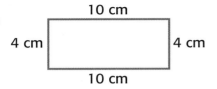
10 cm
4 cm
4 cm
10 cm

10.
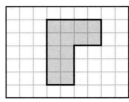

11.

12. Algebra A fountain has 3 sides. It's perimeter is 120 feet. If one side is 40 feet and another is 50 feet, what is the length of the third side?

13. Mila's house has a deck that has 6 sides. The sides have equal lengths. If each side of the deck is 12 feet long, what is its perimeter?

14. Geometry Two squares are shown. The length of each side is 8 feet. Find the total perimeter if the squares are pushed together to make a rectangle.

8 feet

15. Algebra The figure shown below has a perimeter of 21 feet. Find the length of the missing side.

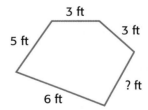
3 ft
3 ft
5 ft
? ft
6 ft

Data File

The Boston Red Sox, a professional baseball team, play at Fenway Park. The park opened in 1912.

16. Three bases and home plate make a diamond that is 90 feet on each side. What is the perimeter of the diamond?

17. There are three square bases in the game of baseball. Each side is 15 inches long. What is the total perimeter of all three bases?

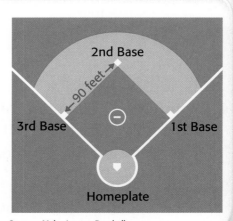

2nd Base
90 feet
3rd Base
1st Base
Homeplate

Source: Major League Baseball

H.O.T. Problems

18. OPEN ENDED Draw and label a figure that has a perimeter of 24 inches.

19. **WRITING IN ►MATH** How can you find the perimeter of a rectangle if you know the length and width? Explain.

2 ft

4 ft

TEST Practice

20. Choose the best unit for measuring the length of a staple. (Lesson 9-4)

  A millimeter

  B centimeter

  C meter

  D kilometer

21. What is the perimeter of the figure? (Lesson 9-5)

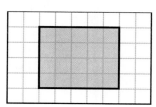

  F 11 units

  G 12 units

  H 18 units

  J 20 units

Spiral Review

Choose the most appropriate unit to measure each length. Write *millimeter, centimeter, meter,* or *kilometer*. (Lesson 9-4)

22. length of a computer

23. distance around the world

24. thickness of cardboard

25. height of a building

26. Three is subtracted from a number. Next, the difference is multiplied by 2. Then, 4 is added to the product. Finally, 9 is subtracted to give a difference of 9. Find the number. (Lesson 9-3)

Tell whether + or − makes each number sentence true. (Lesson 8-2)

27. $15 = 18 \bullet 3$

28. $4 \times 7 = 2 \bullet 26$

29. $4 \bullet 59 = 7 \times 9$

Write the fact family for each set of numbers. (Lesson 6-2)

30. 3, 4, 12

31. 5, 7, 35

32. 6, 8, 48

Area is the number of square units needed to cover a figure without overlapping. You can use grid paper to explore area.

1 square unit

MAIN IDEA

I will estimate the area of a figure.

You Will Need
grid paper
connecting cube
geoboard
rubber boards

New Vocabulary

area

Math Online

macmillanmh.com

• Concepts in Motion

ACTIVITY

1 **Estimate area**

Step 1 **Estimate**

How many square units do you think will cover the side view of the shape?

Step 2 **Trace**

Trace one side of the connecting cube onto grid paper.

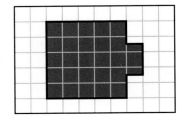

Step 3 **Determine Area**

One whole square is 1 square unit.

Each of these is $\frac{1}{2}$ square unit.

Count the number of whole square units. How many half-square units are there? Estimate the area. How does the estimate compare with your first estimate?

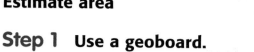

ACTIVITY

2 Estimate area

Step 1 Use a geoboard.

Use a rubberband to make a rectangle on a geoboard.

Step 2 Estimate.

Use what you learned in Activity 1 to estimate the area of the rectangle.

Step 3 Determine the area.

Count the squares in the rectangle.

Hands-On Activity

Think About It

1. Is it easier to find the exact area of a rectangle on grid paper or to estimate? Explain.

2. How did you make your estimate for the area of the rectangle? How close is it to the actual area?

CHECK What You Know

Estimate and determine the area in square units of each figure.

3.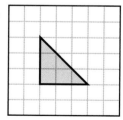

4.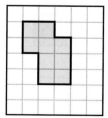

5.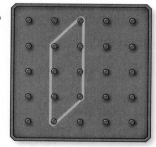

6. Make a design on a geoboard. Find the area.

7. Make a design on a piece of grid paper. Estimate the area.

8. **WRITING IN ►MATH** Explain the difference between area and perimeter.

Explore 9-6 Measure Area **397**

Measure Area

MAIN IDEA

I will determine the area of a figure.

New Vocabulary

area

Math Online

macmillanmh.com
• Extra Examples
• Personal Tutor
• Self-Check Quiz

GET READY to Study

While in art class, Halley drew figures on grid paper. One of the figures is shown at the right. Estimate the area of the figure.

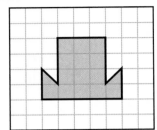

Area is the number of square units needed to cover a figure without overlapping. In the Explore Activity, you estimated area. You can also find the exact area of a figure.

EXAMPLES Determine Area

1 **ART What is the area of the figure that Halley drew?**

Count the number of whole squares. There are 14 whole squares.

There are 2 half squares. Notice that two halves equal one whole.

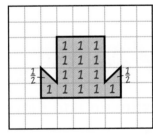

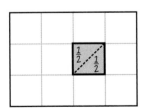

14 squares units + 1 square unit = 15 square units

So, the area is 15 square units.

2 **Find the area of the figure.**

Count the number of whole squares.

The area of this figure is 4 square units.

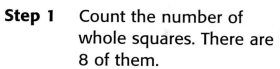

Real-World EXAMPLE Determine Area

3 **Geometry** Phillip created a geoboard design shown at the right. What is the area of the figure?

Remember

Two half squares equal one whole square.

Step 1 Count the number of whole squares. There are 8 of them.

Step 2 Count the half squares. There are 8 halves. Eight halves equal 4 whole squares.

Step 3 Add.

8 square units + 4 square units = 12 square units

So, the area of the design is 12 square units.

CHECK What You Know

Find the area of each figure. See Examples 1–3 (pp. 398–399)

1.

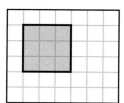

2.

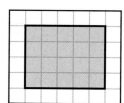

3.

4. Denitra plans to cover a desk with decorative tiles. What is the area of the space she will cover?

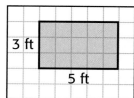

3 ft

5 ft

5. The frame is covered with squares of colored glass. What is the area covered by the colored glass?

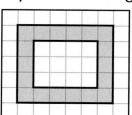

6. **Talk About It** Explain how to find the area of a rectangle.

Find the area of each figure. See Examples 1–3 (pp. 398–399)

7.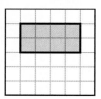

8.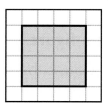

9.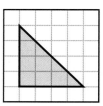

10.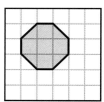

11.

12.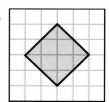

Find the area and perimeter of each figure.

13.

14.

15.

Find the area. Use a geoboard and rubber bands if needed.

16. Luisa is helping to put tile in a hallway. How many square tiles will be needed to fill the area?

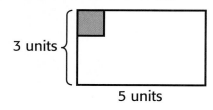
3 units
5 units

17. Elaine is finding the area of her closet. The size is shown. What is the area of her closet?

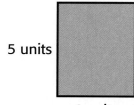
5 units
4 units

H.O.T. Problems

18. OPEN ENDED Draw two rectangles on grid paper that have different lengths and widths but the same area.

19. CHALLENGE A room is 12 units wide by 24 units long. Find the area and the perimeter of the floor of the room.

20. WRITING IN ►MATH Write how to find the area of a rectangle that is 5 units long and 7 units wide.

21. What is the perimeter of the figure? (Lesson 9-5)

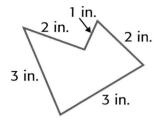

1 in.

2 in.

2 in.

3 in.

3 in.

A 9 inches   **C** 12 inches

B 11 inches   **D** 11 meters

22. What is the area of this figure? (Lesson 9-6)

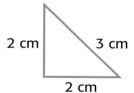

= 1 square unit

F 2 square units

G 4 square units

H 6 square units

J 8 square units

Spiral Review

Find the perimeter of each figure. (Lesson 9-5)

23.

4 in.   6 in.

7 in.

24.

2 cm   3 cm

2 cm

Choose the better estimate. (Lesson 9-4)

25. length of a finger
6 mm or 6 cm

26. height of a van
2 m or 2 km

Multiply. (Lesson 4-2)

27.

4 groups of 2

28.

3 rows of 2

Subtract. Check your answer. (Lesson 3-7)

29. 6,302
      − 5,294

30. 2,001
      − 1,640

31. 4,500
      − 2,406

9-7 Problem-Solving Investigation

MAIN IDEA I will choose the best strategy to solve a problem.

P.S.I. TEAM +

CASSANDRA: **My family is making a sandbox. It has the shape of a rectangle with a width of 4 feet and a length of 6 feet. The plastic that goes under the sand costs $2 for each square foot.**

YOUR MISSION: **Determine if $50 will be enough to pay for the plastic.**

Understand	You know the length and width of the sandbox and the cost of the plastic per square foot. Find the total cost of the plastic.
Plan	Draw a picture to help you solve the problem.
Solve	The drawing shows that the area of the sandbox is 24 square feet. Multiply 24 square feet by $2 per square foot to find the total cost of the plastic.

6 ft

4 ft   4 ft

6 ft

$24 \times \$2 = \$48$ $48 is less than $50

So, $50 will be enough to pay for the plastic.

Check	Look back. Use addition to check. $24 + $24 < $50. So, the answer is correct.

Use any strategy shown below to solve. Tell what strategy you used.

> PROBLEM-SOLVING STRATEGIES
> • Look for a pattern.
> • Make a table.
> • Act it out.

1. **Algebra** Min is reading a book. It has 75 pages. She reads 6 pages every day for 5 days starting on Sunday. After day 5, how many pages will she have left to read?

2. Ms. Dunn bought 6 sheets of plywood, a box of nails, and a hammer from the hardware store. How much did she spend?

Item	Cost
Plywood........	$30 each
Box of nails.....	$5 each
Hammer........	$10 each

3. Vic and Derrick went to the mall. They visited 14 stores on the first floor and 17 stores on the second floor. They went back to 4 of the same stores on the first floor. How many different stores did they visit?

4. **Measurement** Mr. Carson is measuring the length and width of his kitchen counter. Find the measurements for the missing lengths.

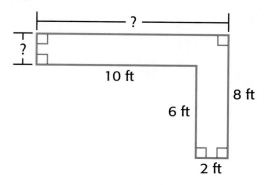

5. Three friends are fishing. Francisco caught 3 times as many fish as Haley. Haley caught 7 less fish than Petra. Petra caught 15 fish. How many fish did Francisco and Haley catch?

6. **Algebra** Malika is shooting baskets. She makes a basket every 6th shot. If she makes a total of 9 baskets, how many shots did she take?

7. Samantha and Edward need to pick 25 tomatoes from their garden. How many more tomatoes do they need to pick?

Vegetables Picked	
Samantha	**Edward**
18 tomatoes	13 peppers
	2 cucumbers
	4 tomatoes

8. **WRITING IN MATH** Can two figures have the same area but different perimeters? Explain with an example.

Problem Solving in Science

Roller Coaster Physics

When the Millennium Force opened at Cedar Point Amusement Park in Ohio in 2000, it broke 10 world records for roller coasters. This roller coaster takes its riders 310 feet into the air and twists them into two 122 degree turns. In 2005, the Millennium Force was voted the best steel roller coaster in the world in a survey of roller coaster fans.

A ride on a roller coaster is an incredible thrill. Did you ever think about how roller coasters work? The train of the roller coaster is pulled up the lift hill along the track with a steel cable. At the top of the hill, the energy of the train plunges the cars down the first downward slope. This energy carries the train up the second hill. As the train goes up and down the hills, this energy conversion continues.

Millennium Force FACTS

Lift height	310 feet
Track length	6,600 feet
Vertical drop	300 feet
Second hill height	169 feet
Third hill height	182 feet

Source: Cedar Point Amusement Park

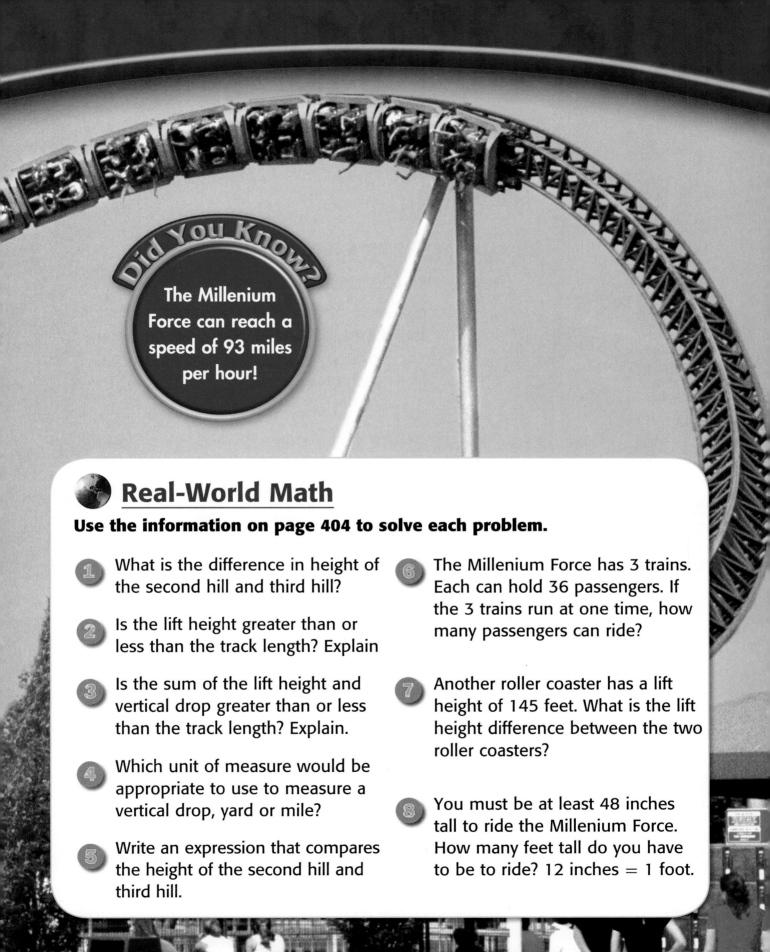

The Millenium Force can reach a speed of 93 miles per hour!

🌎 Real-World Math

Use the information on page 404 to solve each problem.

1. What is the difference in height of the second hill and third hill?

2. Is the lift height greater than or less than the track length? Explain

3. Is the sum of the lift height and vertical drop greater than or less than the track length? Explain.

4. Which unit of measure would be appropriate to use to measure a vertical drop, yard or mile?

5. Write an expression that compares the height of the second hill and third hill.

6. The Millenium Force has 3 trains. Each can hold 36 passengers. If the 3 trains run at one time, how many passengers can ride?

7. Another roller coaster has a lift height of 145 feet. What is the lift height difference between the two roller coasters?

8. You must be at least 48 inches tall to ride the Millenium Force. How many feet tall do you have to be to ride? 12 inches = 1 foot.

Math Activity for 9-8
Measure Temperature

A thermometer is used to measure temperature. The thermometer used in this activity measures degrees Fahrenheit (°F).

MAIN IDEA

I will measure temperature with a thermometer.

You Will Need
3 thermometers (°F)
oven mitt
3 microwave-safe cups
microwave
water
measuring cup
3 ice cubes

Math Online

macmillanmh.com
• Concepts in Motion

ACTIVITY   Use a Thermometer

Step 1   Measure Temperature

Pour $\frac{1}{2}$ cup of water in each of the three cups. Label the cups A, B, C. Place a thermometer in each cup. Copy the table shown. Measure and record each temperature.

	Temperature 1	Temperature 2
A		
B		
C		

Step 2 Change Temperature

Remove the thermometers. Place the ice cubes in Cup A. Place Cup B in the microwave for 45 seconds. Use the oven mitt to carefully remove the cup from the microwave.

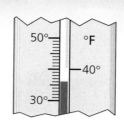

50°   °F
       40°
30°

Step 3 Measure Temperature

Put a thermometer in each cup. Measure and record the temperature of each cup.

Think About It

1. What happened to the temperature of the water in Cup A after ice was added to it?

2. What happened to the temperature of the water in Cup B after it was placed in the microwave?

3. What happened to the temperature of the water in Cup C when it was measured the second time?

4. Describe what happens to the liquid in the thermometer as the water gets hotter and colder.

CHECK What You Know

5. Copy the table below. Record the temperature of Cup B after each time shown to complete the table.

Time (min)	Temperature (°F)
2	■
4	■
6	■

6. **WRITING IN MATH** Predict what would happen to the temperature of the water in Cup A if it was left overnight.

Measure Temperature

°Fahrenheit

GET READY to Learn

Mike lives in Louisville, Kentucky. On a July day, he checks the outside temperature. What is the temperature?

The instrument above is a thermometer. A **thermometer** is an instrument that measures how hot or cold something is (the temperature) in degrees. One unit of measure for temperature is called **degrees Fahrenheit (°F)**.

Today in Louisville
Currently At 9:28 A.M.

83°F

Winds: ENE at 3 mph

Mostly Sunny

Real-World EXAMPLE Measure Temperature

① **MEASUREMENT** Write the temperature shown on Mike's thermometer in degrees Fahrenheit (°F).

The thermometer is labeled Fahrenheit. Find the number next to the top of the red line.

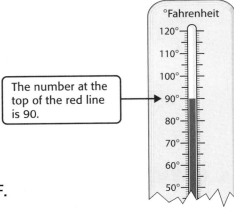

°Fahrenheit

The number at the top of the red line is 90.

So, the temperature is 90°F.

EXAMPLE Compare Temperatures

2 MEASUREMENT A thermometer shows 47°F. Tell whether the temperature is *hotter* or *colder* than room temperature.

Room temperature is about 68°F. The temperature shown is below 68°F.

So, the temperature is *colder* than room temperature.

Remember

As the temperature rises or gets *hotter*, the red line goes up. As the temperature falls or gets *colder*, the red line goes down.

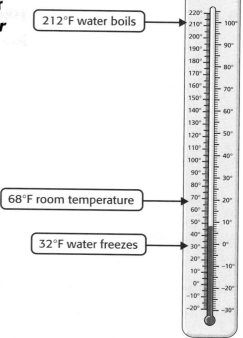

212°F water boils

68°F room temperature

32°F water freezes

CHECK What You Know

Write each temperature in degrees Fahrenheit (°F). See Example 1 (p. 408)

1.

80° °F
70°
60°

2.

90° °F
85°
80°

3.

50° °F
40°
30°

Compare each temperature to the given temperature. Write *hotter* or *colder*. See Example 2 (p. 409)

4. room temperature

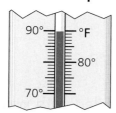

90° °F
80°
70°

5. water freezes

40° °F
30°
20°

6. water boils

20° °F
10°
0°

7. On Monday, the temperature was 87°F. On Tuesday, it was 85°F. Did the temperature get hotter or colder on Tuesday? Explain.

8. **Talk About It** Explain how to read a thermometer.

Write each temperature in degrees Fahrenheit (°F). See Example 1 (p. 408)

9.

10.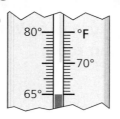

11.

12.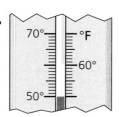

13.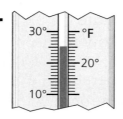

14.

Compare each temperature to the given temperature.
Write *hotter* or *colder*. See Example 2 (p. 409)

15. room temperature

16. room temperature

17. water boils

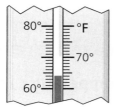

18. water boils

19. water freezes

20. water freezes

Tell which temperature is greater.

21. 29°F or 40°F

22. 116°F or 106°F

23. 32°F or 23°F

24. Use a thermometer to measure the outside temperature in the morning, at lunch, and at the end of the school day. Record the temperatures. Write a statement that compares them.

25. An ice cube tray is filled with water and placed in a freezer. The temperature in the freezer is 40°F. Will the water turn into ice? Explain how you know.

H.O.T. Problems

26. OPEN ENDED Describe a real world situation where you see a thermometer being used.

27. **WRITING IN** ►**MATH** Write a problem that compares temperatures in °F. Solve.

TEST Practice

28. Trish set the room temperature at 68°F. Which thermometer shows this temperature? (Lesson 9-8)

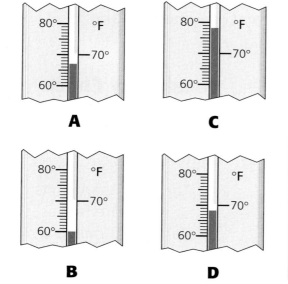

A

B

C

D

29. Below there are 5 groups of marbles. In each group there are 6 marbles. How many marbles are there in all? (Lesson 9-7)

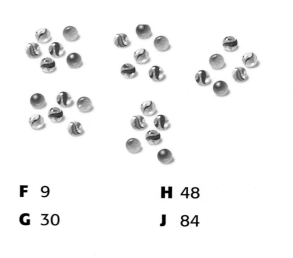

F 9   **H** 48

G 30   **J** 84

Spiral Review

30. Edgar talked to Tom for twice as long as he talked to Rhonda. He talked to Rosita for ten minutes longer than he talked to Tom. Edgar talked to Tom for 20 minutes. How long did he talk to Rhonda and Rosita? (Lesson 9-7)

31. Find the area of the figure at the right. (Lesson 9-6)

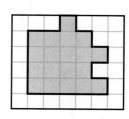

Divide. (Lesson 6-7)

32. 50 ÷ 10   **33.** 90 ÷ 10

34. 10)‾20   **35.** 10)‾30

36. Is the product of 12 × 11 greater than the product of 11 × 12? Explain. (Lesson 5-8)

Study Guide and Review

FOLDABLES Study Organizer — GET READY to Study

Be sure the following Key Vocabulary words and Key Concepts are written in your Foldable.

Measurement	Tool	Unit	Estimate	Measure
Length				
Perimeter				
Area				
Temperature				

Key Concepts

• Estimate and measure **length** using customary and metric units. (p. 375)

Customary Units of Length			
inch	foot	yard	mile

Metric Units of Length	
millimeter	centimeter
meter	kilometer

• **Perimeter** is the distance around a figure. (p. 392)

• **Area** is the number of square units needed to cover a figure without overlapping. (p. 398)

Area = 12 square units

Key Vocabulary

area (p. 398)

length (p. 375)

meter (p. 386)

thermometer (p. 408)

perimeter (p. 392)

Vocabulary Check

Choose the vocabulary word that completes each sentence.

1. The customary units of ____?____ are inches, feet, yards, and miles.

2. To put a fence around your backyard, you would need to find the ____?____.

3. The number of square units needed to cover a figure without overlapping is called the ____?____.

4. A ____?____ is an instrument that measures the temperature of something.

5. A door's width measures about one ____?____.

Lesson-by-Lesson Review

9-1 **Length to the Nearest Half Inch** (pp. 375–377)

Example 1
What is the length of the piece of yarn to the nearest half inch?

The yarn is about the length of one paper clip and half of another.

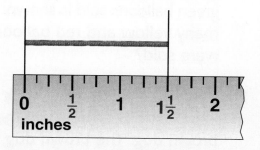

To the nearest half inch, the yarn is $1\frac{1}{2}$ inches long.

6. Estimate the length. Then measure to the nearest half inch.

7. Tavio needs a piece of string that is $3\frac{1}{2}$ inches long and another one that is 4 inches longer. What is the total length of string?

9-2 **Customary Units of Length** (pp. 378–381)

Example 2
What unit would you use to measure the width of a door? Write *inch*, *foot*, *yard*, or *mile*.

An inch and foot are too short. A mile is too long. So, a door is about 1 yard.

Choose the most appropriate unit to measure each length. Write *inch*, *foot*, *yard*, or *mile*.

8. the distance between cities

9. the height of a bike

Measure each length. Use *inch*, *foot*, or *yard*.

10. crayon   **11.** chalkboard

Choose the better estimate.
12. length of a fork
6 inches or 6 yards

9-3 Problem-Solving Strategy: Work Backward (pp. 382–383)

Example 3
Wesley doubles the distance he runs each week. After 3 weeks he runs 8 miles. How many miles did Wesley run the first week?

You know the number of hours Wesley ran after 3 weeks.

Use the *work backward* strategy to find the number of hours he ran the first week.

miles the third week	half	miles the second week
8	÷ 2 =	4

miles the second week	half	miles the first week
4	÷ 2 =	2

So, Wesley ran 2 miles the first week.

13. A balloon store sold three times as many red balloons as yellow balloons. They sold 7 more yellow balloons than green balloons. The number of green balloons sold is shown. How many yellow and red balloons were sold?

14. There are 3 dogs. The black dog is 5 pounds heavier than the brown dog. The brown dog is three times as heavy as the white dog. The white dog is 8 pounds. What is the weight of the other two dogs?

9-4 Metric Units of Length (pp. 386–389)

Example 4
Olinda will travel to another state. What metric unit should be used to measure the distance she will travel?

The millimeter, centimeter, and meter are too short.

So, use the kilometer to measure the distance Olinda will travel.

Choose the most appropriate unit to measure each length. Write *millimeter, centimeter, meter* or *kilometer*.

15. length of a staple

16. length of a dragonfly

Choose the better estimate.

17. length of a football field
90 cm or 90 m

9-5 **Perimeter** (pp. 392–395)

Example 5
Find the perimeter of the triangle.

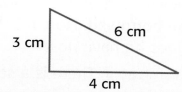

To find the perimeter add the length of the sides.

6 cm + 3 cm + 4 cm = 13 cm

So, the perimeter of this triangle is 13 centimeters.

Find the perimeter of each figure.

18.

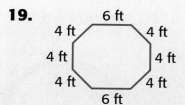

19.

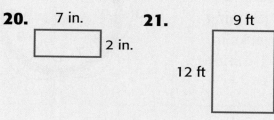

20.   7 in.   **21.**   9 ft

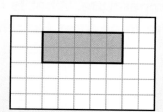

9-6 **Measure Area** (pp. 398–401)

Example 6
What is the area of the patio?

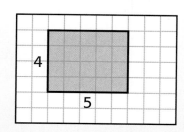

Count the number of squares the figure covers.

The figure covers 20 squares.

So, the area of the patio is 20 square units.

22. Find the area of the figure.

23. Stephen is using a shoebox for an art project. What is the area of the bottom of the shoebox?

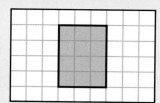

9-7 **Problem-Solving Investigation: Choose a Strategy** (pp. 402–403)

Example 7
Every 10 seconds Winston can do 8 jumping jacks. How many jumping jacks can he do in 1 minute? (*Hint*: 60 seconds = 1 minute)

60 seconds = 1 minute

60 seconds ÷ 10 seconds = 6

Draw a picture.

So, he can do 48 jumping jacks in 1 minute.

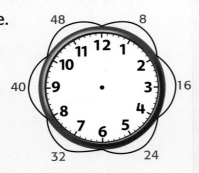

24. Victorio has $37. His aunt gives him $20 for his birthday. He owes his mother $15 for money he borrowed. How much money does he have now?

25. **Geometry** There is a sculpture of wire shapes. There are 4 triangles, 2 squares, and 5 circles. What is the total length of wire used in meters?

Shapes in Sculpture	
Shape	**Amount of Wire Needed**
Triangle	25 cm
Square	50 cm
Circle	100 cm

9-8 **Measure Temperature** (pp. 408–411)

Example 8
The thermometer shows the outside temperature. What is the temperature in degrees Fahrenheit?

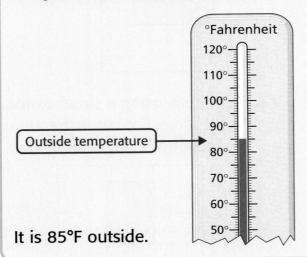

It is 85°F outside.

Write each temperature in degrees Fahrenheit (°F).

26.

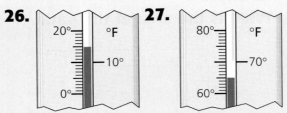

27.

28.

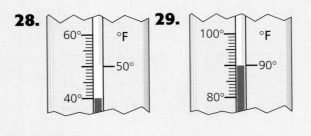

29.

For Exercises 1–2, decide whether each statement is *true* or *false*.

1. A centimeter is about the width of your index finger.

2. Area is the distance around a figure.

Measure each object to the nearest half inch.

3.

4.

Choose the most appropriate unit to measure each length. Write *inch*, *foot*, *yard*, or *mile*.

5. length of your bedroom

6. length of your finger

Choose the better estimate.

7. the height of a ladder
2 millimeters or 2 meters

8. length of a bridge
100 km or 1 km

9. **MULTIPLE CHOICE** Which of the following is longer than a centimeter?

A ladybug   C pen

B staple   D thumbtack

10. Chia starts school at 8:30. It takes her 30 minutes to get dressed and 15 minutes to eat. Her walk to school is 5 minutes long. What time does she get up?

11. Write the temperature in degrees Fahrenheit (°F).

Find the perimeter of each figure.

12. 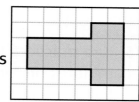 9 ft, 12 ft, 11 ft

13. 9 in., 13 in., 13 in., 5 in.

Find the area of the figure.

14. The diagram shows the shape of a room. What is the area?

15. **MULTIPLE CHOICE** Choose the best estimate for the length of a soccer field.

F 90 km   H 90 inches

G 90 m   J 90 miles

16. **WRITING IN ►MATH** Can you find the perimeter of your desk if you know the length and width? Explain.

PART 1 **Multiple Choice**

Read each question. Then fill in the correct answer on the answer sheet provided by your teacher or on a sheet of paper.

1. Which is the most appropriate unit to measure the length of a shoe?

A centimeter **C** meter

B liter **D** kilometer

2. What is the length of the pencil to the nearest inch?

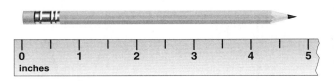

F 3 inches **H** 5 inches

G 4 inches **J** 6 inches

3. Kirti practices the piano for 52 minutes on Monday. She practices for 39 minutes on Tuesday. Which can be used to find how much more she practices on Monday?

A 52 + 39 **C** 52 × 39

B 52 − 39 **D** 52 ÷ 39

4. Which is the most appropriate unit to measure the length of a river?

F millimeter **H** yard

G inch **J** mile

5. Which list shows the numbers of eggs Kim could buy if she buys complete cartons?

A 12, 16, 20 **C** 12, 18, 24

B 12, 18, 22 **D** 12, 18, 26

6. The product of 6 and another factor is 42. What is the missing factor?

F 5 **H** 8

G 7 **J** 9

7. The shaded part of the figure below represents Molly's family room. What is the area of the room?

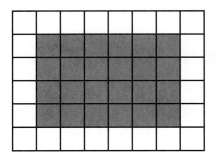

A 18 square units

B 24 square units

C 36 square units

D 48 square units

8. What is the length of the triangle to the nearest centimeter?

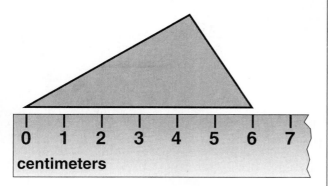

centimeters

F 4 centimeters   **H** 6 centimeters

G 5 centimeters   **J** 7 centimeters

---

9. What is the sum of 33 + 237 rounded to the nearest ten?

A 260   **C** 275

B 270   **D** 280

---

10. What is the area of a rectangle that is 4 square units long and 3 square units wide?

F 7 square units

G 8 square units

H 12 square units

J 15 square units

Preparing for Standardized Tests
For test-taking strategies and practice, see pages R50–R63.

PART 2   Short Response

Record your answers on the answer sheet provided by your teacher or on a sheet of paper.

11. Find the perimeter of the rectangle.

52 feet

13 feet

---

12. Measure to the nearest half inch.

PART 3   Extended Response

Record your answers on the answer sheet provided by your teacher or on a sheet of paper.

13. What is the perimeter of a house that measures 34 feet long and 24 feet wide? Explain your strategy.

NEED EXTRA HELP?													
If You Missed Question...	1	2	3	4	5	6	7	8	9	10	11	12	13
Go to Lesson...	9-4	9-2	9-7	9-2	5-8	5-4	9-6	9-4	1-8	9-6	9-5	9-1	9-5

Measure Capacity, Weight, Volume, and Time

BIG Idea What are capacity, weight, and volume?

The amount a container can hold is called its **capacity**. **Weight** measures how heavy an object is. **Volume** measures the number of cubic units needed to fill a space.

Example The table shows the capacity of different foods.

Capacity of Containers	
Food	**Capacity**
Small jar of jam	1 cup
Large jar of pickles	1 pint
Jar of pasta sauce	1 quart

- Estimate and measure capacity, weight, volume, and time.
- Choose the most appropriate unit of measure.
- Tell time.
- Solve a problem by guessing and checking.

Key Vocabulary

capacity

weight

volume

Math Online > **Student Study Tools** at **macmillanmh.com**

FOLDABLES®
Study Organizer

Use your Foldables to help you organize information about capacity, weight, volume, and time. Begin with one sheet of $8\frac{1}{2}"$ by $11"$ paper.

① **Make** a shutter fold.

② **Fold** the shutter fold in half like a hamburger.

③ **Open** the paper and cut along the two inside valley folds.

④ **Label** as shown. Record what you learn.

| Capacity | Weight |
| Volume | Time |

Chapter 10 Measure Capacity, Weight, Volume, and Time **421**

You have two ways to check prerequisite skills for this chapter.

Option 2

Math Online ▷ Take the Chapter Readiness Quiz at macmillanmh.com.

Option 1

Complete the Quick Check below.

QUICK Check

Identify the object that holds more. (Prior grade)

1.

2.

3. Which should hold more water, a swimming pool or a bathtub?

Identify which object weighs more. (Prior grade)

4.

5.

6. Consuela took two equal-sized glasses of milk to the table. One of the glasses was full, and the other was half full. Which glass weighed less?

Write the time shown on each clock. (Prior grade)

7.

8.

9. The mall opens at 8:00 in the morning. Suppose a customer began shopping when it opened and left three hours later. What time did the customer leave?

The amount a container can hold is called its **capacity**. Common containers and units for capacity are shown.

cup

A yogurt container is about 1 cup.

pint

A can of vegetables is about 1 pint.

quart

A bottle of ketchup is about 1 quart.

gallon

A large jug of milk is a gallon.

MAIN IDEA

I will estimate and measure capacity.

You Will Need
cup, pint, quart, and gallon containers; a variety of containers; water

New Vocabulary

capacity

Math Online

macmillanmh.com
• Concepts in Motion

ACTIVITY

1 **Estimate and measure capacity.**

Step 1 **Cup, Pint, Quart, and Gallon**

• Estimate how many cups a pint, quart, and gallon will hold.

• Use the cup to fill the pint, quart, and gallon.

• How many cups are in a pint, quart, and gallon? Record the results in a table.

Units of Capacity			
	Pint	**Quart**	**Gallon**
Cups	2	▉	▉
Pints		▉	▉
Quarts			▉

Step 2 **Pint, Quart, and Gallon**

• Make an estimate. How many pints do you think a quart container holds? a gallon container? How many quarts do you think a gallon container holds?

• Use the empty can (about 1 pint) to fill each. Use the empty ketchup bottle (about 1 quart) to fill the gallon container.

- How many pints are in a quart? in a gallon? Record the results.

- How many quarts are in a gallon?

ACTIVITY

2 **Measure the capacity of various containers.**

Use the containers that are about a cup, pint, quart, and gallon to measure the capacity of different containers from your teacher. Record the results in a table.

Think About It

1. In Activity 1, how close were your estimates to the exact answers?

2. Name some containers from around your home that would be about the same size as a cup, pint, quart, and gallon.

3. What happens to the number of containers needed as the capacity of the container gets larger? smaller?

✓ CHECK What You Know

Compare. Use >, <, or =.

4. cup ● gallon

5. gallon ● pint

6. quart ● cup

7. cup ● pint

8. quart ● gallon

9. 2 cups ● pint

Copy and complete. Use your table.

10. 2 pints ● 2 gallons

11. 4 pints ● 4 cups

12. 9 quarts ● 9 cups

13. 2 cups ● 1 gallon

14. 1 pint ● 10 cups

15. 2 quarts ● 2 pints

16. **WRITING IN ►MATH** How do you know that 10 gallons is equal to 40 quarts?

Customary Units of Capacity

MAIN IDEA

I will estimate and measure customary units of capacity.

New Vocabulary

capacity

cup (c)

pint (pt)

quart (qt)

gallon (gal)

Math Online

macmillanmh.com

• Extra Examples
• Personal Tutor
• Self-Check Quiz

GET READY to Learn

Armando is filling his younger sister's wading pool. How much water do you think he used to fill the pool?

The amount a container can hold is its **capacity**. The customary units of capacity are the **cup (c)**, **pint (pt)**, **quart (qt)**, and **gallon (gal)**.

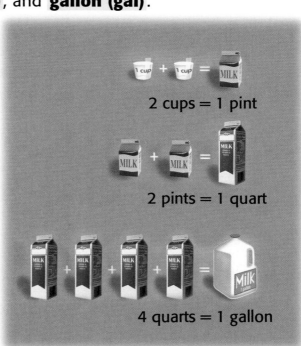

2 cups = 1 pint

2 pints = 1 quart

4 quarts = 1 gallon

Real-World EXAMPLE   Choose Units of Capacity

1) **POOLS** Choose the most appropriate unit Armando should use to measure the amount of water he used to fill the pool.

A cup, pint, and quart are too small. The most appropriate unit to use is the gallon.

Lesson 10-1 Customary Units of Capacity   **425**

2 **SOUP** **Choose the better estimate for the capacity of a soup bowl, 2 cups or 2 gallons.**

Think about a cup and a gallon.

It makes sense that the capacity of a soup bowl is about 2 cups, not 2 gallons.

So, the better estimate for the capacity of a soup bowl is 2 cups.

✓ CHECK What You Know

Choose the most appropriate unit to measure each capacity.
Write *cup, pint, quart,* or *gallon.* See Example 1 (p. 425)

1.

2.

3.

Choose the better estimate. See Example 2 (p. 426)

4. bucket of water
3 c or 3 gal

5. flower vase
2 c or 2 gal

6. Megan has 12 cups of punch. Does she have more or less than a quart?

7. Antonio is making hot chocolate for himself. Should he use 1 cup or 1 quart of milk? Explain.

8. Talk About It How do you decide which capacity unit to use?

**Choose the most appropriate unit to measure each capacity.
Write *cup*, *pint*, *quart*, or *gallon*.** See Example 1 (p. 425)

9.

10.

11.

Choose the better estimate. See Example 2 (p. 426)

12. watering can
2 qt or 2 pt

13. fish aquarium
5 gal or 10 qt

14. punch bowl
1 pt or 6 c

15. Choose the better estimate for the capacity of a pickle jar, 4 cups or 4 gallons.

16. Is the better estimate for the amount of water a person drinks each day 8 cups or 8 gallons?

17. Jackson showers for 10 minutes. Has he used about 15 cups or 150 gallons of water?

18. Would it be faster to fill a sink with a gallon container or a quart container?

19. Would 7 pints or 7 cups hold more?

20. Would 24 pints be more or less than 2 gallons?

H.O.T. Problems

21. OPEN ENDED Name three items from the grocery store that are measured by capacity.

22. CHALLENGE What two units of capacity does the table show?

First Unit	1	2	3	4	5	6
Second Unit	4	8	12	16	20	24

23. **WRITING IN MATH** Sumey and Tina are each filling the same size pails with water. Sumey has a pint container. Tina has a quart container. Who will fill the pail first? Explain.

24. Which container below can hold about 1 cup of liquid? (Lesson 10-1)

A

B

C

D

25. Fillipa has $5 in her wallet. She spent $4 on a poster and $7 on a watch. How much did Fillipa have in her wallet to begin with? (Lesson 8-3)

F $11

G $12

H $16

J $20

26. Angelo has 1 quart of orange juice. How many cups does he have? (Lesson 10-1)

A 2 cups   **C** 6 cups

B 4 cups   **D** 8 cups

Spiral Review

27. Measurement Write the temperature shown at the right in degrees Fahrenheit (°F). (Lesson 9-8)

28. Mr. Orta bought a hot dog at the fair for $3 and a beverage for $2. Now he has $5 left. How much money did he start with? (Lesson 9-7)

Choose the better estimate. (Lesson 9-2)

29. the thickness of a pillow
5 in. or 5 ft

30. the length of a tennis racket
2 ft or 2 yd

31. Algebra What will be the measure of one side of the next square if the pattern continues? (Lesson 5-3)

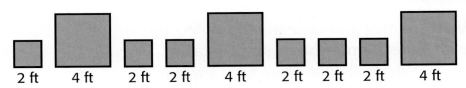

| 2 ft | 4 ft | 2 ft | 2 ft | 4 ft | 2 ft | 2 ft | 2 ft | 4 ft |

Capacity Guess

Estimate Capacity

Get Ready!

Players: 2 or more

Get Set!

Place a variety of empty containers on a table.

Go!

- Player 1 chooses an empty container and gives it to Player 2.

- Player 2 estimates its capacity and guesses which measuring container to use to check the capacity.

- Player 2 fills the measuring container with water and then pours the water into the empty container.

You will need: Labeled measuring containers (cup, pint, quart, and gallon).

Empty containers of different sizes and shapes.

- If the capacity is incorrect, Player 1 will get to estimate the capacity and check.

- Continue, switching roles each time. Each player scores one point for each correct answer.

Problem-Solving Strategy

MAIN IDEA I will solve a problem by using the guess and check strategy.

Octavia's class is making homemade clay. The same amount of flour is used to make red clay and green clay. They use 5 more cups of flour to make blue clay. The total amount of flour used is 29 cups. How much flour is used for each color of clay?

Understand	**What facts do you know?**
	• The total amount of flour is 29 cups.
	• The red and green clay use the same amount of flour.
	• The blue clay uses 5 more cups of flour than the other two colors.
	What do you need to find?
	• The amount of flour used for each color of clay.
Plan	You can use the *guess and check* strategy. Guess different combinations of numbers and check to see if they fit the facts in the problem.
Solve	The red and green clay use the same amount of flour. The blue clay uses 5 more cups than the others. The total flour used in 29 cups.

Guess			Check
Red	**Green**	**Blue**	**Total**
10	10	15	35 no
9	9	14	32 no
8	8	13	29 yes

So, the green and the red clay use 8 cups of flour each, and the blue clay uses 13 cups of flour.

Check	Look back.
	8 + 8 + 13 = 29 and 13 is 5 more 8.
	So, the answer is correct. ✔

ANALYZE the Strategy

Refer to the problem on the previous page.

1. How did the answer from your first guess affect your second guess?

2. Suppose the total number of cups of flour is 29, but each color of clay uses a different amount. The blue clay uses 5 more cups of flour than the green clay. How much flour does each color use?

3. You know that the sum of 9 cups, 9 cups, and 11 cups is 29 cups. Explain why it is an incorrect answer guess.

4. Explain when you would use the *guess and check* strategy to solve a problem.

PRACTICE the Strategy

EXTRA PRACTICE
See page R27.

Solve. Use the *guess and check* strategy.

5. A toy store sold $67 worth of bean-bag animals. How many of each size did they sell?

$6

$5

6. Dylan shares apple juice with his friends. He drinks 1 cup. His friends each drink one more cup than Dylan. There were 13 cups of juice in all. How many friends are there?

7. There are 30 apples in a basket. Half are red. There are 5 more green apples than yellow apples. How many of each color are there?

8. Dawnita has 8 coins. The total is $1. What are the coins?

9. Kira is thinking of two numbers. Their difference is 12 and their sum is 22. What are the numbers?

10. For lunch, Nadia bought 2 items. She spent exactly 70¢. What did she buy?

Food	Cost
Raisins	35¢
Apple	25¢
Granola bar	45¢
Grilled cheese	85¢

11. **WRITING IN ►MATH** Camden, Holden, and Nick have 10 pencils in all. They each have more than 1 pencil. Camden has the most pencils. Holden has the fewest pencils. How many pencils does each boy have? Explain how you would use the *guess and check* strategy to solve.

Lesson 10-2 Problem-Solving Strategy: Guess and Check   **431**

10-3 Metric Units of Capacity

MAIN IDEA

I will estimate and measure metric units of capacity.

New Vocabulary

liter (L)

milliliter (mL)

Math Online

macmilllanmh.com
• Extra Examples
• Personal Tutor
• Self-Check Quiz

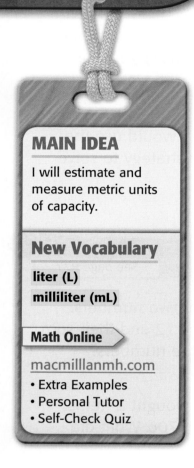

Hands-On Mini Activity

In this activity, you will explore the metric units of capacity.

Step 1 Use an eyedropper to find how many milliliters of water fit on the surface of a penny, and a bottle cap. Count every 10 drops as 1 milliliter.

Step 2 Use an empty 1 liter plastic soda bottle to find how many liters of water fit in a bucket.

You have learned that capacity refers to the amount a container can hold. The metric units of capacity are the **milliliter (mL)** and **liter (L)**.

milliliter (mL)

liter (L)

A dropper holds about 1 milliliter of liquid. This is about 10 drops. Use this unit to measure containers of small capacity.

This water bottle holds about 1 liter of liquid. Use this unit to measure larger containers of capacity.

Metric Units of Capacity
1 liter (L) = 1,000 milliliters (mL)

Remember

Milliliter–smaller unit
Liter–larger unit

Real-World EXAMPLE · **Choose Metric Units**

1 **BIRDS** Choose the unit that should be used to measure the amount of water a bird drinks each day.

A liter is too large. A bird would drink a small amount of water. So, the milliliter should be used.

Real-World EXAMPLE · **Estimate Capacity**

2 **FISH** Choose the better estimate for the amount of water in the aquarium, **50 mL or 5 L**.

50 mL is a small amount. It is not reasonable. Since 5 L is a larger amount of water 5 L is reasonable.

CHECK What You Know

Choose the most appropriate unit to measure each capacity. Write _milliliter_ or _liter_. See Example 1 (p. 433)

1. bucket

2. wading pool

3. spoon

Choose the better estimate. See Example 2 (p. 433)

4.

3 mL or 3 L

5.

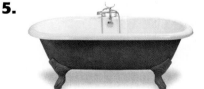

40 L or 4 mL

6.

50 mL or 50 L

7. Karl is using a spoon to measure honey for his tea. Is each spoonful 5 mL or 5 L?

8. **Talk About It** Describe some items sold in a grocery store that come in liter containers.

Lesson 10-3 Metric Units of Capacity **433**

Choose the most appropriate unit to measure each capacity. Write *milliliter* or *liter*. See Example 1 (p. 433)

9. pot of soup

10. juice box

11. pitcher of lemonade

12. bottle of glue

13. water bottle

14. fish tank

Choose the better estimate. See Example 2 (p. 433)

15.

250 L or 250 mL

16.

100 mL or 100 L

17.

10 mL or 10 L

18.

2 mL or 2 L

19.

5 mL or 5 L

20.

200 mL or 200 L

Real-World PROBLEM SOLVING

Food A punch recipe is shown.

21. How many liters of punch will this recipe make?

22. How many liters of punch did the guests drink if there were 650 milliliters left after the party?

23. Mona made 5 liters of punch. She used seven 500-milliliters bottles of grape juice. How many milliliters of grape juice did she use?

Party Punch

3 L pineapple juice
1 L apple juice
1500 mL soda water
500 mL grape juice

Pour into a large punch bowl and chill.

H.O.T. Problems

24. OPEN ENDED Name an item that has a capacity of 1 liter.

25. **WRITING IN MATH** Think about a liter and a milliliter. What do you think the prefix "milli-" means? Explain.

Choose the most appropriate unit to measure each capacity. Write *cup, pint, quart,* or *gallon*. (Lesson 10-1)

1.

2.

Choose the better estimate. (Lesson 10-1)

3.

2 c or 1 qt

4.

2 gal or 3 qt

Solve. Use the *guess and check* strategy. (Lesson 10-2)

5. Find the number.
- It is less than 30 and greater than 10.
- It is a multiple of 3.
- It is an odd number.
- The sum of its digits is 9.

6. MULTIPLE CHOICE Eva went on vacation 6 days longer than Suzie. Which of these shows the number of days that could have been Eva's and Suzie's vacations? (Lesson 10-2)

A Eva 9, Suzie 6

B Eva 12, Suzie 7

C Eva 15, Suzie 9

D Eva 17, Suzie 9

7. Tanya bought 4 items on sale. She spent $90 and saved $30. The table below shows the original and sale prices of each item. How many of each item did she buy? (Lesson 10-2)

Item	Original	Sale
T-shirt	$20	$10
Jeans	$40	$35
Mirror	$15	$13
Blanket	$25	$10

8. MULTIPLE CHOICE Which can hold more than one liter? (Lesson 10-3)

F eye dropper

G cereal bowl

H bath tub

J water balloon

Choose the most appropriate unit to measure each capacity. Write *milliliter* or *liter*. (Lesson 10-3)

9. spoon

10. water bottle

11. **WRITING IN ►MATH** Explain how you know that 1 gal is greater than 1 mL. (Lesson 10-3)

10-4 Problem-Solving Investigation

MAIN IDEA I will choose the best strategy to solve a problem.

P.S.I. TEAM +

CHARLIE: Ria, Marcela, and I were making lemonade for our lemonade stand. I made 3 more liters than Marcela. Ria made 7 liters. Marcela made one less liter of lemonade than Ria.

YOUR MISSION: Find how many liters of lemonade were made in all.

Understand	You know Charlie made 3 liters more than Marcela. Marcela made 1 less than Ria, who made 7. Find how many liters they made in all.
Plan	Organize the data into a table. Then solve.
Solve	The table shows the facts in the problem.

Student	Liters of Lemonade
Ria	7
Marcela	$7 - 1 = 6$
Charlie	$6 + 3 = 9$

$7 + 6 + 9 = 22$
So, the students made 22 liters of lemonade altogether.

Check	Look back. Draw a picture.

Ria   **Marcela**   **Charlie**

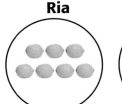 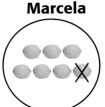 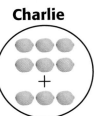

So, the answer is correct. ✓

EXTRA **PRACTICE**
See page R27.

Use any strategy shown below to solve. Tell what strategy you used.

PROBLEM-SOLVING STRATEGIES

- Look for a pattern.
- Make a table.
- Work backward.
- Guess and check.

1. Sumi has 4 cups of punch left. Holly drank 3 cups, Brandon drank 2 cups, and 7 cups were spilled. How many pints of punch did Sumi have to start with?

2. Dustin rides his bike 15 blocks home from school. Each block takes about 2 minutes. About how long will it take him to get home? Explain.

3. Eight oranges make 1 liter of orange juice. How many more oranges would be needed to make a total of 8 liters of orange juice?

4. At an amusement park, the Ghost Castle ride can hold 4 children in each car. If there are 43 children on the ride, how many cars are needed? Explain.

5. Candice walked 2 kilometers in 1 hour. How long did it take her to walk 500 meters?

6. There are 36 frogs in a pond. Four more are on lily pads than are on the banks of the pond. There are a total of 58 frogs. How many frogs are on lily pads and on the banks?

7. How many meters of windows are there?

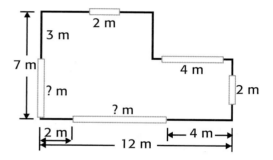

8. **Algebra** Ilan mowed 15 lawns in 5 days. At this rate, how many days will it take him to mow 33 lawns?

9. Ciro played at the park with 5 friends for 30 minutes, 7 friends for 1 hour, and 2 friends for 15 minutes. If he arrived at the park at 1:00 P.M., what time did he leave?

10. **WRITING IN ►MATH** Heath builds wood boxes. He has 12 pieces of wood that are each 2 meters long. Each box uses 8 pieces of wood that are 1-meter long each. How many boxes can he make? Explain.

Hands-On Mini Activity

Step 1 Use a balance scale and weights to find objects that weigh about 1 ounce.

Step 2 Use a balance scale and weights to find objects that weigh about 1 pound.

A scale is often used to measure how heavy something is, or its **weight**. The customary units of weight are **ounce (oz)**, **pound (lb)**, and **ton**.

A golf ball weighs about 1 ounce.

A soccer ball weighs about 1 pound.

A car weighs about 1 ton.

A soccer ball has about the same weight as 16 golf balls.
A car has about the same weight as 2,000 soccer balls.

1 **FOOD** Choose the unit to use to measure the weight of a slice of bread. Write *ounce, pound,* or *ton.*

A pound and ton are too large. The most appropriate unit to use is the ounce.

So, use ounces to measure the slice of bread.

---

2 **BOOKS** Choose the better estimate for the weight of these books, 3 ounces, 3 pounds, or 3 tons.

Think about an ounce, a pound, and a ton. It makes sense that the weight of the books is about 3 pounds, not 3 ounces or 3 tons.

 CHECK **What You Know**

Choose the most appropriate unit to measure the weight of each object. Write *ounce, pound,* or *ton.* See Example 1 (p. 439)

1.    **2.**   **3.**

Choose the better estimate. See Example 2 (p. 439)

4.    compact disc
2 oz or 2 lb

5.    pineapple
5 lb or 5 tons

6. A price tag was placed over the unit of weight on a television box. Is the unit an ounce or pound? Explain.

7. **Talk About It** Explain how you can use a balance scale and golf balls to measure the weight of items.

Choose the unit you would use to measure the weight of each object. Choose from *ounce, pound,* or *ton*. See Example 1 (p. 439)

8.

9.

10.

11. bag of soil

12. paper clip

13. walrus

Choose the better estimate. See Example 2 (p. 439)

14.

pair of socks
2 lb or 2 oz

15.

elephant
5 tons or 5 oz

16.

light bulb
3 lb or 3 oz

17. Marshal's bag of peanuts weighs 5 pounds. His box of crackers weighs 6 ounces. Which weighs more?

18. Cam bought 5 bananas. Would they weigh 5 pounds or 5 tons in all?

19. An ostrich egg weighs 4 pounds, and an Emperor Penguin's egg weighs 24 ounces. Which egg weighs more?

20. One bag of grapefruit weighs 3 pounds 8 ounces. A bag of oranges weighs 2 pounds 2 ounces. What is the difference between the weights?

Data File

Many animals are native to South Carolina.

21. Which weighs more, an American alligator or a Whitetail deer?

22. Which animal weighs about twice as much as a soccer ball? Explain.

South Carolina Wildlife

Animal	Weight	Length
American alligator	600 lb	8 feet
Eastern cottontail	2 lb	15 inches
Whitetail deer	30 lb	6 feet

Source: Columbia Audubon

H.O.T. Problems

23. OPEN ENDED Name four objects that would weigh more than 1 pound.

24. CHALLENGE How many ounces are in 3 pounds?

25. **WRITING IN ▶MATH** Explain what would happen to a balance scale if a 2-pound weight was on one side and a 24-ounce weight was on the other side.

TEST Practice

26. Mr. Mason's class has 9 students. Each student has 3 sharpened pencils and 2 unsharpened pencils. What is the total number of pencils? (Lesson 10-4)

A 18

B 27

C 45

D 50

27. Which of the following objects weighs more than 1 pound?
(Lesson 10-5)

F

H

G

J

Spiral Review

28. Seven boys and 4 girls are playing a game. Each person has 3 points. How many points are there in all? (Lesson 10-4)

Choose the most appropriate unit to measure each capacity. Write *milliliter* or *liter*. (Lesson 10-3)

29. bowl of soup

30. fish aquarium

31. hamster waterbottle

Algebra Find and extend the rule for each table. Then copy and complete. (Lesson 8-4)

32.

Rule: ▨				
Input	2	5	7	8	9
Output	20	▨	70	▨	▨

33.

Rule: ▨				
Input	6	4	3	8	9
Output	▨	16	▨	32	▨

Problem Solving in Science

Lengths, Heights, and Weights *Oh My!*

Animals come in different shapes and sizes. For example, a ruby-throated hummingbird is a little more than $3\frac{1}{2}$ inches long and weighs less than an ounce. On the other end of the scale, a rhinoceros can weigh 2,200 pounds. A beetle is less than an inch long, but a giraffe can be more than 18 feet tall.

Animals also have different characteristics to help them live in their environments. Elephants can hold up to 3 gallons of water in their trunks. A pelican's stomach can hold up to one gallon of food at a time.

Amazing Animals

Animal	Size	Weight (lb)
Zebra	6 ft	530
Ostrich	9 ft	345
Alligator	9 ft	1,000
Tree frog	13 in.	6

 Real-World Math

Use the information on page 442 to solve each problem.

1. What is the length of a ruby-throated hummingbird rounded to the nearest inch?

2. What is the most appropriate unit to measure the height of a rhinoceros?

3. How many yards tall is a giraffe?

4. About how many feet longer is an alligator than a tree frog?

5. How many pints of food can a pelican hold in its stomach?

6. How many more quarts of water can an elephant hold in its trunk than a pelican can hold food in its stomach?

7. How many ounces does a tree frog weigh? (*Hint:* 16 ounces = 1 pound)

8. An alligator is about 8 inches when it is born. Is a tree frog twice that length? Explain.

9. Which animal is almost two times the weight of the zebra?

MAIN IDEA

I will estimate and measure units of mass.

New Vocabulary

mass
gram (g)
kilogram (kg)

Math Online

macmillanmh.com
• Extra Examples
• Personal Tutor
• Self-Check Quiz

> **GET READY to Learn**

Virginia's dad bought a new baseball bat. What do you think is the mass of the baseball bat?

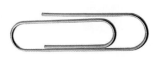

Mass is the amount of material in an object. The most common metric units used to measure the mass of an object are the **gram (g)** and **kilogram (kg)**.

A paper clip has the mass of about 1 gram.

A baseball bat has the mass of about 1 kilogram.

=

1,000 paper clips have about the same mass as one baseball bat.

Metric Units of Mass
1,000 grams (g) = 1 kilogram (kg)

> **Real-World EXAMPLE** Choose Units of Mass

① **SNACKS** **Choose the unit you would use to measure the mass of a pretzel.**

Pretzels are small. It would be more reasonable to measure their mass in grams.

Hands-On Mini Activity

Materials: balance scale, 1-kilogram mass

Step 1 Choose three small objects. Estimate if each one is *less than*, *more than*, or *about* 1 kilogram. Record your estimates in a table like the one shown.

Less than 1 kilogram	1 kilogram	More than 1 kilogram

Step 2 Check your estimates using the balance scale and a 1-kilogram mass.

1. Choose one of the objects. Do you think that it is *less than*, *more than*, or *about* 1 kilogram? Explain.

2. Identify two more objects that would be about 1 gram.

Remember

A kilogram is a little more than 2 pounds.

You can use what you know about the gram and kilogram to estimate and compare mass.

Real-World EXAMPLE Estimate Mass

2 VEGETABLES Choose the better estimate for the mass of a large squash, 500 grams or 500 kilograms.

Think about a gram and a kilogram.

Since a large squash does not have the same mass as 500 baseball bats, we can see that the estimate should be 500 grams. So, the best estimate is 500 grams.

Choose the most appropriate unit to measure each mass.
Write *gram* or *kilogram*. See Example 1 (p. 444)

1. toothbrush   **2.** orange   **3.** shovel

Choose the better estimate. See Example 2 (p. 445)

4.

5 g or 5 kg

5.

50 g or 5,000 g

6.

4 g or 4 kg

7. *Talk About It* Does a large object always have a greater mass than a small object? Explain.

Practice and Problem Solving

EXTRA PRACTICE
See page R28.

Choose the most appropriate unit to measure each mass.
Write *gram* or *kilogram*. See Example 1 (p. 444)

8. teddy bear   **9.** lawn mower   **10.** child

11. bag of pretzels   **12.** pair of sunglasses   **13.** pencil

Choose the better estimate. See Example 2 (p. 445)

14.

15 g or 15 kg

15.

900 g or 900 kg

16.

2 g or 2 kg

17. large ball
500 g or 50 kg

18. apple
160 g or 160 kg

19. cordless phone
200 g or 200 kg

H.O.T. Problems

20. OPEN ENDED A bag of potatoes has a mass of about 3 kilograms. Name two other items that have about the same mass. Explain your reasoning.

21. WHICH ONE DOESN'T BELONG? Identify the unit that does not belong. Explain.

liters   grams   meters   milliliters

22. WRITING IN ►MATH Explain how to change from kilograms to grams.

TEST Practice

23. Elton has 3 more rocks in his rock collection than Curtis. Together they have 35 rocks. How many rocks does each boy have? (Lesson 10-4)

A Elton has 17 and Curtis has 18.

B Elton has 19 and Curtis has 16.

C Elton has 20 and Curtis has 18.

D Elton has 16 and Curtis has 19.

24. There are 1,000 grams in 1 kilogram. How many grams are in 7 kilograms? (Lesson 10-6)

F 70 grams

G 700 grams

H 1,000 grams

J 7,000 grams

Spiral Review

Choose the most appropriate unit to measure each weight. Write *ounce, pound,* or *ton.* (Lesson 10-5)

25. truck   **26.** bumble bee   **27.** dog

28. The soup that Alaina is making needs 8 liters of water. She only has a 500-milliliter container. How can she use that container to measure 8 L? (Lesson 10-4)

Estimate. Round to the nearest hundred. (Lesson 3-2)

29. 769
        − 389

30. 457
        − 253

31. 810
        − 729

Volume is the number of cubic units needed to fill a space. Volume is measured in **cubic units**.

MAIN IDEA

I will use models to explore volume.

You Will Need
cubes, small containers

New Vocabulary

volume
cubic unit

Math Online

macmillanmh.com
• Extra Examples
• Personal Tutor
• Self-Check Quiz

ACTIVITY

1 Estimate and measure volume.

Step 1 Estimate the volume.
A small paper clip box is shown. How many cubes do you think will fit into the box?

Step 2 Fill the box.
Place the cubes in the box in rows as shown until it is full.

Step 3 Find the volume.
Remove the cubes. Count to find how many filled the box. This is the volume. Compare this to your estimate.

ACTIVITY

2 **Estimate and measure volume.**

Step 1 **Estimate the volume.**

Estimate the volume of the figure.

Step 2 **Build the figure.**

Use cubes to build the figure.

Step 3 **Find the volume.**

Count the number of cubes it took to build this figure. How does the actual volume compare to your estimate?

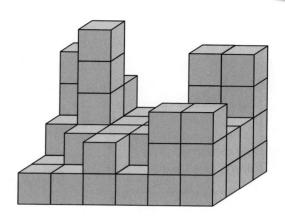

Think About It

1. Do three-dimensional figures with the same volume always have the same shape? Explain.

2. Can you use the same number of cubes to make different figures that have different volumes? Explain.

3. Do you think you can find the exact volume of a round object using cubes? Explain.

CHECK What You Know

Estimate the volume of each figure. Then use unit cubes to build the figure and find its exact volume.

4.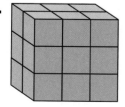

5.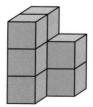

6.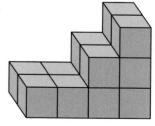

Build two different figures that have the same volume. Use unit cubes.

7. 12 cubic units

8. 26 cubic units

9. 30 cubic units

10. **WRITING IN ►MATH** Explain how objects that are different shapes and different sizes can have the same volume.

10-7 Estimate and Measure Volumes

MAIN IDEA

I will determine the volume of a figure.

New Vocabulary

volume

cubic unit

Math Online

macmillanmh.com

• Extra Examples
• Personal Tutor
• Self-Check Quiz

> ### GET READY to Learn
>
> Jermaine just finished cleaning his fish's aquarium. Now, he has to fill it back up with water. He is trying to determine how many cubic units of water will fill up the fish tank.

Volume is the number of unit cubes a three-dimensional figure holds. Volume is measured in **cubic units**.

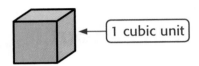

1 cubic unit

To find the volume of a three-dimensional figure, you can count the number of unit cubes needed to fill the figure.

Real-World EXAMPLE   Find Volume

1 **WATER** **How many cubic units of water does the fish tank hold?**

There are 12 cubic units on each layer.

The aquarium will hold 3 layers of 12 cubic units. Add to find how many cubic units are needed.

$12 + 12 + 12 = 36$

So, the aquarium holds 36 cubic units.

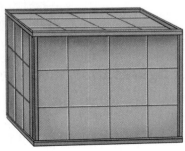

You can use what you know about volume to estimate the number of cubes that will fill a solid.

EXAMPLE Find Volume

2 **Estimate the volume of the figure.**

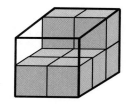

The bottom layer of the figure has a volume of 6 cubic units. The box has two layers.

So, the volume of the box is 12 cubic units.

CHECK What You Know

Use concrete models to find the volume of each figure. See Example 1 (p. 450)

1.

2.

3.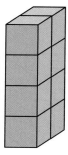

Estimate the volume. Use concrete models to help. See Example 2 (p. 451)

4.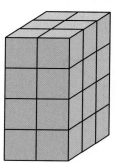

5.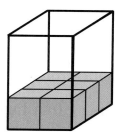

6.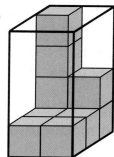

7. This gift has a volume of 16 cubic units. The length is 4 units, and the width is 2 units. What is the height of the gift? Use models if needed.

2 units   4 units

8. Talk About It If you know the volume of a figure, do you know its dimensions? Why?

Use concrete models to find the volume of each figure. See Example 1 (p. 450)

9.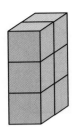

10.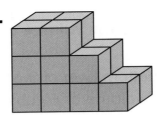

11.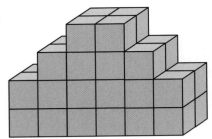

Estimate the volume. Use concrete models to help. See Example 2 (p. 451)

12.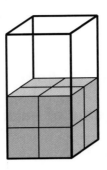

13.

14.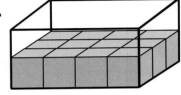

15.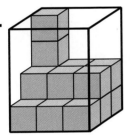

16.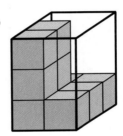

17.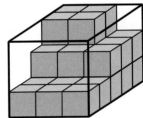

Solve. Use models.

18. Sara wants to know the volume of her music box. She filled it with cubes. It was 6 units long, 5 units high, and 4 units wide. What is the volume of Sara's music box?

19. A book is 8 units long, 6 units wide, and 1 unit high. Use a model to find the volume.

Copy and complete. Use unit cubes.

	Volume			
	Length	**Width**	**Height**	**Unit Cubes**
20.	5	2	▪	20
21.	2	6	3	▪
22.	4	▪	2	32

23. FIND THE ERROR Hunter and Jessica are making figures that have a volume of 6 cubic units. Whose figure is correct? Explain.

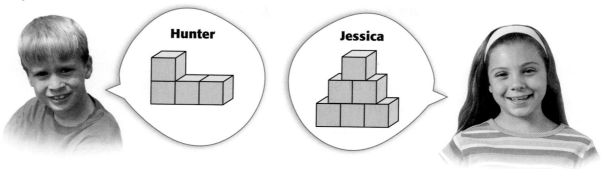

Hunter

Jessica

24. **WRITING IN** ➤**MATH** How are area and volume different?

TEST Practice

25. Which is the best estimate of the mass of a coin? (Lesson 10-6)

A 1 gram

B 100 grams

C 1 kilogram

D 100 kilograms

26. Which of the figures below has a volume of less than 7 cubic units? (Lesson 10-7)

F

H

G

J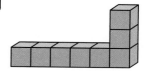

Spiral Review

Choose the better estimate. (Lesson 10-6)

27. wristwatch
60 g or 60 kg

28. tire
35 g or 35 kg

29. rabbit
2 g or 2 kg

Solve. (Lesson 10-5)

30. Tori bought a 10 lb bag of potatoes. April bought ten potatoes that each weighed 10 oz. Whose collection of potatoes weighed more?

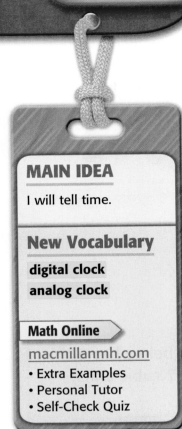

MAIN IDEA

I will tell time.

New Vocabulary

digital clock

analog clock

Math Online

macmillanmh.com
• Extra Examples
• Personal Tutor
• Self-Check Quiz

GET READY to Learn

Jake looked at his watch at the end of soccer practice. What time did soccer practice end?

Jake's watch is digital. A **digital clock** shows the time in numbers.

Real-World EXAMPLE

1 TIME Write the time shown on Jake's digital watch.

The digits before the colon (:) show the hour.

The digits after the colon (:) show the minutes.

Read twelve fifteen   **Write** 12:15

An **analog clock** has an hour hand and a minute hand.

EXAMPLE

2 TIME Write the time shown on the analog clock.

Step 1 Find the hour.

The shorter hand is the hour hand.
It has passed the 5. So, the hour is 5.

Step 2 Count the minutes.

The longer hand is the minute hand. Start at 12.
Count by fives, then count on by ones.
5, 10, 15, 20, 25, 30, 35, 40, 45 then 46, 47
There are 47 minutes.

Read five forty-seven   **Write** 5:47

Write the time shown on each digital or analog clock.

See Examples 1 and 2 (p. 454)

1.

2.

3.

4. If the minute hand is pointing to the number 2, how many minutes past the hour is the clock showing?

5. **Talk About It** Which do you think is more difficult to read, an analog clock or a digital clock? Explain.

Practice and Problem Solving

EXTRA PRACTICE
See page R29.

Write the time shown on each digital or analog clock.

See Examples 1 and 2 (p. 454)

6.

7.

8.

9.

10.

11.

12. If the minute hand is pointing to the number 7, how many minutes is it showing?

13. The clock on the wall showed 8:45. What time did it show 7 minutes later?

H.O.T. Problems

14. **OPEN ENDED** Draw a digital clock with a time on it. Then describe in writing where the hour hand and minute hand would be pointing on an analog clock set at that time.

15. **WRITING IN ►MATH** Does the minute hand or the hour hand move faster on an analog clock? How do you know?

Study Guide and Review

Math Online macmillanmh.com
• STUDY TO GO
• Vocabulary Review

FOLDABLES
Study Organizer
GET READY to Study

Be sure the following Key Vocabulary words and Key Concepts are written in your Foldable.

Key Concepts

- Some customary units of **capacity** are cup, pint, quart, and gallon. (p. 425)

Units of Capacity
1 pint = 2 cups
1 quart = 2 pints
1 gallon = 4 quarts

- Some metric units of capacity are liter and milliliter. (p. 432)

Metric Units of Capacity
1 liter = 1,000 milliliters

- Some customary units of **weight** are ounce, pound, and ton. (p. 438)

Units of Weight
1 pound = 16 ounces

- Some metric units of **mass** are gram and kilogram. (p. 444)

Metric Units of Mass
1 kilogram = 1,000 grams

Key Vocabulary

capacity (p. 425)

gallon (p. 425)

liter (p. 432)

mass (p. 444)

volume (p. 448)

weight (p. 438)

Vocabulary Check

Choose the vocabulary word that completes each sentence.

1. ____?____ is the amount of matter in an object.

2. The number of unit cubes a figure holds is its ____?____.

3. The amount a container can hold is its ____?____.

4. Ounces and pounds measure ____?____ and grams and kilograms measure ____?____.

5. Cup, pint, quart, and gallon are all units of ____?____.

6. The most appropriate metric unit to measure the capacity of a fish tank is ____?____.

Lesson-by-Lesson Review

10-1 Customary Units of Capacity (pp. 425–428)

Example 1
Choose the better estimate for the capacity of a glass of juice, 1 cup or 1 gallon.

Think of a cup and a gallon.

about a cup   about a gallon

It makes sense that a glass of juice is about 1 cup, not 1 gallon.

Choose the better estimate.

7.   1 c or 1 qt

8.   2 qt or 2 pt

9.   2 gal or 2 c

10-2 Problem-Solving Strategy: Guess and Check (pp. 430–431)

Example 2

The value of an even number is more than 10 and less than 20. The sum of its digits is 5. What is the number?

Guess numbers between 10 and 20.

Guess		Check	
Number	Even?	Sum of Digits	
11	no	2	No
16	yes	7	No
14	yes	5	Yes

So, the number is 14.

Solve. Use the *guess and check* strategy.

10. David picked 40 raspberries for his friends. He gave 8 of his friends the same number of raspberries. How many raspberries did each friend receive?

11. Andrew has a combination of 8 quarters, dimes, and nickels that add up to a value of 95¢. How many of each coin does he have?

10-3 **Metric Units of Capacity** (pp. 432–434)

Example 3
Choose the better estimate for the amount water in an ice cube, 10 mL or 10 L.

Think of the size of a mL and L.

milliliter (mL)   liter (L)

So, the better estimate for the amount of water in an ice cube is 10 mL.

Choose the better estimate.

12.    13.

2 L or 2 mL   400 L or 400 mL

Choose the most appropriate unit to measure each capacity. Write *milliliter* or *liter*.

14. bottle of liquid soap   15. paper cup

10-4 **Problem-Solving Investigation: Choose a Strategy** (pp. 436–437)

Example 4

On Monday, a soccer team practiced two hours longer than on Tuesday. On Wednesday, they practiced for twice as long as Monday. They practiced 6 hours on Wednesday. How long was practice on Monday and Tuesday?

Wednesday half Monday
   6   ÷   2   =   3

Monday less Tuesday
   3   −   2   =   1

So, they practiced 3 hours on Monday and 1 hour on Tuesday.

Use any strategy to solve.

16. Jada's goal is to make 42 bracelets. She has already made 7 bracelets. She plans on making an equal number of bracelets each day for the next 5 days. How many bracelets should she make each day to meet her goal?

17. **Measurement** Soccer practice begins at 4:45 P.M. It ends 90 minutes later. What time does soccer practice end?

10-5 Customary Units of Weight (pp. 438–441)

Example 5
Choose the better estimate for the weight of a shoe, 5 ounces or 5 pounds.

Think of an ounce and a pound.

about an ounce   about a pound

5 pounds does not make sense.

So, a shoe weighs about 5 ounces.

Choose the most appropriate unit to measure the weight of each object. Write *ounce, pound,* or *ton*.

18.    19.

Choose the better estimate.

20. polar bear
    9 tons or
    900 lb

21. empty milk jug
    4 oz or 4 lb

10-6 Metric Units of Mass (pp. 444–447)

Example 6
Father put some strawberries in a bag for a snack. Choose the unit you would use to measure their mass.

You need to determine whether to use *grams* or *kilograms*.

Strawberries are small and it would not make sense to measure their mass with kilograms.

So, use grams to measure their mass.

Choose the most appropriate unit to measure each mass. Write *gram* or *kilogram*.

22. bulldozer   23. person

24. dog   25. greeting card

Choose the better estimate.

26.

80 kg or 800 kg

10-7 Measurement: Volume (pp. 450–453)

Example 7

Find the volume of the figure shown below.

There are 6 cubic units on each layer. There are 3 layers of 6 cubic units. Add to find how many cubic units will fill the figure.

6 + 6 + 6 = 18

So, the figure shows 18 cubic units.

Find the volume of each figure.

27.

28.

Solve. Use models.

29. Trina wants to know the volume of her bookshelf. She filled it with cubes. It was 10 units long, 6 units high, and 4 units wide. What is the volume of Trina's bookshelf?

10-8 Tell Time (pp. 454–455)

Example 8
Write the time shown on the analog clock.

Find the hour.

The hour hand has passed the 6. So, the hour is 6.

Count the minutes.

Start at 12. Count by fives. At the 4, the minute hand shows 20. Each mark between the hours shows minutes. There are 3 minutes past the 4.

3 + 20 minutes or 23 minutes.

So, the time is 6:23.

Write the time shown on each digital or analog clock.

30.

31.

32.

33.

34. If the minute hand is pointing to the number 9, how many minutes is it showing?

35. The clock on the wall shows 11:15. What time does it show 9 minutes later?

Chapter Test
Lesson 10-1 to 10-8

Tell whether each statement is *true* or *false*.

1. Capacity can be measured in inches.

2. Cubic units are used to measure volume.

Choose the most appropriate unit to measure each capacity. Write *cup, pint, quart,* or *gallon*.

3.

4.

5. **MULTIPLE CHOICE** What is the volume of the figure shown?

   A 7 cubic units   C 10 cubic units

   B 8 cubic units   D 16 cubic units

6. There are three dogs in the park. Fido is 5 pounds heavier than Buttercup. Buttercup is three times as heavy as Jupiter. Jupiter weighs 8 pounds. How much do Fido and Buttercup each weigh?

Choose the most appropriate unit to measure each capacity. Write *milliliter* or *liter*.

7. trash can

8. juice glass

Choose the better estimate.

9.

   9 g or 9 kg

10.

    32 oz or 2 tons

11. **MULTIPLE CHOICE** Art class began at 11:45. It ended 45 minutes later. What time did art class end?

    F 11:30   H 1:15

    G 12:30   J 1:45

Choose the most appropriate unit to measure each mass. Write *gram* or *kilogram*.

12.

13.

Solve.

14. Kim ordered 3 gallons of punch for the party. Should each person get a cup that holds 8 ounces or 8 quarts?

15. **WRITING IN ▶ MATH** Write a real-world problem that compares ounces and pounds. Solve.

Read each question. Then fill in the correct answer on the answer sheet provided by your teacher or on a sheet of paper.

1. Water freezes when it reaches the temperature shown on the thermometer. At what temperature does water freeze?

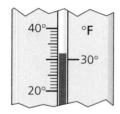

A 30°F   **C** 42°F

B 32°F   **D** 48°F

2. Which is most likely to be the mass of a bicycle?

F 15 centimeters **H** 15 liters

G 15 grams   **J** 15 kilograms

3. Which is most likely to be the capacity of a cocoa mug?

A 10 milliliters   **C** 2 liters

B 100 milliliters   **D** 5 liters

4. Rosie wants to place 12 balls of yarn into 4 baskets. Each basket will have the same number of balls of yarn. How many balls of yarn will be in each basket?

F 3   **H** 5

G 4   **J** 6

5. Ruthie left home at 7:45. She walked 25 minutes to get to school. What other information is needed to determine whether Ruthie arrived at school on time?

A how long school lasts

B how fast Ruthie walks

C the distance to school

D the time school begins

6. The table shows how many crayons come in different numbers of boxes. How many crayons come in one box?

Number of Boxes	Number of Crayons
2	24
4	48
6	72
8	96

F 8   **H** 12

G 10   **J** 14

Preparing for Standardized Tests

For test-taking strategies and practice, see pages R50–R68.

7. The clocks show the times at which Morena began and finished soccer practice. How long did soccer practice last?

Began   Finished

A 1 hour   **C** 2 hours

B $1\frac{1}{2}$ hours   **D** $2\frac{1}{2}$ hours

8. What is the volume of the solid?

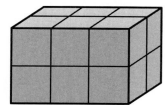

F 1 cubic unit   **H** 8 cubic units

G 2 cubic units   **J** 12 cubic units

9. Which is the most appropriate unit of measurement to use to find the capacity of a bathtub?

A milliliter   **C** cup

B gallon   **D** yard

PART 2   Short Response

Record your answers on the answer sheet provided by your teacher or on a sheet of paper.

10. Which unit would you use to measure the weight of a dog? Explain.

---

11. Order the following units of capacity from least to greatest.

pint, gallon, quart, cup

PART 3   Extended Response

Record your answers on the answer sheet provided by your teacher or on a sheet of paper.

12. What is the volume of the figure below? Describe a different figure that has the same volume but a different length, width, and height.

NEED EXTRA HELP?												
If you missed question...	1	2	3	4	5	6	7	8	9	10	11	12
Go to Lesson...	9-8	10-6	10-3	4-3	10-8	8-7	10-8	10-7	10-1	10-5	10-1	10-7

Identify Geometric Figures and Spatial Reasoning

BIG Idea How do two-dimensional and three-dimensional figures differ?

Two-dimensional figures, or plane figures, have length and width. **Three-dimensional figures**, or solid figures, have length, width, and height.

two-dimensional figures   three-dimensional figures

What will I learn in this chapter?

- Identify, describe, and classify two- and three-dimensional figures.
- Solve problems by solving a simpler problem.
- Identify and extend geometric patterns.
- Identify congruent figures.
- Locate and name whole numbers on a number line.

Key Vocabulary

congruent

polygon

symmetry

three-dimensional figure

two-dimensional figure

Math Online > **Student Study Tools**
at macmillanmh.com

FOLDABLES®
Study Organizer

Make this Foldable to help you organize information about geometric figures. Begin with one sheet of 11″ × 17″ paper.

1 **Fold** the paper so that both ends meet in the center.

2 **Fold** twice in the opposite direction as shown.

3 **Open** the paper and cut to make six flaps.

4 **Label** as shown. Record what you learn in the chapter.

Two-Dimensional Figures	Three-Dimensional Figures
Geometric Patterns	Congruent Figures
Symmetry	Number Lines

Chapter 11 Identify Geometric Figures and Spatial Reasoning **465**

You have two ways to check prerequisite skills for this chapter.

Option 2

Math Online ⟩ Take the Chapter Readiness Quiz at macmillanmh.com.

Option 1

Complete the Quick Check below.

QUICK Check

Identify the figure that does not belong with the other three. Explain. (Prior grade)

1.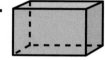

figure A   figure B   figure C   figure D

2.

figure F   figure G   figure H   figure J

3. Marilyn has a box, a can of soup, and a sheet of paper. Which object does not belong with the other two? Explain.

Tell how each pair of figures differ. (Prior grade)

4.    **5.**

6.    **7.**

8.    **9.**

10. Jeffrey and Hana each drew a different figure that has 8 sides. Draw an example of what the figures could look like.

11-1 Three-Dimensional Figures

MAIN IDEA

I will identify, classify, and describe three-dimensional figures.

New Vocabulary

three-dimensional figure

cube   cone

rectangular prism

pyramid   face

cylinder   edge

sphere   vertex

Math Online

macmillanmh.com

- Extra Examples
- Personal Tutor
- Self-Check Quiz

> **GET READY to Study**
>
> We see all of these objects around us everyday. They are three-dimensional figures.

A **three-dimensional figure** is a solid figure that has length, width, and height.

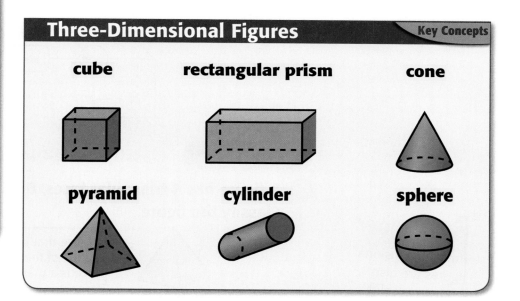

Three-Dimensional Figures   Key Concepts

cube   rectangular prism   cone

pyramid   cylinder   sphere

EXAMPLES   Identify Three-Dimensional Figures

1 **Identify each three-dimensional figure.**

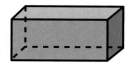

The figure is a rectangular prism.

A soup can looks like a cylinder.

A three-dimensional figure can be classified by its faces, edges, and vertices.

A **face** is a flat surface.

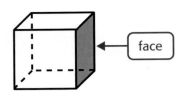

An **edge** is where 2 faces meet.

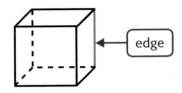

A **vertex** is the point, where 3 or more edges meet. When there is more than one vertex they are called vertices.

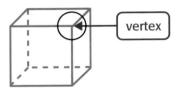

EXAMPLE   Classify Three-Dimensional Figures

2 **A figure has 4 triangular faces, 8 edges, and 5 vertices. Classify the figure.**

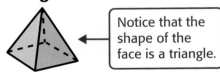

Notice that the shape of the face is a triangle.

So, the figure is a pyramid.

Remember

Three-dimensional figures are also known as solid figures.

EXAMPLE   Describe Three-Dimensional Figures

3 **Describe the figure shown.**

The figure has 6 faces, 12 edges, and 8 vertices.

CHECK What You Know

Identify each three-dimensional figure. See Example 1 (p. 467)

1.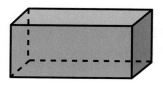

2.

3.

Classify each three-dimensional figure. See Example 2 (p. 468)

4. This figure has 1 circular face.

5. This figure has 2 circular faces.

Describe each three-dimensional figure. Use the terms *faces*, *edges*, **and** *vertices*. See Example 3 (p. 468)

6.

7.

8. Brandi and Ling are playing the drums. What solid figure do the drums represent?

9. Talk About It — How are the three-dimensional figures of a cone and cylinder different? How are they alike?

Practice and Problem Solving

EXTRA PRACTICE See page R29.

Identify each three-dimensional figure. See Example 1 (p. 467)

10.

11.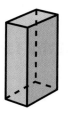

12.

13.

14.

15.

Classify each three-dimensional figure. See Example 2 (p. 468)

16. This figure has 6 faces. It is not a cube.

17. This figure has 1 face.

18. This figure has 0 faces 0 edges, and 0 vertices.

19. This figure has 4 triangular faces and 1 square face. It also has 8 edges and 5 vertices.

Describe each three-dimensional figure. Use the terms *faces, edges,* **and** *vertices.* See Example 3 (p. 468)

20.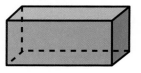
21.
22.
23.

24. Taye bought a box of cereal. What three-dimensional figure is a box of cereal?

25. Max threw his basketball through the hoop. What three-dimensional figure is a basketball?

26. A piece of clay is rolled into a sphere, then cut in half. How many faces does each half have?

27. If a cube is cut in half as shown, what three-dimensional figures are made?

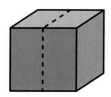

H.O.T. Problems

28. OPEN ENDED Name three real-world objects that you would find at home or in your classroom that resemble a cylinder.

REASONING Find the total area of all of the faces of each figure.

29.

30.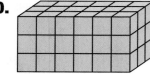

31. **WRITING IN ►MATH** What do a cube and a rectangular prism have in common?

32. A room is in the shape of a rectangle. Two walls measure 10 feet each and another wall measures 6 feet. What is the length of the other wall? (Lesson 9-5)

   A 6 feet

   B 8 feet

   C 10 feet

   D 12 feet

33. Which figure is a cylinder? (Lesson 11-1)

Spiral Review

Write the time shown on each analog or digital clock. (Lesson 10-8)

34.

35.

36.

Solve.

37. A box of crackers is 6 units long, 2 units wide, and 10 units high. What is the volume of the cracker box? (Lesson 10-7)

38. By the end of the game, the Wildcats had doubled their score, and the Cheetahs had increased their score by 3 points. What was the total number of points scored by the end of the game? (Lesson 10-4)

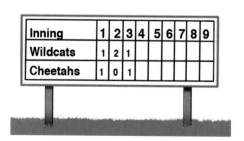

Inning	1	2	3	4	5	6	7	8	9
Wildcats	1	2	1						
Cheetahs	1	0	1						

Subtract. (Lesson 3-6)

39. 607 − 349

40. 800 − 725

41. $300 − $244

Algebra Find the missing numbers. (Lesson 1-1)

42. 175, 225, _____, 325, _____, _____

43. 520, 490, _____, 430, 400, _____, _____

11-2 Two-Dimensional Figures

GET READY to Study

Mary noticed that all stop signs are figures that have eight sides. The shape of the stop sign is a polygon.

MAIN IDEA

I will identify and classify two-dimensional geometric figures.

New Vocabulary

two-dimensional figure

polygon

triangle

quadrilateral

pentagon

hexagon

octagon

Math Online

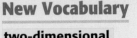

macmillanmh.com

• Extra Examples
• Personal Tutor
• Self-Check Quiz

A **two-dimensional figure**, or plane figure, has length and width. A **polygon** is a closed two-dimensional figure of 3 or more line segments and angles.

triangle
three sides
three angles

quadrilateral
four sides
four angles

pentagon
five sides
five angles

hexagon
six sides
six angles

octagon
eight sides
eight angles

Real-World EXAMPLES   Describe and Identify Two-Dimensional Figures

TRAFFIC SIGNS Describe and identify each polygon.

1

There are 8 sides and 8 angles. So, it is an octagon.

2

There are 5 sides and 5 angles. So, it is a pentagon.

Here are some examples and non-examples of polygons.

Polygons	
Examples	**Non-Examples**
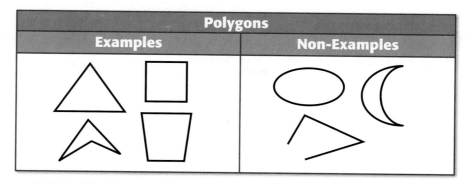	

 **EXAMPLE** Classify Two-Dimensional Figures

3 **A polygon has 6 sides and 6 angles. Classify the figure.**

The figure is a hexagon.

CHECK What You Know

Describe each two-dimensional figure. Use the terms *sides* and *angles*. Then identify the figure. See Examples 1 and 2 (p. 472)

1.

2.

3.

Classify each two-dimensional figure. See Example 3 (p. 473)

4. polygon with 3 sides and 3 angles

5. polygon with 6 angles

6. Bryson used a square pattern block and a triangular pattern block. He placed the edges together.

7. Talk About It Explain why the shape of the tambourine is not a polygon.

Describe each two-dimensional figure. Use the terms *sides* and *angles*. Then identify the figure. See Examples 1 and 2 (p. 472)

8.

9.

10.

11.

12.

13.

14.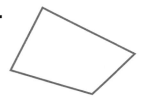

15.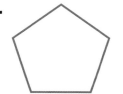

16.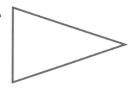

Classify each two-dimensional figure. See Example 3 (p. 473)

17. polygon with 4 sides and 4 angles

18. polygon with 8 sides and 8 angles

19. polygon with less angles than a quadrilateral

20. polygon with 3 long sides and 5 short sides

21. What type of polygon is a square?

22. Dario's shirt has a polygon printed on it. The polygon has 6 sides. What is the name of this polygon?

23. Is the figure shown a polygon? Explain.

24. What three-sided polygon do you get when you fold a square in half, corner to corner?

Identify each face as a two-dimensional figure.

25.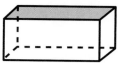

26.

27.

Real-World PROBLEM SOLVING

Music The *triangle* is a musical instrument. It is struck with a metal stick, giving a high-pitched, ringing tone.

28. Why is this instrument called a triangle?

29. Is the shape of this instrument really a triangle? Explain your reasoning.

30. Is a triangle a two-dimensional or three-dimensional figure? Explain.

31. If you placed a triangle on top of a quadrilateral, what other two-dimensional shape can be made?

H.O.T. Problems

32. **OPEN ENDED** Describe an object in your classroom that shows at least 2 polygons.

33. **FIND THE ERROR** Jamie and Pilar are classifying the figure shown. Who is correct? Explain.

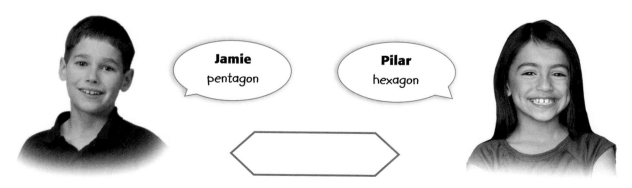

Jamie
pentagon

Pilar
hexagon

34. **WRITING IN ▶MATH** Compare a hexagon and an octagon. How are they different?

11-3 Problem-Solving Strategy

MAIN IDEA I will solve a problem by solving a simpler problem.

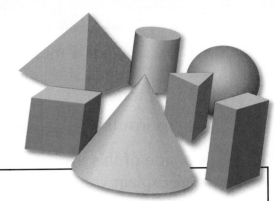

Julianna has 3 three-dimensional figures. The figures have a total of 17 flat faces, 32 edges, and 21 vertices. Two of the figures have 6 equally-shaped flat faces each. What are the names of the 3 figures?

Understand	**What facts do you know?**
	• There are 3 three-dimensional figures.
	• You know the number of faces, edges, and vertices.
	• Two figures have 6 faces of equal size each.
	What do you need to find?
	• The names of the three figures.
Plan	Solve a simpler problem to find the names of the figures.
Solve	A three-dimensional figure with 6 faces of equal size is a cube.
	One cube has 6 faces, 12 edges, and 8 vertices.

6 faces + 6 faces = 12 faces
12 edges + 12 edges = 24 edges
8 vertices + 8 vertices = 16 vertices

	To find the third figure, subtract the total number of faces, edges, and vertices for two cubes.
	17 faces − 12 faces = 5 faces
	32 edges − 24 edges = 8 edges
	21 vertices − 16 vertices = 5 vertices

A pyramid has 5 faces, 8 edges, and 5 vertices. So, Julianna has 2 cubes and 1 pyramid.

Check	Look back. The answer makes sense with the facts in the problem. So, the answer is correct. ✔

476 Chapter 11 Identify Geometric Figures and Spatial Reasoning

Refer to the problem on the previous page.

1. Explain how the *solve a simpler problem* strategy is helpful.

2. Suppose the two figures did not have 6 faces of *equal* size. Could they be other three-dimensional figures? Explain.

3. Explain another strategy you could have used to solve the problem.

4. There are 2 three-dimensional figures. They have a total of 3 flat faces and 1 curved side each. What are the two figures? Write the steps you would take to solve this problem.

PRACTICE the Strategy

EXTRA PRACTICE
See page R30.

Solve. Use the *solve a simpler problem* strategy.

5. During round one of a game, Elio, Nida, and Geoffrey each scored 4 points. In round two, they each scored twice as many points. Find the total number of points.

6. A frame is 2 inches longer and 2 inches wider than the photo below. What is the perimeter of the frame?

4 in.

6 in.

7. Leon wants to buy two gallons of milk. One gallon costs $3. A half-gallon costs $2. Should Leon buy two one-gallon jugs of milk or four half-gallon jugs to spend the least amount of money? Explain.

8. There were 17 bottles, 6 mugs and 5 glasses on a shelf. Rob uses 2 bottles. Anne uses 1 mug and 1 glass. How many items are left?

9. Both Duane and Rajeev followed the homework schedule in the table. What is the total number of hours they spent on homework?

Homework Schedule	
Day	**Time (minutes)**
Monday	45
Tuesday	30
Wednesday	15

10. Lannetta will buy balloons for a party. She invited 6 friends from school, 3 friends from soccer, and 2 cousins. How many balloons will she need to buy if everyone gets two?

11. **WRITING IN MATH** Explain when you would use the *solve a simpler problem* strategy.

Identify and Extend Geometric Patterns

11-4

GET READY to Learn

Maddie's mom is installing a tile floor in their kitchen. What colors of tiles will continue the pattern in the next row?

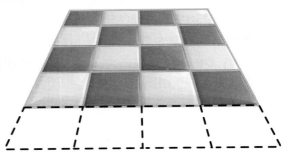

MAIN IDEA

I will identify geometric patterns and use them to make predictions and solve problems.

Math Online

macmillanmh.com
• Extra Examples
• Personal Tutor
• Self-Check Quiz

Identifying geometric patterns can help you make predictions and solve problems.

Real-World EXAMPLE

Identify and Extend Patterns

TILES The drawing shows the pattern that Maddie's mom is using to install the tile floor.

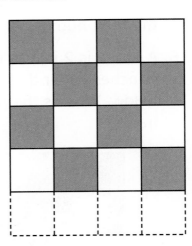

① **Identify the geometric pattern.**

Maddie's mom installs a row of blue, white, blue, white tiles. This is followed by a row of white, blue, white, blue tiles. She continues this pattern.

② **What colors of tiles will continue the pattern in the next row?**

By extending the pattern, the next row will be blue, white, blue, white tiles.

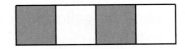

Real-World EXAMPLE Apply Patterns

③ **SCHOOL** Mrs. Pembroke shows her students the pattern below. How many red pattern blocks will be used if the pattern is extended until there are a total of 11 polygons?

Apply the pattern to solve.

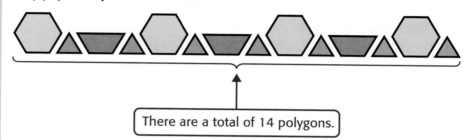

There are a total of 14 polygons.

So, 3 red pattern blocks will be used to extend the pattern to a total of 11 polygons.

CHECK What You Know

Identify and extend each pattern. See Examples 1 and 2 (p. 478)

1.

2.

Apply each pattern. See Example 3 (p. 479)

3. How many triangles will be used if this pattern continues until there are a total of 30 polygons?

4. Marvin will extend a pattern. How many polygons will he use by the time he uses 5 pentagons?

5. **Talk About It** Explain why the pattern of circles at the right can be described as an ABBA pattern.

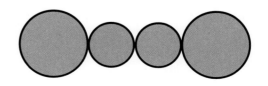

Practice and Problem Solving

Identify and extend each pattern. See Examples 1 and 2 (p. 478)

6. 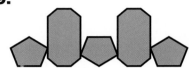 **7.**  **8.**

9.  **10.** **11.**

Apply each pattern. See Example 3 (p. 479)

12. Predict how many red pattern blocks will be used if the pattern is extended until there are a total of 13 polygons.

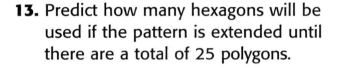

13. Predict how many hexagons will be used if the pattern is extended until there are a total of 25 polygons.

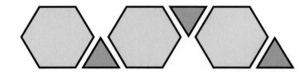

Solve.

14. A banner has a pattern of 4 triangles and 2 squares. Predict the total number of polygons if this pattern is shown 5 times.

15. Susie painted a row of 15 circles. The pattern was 2 red circles and 3 blue circles. Then she added a green circle after each blue circle. How many green circles were there?

16. A pattern shows 2 triangles and 1 square. If a hexagon is placed between each triangle, how many triangles will there be if the pattern is extended until there are 17 shapes?

17. Measurement Each side of each polygon is 1 inch. If the pattern is extended, how many polygons will there be to make a total perimeter of 32 inches?

H.O.T. Problems

18. OPEN ENDED Create a pattern using 3 different shapes.

19. WRITING IN ►MATH Explain where you see geometric patterns in real-world objects.

TEST Practice

20. The triangles below decrease in size from left to right. If the pattern is extended, what is the measure of the height of the next triangle? (Lesson 11-4)

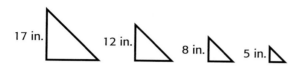

17 in.　12 in.　8 in.　5 in.

A 4 in.

B 3 in.

C 2 in.

D 1 in.

21. All of these figures are *pulanes*.

None of these figures are *pulanes*.

Which of these is NOT a *pulanes*?
(Lesson 11-2)

F **H**

G **J**

Spiral Review

Describe each two-dimensional figure. Use the terms *sides* and *angles*. Then identify the figure. (Lesson 11-2)

22.

23.

24.

Identify each three-dimensional figure. (Lesson 11-1)

25.

26.

27.

Guess the Figure
Identify Three-Dimensional Figures

Get Ready!

Players: 2 players

Get Set!

- Cut a hole in the box large enough for a hand to fit through.

- Place the figures into the box.

Go!

- Player 1 puts his or her hand in the box and chooses a figure without removing it.

- Player 1 names the figure and then removes it. If correct, he or she receives one point. If not correct, he or she places the figure back in the box.

You will need:
three-dimensional objects, a shoe box

- Player 2 takes a turn.

- Play continues until all figures are gone.

- The player with the most points wins.

Identify each three-dimensional figure. (Lesson 11-1)

1.

2.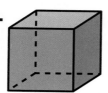

3.

4.

Describe each two-dimensional figure. Use the terms *sides* and *angles*. Then identify the figure. (Lesson 11-2)

5.

6.

7. **MULTIPLE CHOICE** The quadrilaterals below increase in size from left to right. If the pattern extends, what will be the measure of the width of the next quadrilateral?

(Lesson 11-4)

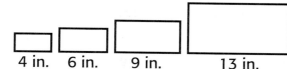

4 in.   6 in.   9 in.   13 in.

A 17 in.   **C** 19 in.

B 18 in.   **D** 20 in.

8. **MULTIPLE CHOICE** Which of these is a rectangular prism? (Lesson 11-1)

F

H

G

J

Solve.

9. Maurice is placing a border around a rectangular bulletin board. The perimeter of the bulletin board is 10 feet. One side measures 2 feet. What will be the length of the border for each side? (Lesson 11-3)

10. Tanika is making a pattern with chalk. The pattern is 1 pink, 3 yellow, 2 blue, and 1 green. If she extends this pattern until there are 30 pieces of chalk, how many yellow pieces will she use? (Lesson 11-4)

11. **WRITING IN ►MATH** What is the connection between the number of sides and number of angles in a polygon? (Lesson 11-2)

Identify Congruent Figures

GET READY to Learn

The surfaces of the two soccer balls are made up of hexagons and pentagons. What do you notice about the hexagons?

MAIN IDEA

I will identify congruent two-dimensional figures.

New Vocabulary

congruent

Math Online

macmillanmh.com
• Extra Examples
• Personal Tutor
• Self-Check Quiz

Figures that have the same size and same shape are **congruent**.

EXAMPLES Identify Congruent Figures

Tell whether the two-dimensional figures in each pair are congruent.

①

The figures have the same size and shape. So, they are congruent.

②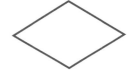

The figures are not the same size or shape. So, they are not congruent.

CHECK What You Know

Tell whether each pair of figures is congruent. Write *yes* or *no*.

See Examples 1 and 2 (p. 484)

1.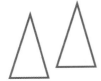

2.

3.

4. One rectangle is 6 inches by 12 inches. Another rectangle is 6 inches by 12 inches. Are the rectangles congruent? Explain.

5. **Talk About It** Can a rectangle and a square be congruent? Explain your reasoning.

Tell whether each pair of figures is congruent. Write *yes* or *no*.

See Examples 1 and 2 (p. 484)

6.

7.

8.

9.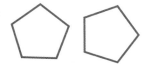

10.

11.

12. One pool measures 16 feet by 12 feet. Another pool measures 12 feet by 6 feet. Are the pools congruent figures? Explain.

13. All of the rectangular doors in Ani's house are the same size. Ani says they are congruent. Is she correct? Explain.

14. A rectangle has two sides that measure 3 feet and 5 feet. What are the measurements of the other two sides?

15. The sides of a square are each 9 feet long. If there is another square that is congruent to the one described, how long are its sides? Explain how you know.

H.O.T. Problems

16. **OPEN ENDED** Draw two triangles that are not congruent.

17. **WHICH ONE DOESN'T BELONG?** Which figure is not congruent to the others? Explain.

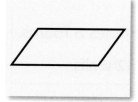

Figure A

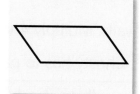

Figure B

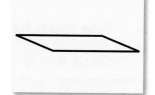

Figure C

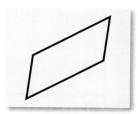

Figure D

18. **WRITING IN ►MATH** Do figures have to be in the same position to be congruent? Explain. Draw a picture to support your answer.

Problem-Solving Investigation

MAIN IDEA I will choose the best strategy to solve a problem.

P.S.I. TEAM ✚

CHRISTY: I am painting all of the walls in my house. Each room in my house is a rectangular solid. There are 8 rooms. How many walls do I have to paint?

YOUR MISSION: Find the total number of walls.

Understand	**What facts do you know?** • Christy is painting all of the walls in 8 rooms. • Each room in her house is a rectangular solid. **What do you need to find?** • The number of walls Christy has to paint.
Plan	Choose an operation to find the number of walls Christy has to paint.
Solve	Find the number of walls in each room. 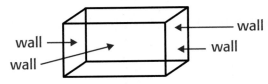 One room has 4 walls. Multiply 4 walls times 8 rooms. Since $4 \times 8 = 32$, Christy has to paint 32 walls.
Check	Look back. Add the walls to check your answer. $4 + 4 + 4 + 4 + 4 + 4 + 4 + 4 = 32$ The answer is correct.

Use any strategy shown below to solve. Tell what strategy you used.

PROBLEM-SOLVING STRATEGIES
- Look for a pattern.
- Choose an operation.
- Make a table.
- Work backward.

1. Brady is planning a party. He sends invitations to 3 friends from his soccer team, 5 friends from school, and 9 neighbors. Seven friends tell him they cannot come. How many friends will come to the party?

2. The pattern shown repeats seven more times. How many triangles will there be in all?

3. Tess ran 4 blocks to get to her friend's house. Then she ran twice as far to the grocery store. How many blocks was the total trip?

4. Maxwell runs 2 miles each day for a week. Jordan runs twice as much as Maxwell. At the end of 7 days, how many miles have Jordan and Maxwell run altogether?

5. Michelle bought 2 orange juices and 1 bottle of water. Carly bought a bottle of water, milk, and a soda. Who spent more money?

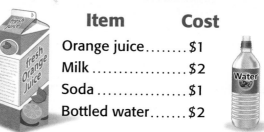

Item	Cost
Orange juice	$1
Milk	$2
Soda	$1
Bottled water	$2

6. Cameron collected 80 rocks in the last 5 years. In the second year, he collected 23 more rocks than he did the first year. He collected 5 rocks each in his third and fourth years. His fifth year he collected 7 rocks. How many did he collect the first year?

7. A store is having a sale on fruit at half the original price. Tyson buys 1 cantaloupe, 2 mangos, and 1 apple. How much money did he spend?

Fruit Original Price

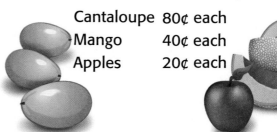

Cantaloupe	80¢ each
Mango	40¢ each
Apples	20¢ each

8. **WRITING IN ►MATH** Look back at Exercise 4. Change the wording so that the *work backward* strategy would have to be used.

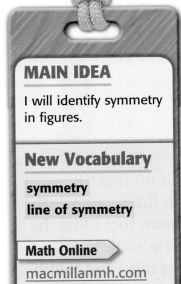

MAIN IDEA

I will identify symmetry in figures.

New Vocabulary

symmetry
line of symmetry

Math Online

macmillanmh.com
• Extra Examples
• Personal Tutor
• Self-Check Quiz

Hands-On Mini Activity

Step 1 Trace a hexagon pattern block. Then trace it again to make one figure as shown.

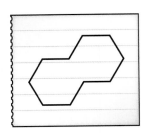

Step 2 Cut out the figure. Fold it in half. Open the figure and trace the fold with a pencil. This is a **line of symmetry**.

Step 3 Fold the figure in a different way to find a different line of symmetry.

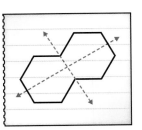

1. Use the pattern blocks at the right to create a figure with symmetry.

2. Are there any other lines of symmetry? Explain.

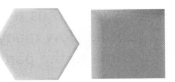

Some figures like the one above are exact matches when cut in half. This is called **symmetry**. The dashed line is called the **line of symmetry**.

EXAMPLES Line of Symmetry

Tell whether each figure has line symmetry. Write *yes* or *no*. If yes, tell how many lines of symmetry it has.

1

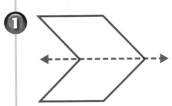

Yes; The figure has one line of symmetry.

2

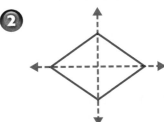

Yes; The figure has two lines of symmetry.

When two halves of a figure do not match, the figure does *not* have line symmetry.

Real-World EXAMPLE

③ **BIRD HOUSES** A family of birds lives in this birdhouse. Does the birdhouse have line symmetry?

The two halves of the birdhouse do not match.

The birdhouse does not have line symmetry.

CHECK What You Know

Tell whether each figure has line symmetry. Write *yes* or *no*. If yes, tell how many lines of symmetry the figure has.

See Examples 1–3 (pp. 488–489)

1.

2.

3.

4.

5.

6.

7. Draw the letter T, and draw any lines of symmetry.

8. **Talk About It** Draw 3 examples of objects that show symmetry.

Tell whether each figure has line symmetry. Write *yes* or *no*. If yes, tell how many lines of symmetry the figure has.

See Examples 1–3 (pp. 488–489)

9.

10. M

11.

12.

13.

14.

15. Name three letters that have line symmetry.

16. Name three digits that have line symmetry.

17. Explain why Circle A has more than 1 line of symmetry and Circle B does not.

18. Explain why this line is not the figure's line of symmetry.

A

B

H.O.T. Problems

19. **OPEN ENDED** Draw a picture of half of a figure that shows symmetry. Have a classmate draw the other half.

20. **CHALLENGE** Look at this picture. How would you test this object to make sure that it shows symmetry?

21. **WRITING IN ►MATH** Can a figure have more than one line of symmetry? Explain.

EXAMPLE

1) Becky has a rectangle. She wants to find its line of symmetry.

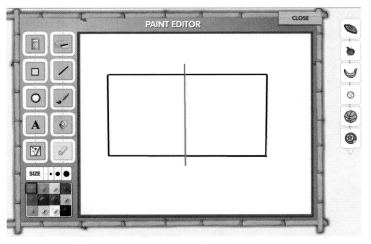

Use the *Math Tool Chest Paint Editor* to draw a rectangle and its line of symmetry.

- Choose the line tool to create a rectangle.
- Choose the line tool to draw one line of symmetry on the rectangle.
- Does the rectangle have other lines of symmetry? Draw them.

CHECK What You Know

Use technology to draw each polygon. Then draw one or more lines of symmetry.

1. triangle

2. rectangle

3. circle

4. pentagon

5. hexagon

6. octagon

7. Name five capital letters of the alphabet that have line symmetry. Draw their line of symmetry.

8. **Talk About It** Explain how to find the line of symmetry in a geometric shape.

Whole Numbers on a Number Line

GET READY to Learn

The timeline shows when Felipe worked on his science project. What did he do on week 5?

Point	Project
A	Plant Seeds
C	Add plant food
E	Repot plants

A   B   C   D   E   F

1   2   3   4   5   6

Week

MAIN IDEA

I will locate and name points on a number line.

New Vocabulary

number line

Math Online

macmillanmh.com
• Extra Examples
• Personal Tutor
• Self-Check Quiz

A timeline is an example of a number line. A **number line** is a line that represents numbers as points.

Real-World EXAMPLE   Locate Points on a Number Line

1 **SCIENCE Use the number line to find what Felipe did on week 5.**

On the number line, week 5 is represented by point E.

Use the table to find that Felipe repotted the plants on week 5.

EXAMPLE   Name Points on a Number Line

2 **What number on the number line does point A represent?**

Locate A on the number line. Each jump is +5.

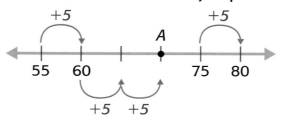

$$55 + 5 + 5 + 5 = 70$$

So, point A represents 70.

---

1. What point on the number line represents 96? See Example 1 (p. 492)

78   84   90   108

2. What number on the number line does point A represent?

See Example 2 (p. 492)

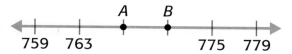

759   763   775   779

3. A number line shows intervals of 4. What number comes right before 32?

4. **Talk About It** Would you use intervals of 1 or of 20 to make a number line from 0 to 100? Explain.

Practice and Problem Solving

EXTRA **PRACTICE**
See page R31.

Tell what point on the number line represents each number. See Example 1 (p. 492)

5. 71

62   65   77

6. 592

602   612   622   632

Tell what number on the number line each letter represents. See Example 2 (p. 492)

7. Point A = ▩

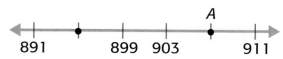

891   899   903   911

8. Point D = ▩

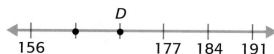

156   177   184   191

9. Point K = ▩

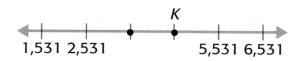

1,531 2,531   5,531 6,531

10. Point R = ▩

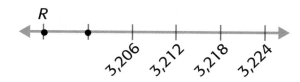

3,206   3,212   3,218   3,224

H.O.T. Problems

11. OPEN ENDED Create a number line that shows 3 even numbers.

12. CHALLENGE Would 302 be a point on the number line shown? Explain.

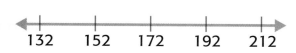

132   152   172   192   212

13. **WRITING IN ►MATH** Describe a real-world example of a number line.

GET READY to Learn

A map of a zoo is shown. From the entrance, the zebras are located 2 units right and 6 units up. This can be written as (2, 6). Where are the giraffes located?

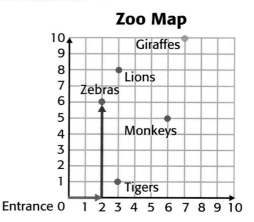

Zoo Map

The map is an example of a **coordinate grid**. An **ordered pair** of numbers like (2, 6) names a location on the grid.

Real-World EXAMPLE Write an Ordered Pair

1 **MAPS** Write the ordered pair for the location of the giraffes.

Step 1 Start at (0, 0). Move right 7 spaces since the giraffes are above the 7. This is the first number of the ordered pair.

Step 2 Move up until you reach the giraffes. The second number in the ordered pair is 10.

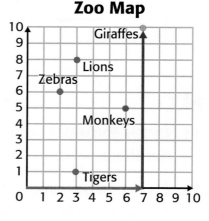

Zoo Map

So, the ordered pair is (7, 10).

2 **SCHOOL** A map of Mrs. Cardona's classroom is shown. What is located at (7, 3)?

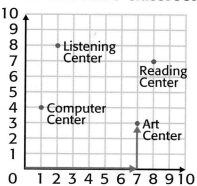

Mrs. Cardona's Classroom

Start at (0, 0). Move 7 spaces to the right. Then, move 3 spaces up. The Art Center is at (7, 3).

CHECK What You Know

Write the ordered pair for the location of each item on the grid. See Example 1 (p. 494)

1. soccer ball

2. game

3. baseball

4. scooter

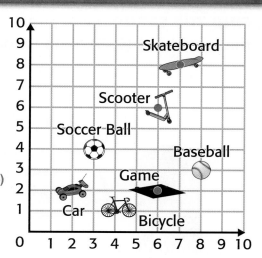

Name the toy at each location. See Example 2 (p. 495)

5. (4, 1)

6. (6, 6)

7. (2, 2)

8. (8, 3)

9. Explain how you would move from the game to the soccer ball.

10. Refer to the grid in Exercises 1–8. If each grid line shows 5 feet, how far is the game from the scooter?

11. **Talk About It** Explain how to locate (0, 2) on a grid.

Write the ordered pair for the location of each item on the grid. See Example 1 (p. 494)

12. swing

13. slide

14. water fountain

15. sandbox

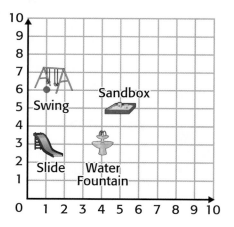

Name the place at each location.

See Example 2 (p. 495)

16. (1, 3)

17. (4, 1)

18. (5, 5)

19. (3, 1)

20. (4, 3)

21. (2, 4)

Grocery Store

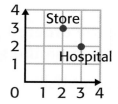

H.O.T. Problems

22. FIND THE ERROR Keith and Suzanne are describing the location of (2, 3). Who is correct? Explain.

Keith
Hospital

Suzanne
Store

23. OPEN ENDED Describe the similarities between maps and grids.

24. WRITING IN ►MATH Does the ordered pair (1, 3) give the same location as (3, 1)? Explain.

25. CHALLENGE Connect the points as you locate each set of ordered pairs on the grid to create figure A. Tell what geometric figure is made.

Figure A: (1,2), (3,5), (10,5), (8,2)

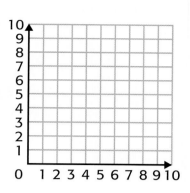

![TEST Practice]

26. Which number on the number line does point *X* represent?

(Lesson 11-8)

A 50   **C** 54

B 52   **D** 59

27. Which point on the number line represents 120? (Lesson 11-8)

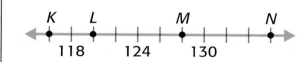

F K   **H** M

G L   **J** N

Spiral Review

Tell what point on the number line represents each number.

(Lesson 11-8)

28. 125

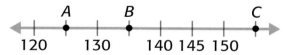

29. 862

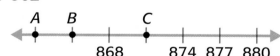

30. 283

31. 484

32. Cornelio jumped 32 inches. Darien jumped 3 feet 1 inch. Cornelio said he jumped farther. Is his answer reasonable? Explain. (Lesson 3-4)

Compare. Use >, <, or =. (Lesson 1-6)

33. 475 ⬤ 478   **34.** 3,392 ⬤ 3,299   **35.** 2,381 ⬤ 12,000

Problem Solving in Art

GARDENS UNDER GLASS

There are three large gardens close to the United States Capitol building in Washington D.C. The Botanic Garden is right across the street. It has more than 4,000 kinds of plants. It also includes a conservatory.

This conservatory is amazing! It is made mostly of glass and aluminum. The conservatory is divided into 10 rooms.

Did You Know?

The Jungle Room is over 80 feet high and has a walkway that is 96 feet long!

Real-World Math

Use the diagram below to solve each problem.

1. What two-dimensional figure does the shape of the east and west galleries create?

2. Which rooms do not have a line of symmetry?

3. If a tent was placed over Oasis to create a figure with 5 vertices, what figure would be made?

4. Suppose the Oasis room was cut in half diagonally. What polygons would be created?

5. What two-dimensional figure does the shape of the Meditation Garden and Children's Garden create? What do you notice about these two rooms?

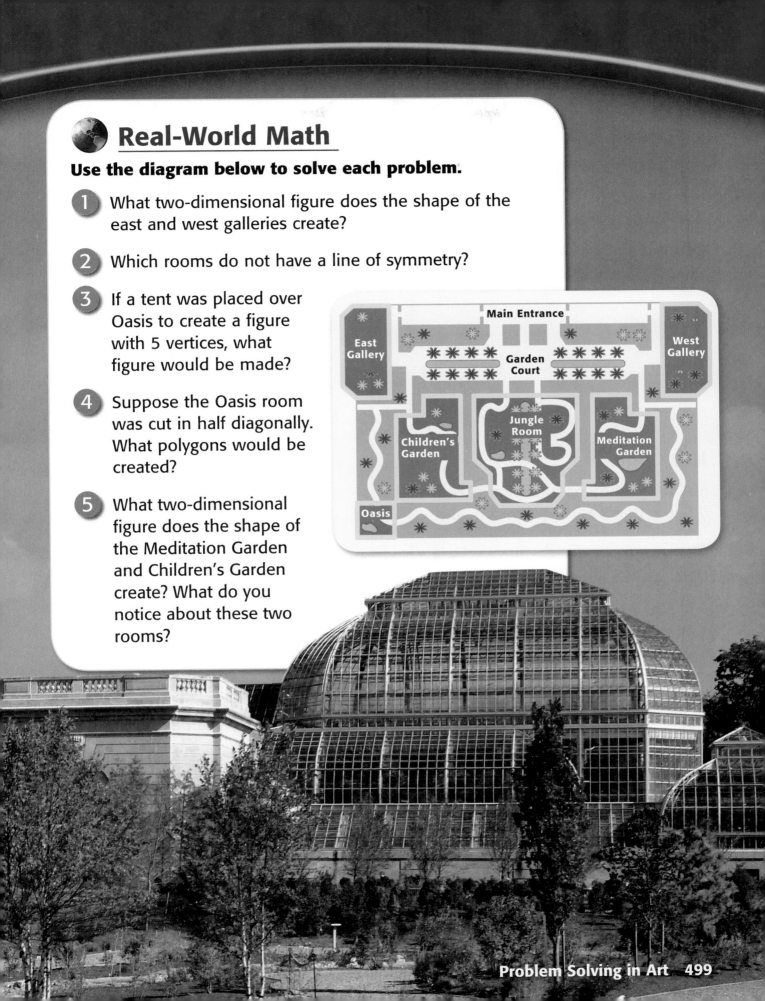

FOLDABLES Study Organizer **GET READY to Study**

Be sure the following Key Vocabulary words and Key Concepts are written in your Foldable.

Key Concepts

- A **three-dimensional figure** is a solid figure that has length, width, and height. (p. 467)

- A **two-dimensional figure** is a plane figure that has length and width. (p. 472)

- Figures that have the same size and same shape are **congruent**. (p. 484)

- A figure has **symmetry** when it is an exact match when cut in half. (p. 488)

Key Vocabulary

congruent (p. 484)

polygon (p. 472)

symmetry (p. 488)

three-dimensional figure (p. 467)

two-dimensional figure (p. 472)

Vocabulary Check

1. A _____?_____ is a solid figure that has length, width, and height.

2. A _____?_____ is a closed, two-dimensional figure of 3 or more line segments.

3. A plane figure is a _____?_____.

4. Figures that have the same size and same shape are _____?_____.

5. A cube is an example of a _____?_____.

6. A figure has _____?_____ when it is an exact match when cut in half.

7. A pentagon is an example of a _____?_____.

Lesson-by-Lesson Review

11-1 **Three-Dimensional Figures** (pp. 467–471)

Example 1
Identify each three-dimensional figure.

The figure is a pyramid.

A party hat looks like a cone.

Identify each three-dimensional figure.

8.   9.

10.   11.

Classify the three-dimensional figure.

12. This figure has 6 faces. All 6 faces are equal in size and shape.

11-2 **Two-Dimensional Figures** (pp. 472-475)

Example 2
The flag of Jamaica is shown. Describe and identify the green polygons.

Each polygon has 3 sides and 3 angles. So, they are triangles.

Example 3
A polygon has 6 sides and 6 angles. Classify the figure.

The figure is a hexagon.

Describe each two-dimensional figure. Use the terms *sides* and *angles*. Then identify the figure.

13.   14.

15.   16.

Classify each two-dimensional figure.

17. polygon with 4 sides and 4 angles

18. polygon with 2 more sides and 2 more angles than a hexagon

19. polygon with the least number of sides of all polygons

11-3 **Problem-Solving Strategy:** Solve a Simpler Problem (pp. 476–477)

Example 4

Mick hit a home run in his Little League baseball game. How far did he run if the distance between the bases and home plate are equal?

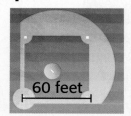

60 feet

Understand

What facts do you know?

- The distances between the bases and home plate are equal.

- The distance between each is 60 feet.

What do you need to find?

- The number of feet around all bases.

Plan   Solve a simpler problem to find the total distance.

Solve   Use a basic multiplication fact and patterns to find the total distance.

$4 \times 6 = 24$

$4 \times 60 = 240$

So, the total distance is 240 feet.

Check   Since $60 + 60 + 60 + 60 = 240$, the answer is correct.

20. Alvin bought 4 packs of trading cards for $2. How much would 6 packs of trading cards cost?

21. Measurement A sidewalk surrounds a rectangular park. What is the total length of the sidewalk?

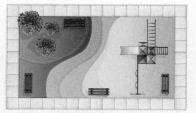

120 yards

180 yards

22. Fina uses 12 red beads and 12 blue beads for every bracelet she makes. If Fina makes 4 bracelets, how many red and blue beads does she use?

23. There are 50 third grade students at Greensdale Elementary School. One out of every 5 third grade students takes the bus to school. How many students take the bus to school?

24. Tasha bought the cat toys shown for $6. If each cat toy is the same price, how much would 5 cat toys cost?

11-4 Identify and Extend Geometric Patterns (pp. 478–481)

Example 5

Jasmine is making a border for a page in her scrapbook. How many stars will be used if this pattern extends until there are a total of 16 shapes?

Identify and extend the pattern until there are 16 shapes. Then count the number of stars.

There are a total of 16 shapes.

So, 8 stars will be used to continue this pattern until there is a total of 16 shapes.

Identify and extend each pattern.

25.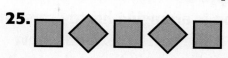

26.

27. How many pentagons will be used if this pattern extends until there is a total of 18 polygons?

28. How many triangles will be used if this pattern extends until there are a total of 25 polygons?

11-5 Identify Congruent Figures (pp. 484–485)

Example 6

Tell whether each pair of figures are congruent.

The figures have the same size and shape. So, they are congruent.

Tell whether each pair of figures are congruent. Write *yes* or *no*.

29.

30.

11-6 **Problem-Solving Investigation: Choose a Strategy** (pp. 486–487)

Example 7
Gordon is arranging his toy cars on a shelf. If the pattern extends, how many red cars will be on the shelf when there are 14 cars on the shelf?

Understand   You know that Gordon is arranging blue, red, and orange toy cars on a shelf. You need to find how many red cars there are when there are 14 cars in all.

Plan   Find and extend a pattern to solve the problem.

Solve   A pattern is blue, red, orange, red. The pattern repeats. Extend this pattern to solve the problem.

B, R, O, R, B, R, O, R, B, R, O, R, B, R

When there are 14 cars, 7 are red.

Check   Use division to check. Half of the cars in the pattern are red. So, if there are 14 cars, 7 cars will be red.

Use any strategy to solve.

31. Silvio walks his dog for 25 minutes each day. How many minutes does Silvio walk his dog in a week?

32. Kendra has $17. She earned $8 for babysitting. She earned $5 for raking leaves. She earned the rest of the money by doing chores. How much money did she earn by doing chores?

33. Pedro has gymnastics lessons four days a week. Each lesson is 45 minutes long. How many minutes does Pedro have gymnastics lessons each month?

34. John has 22 marbles. He has twice as many red marbles as green marbles. The rest of his marbles are yellow. How many yellow marbles does John have?

35. Wesley is half as old as his brother. The sum of their ages is 18. How old are Wesley and his brother?

11-7 Create Lines of Symmetry (pp. 488–490)

Example 8
Tell whether each figure has line symmetry. Write *yes* or *no*. If *yes*, tell how many lines of symmetry the figure has.

Yes; The figure has one line of symmetry.

Yes; The figure has one line of symmetry.

Tell whether each figure has line symmetry. Write *yes* or *no*. If *yes*, tell how many lines of symmetry the figure has.

36.    37.

38.   39.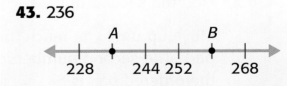

40. Name three polygons that have line symmetry.

41. Name three objects in your desk that have line symmetry.

11-8 Whole Numbers on a Number Line (pp. 492–493)

Example 9
Which number does point *A* represent on the number line?

The interval between lines is 4.

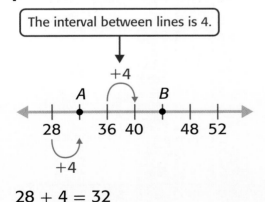

+4
+4

$28 + 4 = 32$

So, point *A* represents 32 on the number line.

Tell what point on the number line represents each number.

42. 49

21 28 35   56

43. 236

A   B
228   244 252   268

44. Tell what number on the number line *Point A* represents.

357 362   372   382

11-9 **Ordered Pairs** (pp. 494–497)

Example 10
A map of Stacy's neighborhood is shown. Write the ordered pair for the location of the park.

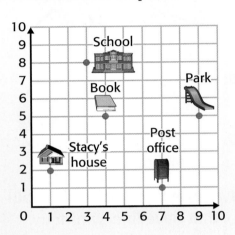

Step 1 Find the first number.

Start at (0, 0). Move 9 spaces to the right since the park is above the 9 on the bottom. This is the first number of the ordered pair.

Step 2 Find the second number.

Move up until you reach the park. The second number in the ordered pair is 5.

So, the ordered pair is (9, 5).

For Exercises 45–48, write the ordered pair for the location of each item in Mr. Bilko's classroom.

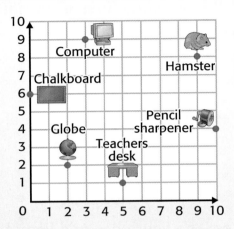

45. globe

46. teacher's desk

47. computer

48. pencil sharpener

For Exercises 49–52, write the ordered pair for the location of each ride at the theme park.

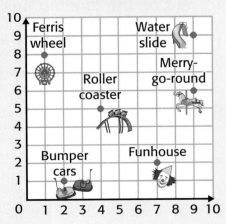

49. roller coaster

50. Ferris wheel

51. water slide

52. fun house

Identify each three-dimensional figure.

1.

2.

Describe each two-dimensional figure. Use the terms *sides* and *angles*. Then identify each figure.

3.

4.

Classify each figure.

5. This figure has 2 faces and 0 edges.

6. This figure has 1 face.

7. **MULTIPLE CHOICE** Which picture best represents a cone?

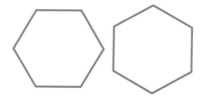

A

B

C

D

8. What point represents 61?

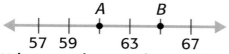

57 59   63   67

9. What number on the number line does point *K* represent?

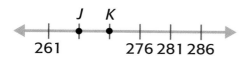

261   276 281 286

10. **MULTIPLE CHOICE** Which point on the number line represents 142?

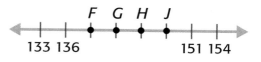

F G H J

133 136   151 154

F *F*   **H** *H*

G *G*   **J** *J*

11. Amanda has a collection of 14 dolls from around the world. There are 3 from Ireland, 2 from Italy, 2 from Mexico, and the rest are from the Americas. How many are from the Americas?

12. Tell whether the pair of figures is congruent. Write *yes* or *no*.

Tell whether each object has a line of symmetry. Write *yes* or *no*. If yes, tell how many.

13.

14.

15. **WRITING IN ►MATH** Draw a figure that has a line of symmetry. Draw its line of symmetry. Explain.

PART 1   Multiple Choice

Read each question. Then fill in the correct answer on the answer sheet provided by your teacher or on a sheet of paper.

1. How many sides does a quadrilateral have?

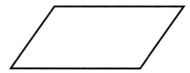

A 2   **C** 6

B 4   **D** 8

2. Each student is given 4 pencils. The pencils are sold in packages of 12. How many packages are needed for 120 pencils in all?

F 7   **H** 9

G 8   **J** 10

3. Which figure is a pentagon?

A   **C**

B   **D**

4. Which pair of figures is congruent?

F

G

H

J

5. What is the length of the cricket to the nearest centimeter?

A 7 centimeters   **C** 5 centimeters

B 6 centimeters   **D** 4 centimeters

6. What is the best estimate of the difference rounded to the nearest ten?

$$88 - 47$$

F 20   **H** 40

G 30   **J** 50

7. What is the perimeter of the rectangle?

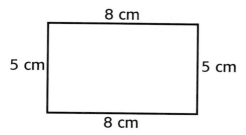

8 cm

5 cm   5 cm

8 cm

A 13 cm   **C** 35 cm

B 26 cm   **D** 40 cm

8. Which of the following best describes a three-dimensional figure with square faces?

F cone   **H** cube

G sphere   **J** pyramid

9. Which of the following describes the shape of the soup can?

A circle   **C** cylinder

B cone   **D** prism

PART 2   Short Response

Record your answers on the answer sheet provided by your teacher or on a sheet of paper.

10. Lorenzo draws a shape that has 6 sides and 6 angles. What shape does he draw?

---

11. Three friends were playing a basketball game. Lance won by 15 points. Marisa had 10 points more than Hanna. If Hanna had 20 points, how many points did Lance and Marisa have?

PART 3   Extended Response

Record your answers on the answer sheet provided by your teacher or on a sheet of paper.

12. Give one example of a two-dimensional figure and one example of a three-dimensional figure. Explain how they are alike and how they are different.

NEED EXTRA HELP?												
If You Missed Question...	1	2	3	4	5	6	7	8	9	10	11	12
Go to Lesson...	11-2	5-8	11-2	11-5	9-4	3-2	9-5	11-1	11-1	11-2	9-3	11-1; 11-2

CHAPTER 12 Organize, Display and Interpret Data

BIG Idea What is Data?

Data is information that can be displayed in graphs.

Example The city of Charleston, South Carolina, has many historical sites to visit. The **pictograph** shows the number of students who prefer each destination.

Favorite Historical Sites in Charleston	
Battery	✋ ✋ ✋ ✋
Rainbow Row	✋ ✋
Waterfront Park	✋ ✋ ✋ ✋ ✋
USS Yorktown	✋
Key: ✋ = 2 students	

What will I learn in this chapter?

- Collect, organize, record, and display data.
- Display and read data in a pictograph, bar graph, and line plot.
- Identify whether events are certain, likely, unlikely, or impossible.
- Solve problems by making a list.

Key Vocabulary

tally chart	probability
bar graph	pictograph

Math Online > **Student Study Tools**
at macmillanmh.com

FOLDABLES®
Study Organizer

Make this Foldable to help you organize information about data, graphs, and probability. Begin with one sheet of $8\frac{1}{2}'' \times 11''$ paper.

① **Fold** one sheet of paper into three equal sections.

② **Fold** a 2-inch pocket along the long edge. Glue the outside edges.

③ **Label** each pocket as shown. Record what you learn on index cards.

Tally Chart Pictographs   Line Plots   Identity Probability

You have two ways to check prerequisite skills for this chapter.

Option 2

Math Online > Take the Chapter Readiness Quiz at macmillanmh.com.

Option 1

Complete the Quick Check below.

QUICK Check

Refer to the picture graph. (Prior grade)

1. How many students said they like summer?

2. How many more students said they like winter than fall?

3. How many students said they like spring or winter?

4. What is the total number of students?

Favorite Season	
Spring	● ● ●
Summer	● ● ● ● ● ●
Fall	🍂 🍂 🍂
Winter	🔔 🔔 🔔 🔔 🔔

Each item = 2 students

Find each sum. (Lesson 2-1)

5.
```
  3
  2
  5
+ 6
```

6.
```
  8
  7
  4
+ 1
```

7.
```
  5
  6
  9
+ 2
```

Identify the color on which each spinner is likely to land. (Prior grade)

8.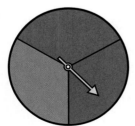

9.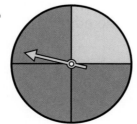

10.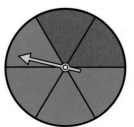

Math Activity for 12-1
Make a Pictograph

A tally chart is a table that organizes data using tally marks. Data displayed in a tally chart can also be displayed in a pictograph.

A pictograph is a graph that uses pictures to *display* or show data. Each picture in a pictograph is the same.

MAIN IDEA

I will collect, organize, record, and display data in pictographs.

You Will Need
notebook paper, 8 small sticky notes, scissors

ACTIVITY   Make a Pictograph

Step 1   **Collect data.**

Make a tally chart like the one shown. Ask 15 people to name their favorite type of fruit. Mark each response with a tally.

Favorite Fruits		
Fruit	Tally	Number
Banana		
Orange		
Strawberry		
Apple		

Step 2   **Organize and record data.**

Count the tally marks in each row and write the number in the last column.

Favorite Fruits		
Fruit	Tally	Number
Banana	⦀⦀	6
Orange	⦀⦀	4
Strawberry	⦀	2
Apple	⦀	3

Step 3 Display the results.

- Fold a piece of paper into fourths lengthwise, and label each section.
- Draw an empty fruit basket on each sticky note.
- Put the fruit baskets on the pictograph to show how many people like each fruit.
- Make a key.
- Each basket represents 2 students.

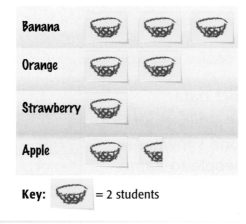

Think About It

1. What does half of a fruit basket represent?

2. How can you use the pictograph to find which fruit is most popular?

3. How many fruit baskets would be used to show 11 students?

4. How are a tally chart and pictograph similar? How are they different?

✓ CHECK What You Know

Collect data from 10 students. Organize and record it in a tally chart. Then use the data to make a pictograph.

5. favorite type of music   **6.** favorite color   **7.** favorite vegetable

8. **WRITING IN ►MATH** Explain how to use the pictograph above to find how many students like strawberries.

Pictographs

GET READY to Learn

MAIN IDEA

I will collect, organize, and display data in pictographs.

New Vocabulary

pictograph

tally chart

Math Online

macmillanmh.com

- Extra Examples
- Personal Tutor
- Self-Check Quiz

Anita is organizing her stuffed animal collection. She records the data that she collected in a tally chart.

My Stuffed Animals		
Stuffed Animal	**Tally**	**Number**
Teddy Bear	ⅢⅠ Ⅰ	6
Cat	ⅢⅠ	4
Dog	ⅢⅠ ⅠⅠ	7
Turtle	ⅠⅠ	2

Anita can use the **tally chart** to make a pictograph. Remember that a **pictograph** is a graph that compares data using picture symbols.

Real-World **EXAMPLE** Display Data in a Pictograph

1 **TOYS** Display the data Anita collected in a pictograph.

Step 1 Make a table with a title, a key, and labels.

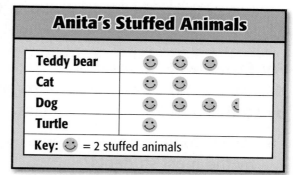

Anita's Stuffed Animals	
Teddy bear	☺ ☺ ☺
Cat	☺ ☺
Dog	☺ ☺ ☺ ☺
Turtle	☺
Key: ☺ = 2 stuffed animals	

Step 2 Choose a symbol that represents the data.

Each ☺ symbol represents 2 stuffed animals.

Step 3 Use ☺ to show the number of stuffed animals.

The pictograph above displays the data Anita collected.

1. Display the set of data in a pictograph. Let each symbol represent two cows.
See Example 1 (p. 515)

Eastwind Cattle Farm		
Type of Cow	Tally	Number of Cows
Holstein	卌 |||	8
Jersey	||||	4
Black Angus	卌	5

2. A pictograph shows 2 ♪ symbols. Each represents 3 people who enjoy rock music. How many people enjoy rock music?

3. **Talk About It** Explain why a pictograph must have a key.

Display each set of data in a pictograph. See Example 1 (p. 515)

4.

Helmets Sold On Saturday	
Type of Helmet	Number Sold
Road	8
Mountain	6
Skate	9

5.

Fish Caught On Sunday	
Type of Fish	Number of Fish
Trout	10
Bass	8
Catfish	17

For Exercises 6–9, use the pictograph that shows the continent reports written by students in Mr. Saxton's Class.

6. Which continent was reported on by the most students?

7. Which continent was reported on by 5 students?

8. Which two continents were reported on by the same number of students?

9. How many students reported on Australia?

Continent Reports by Mr. Saxton's Class	
Asia	●
Europe	● ● ● ● ●
Australia	● ●
South America	● ● ◗
Africa	●
Key: ● = 2 students	

For Exercises 10-13, use the pictograph that shows how many letters a class sent to their pen pals this week. See Example 1 (p. 515)

Pen Pal Letters Sent This Week

Monday	✉ ✉
Tuesday	✉ ✉ ✉
Wednesday	✉
Thursday	✉
Friday	✉ ✉ ✉ ✉ ✉

Key: ✉ = 3 letters

10. How many letters were sent on Tuesday?

11. Which day did the class send the most letters?

12. Which days did the class send the fewest letters?

13. How many letters were sent this week?

14. A pictograph key shows that each 🎥 symbol equals 4 movies. How many symbols would need to be drawn to show 12 movies?

15. A pictograph key shows that each ☀ symbol equals 10 days. Draw the symbols needed to show 25 days.

16. Collect data to find the number of students that have a blue, red, green, or multi-colored toothbrush. Make a pictograph to display the results.

17. Collect data to find the number of students who have birthdays in each season. Organize the data in a tally chart. Then display your results in a pictograph.

H.O.T. Problems

18. **OPEN ENDED** Describe a real-world example of data that can be shown in a pictograph.

19. **CHALLENGE** Find the total number of minutes Alex spent riding on the bus.

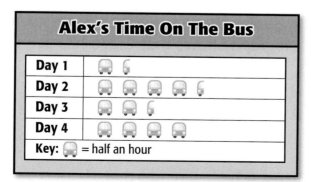

Alex's Time On The Bus

Day 1	🚌 🚌
Day 2	🚌 🚌 🚌 🚌 🚌
Day 3	🚌 🚌 🚌
Day 4	🚌 🚌 🚌 🚌

Key: 🚌 = half an hour

20. **WRITING IN ►MATH** Explain what would happen to the pictograph above if the key showed that each 🚌 = 1 hour.

GET READY to Learn

MAIN IDEA

I will interpret data in pictographs.

Math Online

macmillanmh.com
• Extra Examples
• Personal Tutor
• Self-Check Quiz

Antoine asked his friends how many movies they saw during summer vacation. The pictograph shows the results.

Movies Seen During Summer Vacation

Zack	
Carla	
Grace	
Ivan	
Ricardo	

Key: = 2 movies

You can read and interpret data from a pictograph. To interpret data, write a sentence using the data found in the pictograph.

Real-World EXAMPLE Interpret Pictographs

1 **MOVIES** Use the pictograph above. Who saw two more movies than Grace during summer vacation?

The key shows that each symbol means 2 movies. The pictograph shows that Grace has seen 6 movies.

+ + or 2 + 2 + 2 = 6

To add two more movies, show one more symbol.

+ + + = 2 + 2 + 2 + 2 = 8.

The pictograph shows that Carla has seen 8 movies.

So, Carla saw two more movies than Grace.

Remember

Each pictograph must have a key.

2 **RECYCLING** Every week, a school recycles 55 pounds of paper, 30 pounds of cans, and 25 pounds of plastic. Display the data in a pictograph. Then write a sentence that interprets the data.

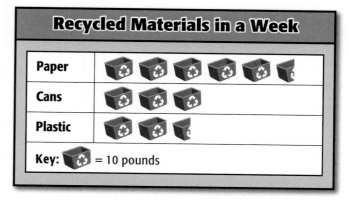

Every week, the school recycles the same amount of paper as cans and plastic together.

CHECK What You Know

For Exercises 1–3, use the pictograph that shows how many gallons of milk were sold. See Example 1 (p. 518)

1. Which store sold the most milk?

2. Which store sold six gallons more than the small grocery store?

3. Suppose each gallon costs $2 How much money was spent on milk at the large grocery store?

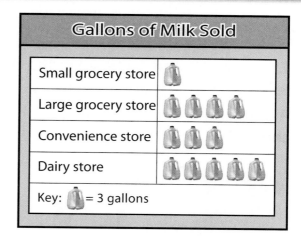

4. A clown makes animal balloons. In one hour, she made 3 giraffes, 12 dogs, and 18 monkeys. Display the data in a pictograph. Then write a sentence that interprets the data.

   See Example 2 (p. 519)

5. **Talk About It** How can you use repeated addition to help you interpret pictographs?

For Exercises 6–12, refer to the pictographs. See Example 1 (p. 518)

Third Grade Shoe Size	
2	👟
4	👟 👟 👟
6	👟 👟 👟 👟 👟
8	👟
Key: 👟 = 4 students	

Chores Per Week	
Latisha	✦ ✦
Kelley	✦ ✦ ✦ ✦
Gabe	✦
Juan	✦ ✦ ✦
David	✦ ✦ ✦
Key: ✦ = 2 chores	

6. What is most common shoe size?

7. What is the second most common shoe size?

8. How many students were asked for their shoe size?

9. Based on this information, should a shoe company make the same number of every size? Explain.

10. Name two students that have 10 chores altogether.

11. If each child earned $1 for each chore, how much money would Latisha earn?

12. Using the key, draw the number of symbols to show how many chores you have each week.

For Exercises 13 and 14, display the data in a pictograph. Then write a sentence that interprets the data. See Example 2 (p. 519)

13. A barn had 4 of each animal shown below and 8 pigs.

14. Ask 10 people to find which of these fruits they like best.

15. A pictograph entitled Backyard Animals has three 🐦 symbols. Each symbol equals 2 animals. Write a sentence about the data.

16. A key shows that each ● symbol means 5 balls. How many symbols would there be to represent 10 balls?

H.O.T. Problems

17. OPEN ENDED A pictograph shows how many points each team scored. The team with the most points scored twice as many as another team. Create a pictograph showing what the pictograph could look like.

18. WRITING IN ►MATH Is it possible to interpret a pictograph without a key? Explain.

TEST Practice

19. The graph shows how many necklaces were sold. Which information is needed to complete the graph? *(Lesson 12-1)*

Necklaces Sold	
Beaded	◯ ◯ ◯ ◯
	◯ ◯
Diamond	◯ ◯ ◯ ◯ ◯
Key: ◯ = 1 necklace	

  A The type of necklace that sold 2.

  B The total number of necklaces that were sold.

  C The number of diamond necklaces that were sold.

  D The location of the store.

20. The graph shows information about pets owned by students. How many more students own birds than fish? *(Lesson 12-2)*

Pets Owned By Students	
Gerbil	🐟 🐟
Fish	🐟
Bird	🐟 🐟 🐟
Key: 🐟 = 4 students	

  F 3

  G 4

  H 6

  J 8

Spiral Review

21. A pictograph shows four 〰 symbols. Each symbol equals 5 miles each student swam. How many miles did the students swim altogether? *(Lesson 12-1)*

Measurement **Find the volume of each figure.** *(Lesson 10-7)*

22.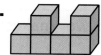

23.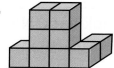

12-3 Problem-Solving Strategy

<u>**MAIN IDEA**</u> I will solve a problem by making a list.

Kia, Kirk, and Shonda are lining up to come in from recess. They are deciding the order they should line up. How many different ways can they line up?

Understand	**What facts do you know?**
	• There are three students.
	What do you need to find?
	• Find how many different ways they can line up.
Plan	Arrange the different combinations in an organized list. Then use the list to solve the problem.
Solve	• Start with Kia. Create different combinations with her first.

Possible Ways to Line Up

	First	Second	Third
1.	Kia	Kirk	Shonda
2.	Kia	Shonda	Kirk
3.	Kirk	Shonda	Kia
4.	Kirk	Kia	Shonda
5.	Shonda	Kirk	Kia
6.	Shonda	Kia	Kirk

• Repeat this method of making a list with each of the other students being first.

• Count all the different combinations.

There are 6 possible ways for the students to line up.

Check	Look back. None of the ways repeat. So, the answer makes sense. ✓

ANALYZE the Strategy

Refer to the problem on the previous page.

1. Explain why the *make a list* strategy was helpful in solving this problem.

2. Explain how to organize all of the combinations in a table.

3. If there were four students, what is the number of combinations?

4. How do you know your answer to Exercise 3 is correct?

PRACTICE the Strategy

EXTRA PRACTICE
See page R33.

Solve. Use the *make a list* strategy.

5. Aleta has black pants and tan pants. She also has a striped shirt, a plaid shirt, and a flowered shirt. How many different outfits can Aleta make?

6. Garcia orders one scoop each of vanilla, chocolate, and strawberry. In how many different ways can his ice cream cone be made?

7. How many lunches can Malia make if she chooses one main item and one side item from the menu shown?

Main Dishes
Pizza
Hamburger
Taco

Side Dishes
Bread
Fruit
Veggies

8. Adele will make a fan out of three different colors of paper. How many color combinations can Adele make if she uses blue, red, and green?

9. List all the possible sandwiches that can be made using one type of bread, meat, and cheese.

Bread   Meat   Cheese

wheat   ham   swiss
rye   beef   provolone

10. Mr. Castillo asked his students to make as many different number combinations as they could using the numbers 5, 7, and 8 without repeating numbers. How many numbers can be made?

11. There are some coins in a jar. The sum of the coins is 13¢. What are the possible coin combinations Amber could have?

12. **WRITING IN ►MATH** Give an example of a problem for which you would use the *make a list* strategy to solve.

Catch Me If You Can!

Make a Graph

Get Ready!

Players: 2 players

Get Set!

- Divide one spinner into 3 equal parts. Label the parts 1, 2, and 4.

- Divide the other spinner into 4 equal parts. Label the parts 1, 2, 3, and 5.

- Make the game board shown.

Go!

- Player 1 spins each spinner and finds the product of the two numbers.

- Player 1 then colors in one square on the graph paper above the product.

- Player 2 takes a turn.

- The game continues, taking turns, until one bar reaches the top.

You will need: 2 spinners, grid paper, crayon

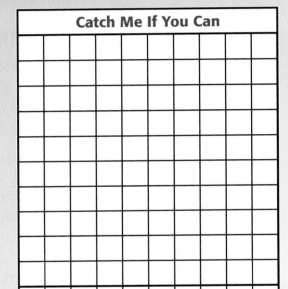

Catch Me If You Can									
1	2	3	4	5	6	8	10	12	20

Products

Display each set of data in a pictograph.
(Lesson 12-1)

1.

Favorite Exercise		
Type of Exercise	Tally	Number of Students
Sit-ups	𝍤𝍤𝍤𝍤𝍤 𝍤𝍤𝍤𝍤𝍤	10
Push-ups	𝍤𝍤𝍤𝍤𝍤 ||	7
Jumping Jacks	𝍤𝍤𝍤𝍤𝍤 |||	8

2.

Weekend Activities	
Activity	Time (hours)
Swim	2
Shop	4
TV	5
Jog	3

3. A pictograph shows 5 ✈ symbols. Each symbol represents 2 times someone has flown in the last year. How many times did all of the people fly in the last year? (Lesson 12-1)

4. MULTIPLE CHOICE How many more students like pepperoni pizza than cheese pizza? (Lesson 12-2)

Favorite Pizza	
Cheese	🍪 🍪
Pepperoni	🍪 🍪 🍪
Vegetable	🍪 🍪 🍪
Key: 🍪 = 2 students	

A 1   **C** 3

B 2   **D** 4

5. MULTIPLE CHOICE The pictograph shows favorite types of movies. How many more people like cartoons than drama? (Lesson 12-2)

Favorite Types of Movies	
Comedy	▢ ▢ ▢
Drama	▢
Cartoon	▢ ▢ ▢ ▢ ▢
Key: ▢ = 3 people	

F 4   **H** 12

G 8   **J** 16

6. Mrs. Torres asked her students to list the four seasons in order of their favorite. How many different lists could she possibly receive? Solve using the *make a list* strategy. (Lesson 12-3)

7. Display the set of data in a pictograph. (Lesson 12-2)

Favorite Place to Read a Book		
Place	Tally	Number of Students
Bed	𝍤𝍤𝍤𝍤𝍤 ||||	9
Outside	|||	3
School	𝍤𝍤𝍤𝍤𝍤 |||	8
Library	||||	4

8. **WRITING IN MATH** Explain why it is important to place a title and labels on pictographs. (Lesson 12-1)

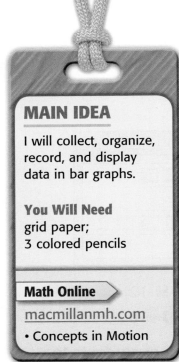

MAIN IDEA

I will collect, organize, record, and display data in bar graphs.

You Will Need
grid paper;
3 colored pencils

Math Online
macmillanmh.com
• Concepts in Motion

Data in a tally chart can also be displayed in a graph. A **graph** is an organized drawing that shows sets of data and how they are related to each other.

A **bar graph** is a graph that uses bars of different lengths and heights to show data.

How Many Pets Do You Have?	
Pets	**Tally**
0						
1												
2				
3					
4 or more	||

ACTIVITY Make a Bar Graph

Step 1 **Draw and label.**

• Draw a rectangle. Separate it into equal rows.

• Label the side and bottom of the graph to describe the information.

• Give the graph a title.

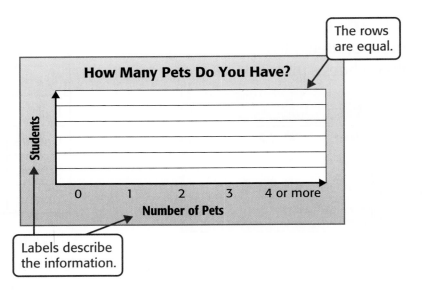

The rows are equal.

Labels describe the information.

Step 2 Choose a scale.

Write a scale on the side of the graph. A **scale** is a set of numbers that represent the data.

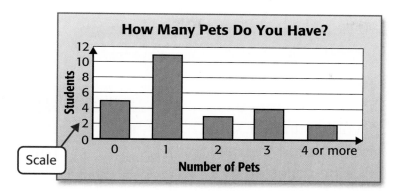

How Many Pets Do You Have?

Students (scale: 0, 2, 4, 6, 8, 10, 12)

Number of Pets: 0, 1, 2, 3, 4 or more

Scale

Step 3 Draw the bars.

Draw vertical bars to match each number from your data.

Think About It

1. How would you decide what scale to use?

2. Why do you think the scale counts by 2?

3. Why do some bars stop between two numbers or two lines?

CHECK What You Know

Display each set of data in a bar graph.

4.

Favorite Meal	
Meal	**Tally**
Breakfast	IIII
Lunch	TTTT TTTT II
Dinner	TTTT TTTT TTTT

5.

Favorite Fruit	
Fruit	**Tally**
Apple	TTTT TTTT TTTT I
Orange	TTTT TTTT
Banana	TTTT III

6. Ask 10 people their favorite color. Display the results in a bar graph.

7. **WRITING IN ►MATH** Refer to Step 2. How will the bar graph change when the scale changes.

Explore 12-4 Make a Bar Graph **527**

Bar Graphs

GET READY to Learn

Desmond asked his friends to name their favorite summer sport. He recorded the data that he collected in a tally chart.

Favorite Summer Sports		
Sport	**Tally**	**Number**
Tennis	ⅠⅠⅠⅠ	4
Swimming	卌 卌	10
Baseball	卌 ⅠⅠ	7
Biking	卌 Ⅰ	6

A **survey** is a way of collecting data by asking a question. You can take data from a tally chart and make a **bar graph**.

Real-World EXAMPLE — Make a Bar Graph

1 SPORTS Make a vertical bar graph to display the results of Desmond's survey.

In a *vertical* bar graph, the bars go up and down. It includes a title, labels, a scale, and bars. There is a space between each bar.

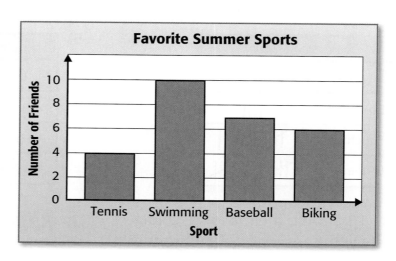

2 **ANIMALS** The bar graph shows how long some animals sleep. Which two animals sleep the most?

In a *horizontal* bar graph, the bars go from left to right.

Remember

On a bar graph, there is a space between each bar.

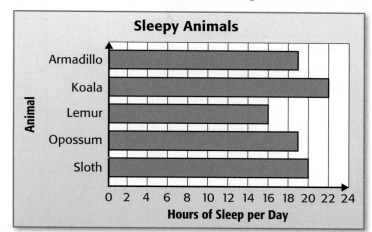

The lengths of the bars for the sloth and the koala are the longest. So, the sloth and the koala sleep the most.

CHECK What You Know

1. Display each set of data below in a vertical bar graph. **See Example 1 (p. 528)**

Favorite Birds to Watch	
Bird	**Tally**
Cardinal	~~||||~~ |||
Robin	||||
Goldfinch	~~||||~~ ||||

2. Display each set of data below in a horizontal bar graph.

See Example 2 (p. 529)

Animal Life Spans	
Animal	**Time (years)**
Lion	10
Hamster	2
Kangaroo	5
Rabbit	7

Source: *Time for Kids Almanac*

For Exercises 3 and 4, refer to Example 2. See Example 2 (p. 529)

3. Which animal sleeps the most?

4. Name one animal that sleeps three hours longer than the lemur.

5. **Talk About It** How are vertical and horizontal bar graphs alike? How are they different?

6. Display the set of data below in a vertical bar graph. See Example 1 (p. 528)

Width of Birds' Nests	
Bird	**Width (ft)**
Bald eagle	8
Blue heron	5
Monk parakeet	3
Stork	6

Source: *Book of World Records*

7. Display the set of data below in a horizontal bar graph. See Example 2 (p. 529)

World Series Wins	
Team	**Wins**
Cardinals	ЖЖ IIII
Giants	ЖЖ
Yankees	ЖЖ ЖЖ ЖЖ ЖЖ ЖЖ I
Dodgers	ЖЖ I

Source: *Book of World Records*

For Exercises 8–15, refer to the graphs.

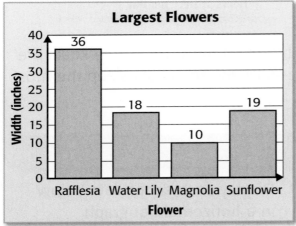

Source: *Book of World Records*

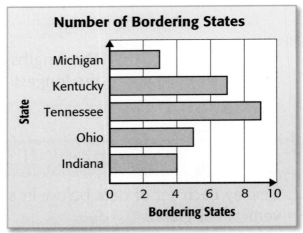

Source: *The World Almanac*

8. How wide is the largest flower?

9. Which two flowers are closest in size?

10. What is the difference between the largest and the smallest flower?

11. Which flower is half the size of the largest flower?

12. How many states border Tennessee?

13. How many more states border Ohio than Michigan?

14. Which states have 5 or fewer states bordering them?

15. Which state has the fewest number of bordering states?

H.O.T. Problems

16. OPEN ENDED Survey 10 people about their favorite baseball team. Display the data in a horizontal bar graph. Write two sentences that interpret the data.

17. WRITING IN ►MATH Why are a title and labels needed in a bar graph?

18. The graph shows how many beads are used in a bracelet. Which data is needed to complete the graph? (Lesson 12-4)

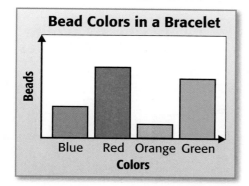

Bead Colors in a Bracelet

Beads

Blue Red Orange Green
Colors

A The colors in a bracelet.

B The scale of the graph.

C The color with the least beads.

D The title of the graph.

19. The pictograph below shows the favorite hiking trails of 24 third graders. How many shoes need to be added to finish the graph?
(Lesson 12-1)

Favorite Hiking Trails	
Blue Trail	👟
Red Trail	👟
Yellow Trail	👟 👟 👟
Key: 👟 = 3 students	

F 3

G 8

H 15

J 19

Spiral Review

20. There are 3 types of wrapping paper. There is also blue ribbon and gold ribbon. How many combinations can be made using 1 type of wrapping paper and 1 type of ribbon? (Lesson 12-3)

21. How many people were surveyed to make the pictograph? (Lesson 12-2)

22. If 48 people were surveyed, how many people would each brush symbol stand for? (Lesson 12-2)

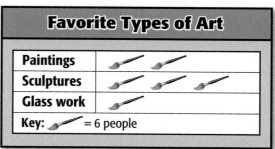

Favorite Types of Art	
Paintings	🖌 🖌
Sculptures	🖌 🖌 🖌
Glass work	🖌
Key: 🖌 = 6 people	

Geometry Describe each three-dimensional figure. Use the terms *faces, edges,* and *vertices.* Then identify the figure. (Lesson 11–1)

23.

24.

25.

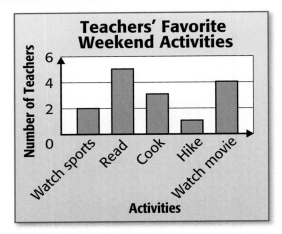

GET READY to Learn

Lauren surveyed the teachers at her school to find out their favorite weekend activities. The bar graph shows the results.

MAIN IDEA

I will interpret data in bar graphs.

Math Online

macmillanmh.com
• Extra Examples
• Personal Tutor
• Self-Check Quiz

You have learned to interpret data in a pictograph. You can also interpret the data in a bar graph.

Real-World EXAMPLES Interpret Bar Graphs

① **HOBBIES** How many more teachers like to read on the weekend than hike?

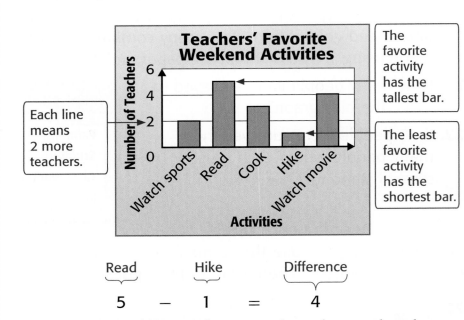

So, 4 more teachers prefer to read on the weekend than hike.

2 **ART** An art contest has 20 drawings. Display the data in a bar graph, and write a sentence to interpret the data.

Step 1 Find the number of drawings of cars.

$6 + 10 + 2 = 18$

$20 - 18 = 2$

There were 2 drawings of cars.

Art Contest	
Drawings	**Number**
Buildings	6
People	10
Animals	2
Cars	▪

Step 2 Make a bar graph to show the data.

Step 3 The data shows that there are the same number of drawings of animals as cars.

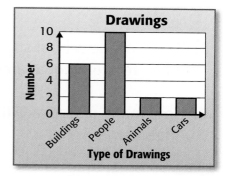

CHECK What You Know

For Exercises 1–2, refer to the bar graph. See Example 1 (p. 532)

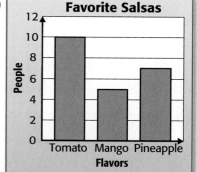

1. How many more people like tomato salsa than pineapple salsa?

2. Write a question that uses the bar graph. Solve.

For Exercise 3, display the data in a bar graph. Then write a sentence that interprets the data.

See Example 2 (p. 533)

3. Three students take ballet lessons. Five students take piano lessons. Ten students play sports.

4. **Talk About It** Jose surveyed five friends about how long they worked on puzzles. The longest time spent on puzzles was 6 hours. How many bars were shown on his graph? Explain.

For Exercises 6–12, refer to the bar graphs. See Example 1 (p. 532)

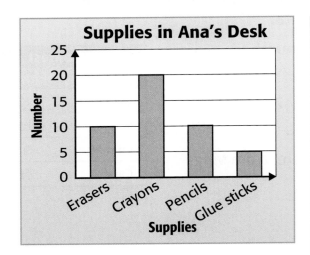

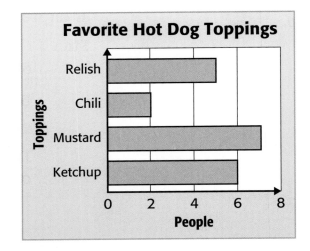

5. Which of the school supplies in Ana's desk are there equal amounts of?

6. How many erasers are there in Ana's desk?

7. Can you find the answer to Exercise 6 just by looking at the bar graph?

8. Explain how to find how many more pencils there are than glue sticks.

9. How does the scale help you find the number of people who like relish?

10. How many more people like relish than chili?

11. Which two toppings have a difference of four?

12. How many people were surveyed?

Display the data in a bar graph. Then write a sentence that interprets the data. See Example 2 (p. 533)

13.

Teachers' Favorite Lunch	
Food	**Tally**
Salad	ЖЖ ЖЖ II
Baked potato	ЖЖ ЖЖ ЖЖ III
Soup	ЖЖ ЖЖ ЖЖ ЖЖ IIII

14.

Races Won	
Name	**Tally**
Julia	ЖЖ I
Martin	IIII
Lin	ЖЖ
Tanya	IIII

15. OPEN ENDED Survey six friends about their favorite recess game, and make a bar graph showing the results.

16. CHALLENGE Draw two bar graphs showing the same data but use different scales.

17. **WRITING IN ►MATH** Explain how you might decide what scale to use in a bar graph.

TEST Practice

18. How many people like the Gators?
(Lesson 12-5)

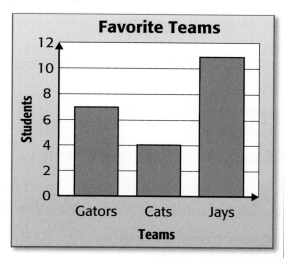

A 12   **C** 6

B 7   **D** 4

19. Which of the shapes listed below has the least number of sides?
(Lesson 11-2)

F quadrilateral

G pentagon

H octagon

J hexagon

20. Estimate 532 + 493 to the nearest thousand.

A 900   **C** 1,000

B 925   **D** 1,030

Spiral Review

21. Display the set of data on the right in a vertical bar graph. (Lesson 12-4)

22. Tracy has to take out the garbage, make her bed, and put her toys away. Show the different orders in which she could do her chores. Use the *make a list* strategy. (Lesson 12-3)

Loaves of Bread	
Type of Bread	**Number of Loaves**
Rye	⊦⊦⊦⊦ ⊦⊦⊦⊦
Wheat	⊦⊦⊦⊦ ⊦⊦⊦⊦ ⊦⊦⊦⊦ ⊦⊦⊦⊦
Marble	⊦⊦⊦⊦ ⊦⊦⊦⊦ ⊦⊦⊦⊦ ⊦⊦⊦⊦ ⊦⊦⊦⊦ ⊦⊦⊦⊦ ⊦⊦⊦⊦ ⊦⊦⊦⊦

23. Geometry A book case is 1 foot deep, 7 feet high and 3 feet wide. Use a model to find the volume. (Lesson 10-7)

Line Plots

GET READY to Learn

Albert surveyed his friends to find out how often they went to a movie theater. The table shows the results.

Movies Per Month			
Wyatt 0	Carina 1	Erika 2	Lazaro 1
Dante 1	Lani 2	Betty 0	Tama 1
Malcolm 2	Kelly 1	Harry 4	Adelmo 1
David 0	Judy 1	Drew 1	Laurie 3

A **line plot** uses Xs above a number line to show how often something happens.

Real-World EXAMPLE **Make a Line Plot**

1 **MOVIES** Make a line plot for the survey results.

Step 1 Draw and label a number line. Include all values of the data. Give it a title that describes the data.

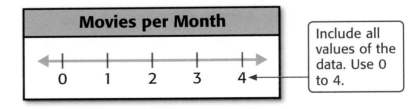

Include all values of the data. Use 0 to 4.

Step 2 Draw an X above the number for each response.

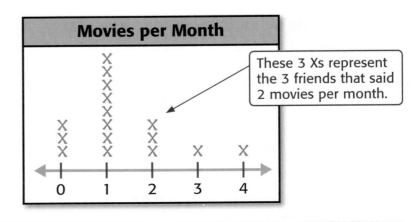

These 3 Xs represent the 3 friends that said 2 movies per month.

Remember

Start with the least number and end with the greatest number you need when numbering a line plot.

2 MOVIES Use Albert's line plot to find how often most students went to the movies.

The most Xs are above number 1. Albert can see that most of his friends went to the movies 1 time per month.

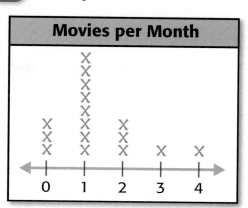

CHECK What You Know

Display each set of data in a line plot. See Example 1 (p. 536)

1.

Third-Grade Shoe Size			
Carlo 2	Arleta 4	Julia 8	Paulo 3
Shen 6	Tamera 5	Ronaldo 3	Desiree 4
William 4	Cole 5	Niles 4	Geraldo 5

2.

Weekly Time Spent on Homework	
Time (hours)	Tally
8				
9	₩₩	
10	₩₩
11	₩₩ |||

For Exercises 3 and 4, refer to the line plot below. See Example 2 (p. 537)

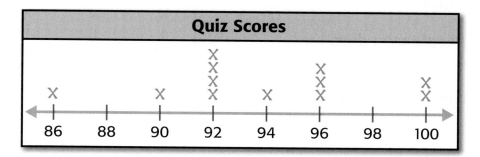

3. How many student's quiz scores are recorded? Explain.

4. What is one conclusion you can draw from this line plot? Explain.

5. **Talk About It** Does a tally table or a line plot make it easier to see how often numbers happen in a set of data? Explain.

Display each set of data in a line plot. See Example 1 (p. 536)

6.

Chores Per Week				
Lena 1	Cirilo 4	Nikita 3	Patrick 2	Raquel 3
Hao 2	Santos 5	Gia 2	Juwan 4	Pia 3
Demitri 2	Tammy 6	Sue 3	Shanti 1	Trey 2
Ayana 3	Jim 5	Maxine 4	Ellis 3	Nate 3

7.

Rollercoaster Rides				
Yuma 1	Barry 3	Rogelio 2	Matt 3	Toni 0
Stella 1	Jen 6	Jodie 0	Jean 3	Vince 3
Charles 0	Thea 3	Tito 1	Eric 1	Stuart 2
Irene 0	Sandra 3	Paul 2	Art 0	Hannah 1

8.

Hours of TV Watched	
Hours of TV	**Students**
0	I
1	IIII
2	
3	II
4	I
5	II

9.

Number of Siblings	
Siblings	**Students**
0	THL II
1	THL
2	THL IIII
3	III
4	II
5 or more	II

For Exercises 10–13, refer to the line plot below. See Example 2 (p. 537)

10. What do the Xs stand for on this line plot?

11. Which animal is owned by the most students?

12. Do more people own dogs or hamsters?

13. What conclusion could you draw?

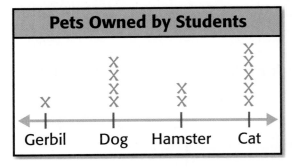

Pets Owned by Students

Gerbil Dog Hamster Cat

H.O.T. Problems

14. OPEN ENDED Give an example of a set of data that is not best displayed in a line plot. Explain.

15. WRITING IN ►MATH Explain how a tally chart and a line plot are alike and different.

16. Which sentence about the data below is true? (Lesson 12–6)

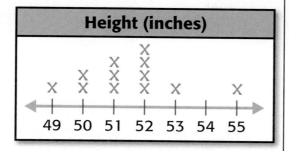

Height (inches)

49 50 51 52 53 54 55

A All members are 55 inches.

B Half of the team members are 52 inches or greater.

C Most members are 51 inches.

D No one is 49 inches tall.

17. Use the graph. What is the difference between the least and most favorite means of travel? (Lesson 12-5)

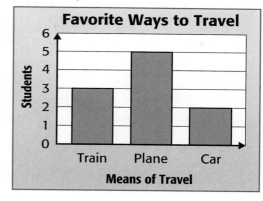

Favorite Ways to Travel

Students

Train Plane Car

Means of Travel

F 2   **H** 5

G 3   **J** 7

Spiral Review

Display each set of data in a vertical bar graph. (Lesson 12-4)

18.

Orchestra Instruments	
Instrument	Tally
Brass	IIII
Woodwind	ΗΗ
Strings	ΗΗ
Percussion	ΗΗ II

19.

Favorite After-School Snack	
Snack	Tally
Apple	III
Granola Bar	II
Smoothie	ΗΗ III
Yogurt	ΗΗ

Measurement **Find the volume of each figure.** (Lesson 10-7)

20.

21.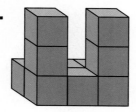

22.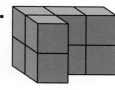

23. Admission to an amusement park is $25. Bus fare is $3 for one way. Find the total cost for Kali and her friend for admission and bus fare to and from the park. (Lessons 2-5 and 2-6)

Problem Solving in Science

EGGS!

When you go to the grocery store, what do the eggs look like? Most of them are usually white and about $1\frac{1}{2}$ inches long. Did you know that eggs come in all sorts of colors, sizes, and shapes? Most birds that lay white eggs, like kingfishers and woodpeckers, lay their eggs in dark holes. Birds that lay their eggs in open areas without a nest lay eggs that are colored like the soil. This keeps predators from finding them.

Did You Know?
Some birds lay their eggs in the nests of other birds so that the other birds will hatch them.

Real-World Math

For Exercises 1–5, use the bar graph to solve each problem.

1 How much longer is an emperor penguin egg than a robin egg?

2 A hummingbird laid 4 eggs. What is the total length of the eggs?

3 What is the second longest egg?

4 How many eggs are being compared?

5 Which egg is half the length of an elephant bird egg?

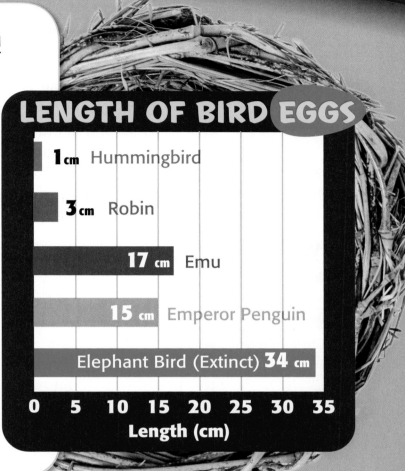

LENGTH OF BIRD EGGS

1 cm	Hummingbird
3 cm	Robin
17 cm	Emu
15 cm	Emperor Penguin
Elephant Bird (Extinct) **34** cm	

0   5   10   15   20   25   30   35
Length (cm)

Identify Probability

GET READY to Learn

Dana has a bag of 8 wristbands. Only one wristband is blue. Avery picks a wristband without looking. How likely is it that Avery will pick a blue wristband?

You can use words to describe probability.

Probability

Key Concept

Words **Probability** describes how likely it is that an event will happen.

Examples

Certain to choose a marble.

Likely to choose red.

Unlikely to choose green.

Impossible to choose yellow.

Real-World EXAMPLES Describe Probability

1 **How likely is it that Avery will pick blue?**

There is only 1 blue wristband out of a total of 8. So, it is *unlikely* that Avery will pick a blue wristband.

2 **How likely is it that Avery will pick green?**

There are 7 green wristbands out of a total of 8. So, it is *likely* that a green one will be picked.

3 **SPINNERS** Andrea spins the spinner. How likely is it that she will spin a multiple of 3?

The numbers 3, 6, 9, and 12 are multiples of 3. So, it is *certain* that Andrea will spin a multiple of 3.

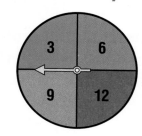

4 **GAMES** Theo and Yasmin are playing a card game. Yasmin needs to draw a 4 to win. The cards shown will be shuffled and placed face down on a table. How likely is it that Yasmin will draw a 4?

There are no 4s. The probability that Yasmin will draw a 4 is *impossible.*

CHECK What You Know

Describe the probability of landing on each color. Write *certain*, *likely*, *unlikely*, or *impossible*. See Examples 1–4
(pp. 542–543)

1. green

2. blue

3. yellow

4. blue, red, or green

5. Jamila is playing a game. She uses a number cube labeled 1, 2, 3, 4, 5, and 6. Describe the probability that she will roll a 7.

6. **Talk About It** Explain the difference between an event that is *certain* and an event that is *likely.*

Describe the probability of choosing each color. Write *certain*, *likely*, *unlikely*, or *impossible*.

See Examples 1–4 (pp. 542–543)

7. purple   **8.** green   **11.** yellow   **12.** red

9. white   **10.** blue or green   **13.** green   **14.** blue

Describe a bag of marbles that represents each statement.

15. Choosing a red marble is *impossible*.

16. Choosing a red marble is *certain*.

17. There are 7 letter tiles in a bag. Five of the tiles are labeled S. One tile is labeled R, and the other is M. Describe the probability of choosing the letter S.

18. Francis asks Dan to choose a marble from a bag of 10 marbles. What is the probability of choosing the color blue if one is blue?

Data File

Maryland's state colors are red, gold, and black.

19. What color is the spinner *likely* to land on?

20. Is the spinner *likely* or *unlikely* to land on gold?

21. How could you change the spinner so that it is *certain* to land on red?

State Colors

H.O.T. Problems

22. FIND THE ERROR Janice and Gus are spinning a spinner. The spinner is evenly divided into 4 sections. The colors are red, yellow, green, and blue. Who is correct? Explain.

Janice
It is unlikely that the spinner will land on orange.

Gus
It is impossible that the spinner will land on orange.

23. **WRITING IN ►MATH** Describe the probability of the following event. Explain.

A cow can fly like a bird.

TEST Practice

24. How many shirt-pant outfits are possible? (Lesson 12-3)

A 2   **C** 6

B 4   **D** 8

25. Lina has 7 cubes in a bag.

She closes her eyes and picks one cube. Describe the probability that she picks a green cube. (Lesson 12-7)

F certain   **H** unlikely

G likely   **J** impossible

Spiral Review

26. How many snack and drink combinations are possible if one snack and one drink is chosen? Explain your reasoning. (Lesson 12-3)

27. Make a line plot for the data: (Lesson 12-6)
5, 7, 2, 1, 5, 2, 8, 9, 3, 5, 7, 3, 9, 7, 2, 10, 4, 4, 3

Cold Snacks
Juice bar
Ice cream cone
Pudding bar

Cold Drinks
Smoothie
Ice water
Ice tea

Problem-Solving Investigation

MAIN IDEA I will choose the best strategy to solve problems.

P.S.I. TEAM +

ELAN: My school is having a book swap. The first day, 8 books were brought to school. The second day, 12 books came. Yesterday, my friends brought 16 books.

YOUR MISSION: Suppose the pattern continues. Find the total number of books after 7 days.

Understand	You know the number of books brought the first three days. Find the total number of books after 7 days.
Plan	Use the *make a table* strategy. Find and extend the pattern to solve the problem.
Solve	The pattern is to add 4 books each day. To find the total, add the number of books from each day.

Day	1	2	3	4	5	6	7
Books	8	12	16	20	24	28	32

+4 +4 +4 +4 +4 +4

$$\begin{array}{r} 8 \\ + 12 \\ \hline 20 \end{array} \quad \begin{array}{r} 20 \\ + 16 \\ \hline 36 \end{array} \quad \begin{array}{r} 36 \\ + 20 \\ \hline 56 \end{array} \quad \begin{array}{r} 56 \\ + 24 \\ \hline 80 \end{array} \quad \begin{array}{r} 80 \\ + 28 \\ \hline 108 \end{array} \quad \begin{array}{r} 108 \\ + 32 \\ \hline 140 \end{array}$$

So, the total number of books is 140.

Check	Look back. The answer makes sense for the facts given in the problem. ✔

Use any strategy shown below to solve. Tell what strategy you used.

PROBLEM-SOLVING STRATEGIES
- Draw a picture.
- Look for a pattern.
- Act it out.
- Make a table.
- Work backward.
- Make an organized list.

1. Horatio has 40 comic books. He keeps 10 comic books for himself and divides the rest equally among his 5 friends. How many comic books does each friend get?

2. Two teams scored a total of 20 points. The Bears scored 6 more points than the Seahawks. How many points did each team score?

3. Five girls signed up for a table tennis tournament. Each girl has to play each of the other girls one time. How many games will the girls play in all? Show your work.

4. Suppose you add 35 to a number then subtract 10. The result is 26. What was the original number?

5. The graph shows the number of people in each car that drove by Niguel's house. What is the total number of people who drove by?

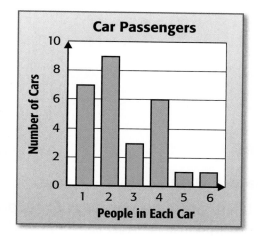

6. The pictograph below shows how many bundles of yarn Mrs. Finn bought. How many bundles did she buy altogether?

Mrs. Finn's Yarn	
Red	🔴 🔴 🔴
Blue	🔴
White	🔴 🔴 ◗
Each 🔴 = 2 bundles of yarn	

7. **WRITING IN ▶MATH** There are 48 oranges in a box and half as many grapefruit. To find the total number of pieces of fruit, would you use the *make a table* strategy? Explain.

Math Online › macmillanmh.com
• STUDY *TO GO*
• Vocabulary Review

FOLDABLES Study Organizer **GET READY to Study**

Be sure the following Key Vocabulary words and Key Concepts are written in your Foldable.

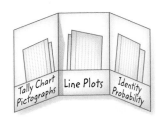

Tally Chart Pictographs | Line Plots | Identity Probability

Key Concepts

Record Data (p. 513)

• There are different ways to record and organize data. Two ways are to use a **tally chart** and a table.

Display Data (p. 513)

• A **pictograph** uses pictures to show data.

• A **bar graph** uses bars to show data.

• A **line plot** uses a number line to show how often something happens.

Probability (p. 542)

Probability describes how likely it is that an event will happen.

• *Certain* to choose a marble.
• *Likely* to choose red.
• *Unlikely* to choose green.
• *Impossible* to choose blue.

Key Vocabulary

bar graph (p. 526)
line plot (p. 536)
pictograph (p. 513)
probability (p. 542)
scale (p. 527)
tally chart (p. 513)

Vocabulary Check

Choose the vocabulary word that completes each sentence.

1. Data displayed on a _____?_____ shows how often an event happens.

2. A _____?_____ displays pictures or symbols to represent data.

3. _____?_____ describes how likely it is that an event will happen.

4. A _____?_____ organizes data using tally marks.

5. In a horizontal _____?_____, the bars go from left to right.

6. A _____?_____ is a set of numbers that represents the data.

Lesson-by-Lesson Review

12-1 **Pictographs** (pp. 515–517)

Example 1

The pictograph shows favorite types of weather. How many people like rain?

Favorite Type of Weather	
Sunny	✹ ✹
Rain	✹
Snow	✹ ✹
Key: ✹ = 4 people	

> The key shows that each symbol represents 4 people.

So, 4 people like rain.

7. Display the set of data in a pictograph.

Vegetables Eaten This Week	
Vegetable	**Tally**
Corn	‖
Broccoli	‖‖
Spinach	‖

8. A pictograph shows 10 symbols. If 30 people were surveyed, how many people are represented by each symbol?

12-2 **Interpret Pictographs** (pp. 518–521)

Example 2

The pictograph shows favorite types of games. How many more people prefer car games to board games?

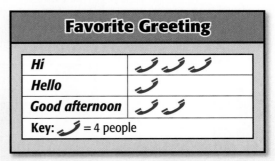

Favorite Games	
Sports	☺ ☺ ☺
Board games	☺
Car games	☺ ☺
Key: ☺ = 5 people	

> One more ☺ means 5 more people prefer card games to board games.

So, 5 more people like car games than board games.

Use the pictograph shown.

Favorite Greeting	
Hi	✍ ✍ ✍
Hello	✍
Good afternoon	✍ ✍
Key: ✍ = 4 people	

9. Which greeting is liked the most?

10. How many people were surveyed?

12-3 **Problem-Solving Strategy: Make a List** (pp. 522–523)

Example 3
Jacob has bologna and turkey. He also has cheddar, provolone, and swiss cheese. How many different sandwiches can he make with one meat and one cheese?

Meat		Cheese
Bologna	–	Cheddar
Bologna	–	Provolone
Bologna	–	Swiss
Turkey	–	Cheddar
Turkey	–	Provolone
Turkey	–	Swiss

Jacob can make 6 sandwiches.

Solve. Use the *make a list* strategy.

11. There are some coins in a jar. They add up to 29¢. How many coin combinations could there be?

12. Mr. Parsons asked his students to make as many different 4-digit number arrangements as they could using the numbers 4, 3, 2, and 1. How many arrangements can be made?

12-4 **Bar Graphs** (pp. 528–531)

Example 4
The vertical bar graph shows campers favorite activity. Which activity do campers prefer most?

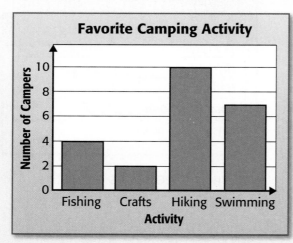

Favorite Camping Activity

The tallest bar is hiking. So, campers prefer hiking.

13. Display the set of data in a horizontal bar graph.

Location of Birds' Nests	
Tree	7
Roof	3
Bush	4
Porch swing	1

14. Display the set of data in a vertical bar graph.

Number of States Visited	
Number of States	**Number of Students**
2	卌
3				
4	卌|
5 or more	卌

Interpret Bar Graphs (pp. 532–535)

Example 5
The bar graph shows the results of a class vote of a field trip. How many more students chose to go to a museum than a farm?

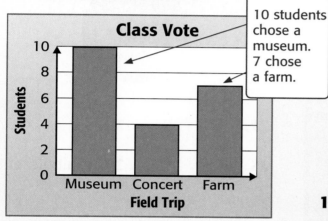

Class Vote

10 students chose a museum. 7 chose a farm.

10 − 7 = 3
So, 3 more students chose a museum than a farm.

Use the bar graph shown.

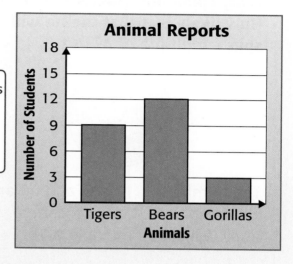

15. How many more students wrote reports about tigers than gorillas?

16. How many reports were written?

12-6 **Line Plots** (pp. 536–539)

Example 6
The number of homeruns scored this season are shown below. Display this set of data in a line plot.

3, 6, 1, 2, 2, 4, 3, 1, 1, 3, 2, 2, 2

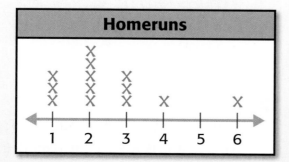

Homeruns

For Exercises 17–20, use the ages of the members of a swim team shown below.

17. 10, 10, 6, 7, 12, 8, 7, 7, 10, 7, 8, 10, 9, 9, 9, 8, 10, 9, 10
Display this set of data in a line plot.

18. How many members are there?

19. **Algebra** Write an expression comparing the number of members who are 6 and 8.

20. How many swimmers are 10 years or older?

12-7 **Identify Probability** (pp. 542–545)

Example 7
Andy spins the spinner shown. How likely is it that the he will spin a multiple of 4?

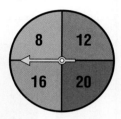

The numbers 8, 12, 16, and 20 are all multiples of 4, so it is *certain* Andy will spin a multiple of 4.

Describe the probability of the spinner landing on each number. Write *certain, likely, unlikely,* or *impossible.*

21. four   **22.** six

23. three   **24.** multiple of 3

12-8 **Problem-Solving Investigation: Choose a Strategy** (pp. 546–547)

Example 8
Ashley has 20 stuffed animals. She will keep 5 and divide the rest equally among 3 friends. How many stuffed animals will each friend get?

You know how many stuffed animals Ashley has and how many she will keep. You know she has 3 friends.

Subtract to find how many animals she has to give away.

$$20 \quad - \quad 5 \quad = \quad 15$$
Total   Ashley keeps   Divide equally

Divide to solve the problem.

$$15 \quad \div \quad 3 \quad = \quad 5$$

So, each friend will get 5 stuffed animals.

Use any strategy to solve.

25. Marcia has the money shown. Will she have enough money to buy an eraser for 35¢ and a note pad for $1.25? Explain.

26. Suppose you add 25 to a number and then subtract 14. The result is 48. What was the original number?

For exercises 1 and 2, tell whether each statement is *true* or *false.*

1. Data from a survey can be recorded on a tally chart.

2. A bar graph does not need a scale.

3. Display the data in a horizontal bar graph.

Pick a Marble	
Color	**Tally**
Yellow	THL THL THL I
Orange	THL III
Green	THL THL II
Blue	THL III

4. MULTIPLE CHOICE The pictograph shows the number of ribbons earned in each gymnastics event. What is the total number of ribbons received?

Gymnast Ribbons	
High bar	🏃 🏃
Parallel bars	🏃 🏃 🏃
Vault	🏃
Each 🏃 = 2 ribbons	

A 5   **C** 11

B 6   **D** 12

5. Display the data from Exercise 4 in a line plot.

Describe the probability of landing on each color. Write *certain*, *likely*, *unlikely*, or *impossible*.

6. blue

7. green

8. purple

9. blue, red, or green

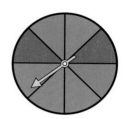

10. MULTIPLE CHOICE The table shows the results of spinning a spinner. What color is the spinner likely to land on?

Color Spin	
Color	**Tally**
Red	THL THL THL
Blue	THL THL THL THL THL THL THL THL
Green	THL
Yellow	III

F red   **H** green

G blue   **J** yellow

11. WRITING IN ▶MATH What does a tally chart of results from an experiment tell you about the probability of each possible outcome?

PART 1 Multiple Choice

Read each question. Then fill in the correct answer on the answer sheet provided by your teacher or on a sheet of paper.

1. The bar graph shows the number of absent students for 5 days. How many students were absent in all?

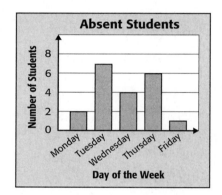

Absent Students

A 1   C 13

B 7   D 20

2. How many students checked out more than 4 books?

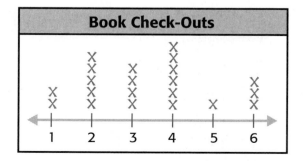

Book Check-Outs

F 4   H 10

G 7   J 23

3. The 2nd graders collected 39 canned goods for the food drive. The 3rd graders collected twice this amount. How many canned goods did the two grades collect altogether?

A 39   C 88

B 78   D 117

4. Tyrell spun the spinner once. On which color is the spinner unlikely to land?

F Green

G Red

H Blue

J Yellow

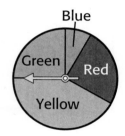

5. Terrance scored 12 points. How many basketballs should be drawn next to Terrance's name?

Points Scored	
Raul	🏀 🏀 🏀 🏀 🏀
Don	🏀 🏀 🏀
Phil	🏀 🏀 🏀 🏀
Terrance	
Key: 🏀 = 2 points	

A 4   C 6

B 5   D 7

6. What conclusion can you draw from the line plot?

Number of Prizes Won

X		X		
X	X	X		X
X	X	X	X	X
1	2	3	4	5

F Most students won 4 or more prizes.

G Most students won 1 to 3 prizes.

H More students won 5 prizes.

J Only 6 students won more than 4 prizes.

7. What rule does the table follow?

Rule: ■	
In	**Out**
7	6
10	9
13	12
16	15

A Add 2. **C** Add 1.

B Subtract 2. **D** Subtract 1.

Preparing for Standardized Tests
For test-taking strategies and practice, see pages R50–R63.

PART 2 Short Response

Record your answer on an answer sheet provided by your teacher or on a sheet of paper.

8. Based on the information in the table below, which students read twice as many books as Linda?

Books Read	
Students	**Books Read**
Benato	8
Linda	4
Colleen	8
Tanisha	7

9. A piggy bank has 6 quarters, 3 dimes, and 1 penny. Which coin is impossible to be picked?

PART 3 Extended Response

Record your answer on an answer sheet provided by your teacher or on a sheet of paper.

10. Ashanti has a bag of marbles. She has 2 red, 8 blue, and 1 green marble. Describe the probability of picking a certain color. Write a sentence for each using likely and unlikely. Explain.

NEED EXTRA HELP?										
If You Missed Question...	1	2	3	4	5	6	7	8	9	10
Go to Lesson...	12-5	12-6	6-3	12-7	12-1	12-6	8-5	12-2	12-7	12-7

CHAPTER 13 Develop Fractions

BIG Idea What is a fraction?

A **fraction** is a number that names part of a whole.

Example Michigan is the largest grower of green Niagra grapes. The fruit plate shown has 4 equal sections. The section with grapes is *one-fourth* or *one of four* sections.

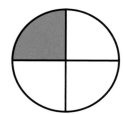 or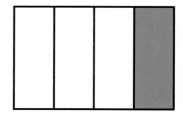

What will I learn in this chapter?

- Use fractions to represent parts of a whole or parts of a set.
- Model fractions and equivalent fractions.
- Compare and order fractions.
- Locate and name fractions on a number line.
- Solve problems by drawing pictures.

Key Vocabulary

fraction

numerator

denominator

equivalent fractions

Math Online > **Student Study Tools** at **macmillanmh.com**

one-fourth
or
one of four sections

FOLDABLES®
Study Organizer

Make this Foldable to help you organize information about fractions. Begin with three sheets of $8\frac{1}{2}$" × 11" paper.

1 **Stack** the paper so that each sheet is one inch higher than the other.

2 **Fold** the sheets upward so that all of the layers are the same distance apart.

3 **Crease** well. Then open and glue together as shown.

4 **Label** as shown. Record what you learn.

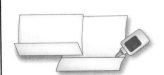

Fractions

Parts of a Whole
Parts of a Set
Find Equivalent Fractions
Compare and Order Fractions
Fractions on a Number Line

You have two ways to check prerequisite skills for this chapter.

Option 2

Math Online > Take the Chapter Readiness Quiz at macmillanmh.com.

Option 1

Complete the Quick Check below.

QUICK Check

Write the number of parts. Then tell whether each figure shows *equal* or *not equal* parts. (Previous grade)

1.

2.

3.

4.

5. Draw a circle that is divided into 6 equal parts.

Tell the number of equal parts. Write *halves, thirds,* or *fourths.* (Previous grade)

6.

7.

8.

9.

10. Jill draws a figure and divides it into fifths. What could her figure look like?

Tell which point represents each given number. (Lesson 1-8)

11. 36

12. 123

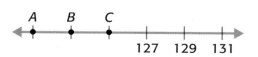

13. How many units are between each mark on the number line?

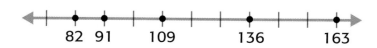

A **fraction** is a number that represents part of a whole or part of a set. You can *construct* or make models to show fractional parts of a whole.

MAIN IDEA

I will make models of fractions.

You Will Need
paper
ruler
scissors

ACTIVITY   Explore Fractions

Step 1   Make a Model

Cut 4 strips of paper that are each 1 inch wide and 8 inches long. Label one strip 1.

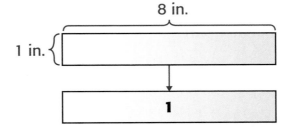

Step 2 Make a Fraction Model

Fold the second strip in half. Cut on the fold. Label each part $\frac{1}{2}$.

| $\frac{1}{2}$ | $\frac{1}{2}$ |

Step 3 Make More Fraction Models

Fold the third strip in half twice. Cut on the folds, and label each part $\frac{1}{4}$.

| $\frac{1}{4}$ | $\frac{1}{4}$ | $\frac{1}{4}$ | $\frac{1}{4}$ |

Step 4

Fold the last strip in half three times. Cut on the folds, and label each part $\frac{1}{8}$.

| $\frac{1}{8}$ | $\frac{1}{8}$ | $\frac{1}{8}$ | $\frac{1}{8}$ | $\frac{1}{8}$ | $\frac{1}{8}$ | $\frac{1}{8}$ | $\frac{1}{8}$ |

Think About It

1. How many strips are labeled $\frac{1}{2}$? $\frac{1}{4}$? $\frac{1}{8}$?

2. How many strips labeled $\frac{1}{2}$ are needed to equal the strip labeled 1?

3. Is $\frac{1}{2}$ greater than or less than $\frac{1}{8}$?

CHECK What You Know

Model each pair of fractions. Then write the greater fraction.

4. $\frac{1}{8}$, $\frac{1}{2}$

5. $\frac{1}{2}$, $\frac{1}{4}$

6. 1, $\frac{1}{4}$

7. $\frac{1}{2}$, 1

8. $\frac{1}{4}$, $\frac{1}{8}$

9. $\frac{1}{8}$, 1

10. **WRITING IN ►MATH** Describe how you would construct fraction models that show $\frac{1}{16}$.

13-1 Parts of a Whole

MAIN IDEA

I will write and read fractions for part of a whole.

New Vocabulary

fraction
numerator
denominator

Math Online

macmillanmh.com
• Extra Examples
• Personal Tutor
• Self-Check Quiz

GET READY to Learn

This rug has 5 equal sections. The sections are red, orange, purple, yellow, and green. What fraction of the rug is red?

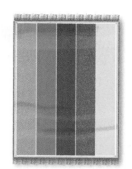

A **fraction** is a number that names part of a whole. You can use a fraction to tell what part of the rug is red.

Real-World EXAMPLE — Write and Read Fractions

1 RUGS **What fraction of the rug is red?**

One Way: Models	Another Way: Draw a Picture
The rug above represents one whole. There are 5 equal parts. A whole can be divided into 5 equal parts, or fifths.	Draw the rug with 5 equal parts. One part is red.

One Way column:

$$1$$

$\frac{1}{5}$ $\frac{1}{5}$ $\frac{1}{5}$ $\frac{1}{5}$ $\frac{1}{5}$

Write $\frac{1}{5}$ ← use numbers

Read one-fifth ← use words

Another Way column:

$\frac{1}{5}$ ← part that is red
← total number of equal parts

So, $\frac{1}{5}$ or one-fifth of the rug is red.

The **numerator** tells the number of equal parts that are being used. The **denominator** tells the total number of equal parts.

$\frac{1}{5}$ ← numerator
← denominator

EXAMPLE   Write and Read Fractions

Remember

The bottom number (denominator) is the total number of equal parts. The top number (numerator) is the number of shaded parts.

2 **What fraction of the figure is green?**

$\dfrac{2}{3}$ ← parts that are green
← total number of equal parts

Write $\dfrac{2}{3}$

Read *two-thirds*

So, $\dfrac{2}{3}$ or two-thirds of the figure is green.

Sometimes a part of a whole cannot easily be named by a fraction because the parts are not equal.

Real-World EXAMPLE

3 **GAMES** **A spinner is shown. What fraction of the spinner is labeled with a 2?**

There are 4 parts. But, the parts are not equal. So, you cannot easily write a fraction.

CHECK What You Know

Write the fraction for the part that is yellow. Then write the fraction for the part that is *not* yellow. Label your answers. See Examples 1–3 (pp. 561–562)

1.

2.

3.

4. What fraction of the pizza has cheese only?

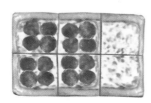

5. **Talk About It** What is a fraction? How does a fraction describe the shaded part of a whole?

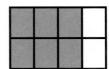
Write the fraction for the part that is blue. Then write the fraction for the part that is *not* blue. Label your answers. See Examples 1–3 (pp. 561–562)

6.

7.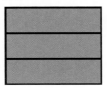

8.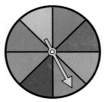

9.

10.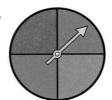

11.

12. What fraction of the honeycomb has bees?

13. What fraction of the pizza has pepperoni?

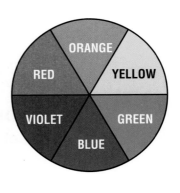

Draw a picture for each fraction. Shade the fraction.

14. $\frac{2}{5}$

15. $\frac{1}{7}$

16. three-eighths

17. two-halves

 Real-World PROBLEM SOLVING

Art The primary colors are red, blue, and yellow. The secondary colors are green, orange, and violet.

18. What fraction of the primary and secondary colors is red?

19. What fraction is blue or orange?

20. What fraction is not violet?

H.O.T. Problems

21. OPEN ENDED Draw a picture for which you cannot easily write a fraction to describe the shaded parts of the whole. Explain your thinking.

22. WRITING IN ▶MATH Explain how to write a fraction to describe part of a whole.

Parts of a Set

MAIN IDEA

I will write and read fractions for part of a set.

Math Online

macmillanmh.com

- Extra Examples
- Personal Tutor
- Self-Check Quiz

 **GET READY to Learn**

Hands-On Mini Activity

Fractions can also be used to name part of a set. You can use counters to help you understand how to describe the fractional parts of a set of objects.

1. What color of the counters is described by the fraction *three out of five*?

2. Use a fraction to describe the fractional part of the counters that are yellow.

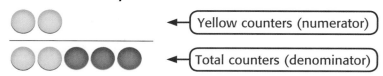

Yellow counters (numerator)

Total counters (denominator)

3. What color of the counters is described by $\frac{3}{5}$?

4. Write the fraction that shows two out of five.

Real-World EXAMPLE   Write and Read Fractions

1 **COUNTERS** What fraction of the set of counters is yellow?

Each counter is a fraction of the set of counters.

There are 3 yellow counters out of a total of 7 counters.

Write $\frac{3}{7}$ use numbers

Read *three-sevenths*   use words

So, three-sevenths or $\frac{3}{7}$ of the counters are yellow.

Write and Read Fractions

DOGS Cody told his four dogs to sit.

2 **What fraction of the set of dogs listened to Cody?**

The yellow counters represent the sitting dogs. The red counter represents the standing dog.

3 dogs, out of a total of 4 dogs, are sitting down.

Write $\dfrac{3}{4}$ ← dogs sitting
← total dogs

Read *three-fourths*
So, $\dfrac{3}{4}$ or three-fourths of the dogs obeyed.

3 **What fraction of the set of dogs did *not* listen to Cody?**

1 dog, out of a total of 4 dogs, is standing.
So, $\dfrac{1}{4}$ or one-fourth of the dogs did *not* obey.

CHECK What You Know

**Write the fraction for the part of the set that is yellow.
Then write the fraction for the part of the set that is *not*
yellow.** See Examples 1–3 (pp. 564–565)

1.

2.

3.

4. Moria has three blue counters, four red counters, and three yellow counters. What fraction of the set of counters is red?

5. **Talk About It** What do the numerator and denominator in a fraction symbol of a set show?

Write the fraction for the part of the set that is yellow.
Then write the fraction for the part of the set that is *not*
yellow. See Examples 1–3 (pp. 564–565)

6.

7.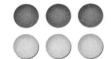

8.

9.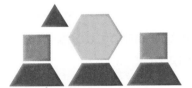

10.

11.

12. What fraction of the set of items shown at the right are living things?

13. There are 3 red paintbrushes and 5 green. What fraction of the set of paintbrushes are red?

Data File

The wild horses of Corolla roam safely in a National Wildlife Refuge along the northern Outer Banks of North Carolina. They are thought to be the oldest breed of horses in this country.

14. There are 6 chestnut-colored horses and 2 gray horses. Write the fraction for the part of the set of horses that are gray.

15. Along the beach stood 2 mares and 1 foal. Write the fraction for the part of the set of horses that are *not* foals.

16. Seven of the horses are grazing. If there are a total of 10 horses, what fraction are *not* grazing?

Source: Outer Banks Conservationists

H.O.T. Problems

17. OPEN ENDED Draw a set of objects that describe a fraction whose numerator is 4.

18. **WRITING IN** ►**MATH** Write a problem describing a fraction of a set. Solve.

TEST Practice

19. Terri is making the pictograph shown. She knows that a total of 20 medals were awarded. How many  symbols does she need to add? (Lesson 12-1)

Medals Awarded

Swimming	
Gymnastics	
Soccer	
Each  = 2 medals	

A 3 **C** 6

B 5 **D** 20

20. Which group shows $\frac{5}{7}$ of the flowers shaded? (Lesson 13-2)

F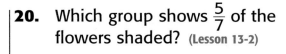

G

H

J

Spiral Review

Write the fraction for the part that is blue. Then write the fraction for the part that is *not* blue. Label your answers. (Lesson 13-1)

21.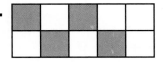

22.

23.

24. Flor drew 4 times as many pictures as Alvaro. Alvaro drew 3 more pictures than Meagan. Meagan drew 2 pictures. How many pictures did Flor and Alvaro each draw? (Lesson 12-8)

25. There are 7 colored tiles in a bag. Six tiles are blue and 1 tile is red. What is the probability of choosing a blue tile? (Lesson 12-7)

Problem-Solving Investigation

<u>**MAIN IDEA**</u> I will choose the best strategy to solve a problem.

P.S.I. TEAM +

CARLOTA: Last night, I played a board game. On one turn, I rolled two number cubes. The sum of the numbers was 9. The difference was 3.

YOUR MISSION: Find the two numbers Carlota rolled.

Understand	Carlota rolled two number cubes. The sum of the numbers was 9. The difference was 3. Find the two numbers rolled.
Plan	Make a table to show all of the possible rolls and their sums.
Solve	The table shows that to get a sum of 9, Carlota must have rolled 5 and 4 or 6 and 3.

$6 + 3 = 9$
$5 + 4 = 9$

Find the difference.

$6 - 3 = 3$
$5 - 4 = 1$

Since $6 + 3 = 9$ and $6 - 3 = 3$, Carlota must have rolled 6 and 3. |
| **Check** | Look back. Since $3 + 6 = 9$ and $6 - 3 = 3$, you know that the answer is correct. ✔ |

+	0	1	2	3	4	5	6
0	0	1	2	3	4	5	6
1	1	2	3	4	5	6	7
2	2	3	4	5	6	7	8
3	3	4	5	6	7	8	9
4	4	5	6	7	8	9	10
5	5	6	7	8	9	10	11
6	6	7	8	9	10	11	12

Use any strategy shown below to solve. Tell what strategy you used.

PROBLEM-SOLVING STRATEGIES
- Draw a picture.
- Look for a pattern.
- Guess and check.
- Make a table.
- Work backward.

1. Measurement Candace practices soccer for 35 minutes each day. She finished soccer practice at 5:30 P.M. What time did she begin?

2. A family of four goes bowling. Find the total cost if each person rents a ball and shoes and plays two games.

BOWLING
$4 per game
$2 shoe rental
$1 ball rental

3. Measurement A moose weighs about 1,300 pounds. A cougar weighs about 250 pounds. How much more does the moose weigh than the cougar?

4. Emil is drawing the pattern shown. He has room for 25 shapes. How many squares will he draw in all?

5. Pablo, Annie, and Terrance each have a pet. One has a cat. One has a turtle, and another has a mouse. Annie and Pablo have a pet with fur. Pablo and Terrance have a pet that lives in an aquarium. Who has the mouse?

6. Suppose you buy the items shown for lunch. If you pay with a $10-bill, how much change will you receive?

Lunch
Soda $1
Pizza $2

7. Algebra There are two numbers whose sum is 8 and quotient is 3. Find the numbers.

8. Lulu's parents took Gavin and two of his friends to the zoo. How much money did they pay for admission and parking?

Zoo Admission Prices	
Adults	$8
Children	$5
Groups of 10 or more	$7 each
Parking	$3

9. Alejandro paid $140 for the cable, phone, and Internet bills. The cable bill was $62. The phone bill was $59. How much was the Internet bill?

10. **WRITING IN MATH** A cake has 4 layers. One layer is orange, one is vanilla, and the others are peach. What fraction of the cake is peach? Explain.

Math Activity for 13-4
Equivalent Fractions

Fraction models can help you find fractions that name the same number, or **equivalent fractions**.

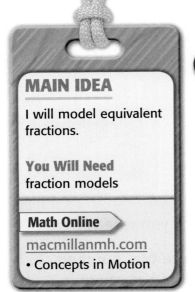

ACTIVITY   Find two fractions equivalent to $\frac{1}{2}$.

Step 1   Model $\frac{1}{2}$.

Start with 1 whole and the $\frac{1}{2}$-fraction tile.

Step 2   Find one fraction equivalent to $\frac{1}{2}$.

Use $\frac{1}{4}$-fraction tiles to equal the length of the $\frac{1}{2}$-fraction tile. Count the number of $\frac{1}{4}$-fraction tiles. So, $\frac{1}{2} = \frac{2}{4}$.

Step 3   Find another fraction equivalent to $\frac{1}{2}$.

Use $\frac{1}{8}$-fraction tiles to equal the length of the $\frac{1}{2}$-fraction tile. Count the number of $\frac{1}{8}$-fraction tiles. So, $\frac{1}{2} = \frac{4}{8}$.

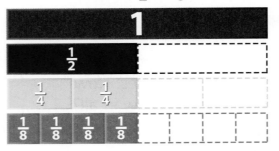

Think About It

1. How many of the $\frac{1}{4}$-fraction tiles are equal to the length of the $\frac{1}{2}$-fraction tile?

2. How many of the $\frac{1}{8}$-fraction tiles are equal to the length of the $\frac{1}{2}$-fraction tile?

3. Write two fractions that name the same amount as $\frac{1}{2}$.

4. Copy and complete $\frac{1}{2} = \frac{\blacksquare}{4} = \frac{\blacksquare}{8}$.

5. Use fraction models to find two fractions equivalent to $\frac{1}{3}$.

CHECK What You Know

Use models to complete the equivalent fractions.

6. $\boxed{\frac{1}{4}}$ = How many $\boxed{\frac{1}{8}}$?

   $\frac{1}{4} = \frac{\blacksquare}{8}$

7. $\boxed{\frac{1}{5}}$ = How many $\boxed{\frac{1}{10}}$?

   $\frac{1}{5} = \frac{\blacksquare}{10}$

8. $\boxed{\frac{1}{3}}$ = How many $\boxed{\frac{1}{6}}$?

   $\frac{1}{3} = \frac{\blacksquare}{6}$

9. $\boxed{\frac{1}{6}}$ $\boxed{\frac{1}{6}}$ = How many $\boxed{\frac{1}{12}}$?

   $\frac{2}{6} = \frac{\blacksquare}{12}$

Use fraction models to determine whether each pair of fractions is equivalent or not equivalent. Write *yes* or *no*.

10. $\frac{1}{2}$ and $\frac{3}{6}$

11. $\frac{1}{4}$ and $\frac{2}{4}$

12. $\frac{3}{4}$ and $\frac{6}{8}$

13. $\frac{3}{3}$ and $\frac{6}{6}$

14. $\frac{3}{5}$ and $\frac{5}{10}$

15. $\frac{2}{3}$ and $\frac{4}{6}$

16. **WRITING IN ►MATH** How do you know whether two fractions are equivalent? Explain how you know whether two fractions are not equivalent.

Model Equivalent Fractions

▶ GET READY to Learn

Kwan has a bookshelf. One of the 3 shelves has books. Kwan says that $\frac{1}{3}$ of the shelves have books. What other fractions name the same number as $\frac{1}{3}$?

Two or more different fractions that name the same amount are called **equivalent fractions**.

EXAMPLE Find Equivalent Fractions

① Complete $\frac{1}{3} = \frac{\blacksquare}{6}$ to find equivalent fractions.

One Way: Fraction Models	Another Way: Picture
Think about the number of equal parts in fraction models. 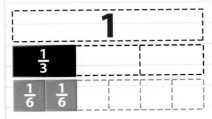 There are 2 sixths in $\frac{1}{3}$. So, $\frac{1}{3} = \frac{2}{6}$.	The rectangle is divided into thirds. One part is shaded. Another equal rectangle is divided into sixths. The same part is shaded. There are 2 sixths in $\frac{1}{3}$. So, $\frac{1}{3} = \frac{2}{6}$.

Complete each number sentence to find equivalent fractions. See Example 1 (p. 572)

1.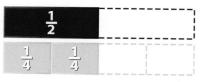

$$\frac{1}{2} = \frac{\blacksquare}{4}$$

2.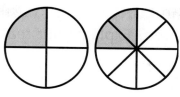

$$\frac{\blacksquare}{4} = \frac{2}{8}$$

3.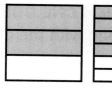

$$\frac{\blacksquare}{3} = \frac{\blacksquare}{6}$$

4. Wayne ate two-fifths of an apple pie. Write another fraction that names $\frac{2}{5}$.

5. **Talk About It** What pattern do you see in the fractions $\frac{1}{2} = \frac{2}{4} = \frac{4}{8}$?

Practice and Problem Solving

EXTRA **PRACTICE**
See page R36.

Complete each number sentence to find equivalent fractions. See Example 1 (p. 572)

6.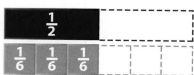

$$\frac{1}{2} = \frac{\blacksquare}{6}$$

7.

$$\frac{1}{5} = \frac{\blacksquare}{10}$$

8.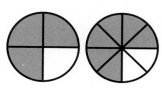

$$\frac{\blacksquare}{4} = \frac{6}{8}$$

9.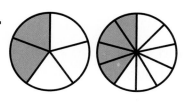

$$\frac{\blacksquare}{5} = \frac{4}{10}$$

10.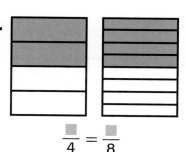

$$\frac{\blacksquare}{4} = \frac{\blacksquare}{8}$$

11.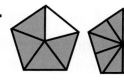

$$\frac{\blacksquare}{5} = \frac{\blacksquare}{10}$$

Algebra Find each missing value. Use models if needed.

12. $\frac{1}{2} = \frac{\blacksquare}{8}$

13. $\frac{1}{3} = \frac{\blacksquare}{12}$

14. $\frac{3}{\blacksquare} = \frac{6}{8}$

15. $\frac{\blacksquare}{5} = \frac{8}{10}$

Write an equivalent fraction for each fraction name.

16. One-sixth of a game spinner is red.

17. Amy read two-thirds of a book.

Lesson 13-4 Model Equivalent Fractions **573**

H.O.T. Problems

18. OPEN ENDED Give an example of two fractions that are not equivalent. Draw a picture to support your answer.

19. WHICH ONE DOESN'T BELONG? Which fraction does not belong with the other three? Explain your reasoning.

$$\frac{4}{8} \qquad \frac{1}{2} \qquad \frac{5}{10} \qquad \frac{3}{5}$$

20. **WRITING IN** ➤**MATH** Explain how you would find a fraction equivalent to $\frac{1}{2}$.

TEST Practice

21. What fraction of the figure is shaded? (Lesson 13-1)

   A $\frac{1}{8}$

   B $\frac{5}{8}$

   C $\frac{6}{8}$

   D $\frac{5}{6}$

22. Which figure is equivalent to $\frac{2}{3}$? (Lesson 13-4)

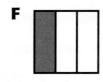

F   H

G   J

Spiral Review

23. In Luanda's neighborhood, 8 children have a scooter and 2 do not. Write a fraction to show how many children do *not* have a scooter. (Lesson 13-3)

24. What fraction of the set of paints is red? (Lesson 13-2)

25. What fraction of the items below is round?

26. Measurement How many seconds are in 2 minutes? (Lesson 10-8)

Model Equivalent Fractions

You can use Math Tool Chest to model equivalent fractions.

Savanna played 4 video games. If Savanna won $\frac{1}{2}$ of the games, how many games did she win?

MAIN IDEA

I will use technology to model equivalent fractions.

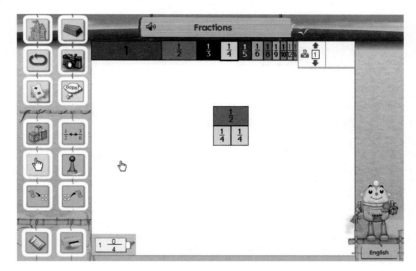

- Stamp out a $\frac{1}{2}$ fraction tile.

- Underneath, stamp out enough $\frac{1}{4}$ fraction tiles to equal $\frac{1}{2}$.

- The model shows that $\frac{1}{2} = \frac{2}{4}$.

How many games did Savanna win?

CHECK What You Know

Use Math Tool Chest to model each fraction and find one equivalent fraction.

1. $\frac{1}{5}$ **2.** $\frac{3}{6}$ **3.** $\frac{3}{9}$ **4.** $\frac{4}{12}$

5. Roberto cut a loaf of bread into 12 slices. Each slice is $\frac{1}{12}$ of the loaf. He used $\frac{1}{3}$ of the loaf to make sandwiches. How many slices of bread did he use?

6. Will was at bat 8 times during a baseball game. He struck out 2 times. What fraction of times at bat did Will strike out?

Fraction Concentration

Find Equivalent Fractions

You will need: 10 index cards

$\frac{1}{2}$	$\frac{1}{3}$	$\frac{1}{4}$	$\frac{1}{5}$
$\frac{3}{4}$	$\frac{2}{5}$	$\frac{2}{3}$	$\frac{3}{9}$
$\frac{4}{5}$	$\frac{4}{8}$	$\frac{4}{10}$	$\frac{2}{12}$
$\frac{1}{6}$	$\frac{6}{10}$	$\frac{2}{8}$	$\frac{3}{5}$
$\frac{2}{10}$	$\frac{8}{10}$	$\frac{4}{6}$	$\frac{6}{8}$

Get Ready!

Players: 2 players

Get Set!

Cut each index card in half. Then label as shown.

Go!

- Shuffle the cards. Then spread out the cards face-down.

- Player 1 turns over any two of the cards.

- If the fractions are equivalent, Player 1 keeps the cards and continues his or her turn.

- If the fractions are not equivalent, the cards are turned over and Player 2 takes a turn.

- Continue playing until all fractions matches are made. The player with the most cards wins.

Write the fraction for the part that is green. Then write the fraction for the part that is *not* green. (Lesson 13-1)

1. 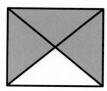 **2.**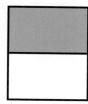

Draw a picture for each fraction. Shade the fraction. (Lesson 13-1)

3. $\frac{1}{6}$ **4.** $\frac{3}{5}$

5. A cake is divided equally into 8 pieces. If two pieces are eaten, what fraction of the cake is left? (Lesson 13-1)

6. MULTIPLE CHOICE What fraction of the figure is shaded? (Lesson 13-1)

A $\frac{1}{2}$

B $\frac{5}{9}$

C $\frac{5}{8}$

D $\frac{3}{8}$

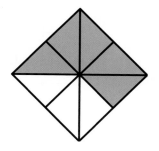

7. Sally has 7 packages of sticky notes. Five of the packages are pink, one is green, and one is blue. Write a fraction to show what part of the set of sticky notes are *not* pink. (Lesson 13-2)

Complete each number sentence to find equivalent fractions. (Lesson 13-4)

8. $\frac{\blacksquare}{6} = \frac{2}{12}$ **9.** $\frac{3}{4} = \frac{\blacksquare}{8}$

10. $\frac{\blacksquare}{6} = \frac{2}{3}$ **11.** $\frac{1}{4} = \frac{\blacksquare}{12}$

12. Write an equivalent fraction for $\frac{2}{6}$. (Lesson 13-4)

13. Algebra Clem is thinking of two numbers. The sum of the numbers is 8. The product is 15. What are the numbers? (Lesson 13-3)

14. MULTIPLE CHOICE What figure shows $\frac{1}{2}$? (Lesson 13-4)

F 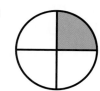 H

G   J

15. **WRITING IN ►MATH** A figure is divided into 12 equal parts. Six parts are shaded. Is more than one-half of the figure shaded? Explain. (Lesson 13-4)

Problem-Solving Strategy

<u>MAIN IDEA</u> I will solve a problem by drawing a picture.

Anessa and her brother have 8 insects in a jar. One-half of the insects are beetles. One is a firefly, and the rest are crickets. How many of the insects are crickets?

Understand	**What facts do you know?**
	• There are 8 insects.
	• One is a firefly.
	• One-half are beetles.
	• The rest are crickets.
	What do you need to find?
	• Find how many of the insects are crickets.
Plan	You can draw a picture to solve the problem.
Solve	• First draw a figure that is divided into 8 equal parts. This shows 8 insects.
	• To show the beetles, shade $\frac{1}{2}$ of the figure. Shade 1 part to show the firefly.
	1 is a firefly.
	There are 4 beetles.
	• There are 3 parts not shaded. This is the number of crickets.
	So, 3 of the insects are crickets.
Check	Look back. 4 beetles + 1 firefly + 3 crickets = 8 insects There are 8 insects. So, the answer is correct. ✔

Refer to the problem on the previous page.

1. Explain why the figure was divided into 8 equal parts.

2. Explain why 4 of the 8 boxes were shaded to show the number of beetles.

3. Suppose Anessa had 10 insects. How many would be crickets?

4. Look back to Exercise 3. Check your answer. How do you know that it is correct? Show your work.

PRACTICE the Strategy

EXTRA PRACTICE
See page R36.

Solve. Use the *draw a picture* strategy.

5. There are 12 books. One-third of the books are Mia's. Two belong to Basilio. The others belong to Tyree. How many books belong to Tyree?

6. Mill Park is 5 miles directly east of Bear Cabin. Nature museum is 5 miles directly south of Bear Cabin. Glacier Lake is 5 miles directly west of Mill Park. Is this possible? Explain.

7. Lucy and Nicole each have an equal-size piece of pizza. The table shows how much of each piece Lucy and Nicole ate. Who ate more?

Lucy	Nicole
$\frac{1}{2}$	$\frac{3}{4}$

8. Four students are standing in line. Ben is ahead of Kendra. Chad is behind Kendra. Tariq is behind Ben. In what order are they standing?

9. There are 36 houses on a street. The table shows what fraction of the houses have a dog or cat. How many of the houses have a dog?

Dogs	Cats
$\frac{3}{4}$	$\frac{1}{4}$

10. Berta rides the elevator 3 floors up from her home to meet her friend Devan. They go down 7 floors, where they meet Tamera. How many floors is Berta from her home?

11. Allison is playing jacks. She tosses 10 jacks on the floor. She then picks up $\frac{2}{5}$ of them. How many jacks are left on the floor?

12. **WRITING IN MATH** Explain what it means to draw a picture to solve a problem. How is a picture helpful in solving a problem?

13-6 Compare and Order Fractions

MAIN IDEA

I will compare and order fractions.

Math Online

macmillanmh.com

- Extra Examples
- Personal Tutor
- Self-Check Quiz

GET READY to Learn

Camila and Pete were reading a book. Camila read $\frac{5}{8}$ of the book while Pete read $\frac{3}{8}$ of the same book. Who read more of the book?

Amount of Book Read	
Camila	$\frac{5}{8}$
Pete	$\frac{3}{8}$

You can compare fractions to see which fraction is greater than (>), which fraction is less than (<), or if they are equivalent fractions.

To compare fractions you can use models or pictures.

Real-World EXAMPLE Compare Fractions

1 **READING** Use models to find if Camila or Pete read more of the book.

Compare $\frac{5}{8}$ and $\frac{3}{8}$. Use >, <, or =.

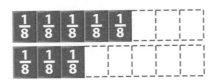

There are 5 one-eighth tiles and 3 one-eighth tiles.

So, $\frac{5}{8}$ is greater than $\frac{3}{8}$.

$\frac{5}{8} > \frac{3}{8}$ or $\frac{3}{8} < \frac{5}{8}$

Camila read more of the book than Pete.

580 Chapter 13 Develop Fractions

You can compare fractions with different denominators.

Real-World EXAMPLE Compare Fractions

2 FOOD Keri and Arturo each have mini pizzas that are the same size. Keri ate $\frac{1}{2}$ of her pizza. Arturo ate $\frac{3}{4}$ of his pizza. Who ate less?

Draw a picture to compare $\frac{1}{2}$ and $\frac{3}{4}$.

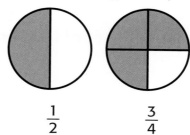

$$\frac{1}{2} \qquad \frac{3}{4}$$

$\frac{1}{2}$ is less than $\frac{3}{4}$.

So, $\frac{1}{2} < \frac{3}{4}$ or $\frac{3}{4} > \frac{1}{2}$.

Keri ate less pizza than Arturo.

Remember

When you compare fractions, be sure the wholes are the same size.

You can also compare fractions.

Real-World EXAMPLE Compare Fractions

3 Order the amount of each ingredient in the recipe from greatest to least.

Use a drawing to help you order the fractions.

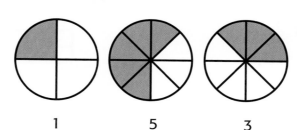

$$\frac{1}{4} \qquad \frac{5}{8} \qquad \frac{3}{8}$$

SNACK MIX

$\frac{1}{4}$ cup walnuts

$\frac{5}{8}$ cup granola

$\frac{3}{8}$ cup yogurt chips

Mix and enjoy

The models show that $\frac{5}{8} > \frac{3}{8} > \frac{1}{4}$.

So, the amounts of each ingredient from the greatest to the least is granola, yogurt chips, and walnuts.

Compare. Use >, <, or =. See Examples 1 and 2 (pp. 580–581)

1.

| $\frac{1}{5}$ | $\frac{1}{5}$ | $\frac{1}{5}$ | $\frac{1}{5}$ | |

| $\frac{1}{5}$ | $\frac{1}{5}$ | $\frac{1}{5}$ | | |

$\frac{4}{5} \bullet \frac{3}{5}$

2.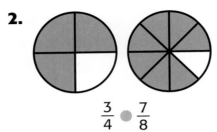

$\frac{3}{4} \bullet \frac{7}{8}$

Use models to order each set of fractions from least to greatest. (See Example 3 p. 581)

3. $\frac{1}{2}, \frac{2}{8}, \frac{3}{4}$

4. $\frac{3}{5}, \frac{4}{10}, \frac{2}{10}$

5. $\frac{1}{4}, \frac{1}{6}, \frac{1}{8}$

6. Alister makes a party mix with $\frac{1}{3}$ cup of raisins and $\frac{2}{3}$ cup of cereal. Are there more raisins or cereal? Explain.

7. **Talk About It** Tell how you know that $\frac{1}{4}$ is less than $\frac{3}{4}$.

▶ Practice and Problem Solving

EXTRA PRACTICE
See page R36.

Compare. Use >, <, or =. See Examples 1 and 2 (pp. 580–581)

8.

| $\frac{1}{3}$ | | |

| $\frac{1}{3}$ | $\frac{1}{3}$ | |

$\frac{1}{3} \bullet \frac{2}{3}$

9.

| $\frac{1}{10}$ | $\frac{1}{10}$ | $\frac{1}{10}$ | $\frac{1}{10}$ | $\frac{1}{10}$ | $\frac{1}{10}$ | | | |

| $\frac{1}{10}$ | $\frac{1}{10}$ | $\frac{1}{10}$ | $\frac{1}{10}$ | | | | | |

$\frac{6}{10} \bullet \frac{4}{10}$

10.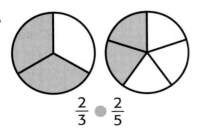

$\frac{2}{3} \bullet \frac{2}{5}$

11.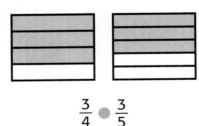

$\frac{3}{4} \bullet \frac{3}{5}$

Use models to order each set of fractions from least to greatest. (See Example 3 p. 581)

12. $\frac{1}{10}, \frac{1}{5}, \frac{4}{5}$

13. $\frac{4}{8}, \frac{7}{8}, \frac{1}{8}$

14. $\frac{5}{8}, \frac{3}{4}, \frac{1}{4}$

15. $\frac{2}{3}, \frac{2}{6}, \frac{3}{6}$

16. $\frac{6}{12}, \frac{1}{3}, \frac{3}{3}$

17. $\frac{3}{4}, \frac{1}{2}, \frac{1}{3}$

Solve.

18. Of the kids at camp, $\frac{5}{8}$ are boys and $\frac{3}{8}$ are girls. Are there more boys or girls? Explain.

19. Measurement Which is greater, $\frac{3}{8}$ of an inch or $\frac{1}{2}$ of an inch?

20. Nat has 6 toy planes. Two of them are red. Are more or less than $\frac{3}{6}$ of the planes red?

21. Kendrick plays soccer, basketball, track, and football. Do more or less than $\frac{2}{4}$ of the sports use a ball?

H.O.T. Problems

22. NUMBER SENSE Is $\frac{1}{4}$ of the smaller waffle the same as $\frac{1}{4}$ of the larger waffle? Explain.

23. OPEN ENDED Write a real-world math problem where you need to compare two fractions.

24. **WRITING IN ►MATH** Two sandwiches are the same size. One is cut into 4 pieces. The other is cut into 8 pieces. How do you know which has the smaller pieces?

TEST Practice

25. Which fraction is more than $\frac{5}{8}$? (Lesson 13-6)

  A $\frac{3}{8}$ **C** $\frac{1}{2}$

  B $\frac{2}{4}$ **D** $\frac{3}{4}$

26. Which set of fractions is ordered from greatest to least? (Lesson 13-6)

  F $\frac{1}{3}, \frac{1}{4}, \frac{1}{5}$ **H** $\frac{1}{4}, \frac{4}{8}, \frac{3}{4}$

  G $\frac{3}{6}, \frac{2}{3}, \frac{3}{3}$ **J** $\frac{5}{8}, \frac{1}{8}, \frac{1}{2}$

Spiral Review

27. A dessert recipe uses $\frac{2}{3}$ cup of berries and $\frac{3}{4}$ cup of grapes. Which is the greater amount, the berries or the grapes? (Lesson 13-5)

Find an equivalent fraction for each fraction. (Lesson 13-4)

28. $\frac{1}{2}$ **29.** $\frac{2}{3}$ **30.** $\frac{3}{5}$

31. Hayden has 6 pets. One-third of them are cats. One is a hamster, and the rest are turtles. How many are turtles? (Lesson 13-1)

13-7 Locate Fractions on a Number Line

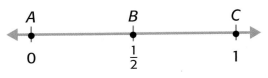
GET READY to Learn

Derek played during one half $\left(\frac{1}{2}\right)$ of a football game. This number line shows a fraction of a whole number.

A   B   C

0        $\frac{1}{2}$ 1

MAIN IDEA

I will locate and name points on a number line using fractions.

Math Online

macmillanmh.com

• Extra Examples
• Personal Tutor
• Self-Check Quiz

In Lesson 11–8, you learned that a number line is a line that can represent whole numbers as points. Fractions can also be represented on a number line.

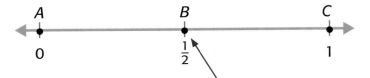

Real-World EXAMPLE   Locate Fractions on a Number Line

1 **GAMES** Use Derek's number line to tell what point represents one half $\left(\frac{1}{2}\right)$ of a football game.

To find a fraction on a number line, divide the line between 0 and 1 by the denominator. The number $\frac{1}{2}$ divides the line into 2 equal parts.

A   B   C

0        $\frac{1}{2}$ 1

> $\frac{1}{2}$ is halfway between 0 and 1. The denominator shows that there are 2 equal parts.

So, point B represents $\frac{1}{2}$ on the number line.

You can find intervals to name fractions.

EXAMPLE   **Locate Fractions on a Number Line**

② **Tell what fraction point _A_ represents on the number line.**

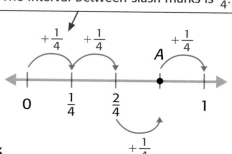

The interval between slash marks is $\frac{1}{4}$.

$\frac{2}{4} + \frac{1}{4} = \frac{3}{4}$.

So, point $A = \frac{3}{4}$.

Remember

A fraction with a numerator of zero is equivalent to 0.

$\frac{0}{4} = 0$

When the numerator and denominator are the same, the fraction equals 1.

$\frac{4}{4} = 1$

CHECK What You Know

Tell what point each fraction represents. See Example 1 (p. 584)

1. $\frac{1}{4}$

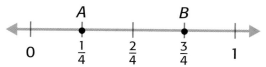

2. $\frac{4}{5}$

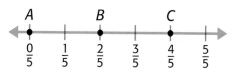

Tell what fraction each point represents. See Example 2 (p. 585)

3. Point $A = \blacksquare$

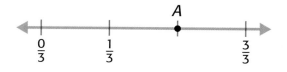

4. Point $X = \blacksquare$

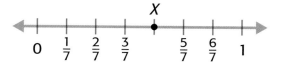

5. Point _H_ represents a fraction half way between $\frac{1}{2}$ and 1 on a number line. Name the fraction.

6. **Talk About It** Explain how a number line can be used to compare fractions.

Practice and Problem Solving

Tell what point each fraction represents. See Example 1 (p. 584)

7. $\frac{1}{3}$

8. $\frac{3}{5}$

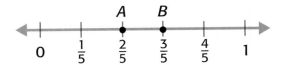

9. $\frac{4}{4}$

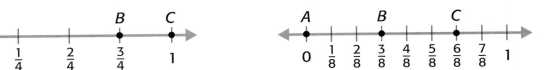

10. $\frac{3}{8}$

Tell what fraction each point represents. See Example 2 (p. 585)

11. Point $B = $ ▨

12. Point $C = $ ▨

13. Point $Y = $ ▨

14. Point $Z = $ ▨

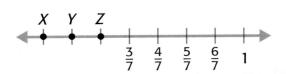

15. Name a fraction that is between $\frac{7}{8}$ and $\frac{5}{8}$ on a number line.

16. Name three fractions between $\frac{1}{6}$ and $\frac{6}{6}$ on a number line.

17. Name two fractions that could be next to $\frac{3}{5}$ on a number line.

18. Name a fraction that is greater than $\frac{3}{7}$.

H.O.T. Problems

19. OPEN ENDED Draw a number line showing five fractions.

20. CHALLENGE Name point *A* on the number line.

21. FIND THE ERROR Johnny and Lars each put three fractions on a number line in the order shown. Who is correct? Explain.

Johnny

$\frac{2}{3}, \frac{3}{3}, \frac{4}{3}$

Lars

$\frac{3}{4}, \frac{4}{3}, \frac{5}{4}$

22. WRITING IN ▸MATH Explain how to name a point on a number line.

TEST Practice

23. Which point best represents $\frac{3}{4}$ on the number line? (Lesson 13-7)

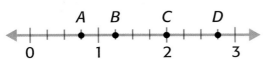

A *A*   **C** *C*

B *B*   **D** *D*

24. Which group shows less than $\frac{4}{7}$ of the cups empty? (Lesson 13-6)

F

G

H

J

Spiral Review

25. Compare $\frac{3}{5}$ and $\frac{5}{5}$. Write >, <, or =. (Lesson 13-6)

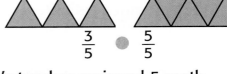

$\frac{3}{5} \bullet \frac{5}{5}$

26. A grocer unpacks a box of soup that holds 4 cartons. Each carton holds 12 cans of soup. How many cans did the grocer unpack? (Lesson 5-8)

27. Kylie's teacher assigned 5 math problems each day. After 10 days how many math problems did her class do? (Lesson 4-6)

Problem Solving in Science

The BUZZ on Insects

Insects have been around for 250 million years...long before the dinosaurs. Today, there are over 800,000 known species of insects!

These species represent $\frac{1}{12}$ of all the animal species on Earth.

Insects have different sizes and shapes but they all have four things in common. Insects have
- three body parts,
- six jointed legs,
- two antennae, and
- an outside skeleton.

🌐 Real-World Math

Use the information on page 588 and 589 to solve each problem.

1. Is the length of the firefly greater than or less than the length of the house fly?

2. Is the length of the lady beetle greater than or less than $\frac{1}{2}$ of an inch?

3. Of Earth's animals species, $\frac{1}{12}$ are insects. What part are not insects? Draw a model to show your work.

4. What part of all animals on Earth are not beetles?

5. Round the number of bugs to the nearest ten thousand.

6. What fraction of the insects shown are longer than $\frac{1}{2}$-inch?

7. Suppose you put one lady beetle in front of another. Which insect would have this same length?

8. Which two insects are the same length?

Six Spotted Green Tiger Beetle
$\frac{5}{8}$-inch

Firefly
$\frac{3}{4}$-inch

Honey Bee
$\frac{4}{8}$-inch

Lady Beetle
$\frac{3}{8}$-inch

House Fly
$\frac{1}{2}$-inch

That's a lot of Insects!

Insect	Number of Species
Beetle	350,000
Butterfly & Moth	170,000
Fly	120,000
Bee & Ant	110,000
Bug	82,000
Grasshopper	20,000
Dragonfly	5,000

Source: Enchanted Learning

Did You Know?

Of all the animals on Earth, $\frac{1}{4}$ are beetles.

Study Guide and Review

FOLDABLES®
Study Organizer **GET READY to Study**

Be sure the following Key Concepts and Key Vocabulary are written in your Foldable.

Fractions
Parts of a Whole
Parts of a Set
Find Equivalent Fractions
Compare and Order Fractions
Fractions on a Number Line

Key Concepts

- A **fraction** names part of a whole or part of a set. (p. 561)

$\frac{2}{3}$ ← numerator
      ← denominator

- **Equivalent fractions** name the same number. (p. 572)

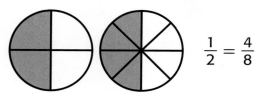

$\frac{1}{2} = \frac{4}{8}$

- Compare and order fractions by drawing a picture. (p. 580)

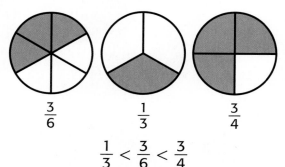

$\frac{3}{6}$ $\frac{1}{3}$ $\frac{3}{4}$

$\frac{1}{3} < \frac{3}{6} < \frac{3}{4}$

Key Vocabulary

denominator (p. 561)

equivalent fractions (p. 572)

fraction (p. 561)

numerator (p. 561)

Vocabulary Check

Choose the vocabulary word that completes each sentence.

1. The number $\frac{1}{2}$ is a(n) _____?_____.

2. In the fraction $\frac{1}{5}$, the 5 is called the _____?_____.

3. In the fraction $\frac{3}{4}$, the 3 is called the _____?_____.

4. The fractions $\frac{1}{3}$ and $\frac{2}{6}$ are _____?_____.

5. The _____?_____ tells the total number of equal parts.

6. A _____?_____ is a number that names part of a whole or part of a set.

Lesson-by-Lesson Review

13-1 Parts of a Whole (pp. 561–563)

Example 1
What fraction of the figure is blue?

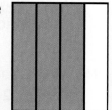

$\dfrac{3}{4}$ ← numerator
← denominator

Write $\dfrac{3}{4}$ ← use numbers

Read *three-fourths* ← use words

So, $\dfrac{3}{4}$ or three-fourths is blue.

Write the fraction for the part that is blue. Then write the fraction for the part that is *not* blue. Label your answers.

7.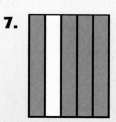

8.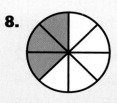

Draw a picture for each fraction. Shade the fraction.

9. $\dfrac{2}{3}$

10. $\dfrac{3}{8}$

13-2 Parts of a Set (pp. 564–567)

Example 2
What fraction of the set is blue?

Write $\dfrac{2}{5}$ ← use numbers

Read *two-fifths* ← use words

So, $\dfrac{2}{5}$ or two-fifths is blue.

Write the fraction for the part of the set that is blue. Then write the fraction for the part of the set that is *not* blue.

11.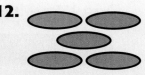

12.

13. What fraction of the set of bananas is peeled?

13-3 **Problem-Solving Investigation:** **Choose a Strategy** (pp. 568–569)

Example 3
Zane saves $6 each month in coins. How long will it take him to save $48 in coins?

To solve the problem, you can use the *guess and check* strategy.

$6 × 7 = $42   no
$6 × 9 = $54   no
$6 × 8 = $48   yes

It will take Zane 8 months to save $48.

14. Sodas cost $2. Salads cost $4. Lia buys 1 soda and 1 salad. Trish orders 1 soda and 2 salads. How much money is spent in all?

15. Mrs. Cook drove 7 hours each day for 2 days. Then she drove 4 hours each day for 2 days. How many hours did she drive in all?

16. Paloma needs 5 wall tiles for each mural she makes. She has 15 tiles. How many murals can she make?

13-4 **Model Equivalent Fractions** (pp. 572–574)

Example 4
Complete $\frac{2}{3} = \frac{\blacksquare}{6}$ to find equivalent fractions.

One Way: Use Fraction Models

$\frac{1}{3}$	$\frac{1}{3}$	

$\frac{1}{6}$	$\frac{1}{6}$	$\frac{1}{6}$	$\frac{1}{6}$	

Another Way: Draw a Picture

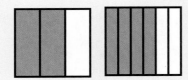

There are 4 sixths in $\frac{2}{3}$. So, $\frac{2}{3} = \frac{4}{6}$.

Complete each number sentence to find equivalent fractions.

17. $\frac{1}{2} = \frac{\blacksquare}{4}$ **18.** $\frac{2}{2} = \frac{\blacksquare}{6}$

19. $\frac{1}{3} = \frac{\blacksquare}{6}$ **20.** $\frac{3}{5} = \frac{\blacksquare}{10}$

Algebra Find each missing value.

21. $\frac{2}{3} = \frac{\blacksquare}{12}$ **22.** $\frac{5}{\blacksquare} = \frac{10}{12}$

Write an equivalent fraction for each fraction name.

23. Jordan put extra cheese on $\frac{1}{5}$ of a pizza.

24. Four-fifths of the quilt squares are red.

13-5 Problem-Solving Strategy: Draw a Picture (pp. 578–579)

Example 5

There are 12 trucks. One-third are red. Two are blue. The rest are green. How many trucks are green?

Divide a figure into 12 equal parts. Shade $\frac{1}{3}$ to show the red trucks and 2 parts to show the blue trucks.

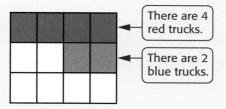

There are 4 red trucks.

There are 2 blue trucks.

There are 6 parts left. So, 6 are green.

25. A music CD tower can hold 16 CDs. One-half of the slots are filled with CDs. How many CDs are in the CD tower?

26. A boat ride has 15 boats. The boats are yellow, purple, and orange. One-fifth are yellow. Five are purple. How many of the boats are orange?

27. Patty and Ruben are playing tic-tac-toe. Patty has Xs in one-third of the 9 squares. Ruben has Os in 2 of the squares. How many squares are empty?

13-6 Compare and Order Fractions (pp. 580–583)

Example 6

Compare $\frac{2}{3}$ and $\frac{3}{5}$.

One Way: Use Fraction Models

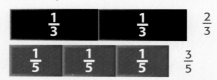

Another Way: Draw a Picture

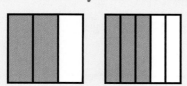

$\frac{2}{3}$ is greater than $\frac{3}{5}$. So, $\frac{2}{3} > \frac{3}{5}$.

Compare. Use >, <, or =.

28.

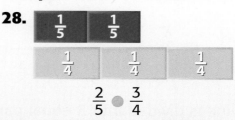

$\frac{2}{5}$ ● $\frac{3}{4}$

29.

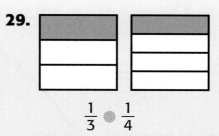

$\frac{1}{3}$ ● $\frac{1}{4}$

30. Lucita walks $\frac{1}{4}$ mile. Sergio walks $\frac{3}{8}$ mile. Who walks farther?

13-7 **Locate Fractions on a Number Line** (pp. 584–587)

Example 7

Tell what point represents $\frac{3}{4}$.

The denominator of $\frac{3}{4}$ shows that the number line is divided into 4 equal parts.

The numerator of $\frac{3}{4}$ shows that the point is the third interval after zero.

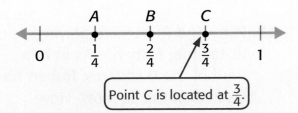

Point C is located at $\frac{3}{4}$.

So, point C represents. $\frac{3}{4}$ on the number line.

Example 8

Tell what fraction point X represents.

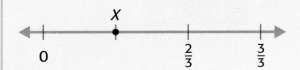

The line is divided into 3 equal parts. The point X represents the first part of the line.

$\frac{1}{3}$ ← part
← total parts

So, point X represents $\frac{1}{3}$ on the number line.

Tell what point each fraction represents.

31. $\frac{3}{6}$

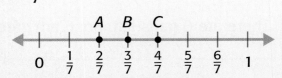

32. $\frac{3}{5}$

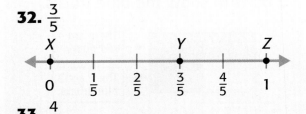

33. $\frac{4}{7}$

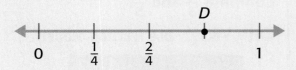

Tell what fraction each point represents.

34. Point D = ▪

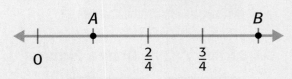

35. Point A = ▪

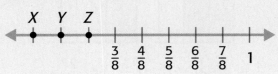

36. Point Y = ▪

For Exercises 1 and 2, tell whether each statement is *true* or *false*.

1. The numerator is the top number in a fraction.

2. The fractions $\frac{3}{5}$ and $\frac{5}{10}$ are equivalent fractions.

Complete each number sentence to find equivalent fractions.

3. $\frac{1}{4} = \frac{\blacksquare}{8}$

4. $\frac{3}{5} = \frac{\blacksquare}{10}$

5. What fraction of the spinner is purple?

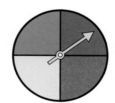

6. There are 2 groups of 5 students and 2 groups of 7 students on a field trip. How many students are there altogether?

7. MULTIPLE CHOICE Which figure is equivalent to $\frac{3}{4}$?

A

B

C

D

8. Of the fish in an aquarium, $\frac{8}{12}$ are orange and $\frac{4}{12}$ are striped. Are more fish orange or striped?

9. MULTIPLE CHOICE Which fraction of the leaves is shaded?

F $\frac{2}{6}$ **H** $\frac{4}{6}$

G $\frac{2}{4}$ **J** $\frac{4}{2}$

Tell what fraction each point represents.

10. Point $A = \blacksquare$

11. Point $X = \blacksquare$

12. Does Tómas practice the piano more or less than the oboe?

Instrument	Time (hour)
Oboe	$\frac{1}{8}$
Piano	$\frac{5}{8}$

13. **WRITING IN ►MATH** There are 9 students. Four-ninths of them are carrying a backpack. Are more people carrying a backpack or not carrying a backpack? Explain.

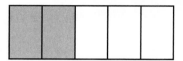

PART 1   Multiple Choice

Read each question. Then fill in the correct answer on the answer sheet provided by your teacher or on a sheet of paper.

1. What fraction of the bar is shaded?

A $\frac{1}{5}$ **C** $\frac{2}{5}$

B $\frac{1}{4}$ **D** $\frac{1}{2}$

2. There are 3 black cats and 1 white cat. What fraction of the cats are white?

F $\frac{1}{4}$

G $\frac{1}{2}$

H $\frac{2}{3}$

J $\frac{3}{4}$

3. Judie had $8,000 in her savings account. She spent $4,537 on a used car. How much does she have left in her savings account?

A $3,463   **C** $4,363

B $3,467   **D** $5,473

4. Casey drew smiley faces on her paper. Which face appears to have a line of symmetry?

F   **H**

G   **J**

5. If $\frac{3}{5}$ of a wheel of cheese has been eaten, what fraction of the cheese has not been eaten?

A $\frac{1}{4}$ **C** $\frac{2}{5}$

B $\frac{3}{8}$ **D** $\frac{2}{4}$

6. Which fraction represents the same part of the whole as $\frac{6}{9}$?

F $\frac{2}{3}$ **H** $\frac{2}{5}$

G $\frac{3}{4}$ **J** $\frac{1}{3}$

7. A humpback whale weighs 2,558 pounds. What is the weight rounded to the nearest hundred?

A 2,530 pounds   **C** 2,600 pounds

B 2,540 pounds   **D** 3,000 pounds

8. A play area has a width of 9 feet and a length of 7 feet. What is the perimeter of the play area?

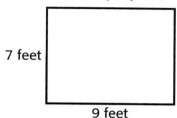

7 feet

9 feet

F 11 feet   **H** 32 feet

G 16 feet   **J** 58 feet

---

9. What fraction represents point *K* on the number line?

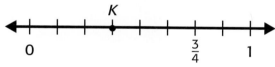

0   $\frac{3}{4}$ 1

A $\frac{3}{4}$ **C** $\frac{3}{8}$

B $\frac{1}{2}$ **D** $\frac{5}{8}$

---

10. What fraction of the figure is shaded?

F $\frac{8}{16}$

G $\frac{3}{5}$

H $\frac{8}{12}$

J $\frac{16}{8}$

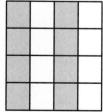

PART 2   Short Response

Record your answers on the answer sheet provided by your teacher or on a sheet of paper.

11. Order the set of fractions $\frac{4}{8}, \frac{5}{8}, \frac{3}{4}$ from greatest to least.

---

12. Angie owns a rule book for 10 board games and card games. Six of the rules are for board games. What fraction of the rules are *not* for board games?

---

13. A pizza is cut into 8 equal slices. Some friends eat 6 pieces. Do they eat $\frac{3}{4}$ or $\frac{2}{3}$ of the pizza?

PART 3   Extended Response

Record your answers on the answer sheet provided by your teacher or on a sheet of paper.

14. Order the set of fractions $\frac{1}{4}, \frac{1}{2}, \frac{1}{3}$ from greatest to least. What happens to the size of a fraction tile when the denominators are greater? What happens when the denominators are less? Explain.

NEED EXTRA HELP?														
If You Missed Question...	1	2	3	4	5	6	7	8	9	10	11	12	13	14
Go to Lesson...	13-1	13-2	3-8	11-7	13-1	13-4	1-8	9-5	13-7	13-1	13-6	13-1	13-4	13-6

CHAPTER 14

Understand Fractions and Decimals

 **How are fractions and decimals alike?**

Fractions and decimals both describe parts of a whole.

Example Nataya uses a rain gauge to measure the amount of rainfall in inches. It shows 0.1, one-tenth.

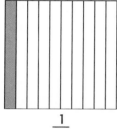 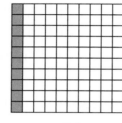

$$\frac{1}{10} \qquad = \qquad 0.1$$

What will I learn in this chapter?

- Understand tenths and hundredths.
- Relate fractions and decimals to money.
- Solve problems by acting them out.

Key Vocabulary

decimal

decimal point

tenths

hundredths

Math Online ▶ **Student Study Tools** at macmillanmh.com

Make this Foldable to help you organize information about fractions and decimals. Begin with one sheet of $8\frac{1}{2}'' \times 11''$ paper.

1 **Fold** the ends of the paper to meet in the middle.

2 **Fold** again in half as shown.

3 **Unfold** and cut along the two inside folds as shown.

4 **Label.** Record what you learn.

Tenths   Hundredths

Fractions, Decimals, Money

Decimals and Money

Chapter 14 Understand Fractions and Decimals **599**

You have two ways to check prerequisite skills for this chapter.

Option 2

Math Online ▷ Take the Chapter Readiness Quiz at macmillanmh.com.

Option 1

Complete the Quick Check below.

QUICK Check

Write the fraction for the part that is shaded. (Lesson 13-1)

1.

2.

3.

4. Three of 5 trees are oak trees. What fraction of the trees are not oak trees?

Write each fraction using numbers. (Lesson 13-1)

5. three-fourths

6. two-tenths

7. one-half

8. Alan will sing 3 verses of a song. The song has 7 verses. Write the fraction for the part of the song he will sing in words and in numbers.

Write the amount of money shown. Use a cent sign then a dollar sign and decimal point. (Lesson 2-5)

9.

10.

11.

12. Write 25¢ using a decimal point.

A **decimal** is a number that uses place value and a **decimal point** to show part of a whole. You can use models to show how fractions and decimals are related.

ACTIVITY Explore Fractions and Decimals

Step 1 Make a Model

Make a grid that is divided into 10 equal parts or columns. Shade 3 parts.

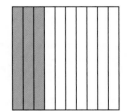

Step 2 Write a Fraction and Decimal

Write the fraction for the shaded part of the grid. Copy and complete the place value chart.

 shaded part
parts in all

Hundreds	Tens	Ones	Tenths
		0	

decimal point

Step 3 Make Another Model

Make a grid that is divided into 10 rows and 10 columns. Shade 30 of the 100 parts.

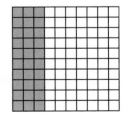

Step 4 Write a Fraction and a Decimal

Write the fraction for the shaded part. Copy and complete the place value chart.

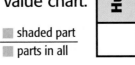 shaded part
parts in all

Hundreds	Tens	Ones	Tenths	Hundredths
		0		

Think About It

1. In Step 1, how many tenths are shaded in the grid?

2. In Step 3, how many hundredths are shaded in the grid?

3. How do you write in words the shaded part of the grids?

4. Do $\frac{3}{10}$ and $\frac{30}{100}$ name the same number? How do you know?

CHECK What You Know

Write a fraction and a decimal for each shaded part.

5.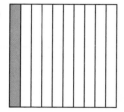

6.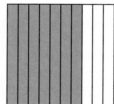

7.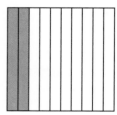

8.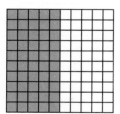

9.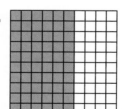

10.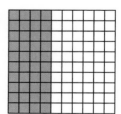

Model each fraction. Then write the fraction as a decimal.

11. $\frac{15}{100}$

12. $\frac{5}{10}$

13. $\frac{7}{10}$

Model each decimal. Then write each decimal as a fraction.

14. 0.25

15. 0.80

16. 0.4

17. **WRITING IN ►MATH** Describe how fractions are like decimals and how they are different from decimals.

14-1 Tenths

GET READY to Learn

Alfredo's mother made a blanket using pieces of fabric. What part of the blanket is blue?

MAIN IDEA

I will learn the meaning of tenths.

New Vocabulary

decimal
decimal point
tenths

Math Online

macmillanmh.com
• Extra Examples
• Personal Tutor
• Self-Check Quiz

Recall that a **decimal** is a number that uses place value and a **decimal point** to show part of a whole. Everything to the right of the decimal point is *part of a whole*. A **tenth** is one of ten equal parts.

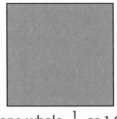

one whole, $\frac{1}{1}$, or 1.0

There are ten tenths in one whole.

one tenth, $\frac{1}{10}$, or 0.1

Real-World EXAMPLE   Write Tenths

1 SEWING What part of the blanket is blue?

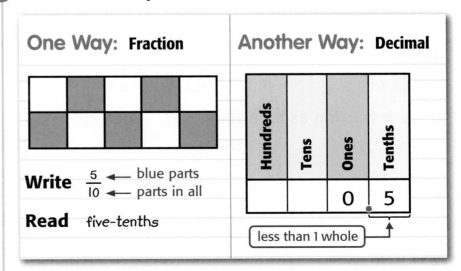

One Way: Fraction	Another Way: Decimal

Write $\frac{5}{10}$ ← blue parts ← parts in all

Read five-tenths

Hundreds	Tens	Ones	Tenths
		0	5

less than 1 whole

So, $\frac{5}{10}$ or 0.5 of the blanket is blue.

You can write a fraction as a decimal and a decimal as a fraction.

Remember

To read a decimal, read the digits to the right of the decimal point as a whole number. Then say its place value. Example: 0.7 is read *seven-tenths*.

EXAMPLES Write Fractions and Decimals

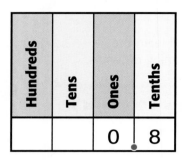

2 Write $\frac{4}{10}$ as a decimal.

$\frac{4}{10}$ is written as 0.4.

Hundreds	Tens	Ones	Tenths
		0	4

So, $\frac{4}{10}$ = 0.4.

3 Write 0.8 as a fraction.

0.8 is eight-tenths.

Hundreds	Tens	Ones	Tenths
		0	8

So, 0.8 = $\frac{8}{10}$.

CHECK What You Know

Write a fraction and a decimal for the part that is shaded. See Example 1 (p. 603)

1.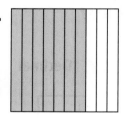

2.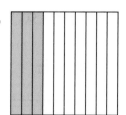

3.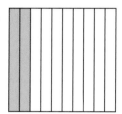

Write each fraction as a decimal. See Example 2 (p. 604)

4. $\frac{7}{10}$

5. $\frac{1}{10}$

6. $\frac{2}{10}$

Write each decimal as a fraction. See Example 3 (p. 604)

7. 0.5

8. 0.9

9. 0.4

10. There are six-tenths of pizza left. What is the decimal for this amount?

11. **Talk About It** What is the number 0.7 in word form? What does the amount mean?

Write a fraction and a decimal for the part that is shaded. See Example 1 (p. 603)

12.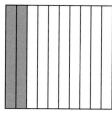

13.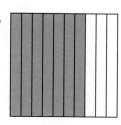

14.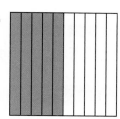

15.

16.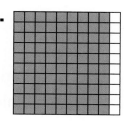

17.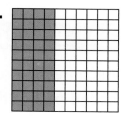

Write each fraction as a decimal. See Example 2 (p. 604)

18. $\frac{2}{10}$

19. $\frac{6}{10}$

20. $\frac{9}{10}$

21. $\frac{3}{10}$

22. eight-tenths

23. five-tenths

Write each decimal as a fraction. See Example 3 (p. 604)

24. 0.4

25. 0.9

26. 0.5

27. 0.8

28. six-tenths

29. three-tenths

30. Four out of 10 chess players are girls. What part of the group are boys? Write as a decimal.

31. Manu needs five-tenths of a cup of flour. Write the amount of flour needed as a fraction.

Data File

The table lists cities of Indiana and their rainfall.

32. Write all of the cities' rainfalls as a fraction.

33. Write Fort Wayne's rainfall as a decimal.

Indiana Cities	Least Rainfall in June (in.)
Evansville	0.8
Bloomington	0.3
South Bend	0.5
Gary	0.1
Fort Wayne	$\frac{3}{10}$

Source: Weather Reports

H.O.T. Problems

34. OPEN ENDED Write any number in the tenths place as a decimal and as a fraction. Explain the meaning of the number.

35. NUMBER SENSE Is the number 0.3 greater than or less than 1? Explain how you know.

36. WRITING IN ►MATH Write about a real-world situation where you see tenths written as decimals.

TEST Practice

37. Which fraction represents the shaded part of the figure?

(Lesson 14-1)

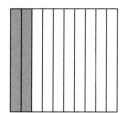

A $\frac{2}{100}$ **C** $\frac{2}{10}$

B $\frac{3}{100}$ **D** $\frac{3}{10}$

38. Dena shaded $\frac{9}{10}$ of the figure. Which decimal equals $\frac{9}{10}$?

(Lesson 14-1)

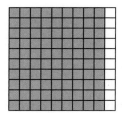

F 10.59 **H** 5.09

G 9.5 **J** 0.9

Spiral Review

Tell what fraction each point represents. (Lesson 13-7)

39. *Point A = ▓*

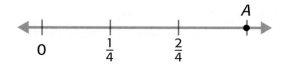

40. *Point X = ▓*

41. Edmund and Lloyd are planning to help cut the grass. Lloyd says he will cut $\frac{1}{2}$ of the grass. Edmund says he will cut $\frac{4}{8}$ of the grass. Who will cut more? Explain. (Lesson 13-4)

42. Algebra Find the missing value in $\frac{5}{7} = \frac{10}{▓}$. (Lesson 13-4)

Fractoes and Decimoes

Match Fractions and Decimals

Get Ready!

Players: 2–3 players

Get Set!

Cut apart the Fractoes and Decimoes tiles.

Go!

- Place the tiles spread out on a table face down.

- Each player chooses five tiles and holds them so the other players cannot see them.

- The first player chooses a tile from his or her hand and places it on the table. A new tile is then chosen.

- The next player places one tile from his or her hand that matches one of the ends of the tiles on the table. A new tile is then chosen.

You will need: Fractoes and Decimoes resource master, scissors.

0.8	eight tenths	0.5	$\frac{5}{10}$
$\frac{4}{10}$			0.7
0.4			
$\frac{1}{10}$			

- If a player is unable to make a match, a new tile must be chosen until a match is made.

- Play continues until one player has placed all his or her tiles and wins the game.

14-2 Hundredths

MAIN IDEA

I will learn the meaning of hundredths.

New Vocabulary

hundredths

Math Online

macmillanmh.com

• Extra Examples
• Personal Tutor
• Self-Check Quiz

GET READY to Learn

Mr. Rivera's class took a survey of 100 people to find out their favorite food. What part of the group favors spaghetti?

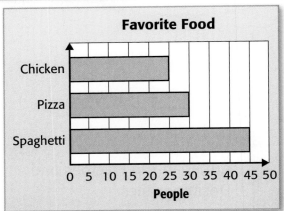

Favorite Food

Decimals can also be written in **hundredths**.

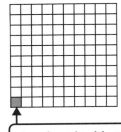

One hundredth. There are 100 hundredths in one whole.

Hundreds	Tens	Ones	Tenths	Hundredths
		0	0	1

No tenths

Real-World EXAMPLE   Write Hundredths

1 FOOD What part of the group favors spaghetti?

Out of the 100 students, 45 students favor spaghetti.

One Way: Fraction

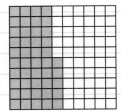

$\frac{45}{100}$ forty-five hundredths

Another Way: Decimal

Hundreds	Tens	Ones	Tenths	Hundredths
		0	4	5

0.45 forty-five hundredths

608   Chapter 14 Understand Fractions and Decimals

Fractions can be written as decimals and decimals can be written as fractions.

EXAMPLE **Write Fractions and Decimals**

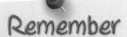

Remember

Place value names to the right of the decimal point end in *ths*.
Example: *tenths*, *hundredths*

② **Write $\frac{48}{100}$ as a decimal.**

$\frac{48}{100}$ is written as 0.48.

Hundreds	Tens	Ones	Tenths	Hundredths
		0	4	8

So, $\frac{48}{100} = 0.48$.

③ **Write 0.75 as a fraction.**

0.75 is seventy-five hundredths.

Hundreds	Tens	Ones	Tenths	Hundredths
		0	7	5

So, $0.75 = \frac{75}{100}$.

✓ CHECK What You Know

Write a fraction and a decimal for the part that is shaded.
See Example 1 (p. 608)

1.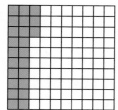

2.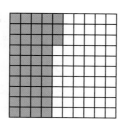

3.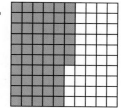

Write each fraction as a decimal. See Example 2 (p. 609)

4. $\frac{56}{100}$

5. $\frac{25}{100}$

6. $\frac{86}{100}$

Write each decimal as a fraction. See Example 3 (p. 609)

7. 0.85

8. 0.34

9. 0.19

10. Andre is reading a book. It has 100 pages. He has read 54 pages. Write the fraction and decimal for the amount that he has read.

11. **Talk About It** Where have you seen a decimal in the hundredths used everyday?

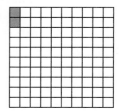
Write a fraction and a decimal for the part that is shaded.

See Example 1 (p. 608)

12.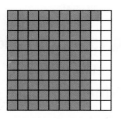

13.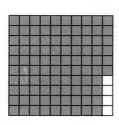

14.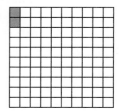

15.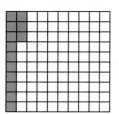

16.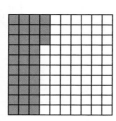

17.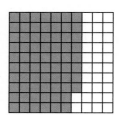

Write each fraction as a decimal. See Example 2 (p. 609)

18. $\frac{23}{100}$

19. $\frac{27}{100}$

20. $\frac{14}{100}$

21. $\frac{10}{100}$

22. $\frac{73}{100}$

23. $\frac{91}{100}$

Write each decimal as a fraction. See Example 3 (p. 609)

24. 0.58

25. 0.24

26. 0.05

27. 0.49

28. 0.55

29. 0.30

For Exercises 30 and 31, write a fraction and a decimal.

30. Sareeta has a box of 100 crackers. She eats 15 crackers. What part of the crackers did she not eat?

31. Yolanda read 100 books. Of these, 35 are fiction. What part of the books are not fiction?

H.O.T. Problems

32. OPEN ENDED Write a decimal with 9 in the hundredths place.

33. WHICH ONE DOESN'T BELONG? Which of these numbers does not belong? Explain.

$\frac{4}{10}$ $\frac{25}{100}$ 0.36   0.58

34. **WRITING IN ►MATH** Explain why 0.38 contains 3 tenths and 8 hundredths.

Write a fraction and a decimal for the part that is shaded. (Lessons 14-1 and 14-2)

1.

2.

3.

4.

Write each fraction as a decimal.

(Lesson 14-1)

5. April has $\frac{2}{10}$ of a dollar.

6. Rhonda ate $\frac{7}{10}$ of a bag of fruit snacks.

Write each decimal as a fraction.

(Lesson 14-1)

7. 0.1　　　　**8.** 0.8

9. MULTIPLE CHOICE Which decimal equals $\frac{4}{10}$? (Lesson 14-1)

A 10.4　　　**C** 0.4

B 5.4　　　　**D** 0.04

Write each fraction as a decimal.

(Lesson 14-2)

10. $\frac{37}{100}$　　　　**11.** $\frac{10}{100}$

Write each decimal as a fraction.

(Lesson 14-2)

12. 0.94　　　　**13.** 0.43

14. MULTIPLE CHOICE Burt shaded $\frac{4}{100}$ of a figure then erased $\frac{1}{100}$. Which fraction equals the amount left shaded? (Lesson 14-2)

F $\frac{3}{100}$　　　**H** $\frac{3}{10}$

G $\frac{2}{100}$　　　**J** $\frac{2}{10}$

15. Jodi played 20 CDs. She played half of them in the car and the rest in her room. What part of the CDs did she play in her room? Write it as a decimal. (Lesson 14-1)

16. The store sold 75 of its 100 airplane models. What fraction of models are left?

17. Arlo counts 100 vehicles in a parking lot. Write a decimal for the fractional part that is other vehicles. (Lesson 14-2)

Parking Lot Vehicles	
Type	**Amount**
Car	65
Mini-van	30
Other	

18. **WRITING IN ▶MATH** Is the number 0.5 greater than or less than 1? Explain. (Lesson 14-1)

CONTINENTS

The land masses of Earth are continents. There are seven continents. The continents make up about $\frac{3}{10}$ of Earth's surface. The rest of Earth's surface, about $\frac{7}{10}$, is water.

The seven continents on Earth are Africa, Antarctica, Asia, Australia, Europe, North America, and South America. The table shows the amount of Earth's land that each continent covers.

Earth's Land Area

Continent	Land Area
Africa	0.2
Antarctica	0.09
Asia	0.3
Australia	0.05
Europe	0.07
North America	0.17
South America	0.12

🌐 Real-World Math

Use the information on page 612 to solve each problem.

1 Write a decimal that shows how much of Earth is water.

2 Copy and complete the tenths model to show what part of Earth is land. Then, write the decimal.

3 Name the continents whose land area is written in tenths. Change those decimals to fractions.

4 Which continent covers the greatest area? Tell how you know.

5 Which continent's land area has a 2 in the hundredths place?

6 Write in words the decimal that represents the total land area for the continent of North America.

7 Name the continents whose land areas are written with a decimal that has no tenths. Write those decimals as fractions.

8 How many parts of a hundredths grid, would you shade to represent the land area of Australia?

14-3 Problem-Solving Strategy

MAIN IDEA I will solve a problem by working backward.

Frannie put some money in the bank to start a savings account. Last month she put in enough money to double that amount. Today, she put in more money and the total amount doubled, again. Now she has $20. How much money did Frannie start with?

Understand	**What facts do you know?** • The money doubled two times. • The total amount at the end is $20. **What do you need to find?** • The amount of money Frannie started with.
Plan	Work backward from what you know, $20, to find the amount Frannie started with.
Solve	• Start with $20. • Find the number that was doubled. • Since the amount was doubled two times, find half of $10. So, the amount of money Frannie started with was $5. $20 ↓ half of $20 is $10. ↓ $10 ↓ half of $10 is $5. ↓ $5
Check	Look back. When you double $5, the result is $5 × 2 or $10. When you double $10, the result is $10 × 2 or $20. So, the answer is correct.

ANALYZE the Strategy

Refer to the problem on the previous page.

1. Explain how the *work backward* strategy helped to solve the problem.

2. Explain when to use the *work backward* strategy.

3. Suppose Frannie ended up with $36 after the amount doubled two times. How much did she start with?

4. How would you check your answer in Exercise 3?

PRACTICE the Strategy

EXTRA PRACTICE
See page R38.

Solve. Use the *work backward* strategy.

5. **Measurement** It took Gerardo one hour to eat lunch. Then he worked at a store for 3 hours. If he finished at 5:00 P.M., what time did he start eating lunch?

6. Flora, Alonso, and Luz went fishing. Find how many fish each caught.

Fish Caught

Name	Fish
Flora	3 more than Luz
Alonso	3 more than Flora
Luz	5 fish

7. Mariah celebrated her birthday in March, 4 months after joining the swim team. Two months after joining the team, she swam in her first swim meet. What month did she swim in her first meet?

8. **Measurement** The table shows the starting times of the three movies at the mall.

Movie	Starting Times		
Finding Freddie	NOON	3:00 P.M.	6:00 P.M.
Eating Oranges	2:00 P.M.	3:30 P.M.	5:00 P.M.
Running Races	2:30 P.M.	5:00 P.M.	7:30 P.M.

Whitney and her brother saw a movie that lasted 2 hours. It took 30 minutes to get home. They got home at 5:00 P.M. Which movie did they see? What was the starting time?

9. Mr. Robbins gave 9 students one pencil each. That afternoon, he gave 5 more students one pencil each. Now he has 15 pencils. How many pencils did he start with?

10. **WRITING IN MATH** Write a real-world problem in which the *work backward* strategy must be used to solve.

Explore

Math Activity for 14-4

Fractions, Decimals, and Money

You can use what you know about money to understand fractions and decimals.

MAIN IDEA

I will relate money to fractions and decimals.

You Will Need
play money

Math Online

macmillanmh.com

• Concepts in Motion

ACTIVITY   Explore Parts of a Dollar

Step 1   **Count coins.**

- Count out enough pennies to make one dollar. Recall that 100 pennies = 1 dollar.
- Take 1 penny from the pile.
- How much money is 1 penny in cents?
- How do you write 1¢ as a fraction?
- How do you write 1¢ as a decimal?

Coin	Amount in Cents	Fraction of a Dollar	Amount as a Decimal
Penny	1¢	$\frac{1}{100}$	$0.01
Nickel			
Dime			
Quarter			
Half-Dollar			

Step 2   **Complete the table.**

Copy the table shown. Repeat Step 1 for each coin listed. Record your answers on your table.

616 Chapter 14

Step 3 Model fractions.

The model shows that 1¢ is $\frac{1}{100}$ of a dollar. Model the fraction of a dollar for 5¢, 10¢, 25¢, and 50¢.

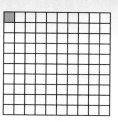

Think About It

1. Write a fraction for the part of a dollar that is 7 pennies.

2. Use graph paper to model the fraction in Exercise 1.

3. Make a model to show 50¢.

CHECK What You Know

4. Copy and complete the table. What patterns do you notice?

Coins	Fraction of a Dollar	Amount as a Decimal	How Many Make $1.00?
Penny			
Nickel			
Dime	$\frac{10}{100}$		
Quarter			
Half-Dollar		$0.50	

Model each amount. Then write as a fraction and a decimal.

5. 47¢　　　　　6. 26¢　　　　　7. 87¢

Write each fraction as cents and as a decimal.

8. $\frac{22}{100}$　　　　　9. $\frac{34}{100}$　　　　　10. $\frac{75}{100}$

11. **WRITING IN ►MATH** You have 2 nickels. Explain how you know what that amount is in decimal form.

14-4 Decimals and Money

> **GET READY to Study**
>
> On Monday, Jamil paid 75¢ or $0.75 for a can of orange juice. What part of a dollar is 75¢?

MAIN IDEA

I will relate money to fractions and decimals.

Math Online

macmillanmh.com
• Extra Examples
• Personal Tutor
• Self-Check Quiz

In the Explore Activity, you related fractions and decimals. In this lesson, you will look at specific parts of a dollar.

Fractions, Decimals, and Money		**Key Concepts**
Money	**Words**	**Numbers**
	one cent one hundredth of a dollar	1¢ or $0.01 $\frac{1}{100}$
	five cents five hundredths of a dollar	5¢ or $0.05 $\frac{5}{100}$
	ten cents ten hundredths of a dollar	10¢ or $0.10 $\frac{10}{100}$
	twenty-five cents twenty-five hundredths of a dollar	25¢ or $0.25 $\frac{25}{100}$
	fifty cents fifty hundredths of a dollar	50¢ or $0.50 $\frac{50}{100}$
	one hundred cents one hundred hundredths of a dollar	100¢ or $1.00 $\frac{100}{100}$

You can write amounts of money as part of a dollar.

Real-World EXAMPLES

1 **MONEY** **Jamil spent 75¢ on a can of orange juice. What part of a dollar is 75¢?**

Write 75¢ as a fraction.

You know that 75¢ can be written as $0.75 or $\frac{75}{100}$.

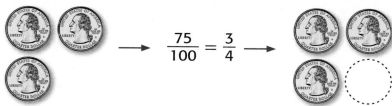

$$\frac{75}{100} = \frac{3}{4}$$

So, 75¢ is $\frac{3}{4}$ of a dollar.

2 **MONEY** **Arthur found 2 dimes on the counter. What part of a dollar is 20¢?**

Write 20¢ as a fraction.

You know that 20¢ can be written as $0.20 or $\frac{20}{100}$.

$$\frac{20}{100} = \frac{2}{10} = \frac{1}{5}$$

So, 20¢ is $\frac{2}{10}$ or $\frac{1}{5}$ of a dollar.

3 **MONEY** **Belinda bought an apple for 50¢. What part of a dollar is 50¢?**

Write 50¢ as a fraction.

You know that 50¢ can be written as $0.50 or $\frac{50}{100}$.

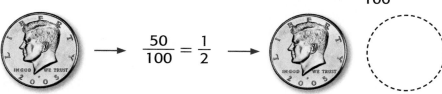

$$\frac{50}{100} = \frac{1}{2}$$

So, 50¢ is $\frac{1}{2}$ of a dollar.

Write the part of a dollar each amount represents. See Examples 1–3 (p. 619)

1.

2.

3.

4.

5. Raymond has 9 dimes and 5 pennies. He wants to buy something that costs $1.00. What part of a dollar does he still need?

6. **Talk About It** Six dimes is $0.60. How can you use $\frac{6}{10} = \frac{\blacksquare}{5}$ to find the part of a dollar for 60¢?

Practice and Problem Solving

EXTRA PRACTICE See page R38.

Write the part of a dollar each amount represents. See Examples 1–3 (p. 619)

7.

8.

9.

10.

11.

12.

13. Cierra saves $\frac{3}{4}$ of every dollar she earns. If she earns $5, how much will she save?

14. Mel spent $\frac{7}{10}$ of a dollar on a pen. He gave the clerk $\frac{3}{4}$ of a dollar. How much change did he receive?

H.O.T. Problems

15. OPEN ENDED Explain how you could represent the decimal 0.6 as money.

16. WHICH ONE DOESN'T BELONG? Identify the form that does not name the part of a dollar for 75¢.

seventy-five hundredths	2 quarters and 2 dimes	$0.75	$\frac{75}{100}$

17. WRITING IN ►MATH Compare money written as a fraction and money written as a decimal. How are they alike? How are they different?

TEST Practice

18. Justino shaded $\frac{1}{100}$ of the figure. Which decimal equals $\frac{1}{100}$?

(Lesson 14-2)

- **A** 1.0
- **B** 0.100
- **C** 0.1
- **D** 0.01

19. How would you write 3 dollars and 45 cents in decimal form?

(Lesson 14-4)

- **F** $345
- **G** $34.5
- **H** $3.45
- **J** $0.345

Spiral Review

20. In line, Enrique said he could not see over Brent's head. Lydia is not last but is taller than Enrique. What is the order of the line if Halley is not first and cannot see over anyone? (Lesson 8-3)

Write each decimal as a fraction. (Lesson 14-2)

21. 0.82   **22.** 0.47   **23.** 0.07

24. If Alek takes a pair of jeans, a pair of shorts, a pair of dress pants, and 3 different T-shirts, how many outfits will he have for his vacation? (Lesson 7-3)

Problem-Solving Investigation

<u>MAIN IDEA</u> I will choose the best strategy to solve a problem.

P.S.I. TEAM +

JULIO: A community garden is 50 yards long and 14 yards wide. It is equally divided into 10 smaller gardens in 2 rows of 5.

▸

YOUR MISSION: Find the perimeter of one of the smaller gardens.

Understand	The garden is 50 yards long and 14 yards wide. There are 10 smaller gardens in 2 rows of 5.
Plan	Draw a picture to represent the whole garden divided into 10 smaller gardens. Find the length of each side of one smaller garden.
Solve	There are 5 gardens along one side. 50 yards ÷ 5 = 10 yards There are 2 gardens along the other side. 14 yards ÷ 2 = 7 yards There are 4 sides to each garden. 10 + 10 + 7 + 7 = 34 yards So, the perimeter of one smaller garden is 34 yards. 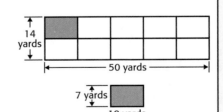
Check	Look back. Check the answer by multiplying. 7 yards × 2 = 14 yards and 10 yards × 5 = 50 yards. So, the answer is correct.

**Use any strategy shown below to solve.
Tell what strategy you used.**

PROBLEM-SOLVING STRATEGIES
• Make an organized list.
• Draw a picture.
• Act it out.
• Use logical reasoning.
• Work backward.

1. There are 136 guests coming to the party. Should 5, 6, or 8 guests be seated at each table so that each table is full and the same number of guests are at each table? Explain.

2. **Measurement** A potato-sack race will be held inside a rectangular area that is 50 meters long and 40 meters wide. Donna has a piece of rope 200 meters long. She will make the rectangle. How much of the rope will be left over?

3. Izzie and her family buy the items shown. They want to make 4 equal payments to pay for the items. If each payment is $213, how much money did they already pay?

Computer Items	
Item	**Cost**
Computer	$676
Printer	$177
Software	$99

4. **Measurement** The third grade class went on a one-day trip to a state park. They arrived at 12:45 P.M. after driving for 3 hours and 15 minutes. What time did they leave?

5. At a zoo, Exhibit One has 3 spiders and 4 camels. Exhibit Two has 3 insects and 10 birds. What is the difference in the number of legs in Exhibit One and Two?

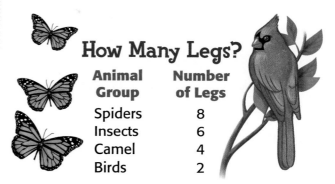

How Many Legs?

Animal Group	Number of Legs
Spiders	8
Insects	6
Camel	4
Birds	2

6. There are 1,440 minutes in one day. What is the total number of minutes in one week?

7. The table shows finishing times for walkers in a 7-mile race. Who walked faster per mile? Explain.

7-Mile Race	
Walker	**Time (min)**
Janna	77
Gilbert	84

8. **WRITING IN ►MATH** How many 8-ounce packages of dog bones are in a box that weighs 200 ounces? Explain the strategy you found was most helpful in solving this problem.

Study Guide and Review

GET READY to Study

Be sure the following Key Vocabulary words and Key Concepts are written in your Foldable.

Tenths | Hundredths

Fractions, Decimals, Money | Decimals and Money

Key Concepts

- A **tenth** is one of ten equal parts. (p. 603)

  Write: 0.5

  Read: five-tenths

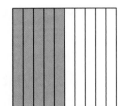

- A **hundredth** is one of one hundred equal parts. (p. 608)

  Write: 0.45

  Read: forty-five hundredths

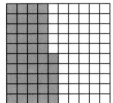

Fractions, Decimals, and Money (p. 618)

$25¢ = 0.25 = \frac{25}{100} = \frac{1}{4}$ of a dollar

$50¢ = 0.50 = \frac{50}{100} = \frac{1}{2}$ of a dollar

Key Vocabulary

decimal (p. 603)
decimal point (p. 603)
hundredths (p. 608)
tenths (p. 603)

Vocabulary Check

Choose the word that completes each sentence.

1. In 6.31, the 3 is in the _____?_____ place.

2. When adding money, you line up the _____?_____.

3. In 3.47, the 7 is in the _____?_____ place.

4. The tenths place is to the right of the _____?_____.

5. $\frac{76}{100}$ reads as seventy-six _____?_____.

6. A dime is one-_____?_____ of a dollar.

7. The number 5.2 is a _____?_____.

8. One quarter is 25 _____?_____ of a dollar.

Lesson-by-Lesson Review

14-1 Tenths (pp. 603–606)

Example 1
What part of the figure is shaded?
Write as a fraction and decimal.

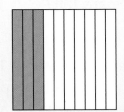

Fraction: $\frac{3}{10}$

Decimal: 0.3

So, $\frac{3}{10}$ or 0.3 of the figure is shaded.

Write a fraction and a decimal for the part that is shaded.

9. 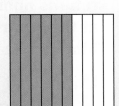 10.

Write each fraction as a decimal.

11. $\frac{2}{10}$ 12. $\frac{9}{10}$

Write each decimal as a fraction.

13. 0.8   14. 0.2

15. Rodney ate 0.5 of a banana and gave away the rest. How much of the banana did he give away?

14-2 Hundredths (pp. 608–610)

Example 2
The table shows how 100 students answered the question, "How do you help at school?" What part of the group answered recycle?

Helping Hands	
Follow the rules	55
Pick up litter	26
Recycle	19

Fraction: $\frac{19}{100}$ **Decimal:** 0.19

So, $\frac{19}{100}$ or 0.19 students said recycle.

Write a fraction and a decimal for the part that is shaded.

16.  17.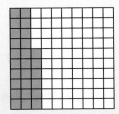

Write each fraction as a decimal.

18. $\frac{37}{100}$ 19. $\frac{14}{100}$

Write each decimal as a fraction.

20. 0.03   21. 0.18

14-3 Problem-Solving Strategy: Work Backward (pp. 614–615)

Example 3
Marlon lives 5 minutes from the library. He walks to the library and returns home by 2:15 P.M. If he is at the library 25 minutes, what time did he leave for the library?

Understand   Bernardo lives 5 minutes from the library. He was there 25 minutes. He got home at 2:15.

Plan   Work backward to find the time he left for the library.

Solve   Go backward from the time he got home.

2:15 − 5 minutes = 2:10

2:10 − 25 minutes = 1:45

1:45 − 5 minutes = 1:40

So, Bernardo left for the library at 1:40 P.M.

Check   Look back. Start with 1:40. Work forward.

1:40 + 5 minutes = 1:45

1:45 + 25 minutes = 2:10

2:10 + 5 minutes = 2:15

So, the answer is correct.

Solve.

22. Three friends have 5 toads each. If 2 friends give away 2 toads each, how many toads are left?

23. Leanne gave away 14 apples. She now has the apples shown below. How many apples did she have to start with?

24. Three is subtracted from a number. The difference is multiplied by 2. Then, 4 is added to the product. Finally, 9 is subtracted to give a difference of 9. What is the number?

25. Maria has 40 mystery books in her collection. She will keep 10 books for herself and divide the rest equally among 5 friends. How many mystery books does each friend get?

14-4 Decimals and Money (pp. 618–621)

Example 4

Estella bought a goldfish for 25¢. What part of a dollar is 25¢?

Write 25¢ as a fraction.

You know that 25¢ can be written as $0.25 or $\frac{25}{100}$.

$\frac{25}{100} = \frac{1}{4}$

So, 25¢ is $\frac{1}{4}$ of a dollar.

Example 5

A store sells stickers for 40¢ each. What part of a dollar is 40¢?

Write 40¢ as a fraction.

You know that 40¢ can be written as $0.40 or $\frac{40}{100}$.

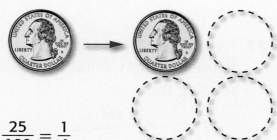

$\frac{40}{100} = \frac{4}{10} = \frac{2}{5}$

So, 40¢ is $\frac{4}{10}$ or $\frac{2}{5}$ of a dollar.

Write the part of a dollar each amount represents.

26.

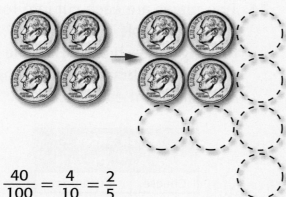

27.

28.

29. Sherman wants to buy 2 tickets for a ride at the fair. The tickets cost $1.00. If he still needs $0.22, what part of a dollar does he have? Write it as a fraction.

30. A keychain costs $1.75. Moira and Tionna will share the cost of the keychain. The table shows how much money they each have. What part of a dollar do they still need?

Coins	Q	D	N	P
Moira	0	1	5	17
Tionna	2	4	0	1

14-5 **Problem-Solving Investigation:** **Choose a Strategy** (pp. 622–623)

Example 6
Marquez has six coins that total $0.75. There is at least one quarter, one dime, and one nickel. What are the other three coins?

Understand
You know the number of coins and their total value. Find the other three coins.

Plan Guess and check to solve.

Solve Find the value of the three coins you know.

$$\$0.25 + \$0.10 + \$0.05 = \$0.40$$

Then, find the difference between the total value of the six coins and the value of the three coins.

$$\$0.75 - \$0.40 = \$0.35$$

Find the three coins that add to $0.35.

Guess			Check
dime	dime	nickel	$0.25 no
quarter	dime	dime	$0.45 no
quarter	nickel	nickel	$0.35 yes

So, 1 quarter and 2 nickels.

Check Look back. The answer makes sense for the situation. ✔

Use any strategy to solve.

31. **Measurement** The window in Jody's room is 2 feet wide by 3 feet high. The perimeter of her brother's window is twice as much. What could be the measurements of her brother's window?

32. **Geometry** The perimeter of the figure is 71 cm. What is the missing measure?

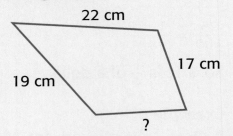

33. Kareem gave 25 football cards to Cindy, 13 to Millie, and 14 to Brad. Kareem now has half the cards he started with. How many cards did he start with?

34. Two pizzas are each cut into four slices. The table shows the topping on each slice. Find the possible combinations choosing one slice from each pizza?

Pizza Slices	
Pizza A	**Pizza B**
Cheese	Mushroom
Cheese	Mushroom
Sausage	Pepperoni
Sausage	Pepperoni

For Exercises 1 and 2, tell whether each statement is *true* or *false*.

1. One cent is one hundredth of a dollar.

2. One place to the right of the decimal point is hundredths place.

Write a fraction and a decimal for the part that is shaded.

3.    **4.**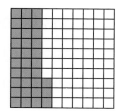

5. Teresa wants to share her 17 dolls with her three friends. Will she be able to share them equally? Explain.

Write the part of a dollar each amount represents.

6. 3 quarters   **7.** 2 pennies

Write each decimal as a fraction.

8. 0.08   **9.** 0.18

10. MULTIPLE CHOICE How would you write 2 dollars and 95 cents in decimal form?

   A $0.295   **C** $29.5

   B $295   **D** $2.95

11. At the beginning of the school year, the third grade class had 29 students. Four students moved away and 6 new students moved in. How many students were in the third grade class at the end of the year?

Write each amount as a fraction and a decimal.

12. 2 dimes, 6 pennies

13. 1 quarter, 3 nickels

Write each fraction as a decimal.

14. $\frac{3}{10}$ **15.** $\frac{97}{100}$

16. MULTIPLE CHOICE What part of a dollar does the jump rope cost?

  F $\frac{25}{100}$ **H** $\frac{50}{100}$

  G $\frac{40}{100}$ **J** $\frac{75}{100}$

17. **WRITING IN ►MATH** Is $2.39 greater than $5? Explain.

PART 1 **Multiple Choice**

Read each question. Then fill in the correct answer on the answer sheet provided by your teacher or on a sheet of paper.

1. What decimal does the model below show?

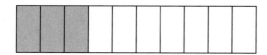

   A 3   **C** 0.3

   B 0.33   **D** 0.03

2. Which fraction is equal to 0.5?

   F $\frac{1}{2}$ **H** $\frac{1}{5}$

   G $\frac{5}{12}$ **J** $\frac{5}{100}$

3. How do you write twenty-seven hundredths as a decimal?

   A 2700   **C** 2.7

   B 27   **D** 0.27

4. Tim placed his money shown below on the table.

How much money does he have?

   F $0.55   **H** $0.65

   G $0.75   **J** $0.85

5. Which is most likely to be the capacity of a cocoa mug?

   A about 10 milliliters

   B about 100 milliliters

   C about 2 liters

   D about 5 liters

6. Destiny bought a bag of 10 marbles. Three marbles were blue and two were green. The rest were yellow. Which fraction shows the part that was green?

   F $\frac{2}{10}$ **H** $\frac{5}{10}$

   G $\frac{3}{10}$ **J** $\frac{7}{10}$

7. Which fraction equals $\frac{1}{2}$?

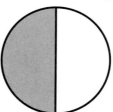

   A $\frac{2}{4}$ **C** $\frac{2}{6}$

   B $\frac{4}{10}$ **D** $\frac{1}{4}$

Preparing for Standardized Tests
For test-taking strategies and practice,
see pages R50–R63.

8. Which object looks like a cone?

F

H

G

J

9. Which number sentence relates to $45 \div 5 = 9$?

A $45 - 5 = 40$

B $9 \times 5 = 45$

C $9 + 5 = 14$

D $45 + 5 = 50$

10. Last year a bike shop had bike sales of $2,355. For the first 6 months of this year, the bike sales are $975. How much more does the shop need to sell to reach last year's sales?

F $1,380

G $2,480

H $2,620

J $3,330

PART 2   Short Response

Record your answer on an answer sheet provided by your teacher or on a sheet of paper.

11. Lonzo shared an apple with his friends. He gave 0.3 to Pat, 0.2 to Sue, 0.1 to Kurt, and the rest to Oliver. How much of the apple did Oliver get? Write it as a decimal and a fraction.

12. What would be the most appropriate metric unit of measurement to use to find the length of a highway?

PART 3   Extended Response

Record your answer on an answer sheet provided by your teacher or on a sheet of paper.

13. A quilt has 10 rows with 10 squares in each row. Each row has 4 blue squares. What decimal represents the total number of blue squares in the whole blanket? Explain.

NEED EXTRA HELP?													
If You Missed Question...	1	2	3	4	5	6	7	8	9	10	11	12	13
Go to Lesson...	14-1	14-2	14-2	1-10	10-3	14-2	13-1	11-1	6-2	3-3	14-1	9-4	13-1

CHAPTER 15

Multiply by One-Digit Numbers

BIG Idea How do I multiply greater numbers?

Models can be used to multiply larger numbers.

Example If 2 prairie dogs each dig 13 holes, there will be 2×13 or 26 holes.

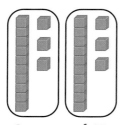

$$\begin{array}{r} 13 \\ \times\ 2 \\ \hline 6 \\ +\ 20 \\ \hline 26 \end{array}$$

$2 \times 3 = 6$
$2 \times 10 = 20$

2 groups of 13

What will I learn in this chapter?

■ Multiply multi-digit numbers.

■ Estimate products.

■ Multiply money.

■ Solve problems by using logical reasoning.

Key Vocabulary

multiples

factors

product

estimate

round

 **Math Online** > **Student Study Tools** at macmillanmh.com

FOLDABLES®
Study Organizer

Make this Foldable to organize information about multiplying by a one-digit number. Begin with one sheet of 11″ × 17″ paper.

1 **Fold** the sheet of paper into thirds as shown.

2 **Fold** the bottom edge up two inches and crease well.

3 **Glue** the outer edges to create three pockets.

4 **Label** as shown. Record what you learn on index cards.

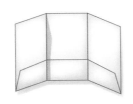

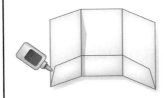

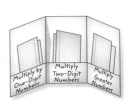

You have two ways to check prerequisite skills for this chapter.

Option 2

Math Online Take the Chapter Readiness Quiz at macmillanmh.com.

Option 1

Complete the Quick Check below.

QUICK Check

**Write a multiplication sentence for each array.
Then multiply.** (Lesson 4-1)

1.

2.

3.

Multiply. (Chapters 4 and 5)

4. $\begin{array}{r} 4 \\ \times 7 \\ \hline \end{array}$

5. $\begin{array}{r} 5 \\ \times 6 \\ \hline \end{array}$

6. $\begin{array}{r} 9 \\ \times 2 \\ \hline \end{array}$

7. 5×5

8. 9×1

9. 10×8

10. There are 2 space shuttles docked at the station. If each shuttle has 5 astronauts, how many astronauts are there altogether?

11. For every pound of play clay Reynaldo makes, he needs 6 cups of flour. How many cups of flour does he need to make 6 pounds of play clay?

Round to the nearest ten. (Lesson 1-8)

12. 78

13. 53

14. 49

Round to the nearest hundred.

15. 125

16. 111

17. 199

15-1 Multiply Multiples of 10, 100, and 1,000

MAIN IDEA

I will multiply multiples of 10, 100, and 1,000.

New Vocabulary

multiples

Math Online

macmillanmh.com

• Extra Examples
• Personal Tutor
• Self-Check Quiz

GET READY to Learn

Four multiplication sentences are modeled below. Notice the pattern of zeros. Describe this pattern.

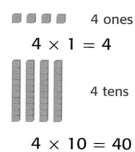

■ ■ ■ ■   4 ones

$4 \times 1 = 4$

4 tens

$4 \times 10 = 40$

4 hundreds

$4 \times 100 = 400$

4 thousands

$4 \times 1,000 = 4,000$

You can use basic facts and patterns of zeros to help you multiply a number mentally by 10, 100, and 1,000.

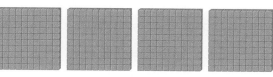

Real-World EXAMPLE

1 **ART** Erik buys 7 boxes of straws for a project. Each box has 100 straws. How many straws does he have?

Find 7×100. Use basic facts and patterns of zeros.

$7 \times 1 = 7$ 7×1 one $= 7$ ones
$7 \times 10 = 70$ 7×1 ten $= 7$ tens
$7 \times 100 = 700$ 7×1 hundred $= 7$ hundred

So, Erik has 700 straws.

Lesson 15-1 Multiply Multiples of 10, 100, and 1,000   **635**

You can also multiply a number mentally by multiples of 10, 100, and 1,000. A **multiple** is the product of a given number and any other whole number.

20 is a multiple of 10.

200 is a multiple of 100.

2,000 is a multiple of 1,000.

 Real-World EXAMPLE

Remember

To multiply by multiples of 10, find the product of the basic fact then place the zeros.

2 **CRAFT** **Bags of 3,000 craft beads are on sale. Ellie bought 5 bags. How many beads did Ellie buy?**

You need to find $5 \times 3,000$.

$5 \times 3 = 15$	5×3 ones = 15 ones
$5 \times 30 = 150$	5×3 tens = 15 tens
$5 \times 300 = 1,500$	5×3 hundreds = 15 hundreds
$5 \times 3,000 = 15,000$	5×3 thousands = 15 thousands

So, $5 \times 3,000 = 15,000$. Ellie bought 15,000 beads.

When the basic fact has a zero in it, you still need to add all the other zeros.

Real-World EXAMPLE

3 **TRAFFIC** **A busy intersection has 5,000 vehicles pass through each day. How many vehicles will pass through in 4 days?**

You need to find $4 \times 5,000$.

$4 \times 5 = 20$ ◄——— 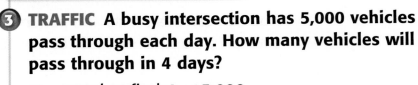 THINK Sometimes the basic fact has a zero.

$4 \times 50 = 200$

$4 \times 500 = 2,000$

$4 \times 5,000 = 20,000$

So, $4 \times 5,000 = 20,000$. About 20,000 vehicles will pass through in 4 days.

Multiply. Use basic facts and patterns. See Examples 1–3 (pp. 635–636)

1. 3 × 1 = ■
3 × 10 = ■
3 × 100 = ■
3 × 1,000 = ■

2. 7 × 4 = ■
7 × 40 = ■
7 × 400 = ■
7 × 4,000 = ■

3. 5 × 6 = ■
5 × 60 = ■
5 × 600 = ■
5 × 6,000 = ■

4. 4 × 90

5. 4 × 4,000

6. 500 × 8

7. Hunter's Pizza Shop sold 3,000 pizzas each month for 6 months. Find the total number of pizzas sold.

8. **Talk About It** Explain the pattern of zeros that you see when you multiply by 10, 100, or 1,000.

Practice and Problem Solving

EXTRA PRACTICE
See page R39.

Multiply. Use basic facts and patterns. See Examples 1–3 (pp. 635–636)

9. 2 × 1 = ■
2 × 10 = ■
2 × 100 = ■
2 × 1,000 = ■

10. 6 × 4 = ■
6 × 40 = ■
6 × 400 = ■
6 × 4,000 = ■

11. 7 × 8 = ■
7 × 80 = ■
7 × 800 = ■
7 × 8,000 = ■

12. 5 × 50

13. 30 × 8

14. 4 × 30

15. 900 × 7

16. 600 × 90

17. 60 × 80

18. Demont's album has 20 pages, and 6 trading cards are on each page. How many cards all?

19. There are 100 houses. Each house has 10 windows. How many windows are there in all?

20. Carmen sold 200 flats of flowers each day for 9 days. Each flat holds 4 flowers. How many flowers did she sell?

21. Carlita has 3 boxes of teddy bears. Each box holds 20 bears. She sells each bear for $4. How much money did she earn?

H.O.T. Problems

22. **OPEN ENDED** Write a multiplication sentence that uses a multiple of 10 and has a product of 24,000.

23. **WRITING IN ►MATH** Write a real-world problem that involves multiplying by a multiple of 10.

Problem-Solving Strategy

__MAIN IDEA__ I will solve a problem by using logical reasoning.

Three friends all have on different shirts. Hallie's shirt is white. Jimar's shirt is not green. Marina's shirt is not red. What is the color of each of their shirts?

Understand	**What facts do you know?**
	• Hallie has on a white shirt.
	• Jimar's shirt is not green.
	• Marina's shirt is not red.
	What do you need to find?
	• The color of each person's shirt.
Plan	Make a table to show what you know. Then use logical reasoning to find the color of each person's shirt.
Solve	Hallie is wearing white. So, write yes by her name under white. Place an X in all the rest of the white column and the other colors for Hallie.

	White	**Red**	**Green**
Hallie	yes	X	X
Marina	X	X	yes
Jimar	X	yes	X

	Marina's shirt is not red and can not be white, so it is green. The color that is left is red. Jimar's shirt must be red.
	So, Hallie is wearing white, Marina is wearing green, and Jimar is wearing red.
Check	Look back. The answer makes sense for the facts given. So, the answer is correct. ✔

Refer to the problem on the previous page.

1. Explain how making a table helped in solving the problem.

2. What does it mean to use logical reasoning?

3. If the colors of shirts changed how would the problem be different? How would it be the same?

4. How would the results be different if Marina's shirt was not green?

PRACTICE the Strategy

EXTRA PRACTICE
See page R39.

Solve. Use logical reasoning.

5. Marilee places her math book next to her reading book and language book. Her language book is next to her science book, which is next to her history book. What is a possible order?

6. Emerson, Thi, Joyce, and Lawanda each have a different pet. Emerson has a cat. Thi does not have a dog or a fish. Joyce does not have a bird or a fish. What pet does each person have?

7. Paulita, Daniel, and Drake each play a different sport. According to the information in the table what sport does each student play?

Sports Students Play	
Student	**Sport**
Paulita	soccer
Drake	not basketball
Daniel	not soccer or football

8. Larrisa, Jo, and Callie went to lunch. They each ordered something different. Larrisa does not like hamburgers. Jo and Callie do not like salad. Who ordered the salad?

LUNCH MENU
* SALAD
* FRUIT PLATE
* HAMBURGER

9. Three friends want to buy the game shown below. Dexter has 5 quarters and 6 dimes. Alma has 6 quarters and 8 dimes. Emmett has 5 coins. If they will receive 10 cents in change, what coins does Emmett have?

10. **WRITING IN ▸MATH** Write two sentences describing how you would use logical reasoning to help solve a real-world situation.

Estimate Products

MAIN IDEA

I will estimate products.

Math Online

macmillanmh.com
- Extra Examples
- Personal Tutor
- Self-Check Quiz

GET READY to Learn

Each of the 26 schools in Fair City sends 6 of their best spellers to the city spelling bee. About how many students go to the spelling bee?

When you do not need an exact answer, you can estimate. One way to estimate is to round.

Real-World EXAMPLE   Estimate by Rounding

① SCHOOL About how many students go to the spelling bee?

Estimate 6 × 26 by rounding to the nearest ten.

Step 1 Round the factor, that is greater than 10, to the nearest ten.

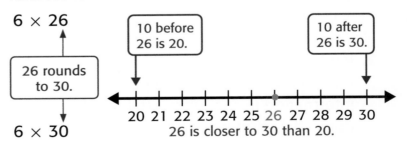

6 × 26

26 rounds to 30.

6 × 30

10 before 26 is 20.

10 after 26 is 30.

20 21 22 23 24 25 26 27 28 29 30
26 is closer to 30 than 20.

Step 2 Multiply mentally.

6 × 30 = 180

So, about 180 spellers go to the spelling bee.

2 **PLAYS** There are 140 students. Each student can invite 3 people to a play. About how many people can be invited?

Estimate 3 × 140 by rounding to the nearest hundred.

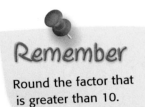

Remember

Round the factor that is greater than 10.

3 × 140

140 rounds to 100.

3 × 100 = 300

| 100 before 140 is 100. | | 100 after 140 is 200. |

100 110 120 130 140 150 160 170 180 190 200

140 is closer to 100 than 200.

Real-World **EXAMPLE**   Use Estimation to Verify Reasonableness

3 **SCHOOL BUS** A school bus holds about 52 students. Will 4 buses be enough to transport 175 students?

Estimate 4 × 52. Then compare to 175.

4 × 52   Round 52 to 50.

4 × 50 = 200   Multiply mentally.

The buses can hold about 200 students. 200 > 175 So, 4 buses will be enough to transport the 175 students.

CHECK What You Know

Estimate. Round to the nearest ten. See Example 1 (p. 640)

1. 47
    × 4

2. 51
    × 8

3. 58
    × 2

Estimate. Round to the nearest hundred. See Example 2 (p. 641)

4. 315
    × 3

5. 189
    × 5

6. 150
    × 6

7. Measurement Jan spends 5 hours each week in math class. Is her estimate reasonable, if each class is 55 minutes long? See Example 3 (p. 641)

8. **Talk About It** Is estimating 878 × 9, to the nearest hundred, greater than the actual product? Explain.

Practice and Problem Solving

EXTRA PRACTICE See page R39.

Estimate. Round to the nearest ten. See Example 1 (p. 640)

9. 27
× 4

10. 17
× 6

11. 36
× 3

12. 28 × 8

13. 32 × 5

14. 43 × 4

Estimate. Round to the nearest hundred. See Example 2 (p. 641)

15. 180
× 9

16. 197
× 6

17. 306
× 3

18. 271 × 4

19. 290 × 7

20. 114 × 8

21. Hiking burns about 288 Calories each hour. About how many Calories can be burned if someone hikes 3 days a week for an hour?

22. A restaurant keeps track of the paper goods it uses for one day as shown below. Is it reasonable to estimate 2,500 napkins, cups, and bags are used in one week? Explain. See Example 3 (p. 641)

23. Corrine uses 27 sheets of paper for a book she makes. About how many sheets would she need if she makes 8 books?

24. Measurement Hector studies about 3 hours each day. Is it reasonable to estimate that is equal to about 100 hours in 4 weeks? Explain.
See Example 3 (p. 641)

Daily Paper Use

bags 532

hamburger wrappers 875

cups 1,091

napkins 913

H.O.T. Problems

25. FIND THE ERROR Libby and Grady are estimating 458 × 4. Who is correct? Explain.

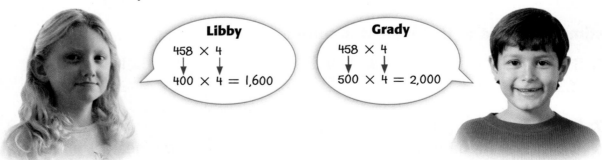

Libby
458 × 4
↓   ↓
400 × 4 = 1,600

Grady
458 × 4
↓   ↓
500 × 4 = 2,000

26. **WRITING IN ►MATH** Explain how you would estimate 77 × 6.

Multiply. Use basic facts and patterns.
(Lesson 15-1)

1. $4 \times 6 = \blacksquare$
   $4 \times 60 = \blacksquare$
   $4 \times 600 = \blacksquare$
   $4 \times 6,000 = \blacksquare$

2. $8 \times 4 = \blacksquare$
   $8 \times 40 = \blacksquare$
   $8 \times 400 = \blacksquare$
   $8 \times 4,000 = \blacksquare$

Multiply. Use mental math.

3. 2×60

4. $3 \times 3,000$

5. **MULTIPLE CHOICE** What number makes this number sentence true? (Lesson 15-1)

$$6 \times \blacksquare = 48,000$$

A 80   C 8,000

B 800   D 80,000

Solve. Use logical reasoning. (Lesson 15-2)

6. Robert, Aiden, and Ramon like different kinds of books. Ramon does not like mysteries. Aiden does not like science fiction or sports stories. Robert loves science fiction. What kind of book does each boy like?

7. Taryn lives down the street from Roger. Dakota lives next door to Roger. Regina lives in between Taryn and Dakota. What order do the kids live in on the block? (Lesson 15-2)

Estimate. Round to the nearest ten. (Lesson 15-3)

8. $\begin{array}{r} 78 \\ \times\ 8 \\ \hline \end{array}$

9. $\begin{array}{r} 23 \\ \times\ 2 \\ \hline \end{array}$

10. Mrs. Henry plans to teach 3 math lessons each week. About how many lessons does she plan to teach in 22 weeks? (Lesson 15-3)

Estimate. Round to the nearest hundred. (Lesson 15-3)

11. $\begin{array}{r} 173 \\ \times\ 5 \\ \hline \end{array}$

12. $\begin{array}{r} 168 \\ \times\ 6 \\ \hline \end{array}$

13. A third grade class has 122 students. Each student needs 4 folders. About how many folders do the third graders need? (Lesson 15-3)

14. **MULTIPLE CHOICE** Kisho made a quilt out of squares by putting the squares in 6 equal rows of 11 squares each. About how many squares did Kisho use? (Lesson 15-3)

F 40   H 60

G 50   J 70

15. **WRITING IN ▶MATH** Explain the steps in estimating the product of 88×3 by rounding to the nearest ten. (Lesson 15-3)

15-4 Multiply by a One-Digit Number

GET READY to Learn

Ken and his two brothers each have 13 marbles. How many marbles do they have in all?

You can use what you know about multiplying smaller numbers to find products like 3 × 13.

Real-World EXAMPLE   Use a Model

① **MARBLES Ken and his two brothers each have 13 marbles. How many marbles do they have in all?**

The array shows 3 × 13. Break the grid into parts.

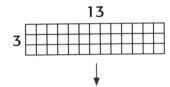

The part shaded orange shows 3 × 10.
The part shaded blue shows 3 × 3.

Find the product of each part, or the **partial products**. Then add the partial products.

$3 \times 10 = 30$
$3 \times 3 = 9$

$30 + 9 = 39$

$3 \times 13 = 39$

So, Ken and his brothers have 39 marbles in all.

You can also use an area model to find a product. Area models do not have grid lines but you can still break the rectangle into parts to find the product.

Real-World EXAMPLE Use a Model

2 **BOOKS** A library has 2 shelves. Each shelf holds 23 books. How many books will fit on the 2 shelves?

Find the product of 2 and 23.

Estimate $2 \times 23 \longrightarrow 2 \times 20 = 40$

Think of 23 as 20 + 3.

	20	+	3
2	$2 \times 20 = 40$		$2 \times 3 = 6$

$$\begin{array}{r} 23 \\ \times\ 2 \\ \hline 6 \\ +40 \\ \hline 46 \end{array}$$

So, 46 books will fit on the shelves.

You can also multiply by a one-digit number without models.

EXAMPLE Use Paper and Pencil

3 **Find 3 × 32.**

Multiply the ones. Then multiply the tens.

Step 1 Multiply ones.

$$\begin{array}{r} 32 \\ \times\ 3 \\ \hline 6 \end{array}$$ ← 3×2 ones = 6 ones

Step 2 Multiply tens.

$$\begin{array}{r} 32 \\ \times\ 3 \\ \hline 96 \end{array}$$ ← 3×3 tens = 9 tens

Check

The model shows that $3 \times 32 = 96$. ✔

	30	+	2
3	$3 \times 30 = 90$		$3 \times 2 = 6$

$$\begin{array}{r} 32 \\ \times\ 3 \\ \hline 6 \\ +90 \\ \hline 96 \end{array}$$

$3 \times 2 = 6$
$3 \times 30 = 90$

Multiply. Use estimation to check. See Examples 1–3 (pp. 644–645)

1. 4 × 22

2. 3 × 21

3. 5 × 11

4. 12
   × 4

5. 41
   × 2

6. 32
   × 3

7. A classroom has 23 desks. Each desk has 3 books on it. How many books are on the desks?

8. **Talk About It** Explain how area models help you multiply.

Practice and Problem Solving

EXTRA PRACTICE
See page R40.

Multiply. Use estimation to check. See Examples 1–3 (pp. 644–645)

9. 21
   × 4

10. 32
   × 2

11. 44
   × 2

12. 13
   × 3

13. 12
   × 4

14. 20
   × 3

15. 43
   × 2

16. 33
   × 3

17. 2 × 23

18. 2 × 33

19. 2 × 22

20. 2 × 14

21. There are 21 bags of bagels with 4 bagels in each. If the scouts ate all but 9, how many did they eat?

22. **Measurement** Fran cut 4 pieces of yarn that measured 15 inches each. How many feet of yarn did she use?

Real-World PROBLEM SOLVING

School The table shows the number of classes and students in grades 3, 4, and 5.

23. Write an expression for the total number of students in the fourth grade.

24. How many more students are in grade 4 than grade 3?

25. Write an expression that compares the total number of students in the 3rd grade and 5th grade. Use < or >.

26. How many students are in all of the grades?

Fair Street School Student Count

Grade	Number of Classes	Students per Class
3	3	23
4	4	22
5	2	31

H.O.T. Problems

27. OPEN ENDED Write a number that when multiplied by 3 is one less than 100.

28. **WRITING IN** ►**MATH** Is the product of 3 and 32 the same as the product of 32 and 3? Explain your reasoning.

TEST Practice

29. Kome made a large mat out of carpet squares by putting the squares in 3 equal rows of 21 squares each. How many squares did Kome use? (Lesson 15-4)

   A 42   **C** 63

   B 62   **D** 84

30. Which product would be a reasonable estimate for this number sentence? (Lesson 15-3)

$$82 \times 9 = \blacksquare$$

   F 70

   G 700

   H 720

   J 810

Spiral Review

Estimate. Round to the nearest hundred. (Lesson 15-3)

31. 125
  × 8

32. 233
  × 4

33. 158
  × 3

34. A number has two digits. The first digit is odd. The difference of the digits is 2 and their sum is 12. What is the number? (Lesson 15-2)

Multiply. Use basic facts and patterns. (Lesson 15-1)

35. $5 \times 300 = \blacksquare$ **36.** $8 \times 9,000 = \blacksquare$ **37.** $400 \times 60 = \blacksquare$

Write the part of a dollar each amount represents. (Lesson 14-4)

38.

39.

40.

41. Mr. Harris bought a tie for $12 and a pair of shoes for $26. He paid with a $50-bill. What is his change? (Lesson 3-3)

15-5 Problem-Solving Investigation

MAIN IDEA I will choose the best strategy to solve a problem.

P.S.I. TEAM +

RYDELL: I need to fill 3 pitchers and 2 punch bowls with punch. It takes 11 cans of punch to fill one pitcher and 24 cans of punch to fill the punch bowl.

YOUR MISSION: Find how many cans of punch are needed in all.

Understand	It takes 11 cans to fill one pitcher. It takes 24 cans of punch to fill one punch bowl. Find how many cans of punch are needed in all.
Plan	Solve a simpler problem. Solve for each part of the problem then add.
Solve	1 pitcher = 11 cans of punch. So, it takes 11 × 3 or 33 cans to fill 3 pitchers. 1 punch bowl = 24 cans. So, it takes 24 × 2 or 48 cans to fill 2 bowls. Now find the total. 33 + 48 = 81 So, 81 cans of punch are needed in all.
Check	Look back. Use addition to check. bowl + bowl + pitcher + pitcher + pitcher 24 + 24 + 11 + 11 + 11 = 81 So, the answer is correct. ✔

Use any strategy shown below to solve. Tell what strategy you used.

PROBLEM-SOLVING STRATEGIES
• Work a simpler problem.
• Make an organized list.
• Draw a picture.
• Act it out.
• Use logical reasoning.

1. Tyron and Freda collected tin cans for recycling. Tyron collected 3 times as many as Freda. The total number collected by their class was 500 cans. Tyron and Freda collected $\frac{1}{5}$ of that. How many cans did they each collect?

2. **Measurement** A log is shown. Suppose a piece that measures 11 inches is cut off. How many 5-inch pieces can be made from the part of the log that is left?

|← 46 inches →|

3. **Measurement** Constance places 6 books on one side of a scale. To balance the scale, Eli places 2 books and his baseball glove on the other side. If each book weighs 3 ounces, how much does Eli's baseball glove weigh?

4. Sofia, Roxana, and Adrian are playing a game. Sofia has 88 points. Roxana has 26 points more than Sofia. Adrian wins with 50 points more than Roxana. How many points does each person have?

5. Alfonso and Juanna are playing a game with one 0–5 number cube and one 5–10 number cube. Each cube is rolled twice. The total of their rolls is 25. What could be the other three numbers rolled if one was a 5?

6. Logan, Rodolfo, Emanuel, and Corbin were waiting for the bus. Logan was next to Emanuel, who was not next to Corbin. Corbin was next to Rodolfo, but not next to Logan. In what order were they standing?

7. **WRITING IN MATH** Alexa ran the distances shown. She ran 2 miles more than the total of these on Sunday. Explain how to find how many miles she ran for the four days.

Day	Distance Ran
Monday	4 miles
Thursday	6 miles
Saturday	8 miles

Multiplication with Regrouping

MAIN IDEA

I will use models to explore multiplication with regrouping.

You Will Need
base-ten blocks

Math Online

macmillanmh.com
• Concepts in Motion

You sometimes need to regroup when adding. Regrouping is also sometimes used when multiplying.

ACTIVITY   **Multiply 2 × 16.**

Step 1 **Model 2 × 16.**
Model 2 groups of 16.
Use 1 ten and 6 ones in each group.

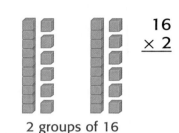

16
× 2

2 groups of 16

Step 2 **Combine the models.**
Combine the ones.
Combine the tens.

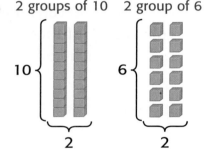

2 groups of 10   2 group of 6

10 {   6 {

2   2

Step 3 **Regroup.**
Regroup 12 ones as 1 ten and 2 ones.

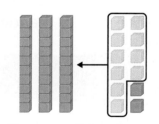

12 ones = 1 ten, 2 ones

Step 4 **Add the partial products.**

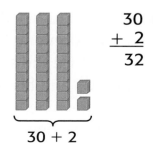

30
+ 2
32

So, 2 × 16 = 32.

30 + 2

Think About It

1. How did you model each factor?

2. Why did you regroup?

3. How did the number of tens and ones change after you regrouped?

4. Will you always have to regroup in multiplication? Explain.

5. If you have 4 groups of 16, what would be the product?

CHECK What You Know

Write a multiplication expression for each model. Then multiply.

6.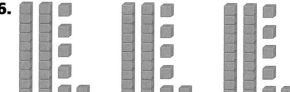

7.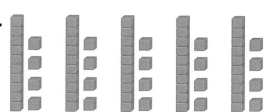

8.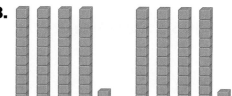

9.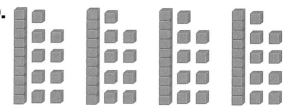

Multiply. Use base-ten models.

10. 5×18 11. 12×6 12. 4×24 13. 17×3

14. **WRITING IN ►MATH** Explain why knowing how to estimate is useful when multiplying larger numbers.

Multiply Two-Digit Numbers

 **to Learn**

A new apartment building will have 5 floors with 13 apartments on each floor. How many apartments will there be in the building?

You can connect multiplication models to paper and pencil.

Real-World EXAMPLE Multiply with Regrouping

① **BUILDINGS How many apartments will the building have?**

Use models to help you find the product of 5 × 13.

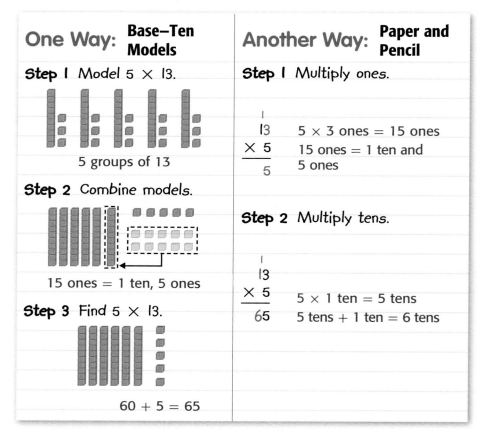

One Way: Base–Ten Models	Another Way: Paper and Pencil
Step 1 Model 5 × 13.	**Step 1** Multiply ones.
5 groups of 13	$\begin{array}{r}1\\13\\\times\ 5\\\hline 5\end{array}$ 5 × 3 ones = 15 ones 15 ones = 1 ten and 5 ones
Step 2 Combine models.	**Step 2** Multiply tens.
15 ones = 1 ten, 5 ones	$\begin{array}{r}1\\13\\\times\ 5\\\hline 65\end{array}$ 5 × 1 ten = 5 tens 5 tens + 1 ten = 6 tens
Step 3 Find 5 × 13.	
60 + 5 = 65	

So, the building will have 65 apartments.

2 **REPTILES** **A female desert tortoise can lay as many as 8 eggs at one time. How many eggs could 12 female desert tortoises lay?**

Find 8 × 12. **Estimate** 8 × 12 ⟶ 8 × 10 = 80

Step 1 Multiply ones.

$$\begin{array}{r} {\scriptstyle 1} \\ 12 \\ \times\ 8 \\ \hline 6 \end{array}$$ 8 × 2 ones = 16 ones

Step 2 Multiply tens.

$$\begin{array}{r} {\scriptstyle 1} \\ 12 \\ \times\ 8 \\ \hline 96 \end{array}$$ 8 × 1 ten = 8 tens
8 tens + 1 ten = 9 tens

> **Remember**
> Do not multiply the regrouped tens again, add them once the tens are multiplied.

The area model shows that 8 × 12 is 96.

	10	+	2
8	8 × 10 = 80		8 × 2 = 16

$$\begin{array}{r} 12 \\ \times\ 8 \\ \hline 16 \\ +80 \\ \hline 96 \end{array}$$

Multiply ones.
Multiply tens.
Add partial products.

So, 12 female tortoises could lay 96 eggs.

Check for Reasonableness

80 is close to 96, so the answer is reasonable. ✔

CHECK What You Know

Multiply. Use models if needed. See Examples 1 and 2 (pp. 652–653)

1. $\begin{array}{r} 13 \\ \times\ 4 \\ \hline \end{array}$

2. $\begin{array}{r} 27 \\ \times\ 3 \\ \hline \end{array}$

3. $\begin{array}{r} 13 \\ \times\ 8 \\ \hline \end{array}$

4. Measurement A construction crew finished 14 miles of highway in 1 week. At this rate, how many miles could they finish in 4 weeks?

5. **Talk About It** What is the greatest number of ones that could be in the ones column without having to regroup? Explain.

Practice and Problem Solving

EXTRA **PRACTICE**
See page R41.

Multiply. Use models if needed. See Examples 1 and 2 (pp. 652–653)

6. 46
× 2

7. 17
× 4

8. 53
× 2

9. 92
× 3

10. 13
× 6

11. 18
× 9

12. 15
× 4

13. 12
× 5

14. 18 × 8

15. 14 × 9

16. 28 × 4

17. 31 × 8

18. Measurement A stick insect can measure 22 inches in length. How many inches long would 3 stick insects measure?

19. Measurement A gecko can grow close to 35 centimeters in length. How many centimeters would 4 geckos measure?

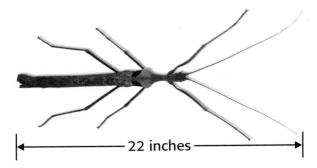

|← 22 inches →|

|← 35 centimeters →|

20. It is suggested to eat 5 servings of grains each day. How many servings is this in a 31-day month?

21. The Pizza Parlor put 65 pieces of pepperoni on each pizza. How many pieces of pepperoni are on 6 pizzas?

Real-World PROBLEM SOLVING

Airships The first airship, which was a blimp, was built more than 80 years ago. Today, television viewers get a bird's-eye view of many events.

22. Rounded to the nearest hundred, what is the blimp's fuel tank capacity?

23. At top speed, how many miles could a blimp travel in 3 hours?

24. What is the maximum height a blimp can rise if it is 5 times its average cruising height?

Blimp Facts
• A blimp cruises at 30–40 miles per hour but, its maximum speed is 65 miles per hour.
• The average cruising height of a blimp is 2,000 feet.
The capacity of a blimp's fuel tank is 426 gallons.

Source: The Columbus Dispatch.

H.O.T. Problems

25. OPEN ENDED Write a multiplication sentence whose product is less than 52.

26. NUMBER SENSE Without multiplying, how do you know that 21 × 3 is greater than 60?

27. WHICH ONE DOESN'T BELONG? Identify the multiplication expression that does not belong. Explain.

| 3 × 33 | 4 × 23 | 5 × 15 | 7 × 18 |

28. **WRITING IN ►MATH** Describe the steps you would take to multiply 76 and 4.

TEST Practice

29. On Thursday, 132 people visited the library. Three times as many people visited over the weekend. How many people visited over the weekend? (Lesson 15-4)

   A 264   **C** 375

   B 300   **D** 396

30. Lee's Pizza Shop puts 75 pepperoni slices on every pizza. If someone buys 6 pizzas, how many pepperoni slices would there be altogether? (Lesson 15-6)

   F 150   **H** 420

   G 300   **J** 450

Spiral Review

31. Milo's grandmother is making 3 baby quilts. Each quilt's squares will be sewn in an array of 8 rows with 4 squares in each row. How many squares will she need to cut out?
(Lesson 15-5)

Estimate. Round to the nearest hundred. (Lesson 15-3)

32. 225
  × 5

33. 168
  × 8

34. 177
  × 9

Identify each two-dimensional figure. (Lesson 11-2)

35.

36.

37.

Multiply Greater Numbers

GET READY to Learn

Griffen is reading a book about pencil making. He learns about a machine that makes 132 pencils a minute. How many pencils are made in 5 minutes?

MAIN IDEA

I will multiply three and four-digit numbers by a one-digit number with regrouping.

Math Online

macmillanmh.com
• Extra Examples
• Personal Tutor
• Self-Check Quiz

You have learned how to multiply two-digit numbers. Use what you know to multiply larger numbers.

Real-World EXAMPLE

① **PENCILS** **How many pencils are made in 5 minutes?**

Find 5 × 132.   **Estimate** 5 × 132 ⟶ 5 × 100 = 500

Step 1 Multiply ones.

$$\begin{array}{r} \overset{1}{132} \\ \times\ 5 \\ \hline 0 \end{array}$$ 5 × 2 ones = 10 ones

Step 2 Multiply tens.

$$\begin{array}{r} \overset{1\ 1}{132} \\ \times\ 5 \\ \hline 60 \end{array}$$ 5 × 3 tens = 15 tens
Add the regrouped amount. 15 + 1 = 16 tens

Step 3 Multiply hundreds.

$$\begin{array}{r} \overset{1\ 1}{132} \\ \times\ 5 \\ \hline 660 \end{array}$$ 5 × 1 hundred = 5 hundreds
Add the regrouped amount. 5 + 1 = 6 hundreds

So, 660 pencils are made in 5 minutes.

Check for Reasonableness

Since 660 is close to 500, the answer is reasonable. ✔

2 DUCKS A duck eats about 1,960 grams of food in 1 week. How much would it eat in 4 weeks?

Find 1,960 × 4.

Step 1 Multiply ones.

$$\begin{array}{r} 1,96\mathbf{0} \\ \times\quad 4 \\ \hline \mathbf{0} \end{array}$$
4 × 0 one = 0 ones

Step 2 Multiply tens.

$$\begin{array}{r} {}^{2} \\ 1,9\mathbf{6}0 \\ \times\quad 4 \\ \hline \mathbf{4}0 \end{array}$$
4 × 6 tens = 24 tens

Step 3 Multiply hundreds.

$$\begin{array}{r} {}^{3\ 2} \\ 1,\mathbf{9}60 \\ \times\quad 4 \\ \hline \mathbf{8}40 \end{array}$$
4 × 9 hundreds = 36 hundreds
Add the regrouped amount.
36 hundreds + 2 hundreds = 38 hundreds

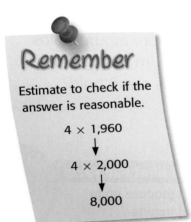

Remember

Estimate to check if the answer is reasonable.

4 × 1,960
↓
4 × 2,000
↓
8,000

Step 4 Multiply thousands.

$$\begin{array}{r} {}^{3\ 2} \\ \mathbf{1},960 \\ \times\quad 4 \\ \hline \mathbf{7},840 \end{array}$$
4 × 1 thousand = 4 thousands
Add the regrouped amount.
4 thousands + 3 thousands = 7 thousands

So, 1,960 × 4 = 7,840.

CHECK What You Know

Multiply. See Examples 1 and 2 (pp. 656–657)

1. 125
  × 5

2. 248
  × 3

3. 1,276
  × 4

4. 1,342
  × 7

5. If there are 365 days in one year, how many days are in 3 years?

6. (**Talk About It**) How is multiplying a three-digit number with regrouping similar to multiplying a two-digit number with regrouping?

Multiply. See Examples 1 and 2 (pp. 656–657)

7. 518
× 2

8. 222
× 5

9. 159
× 3

10. 293
× 7

11. 1,042
× 8

12. 1,513
× 9

13. 2,278
× 3

14. 3,150
× 6

15. 170 × 4

16. 821 × 4

17. 1,122 × 9

18. 1,189 × 5

Algebra Copy and complete each table.

19.

Rule: Multiply by 6.	
Input	**Output**
112	■
821	■
145	■

20.

Rule: Multiply by 4.	
Input	**Output**
38	■
29	■
417	■

21.

Rule: Multiply by ■.	
Input	**Output**
60	120
17	■
75	■

22. Measurement A jet is 232 feet long. What is the length if 7 jets were lined up nose to tail?

23. Each page of a photo album holds 6 pictures. The album has 125 pages. How many photos can it hold?

Data File

The Philadelphia Mint is one of several places in the U.S. where coins are made.

24. How many coins can be produced in 5 minutes?

25. Suppose a coin bag contained 1,575 nickels. How much would the bag of nickels weigh?

26. What is the total number of reeds found on $1 worth of quarters?

Philadelphia Mint Fun Facts

- Machines can produce up to 850 coins per minute.
- A nickel weighs 5 grams.
- A Presidential $1 coin is 2 millimeters thick.
- A quarter has 119 reeds, or ridges, while a dime, which is smaller, has 118.

Source: The United States Mint

H.O.T. Problem

27. **WRITING IN ▶MATH** Is the product of a two-digit number and a one-digit number always a three-digit number? Explain.

High and Low
Find a Product

Get Ready!
Players: 2 or more

Get Set!
Label a number cube 1–6.
Make 2 game sheets like the
one shown.

Go!
- Decide if the product goal for
  the game is HIGH or LOW.

- Player 1 rolls the number
  cube, and records the number
  in any of the factor spaces on
  the game sheet.

- Player 2 rolls the number cube
  and records the number in any
  of the factor spaces on their
  game sheet.

- Play continues until all
  players have filled in the
  factor spaces.

- Each player then finds the
  products of his or her factors.

- The winner is the player
  with the greatest or least
  product, depending on
  the goal.

You will need:
blank number cube, game sheet

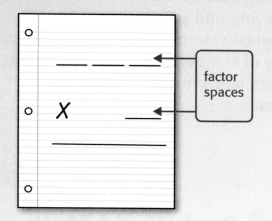

factor
spaces

STAMP COLLECTING

The U.S. government began issuing stamps in 1847. People began saving and collecting them. During the Civil War, unused Union stamps were actually used as money. At first, small engraving and printing companies produced stamps for the government. Many of the same images—American leaders and symbols—appeared on stamps.

In 1924, a new era of stamp making began in response to collectors' great interest in stamps. Stamp makers began producing colorful, exotic stamps with a wide variety of subjects and many different colors of ink. Today, you can see everything from comic-book characters to your favorite musicians on stamps.

Real-World Math

Use the stamps on page 661 to solve each problem.

1 What was the original cost of three Elvis stamps?

2 How many 24¢ stamps could have been bought for less than $1?

3 Find the cost of thirteen 6¢ stamps.

4 If 24 Lincoln stamps were on a page, what is the total value of the page of stamps?

5 If 78 stamp collectors each have one of each of the five stamps, how many stamps are there altogether?

6 What combination of stamps would equal the postage needed for a package that costs 75¢ to mail?

7 Suppose the First Man on the Moon stamp was issued in sheets of 10 stamps each, what was the value of a sheet of Man on the Moon stamps?

Did You Know?

The most valuable U.S. stamp was sold for $930,000.

Study Guide and Review

FOLDABLES
Study Organizer

GET READY to Study

Be sure the following Key Vocabulary words and Key Concepts are written in your Foldable.

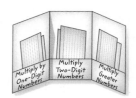

Key Concepts

Multiply by Multiples of 10, 100, and 1,000 (p. 635)

Use basic facts and patterns.

$3 \times 8 = 24$	3×8 ones
$3 \times 80 = 240$	3×8 tens
$3 \times 800 = 2,400$	3×8 hundreds
$3 \times 8,000 = 24,000$	3×8 thousands

Estimate

• You can **estimate** a product by rounding. (p. 640)

3×115 115 rounds to 100.

$\downarrow$

$3 \times 100 = 300$

Multiply by One-Digit Numbers (p. 644)

$$\begin{array}{r} 1 \\ 36 \\ \times\ 2 \\ \hline 72 \end{array} \qquad \begin{array}{r} 2\ 1 \\ 173 \\ \times\ 4 \\ \hline 692 \end{array}$$

Key Vocabulary

estimate (p. 640)

factors (p. 636)

multiples (p. 635)

product (p. 636)

round (p. 640)

Vocabulary Check

Choose the vocabulary word that completes each sentence.

1. A(n) _____?_____ is the product of a given number and any other whole number.

2. _____?_____ are numbers that are multiplied together to get a product.

3. When you find an answer that is close to the exact answer you _____?_____.

4. One way to estimate is to _____?_____.

5. 36, 45, 54, and 63 are _____?_____ of 9.

6. In 70×6, the number 420 is the _____?_____.

Lesson-by-Lesson Review

15-1 Multiply Multiples of 10, 100, and 1,000 (pp. 635–637)

Example 1
Each student needs 200 felt squares. There are 4 students. What is the total number of felt squares needed?

You need to find 4×200. Use a basic fact and patterns of zeros.

$4 \times 2 = 8$
$4 \times 20 = 80$ 4×2 tens
$4 \times 200 = 800$ 4×2 hundreds

So, $4 \times 200 = 800$ felt squares.

Multiply. Use basic facts and patterns.

7. $3 \times 1 = \blacksquare$ **8.** $7 \times 4 = \blacksquare$
$3 \times 10 = \blacksquare$ $7 \times 40 = \blacksquare$
$3 \times 100 = \blacksquare$ $7 \times 400 = \blacksquare$
$3 \times 1{,}000 = \blacksquare$ $7 \times 4{,}000 = \blacksquare$

9. 8×50 **10.** 5×70

11. There are 120 crayons in each box. The teacher dumped all 7 boxes into an art tub. How many crayons are in the tub?

15-2 Problem-Solving Strategy: Use Logical Reasoning (pp. 638–639)

Example 2
Heather has 3 people in her family. Kaneesha does not have 4 or 7 people. Ruby does not have 5 or 7 people. How many people are in each family?
Use a table and logical reasoning.

Girl	3	4	5	7
Heather	yes	✗	✗	✗
Kaneesha	✗	✗	yes	✗
Ruby	✗	yes	✗	✗
Johanna	✗	✗	✗	yes

So, Heather has 3 family members, Kaneesha has 5, Ruby has 4, and Johanna has 7.

Solve. Use logical reasoning.

12. The girls' softball team wants to buy new uniforms for $495. There are 20 girls. How much does each uniform cost?

13. Suppose you meet 5 friends and all of you shook hands with each other when you met. How many handshakes would there be?

14. Tory planted sunflower seeds in 10 rows. If there were 25 seeds in each row, how many seeds were planted?

15-3 **Estimate Products** (pp. 640–642)

Example 3
Jewel's mom spent 4 days baking for the craft fair. If she made 120 baked goods each day, about how many things did she bake in 4 days?

Estimate 4 × 120 by rounding to the nearest hundred.

4 × 120

> 120 rounds to 100.

4 × 100 = 400   Multiply mentally.

Estimate. Round to the nearest ten.

15. 37 × 4   **16.** 27 × 6

17. 38 × 8   **18.** 42 × 5

Estimate. Round to the nearest hundred.

19. 190 × 9   **20.** 187 × 6

21. 371 × 4   **22.** 490 × 7

23. Four different types of cereal are stacked in rows of 21 boxes each. About how many boxes of cereal are there altogether?

15-4 **Multiply by a One-Digit Number** (pp. 644–647)

Example 4
If 2 men caught 24 fish each, how many fish is that altogether?

Find 2 × 24.

Step 1   Multiply ones.

$$\begin{array}{r} 24 \\ \times\ 2 \\ \hline 8 \end{array}$$ 2 × 4 ones = 8 ones

Step 2   Multiply tens.

$$\begin{array}{r} 24 \\ \times\ 2 \\ \hline 48 \end{array}$$ 2 × 2 tens = 4 tens

So, 48 fish were caught.

Multiply. Use estimation to check.

24. $\begin{array}{r} 41 \\ \times\ 2 \\ \hline \end{array}$ **25.** $\begin{array}{r} 22 \\ \times\ 2 \\ \hline \end{array}$

26. 2 × 34   **27.** 6 × 11

28. Measurement Two triangles have a perimeter of 44 centimeters each. What is the total perimeter of the two triangles?

29. When the new office building opens, there will be 32 new desks for each of the 3 floors. How many new desks will need to be ordered?

Problem-Solving Investigation: Choose a Strategy (pp. 648–649)

Example 5
Lorena earned $70 each week for 7 weeks. She spent $125 and saved the rest. How much money was she able to save?

First, find $70 × 7.

$$\begin{array}{r} \$70 \\ \times\ 7 \\ \hline \$490 \end{array}$$ Lorena earned $490.

Then subtract.

$$\begin{array}{r} {\scriptstyle 8\ 10} \\ \$4\cancel{9}\cancel{0} \\ -\$125 \\ \hline \$365 \end{array}\begin{array}{l} \\ \text{earned} \\ \text{spent} \\ \text{saved} \end{array}$$

So, Lorena saved $365.

30. Measurement Each day you run 15 miles. In two weeks, how many miles will you have run?

31. A book has 300 pages. A second book has 3 times as many pages. How many pages are in both books?

32. Measurement It takes Corey's dad about 5 hours to drive to his aunt's house. If he drives about 62 miles each hour, about how many miles does Corey live from his aunt?

15-6 **Multiply Two-Digit Numbers** (pp. 652–655)

Example 6
Eight boys have 12 baseballs each. How many baseballs are there altogether?

Find 8 × 12.

Step 1
Multiply ones.

$$\begin{array}{r} {\scriptstyle 1} \\ 12 \\ \times\ 8 \\ \hline 6 \end{array}$$

Step 2
Multiply tens.

$$\begin{array}{r} {\scriptstyle 1} \\ 12 \\ \times\ 8 \\ \hline 96 \end{array}$$

So, there are 96 baseballs altogether.

Multiply. Use models if needed.

33. $\begin{array}{r} 46 \\ \times\ 2 \\ \hline \end{array}$

34. $\begin{array}{r} 17 \\ \times\ 4 \\ \hline \end{array}$

35. $\begin{array}{r} 35 \\ \times\ 2 \\ \hline \end{array}$

36. $\begin{array}{r} 29 \\ \times\ 3 \\ \hline \end{array}$

37. Measurement Elvio made a long chalk line on the sidewalk. He traced along a one foot ruler 8 times. How many inches long was Elvio's chalk line?

15-7 **Multiply Greater Numbers** (pp. 656–658)

Example 7

Find 6 × 112.

Step 1 Multiply ones.

$$\begin{array}{r} {}^{1} \\ 112 \\ \times\ 6 \\ \hline 2 \end{array}$$ 2 × 6 ones = 12 ones
  12 ones = 1 ten and 2 ones

Step 2 Multiply the tens.

$$\begin{array}{r} {}^{1} \\ 112 \\ \times\ 6 \\ \hline 72 \end{array}$$ 6 × 1 tens = 6 tens
  6 tens + 1 ten = 7 tens

Step 3 Multiply the hundreds.

$$\begin{array}{r} {}^{1} \\ 112 \\ \times\ 6 \\ \hline 672 \end{array}$$ 6 × 1 hundred = 6 hundreds

Example 8
A car factory completes 1,186 cars in one week. How many cars are completed in 4 weeks?

Find 1,186 × 4.

Estimate
1,186 × 4 ⟶ 1,000 × 4 = 4000

$$\begin{array}{r} {}^{3\,2} \\ 1,186 \\ \times\ 4 \\ \hline 4,744 \end{array}$$

So, 4,744 cars are completed in 4 weeks.

Multiply.

38. 418
    × 2

39. 272
    × 5

40. 1,042
    × 8

41. 1,313
    × 9

42. 168 × 5

43. 2,513 × 4

44. The first 2,525 people who went to the game each received the number of coupons shown below. How many coupons were given away?

45. Charlene walked 4,988 steps on Monday. If she walked at least that many steps Tuesday and Wednesday, how many steps would she have walked the three days?

46. **Measurement** Suppose you hiked Mammoth Cave to the end and back. How many miles would you have hiked?

Mammoth cave Facts
• Located in Kentucky.
• Thought to be the longest cave system in the United States.
• Length is 352 miles long.

Chapter Test

For Exercises 1 and 2, tell whether each statement is *true* or *false*.

1. When an exact answer is not needed, estimate to find an answer that is close to the exact answer.

2. You do not need to add any regrouped tens after multiplying the tens.

3. Multiply. Use basic facts and patterns.
$5 \times 6 =$ ▥
$5 \times 60 =$ ▥
$5 \times 600 =$ ▥
$5 \times 6,000 =$ ▥

Multiply.

4. 115
　　× 4

5. 　43
　　× 9

6. 270
　　× 3

7. 421
　　× 2

8. MULTIPLE CHOICE There are 365 days in one year. About how many days are in 7 years?

　A 2,500　　　**C** 2,645

　B 2,600　　　**D** 2,800

Estimate. Round to the nearest ten.

9. 　42
　　× 6

10. 　75
　　× 4

11. Algebra Copy and complete.

| Rule: Multiply by 6. ||
Input	Output
251	▥
332	▥
469	▥
102	▥

12. Jenni and Tom each collected $75 for a fundraiser. They want to raise $250 altogether. How much more does each of them need to collect if they each raise the same amount?

Estimate. Round to the nearest hundred.

13. 289×5　　　**14.** 350×6

Estimate. Round to the nearest ten.

15. In one year there are 7 months that have 31 days. About how many days is that altogether?

16. MULTIPLE CHOICE What number can be multiplied by 3,573 to give the answer 7,146?

$3,573 \times$ ▥ $= 7,146$

　F 2　　　　　**H** 7

　G 6　　　　　**J** 8

17. WRITING IN ►MATH Explain why it is important to estimate the answer

PART 1   Multiple Choice

Read each question. Then fill in the correct answer on the answer sheet provided by your teacher or on a sheet of paper.

1. Amos spent 24 days at camp. He hiked 3 miles each day. How many miles did he hike in all?

   A 72 miles   **C** 27 miles

   B 60 miles   **D** 8 miles

2. Tessa collected $125. Ariana collected 3 times as much. How much money did Ariana collect?

   F $128   **H** $375

   G $275   **J** $500

3. What decimal does the model show?

   A 7   **C** 0.7

   B 0.77   **D** 0.07

4. Trent's bookcase has 6 shelves. Each shelf holds 14 books. About how many books does Trent have in his bookcase?

   F 20   **H** 80

   G 60   **J** 120

5. Sonia has 30 shells. Brenda owns 10 times as many as Sonia. How many shells does Brenda have?

   A 40   **C** 400

   B 300   **D** 3,000

6. Roberta bought 72 eggs. Eggs are sold in cartons of 12. How many cartons of eggs did Roberta buy?

   F 6   **H** 8

   G 7   **J** 9

7. A piggy bank has 6 quarters, 4 dimes, 3 nickels, and 5 pennies. What is the value of the coins?

   A $2.00   **C** $2.50

   B $2.10   **D** $2.55

8. Which expression describes the array shown below?

   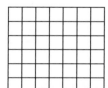

   F 8×6 **H** 6×7

   G 7×8 **J** 6×6

9. Preston emptied his change bag.
How much money does he have?

A $1.30	**C** $1.90
B $1.75	**D** $2.05

10. Dunn Elementary has 300 students.
Each student will bring 5 pencils on
the first day of school. How many
pencils will there be altogether?

F 150	**H** 1,500
G 305	**J** 15,000

11. Troy made 23 paper airplanes.
Lewis made twice as many. About
how many paper airplanes do the
boys have altogether?

A 20	**C** 50
B 46	**D** 60

PART 2   Short Response

**Record your answer on an answer
sheet provided by your teacher or on
a sheet of paper.**

12. Esteban reads the amount shown
each day for 8 days. How many
pages does he read in the 8 days?

45 pages

---

13. What number would make this
number sentence true?

$$5 \times 8{,}000 = \blacksquare$$

PART 3   Extended Response

**Record your answer on an answer
sheet provided by your teacher or on
a sheet of paper.**

14. There are 1,060 steps to the second
floor of the Eiffel Tower in France.
Odell walked up and down 4 times.
How many steps did he walk up
and down altogether? Explain.

NEED EXTRA HELP?														
If you missed question...	1	2	3	4	5	6	7	8	9	10	11	12	13	14
Go to Lesson...	15-4	15-7	14-1	15-4	15-1	7-6	1-10	4-1	1-10	15-1	15-3	15-4	15-4	15-1

Looking Ahead

Let's Look Ahead!

Looking Ahead

Distributive Property

MAIN IDEA

I will use the Distributive Property to make multiplication easier.

New Vocabulary

Distributive Property

▶ **GET READY to Learn**

Chef Cora hard boils six dozen eggs each day to make egg salad for sandwiches. What is the total number of eggs Chef Cora uses each day?

To multiply larger numbers, the Distributive Property is helpful. The Distributive Property combines addition and multiplication.

Distributive Property	Key Concept

Words The **Distributive Property** says that you can multiply the addends of a number and then add the products.

Symbols $6 \times 12 = (6 \times 10) + (6 \times 2)$

Real-World EXAMPLE Use a Model

1 **FOOD** **What is the total number of eggs Chef Cora uses each day?**

There are 12 eggs in one dozen. Find 6×12.
Think of 6×12 as $(6 \times 10) + (6 \times 2)$.

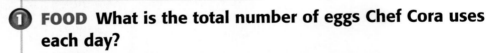

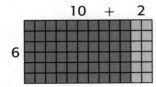
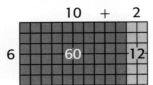

$6 \times 12 = (6 \times 10) + (6 \times 2)$

$\qquad = 60 + 12$

$\qquad = 72$

So, Chef Cora uses 72 eggs each day.

2 **THEATRE** **Twenty third graders are attending a performance at a Children's Theatre. Each admission ticket is $5 and each bus ticket is $3. What is the total cost for the 20 students?**

Remember

The Associative Property of Multiplication states that the grouping of the factors does not change the product.

One Way	Another Way
Find the cost of 20 admissions and 20 bus tickets. Then add.	Find the cost for 1 person. Then multiply by 20.
20 × $5 + 20 × $3	20 × ($5 + $3)
cost of 20 cost of 20 admissions bus tickets	cost for 1 person

$$= \$100 + \$60 \qquad \text{or} \qquad = 20 \times \$8$$
$$= \$160 \qquad\qquad\qquad\qquad = \$160$$

The total cost for 20 students is $160.

The Distributive Property can help you solve more difficult problems mentally.

3 **MAPS** **The World of Maps store displays its maps on 3 shelves. There are 26 world maps on each shelf. How many world maps does the store have to sell?**

Find 3 × 26 mentally using the Distributive Property.

$$3 \times 26 = 3 \times (20 + 6) \qquad \text{Think of 26 as } 20 + 6.$$
$$= (3 \times 20) + (3 \times 6) \qquad \text{Distributive Property}$$
$$= \quad 60 \quad + \quad 18 \qquad \text{Find each product mentally.}$$
$$= \qquad 78 \qquad\qquad \text{Add 60 and 18 mentally.}$$

So, there are 78 world maps to sell.

Use models and the Distributive Property to solve. Record your work. See Examples 1–3 (pp. LA2–LA3)

1. 12
      × 9

2. 22
      × 6

3. 32
      × 7

4. 15
      × 8

5. Mr. Kline bought 4 sheets of postage stamps. Each sheet had 16 stamps. What is the total number of postage stamps Mr. Kline bought?

6. **Talk About It** Explain how the Distributive Property can help when solving difficult problems.

Practice and Problem Solving

Use models and the Distributive Property to solve. Record your work.

See Examples 1–3 (pp. LA2–LA3)

7. 11
      × 8

8. 63
      × 4

9. 55
      × 6

10. 49
       × 9

11. 37 × 5

12. 99 × 9

13. 85 × 5

14. Write a multiplication problem that the model below represents.

15. Trisha is solving a problem using the Distributive Property. Explain her error.

$4 \times 76 = (4 \times 7) + (4 \times 6)$
$= 28 + 24$
$= 52$

16. The grocery store has 3 shelves of milk in the cooler. When the shelves are filled, there are 12 gallon bottles on each shelf. What is the total cost of all the milk?

17. There are 52 weeks in one year. Use the Distributive Property to find the number of weeks in 6 years.

H.O.T. Problems

18. OPEN ENDED Create a multiplication problem in which the Distributive Property is used to solve it.

NUMBER SENSE Rewrite each expression using the Distributive Property. Then solve.

19. $3 \times (40 + 6)$ **20.** $5 \times (60 + 7)$ **21.** $2 \times (20 + 1)$

22. **WRITING IN ►MATH** Explain how you can use the Distributive Property and an area model to solve 3×24.

TEST Practice

23. Which expression means the same as $(6 \times 10) + (6 \times 7)$?
(Lesson LA1)

A $16 + 13$ **C** $16 + 42$

B 6×17 **D** 12×17

24. Choose the correct factor to complete the multiplication sentence. (Lesson LA1)

$4 \times 35 = (4 \times \blacksquare) + (4 \times 5)$

F 3   **H** 30

G 5   **J** 50

Spiral Review

Multiply. (Lesson 15-7)

25. $\begin{array}{r} 1{,}056 \\ \times\ 7 \\ \hline \end{array}$

26. $\begin{array}{r} 3{,}225 \\ \times\ 4 \\ \hline \end{array}$

27. $\begin{array}{r} 9{,}123 \\ \times\ 8 \\ \hline \end{array}$

Use basic facts and patterns of zeros to find each product.
(Lesson 15-1)

28. $800 \times 3 = \blacksquare$ **29.** $40 \times 50 = \blacksquare$ **30.** $7 \times 7{,}000 = \blacksquare$

Solve. (Lesson 15-6)

31. There are 60 families living in an apartment building. Half of the families have 4 people. Seven of the families have 3 people. The rest of the families have 5 people. How many people live in the apartment building?

Estimate Products

MAIN IDEA

I will estimate products by rounding.

GET READY to Learn

Some airplanes may fly as many as 13 hours a day. About how many hours would one airplane fly in 3 weeks?

The word *about* tells you to estimate. When you estimate the product of 2 two-digit factors, it is helpful to round them both.

Real-World EXAMPLE   Estimate Products

 AIRPLANES If an airplane flies about 13 hours each day, about how many hours does it fly in 3 weeks?

There are 21 days in 3 weeks. So, estimate 21 × 13.

Step 1   Round each factor to the nearest ten.

$$
\begin{array}{r} 21 \\ \times\ 13 \end{array} \longrightarrow
\begin{array}{r} 20 \\ \times\ 10 \end{array}
$$

⟵ 21 rounds to 20.
⟵ 13 rounds to 10.

Step 2   Multiply.

$$
\begin{array}{r} 20 \\ \times\ 10 \\ \hline 200 \end{array}
$$

Find the product of the basic fact.
2 × 1 = 2
Place the zeros.

So, one airplane flies about 200 hours in 21 days or 3 weeks. Since both factors were rounded down, the estimate is less than the actual product.

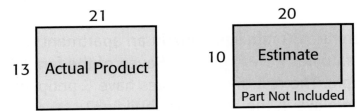

	21
13	Actual Product

	20
10	Estimate
	Part Not Included

H.O.T. Problems

18. OPEN ENDED Create a multiplication problem in which the Distributive Property is used to solve it.

NUMBER SENSE Rewrite each expression using the Distributive Property. Then solve.

19. $3 \times (40 + 6)$ **20.** $5 \times (60 + 7)$ **21.** $2 \times (20 + 1)$

22. **WRITING IN ▶MATH** Explain how you can use the Distributive Property and an area model to solve 3×24.

TEST Practice

23. Which expression means the same as $(6 \times 10) + (6 \times 7)$?
(Lesson LA1)

A $16 + 13$ **C** $16 + 42$

B 6×17 **D** 12×17

24. Choose the correct factor to complete the multiplication sentence. (Lesson LA1)

$4 \times 35 = (4 \times \blacksquare) + (4 \times 5)$

F 3   **H** 30

G 5   **J** 50

Spiral Review

Multiply. (Lesson 15-7)

25. 1,056
$\times 7$

26. 3,225
$\times 4$

27. 9,123
$\times 8$

Use basic facts and patterns of zeros to find each product.
(Lesson 15-1)

28. $800 \times 3 = \blacksquare$ **29.** $40 \times 50 = \blacksquare$ **30.** $7 \times 7,000 = \blacksquare$

Solve. (Lesson 15-6)

31. There are 60 families living in an apartment building. Half of the families have 4 people. Seven of the families have 3 people. The rest of the families have 5 people. How many people live in the apartment building?

Estimate Products

GET READY to Learn

Some airplanes may fly as many as 13 hours a day. About how many hours would one airplane fly in 3 weeks?

MAIN IDEA

I will estimate products by rounding.

The word *about* tells you to estimate. When you estimate the product of 2 two-digit factors, it is helpful to round them both.

Real-World EXAMPLE   Estimate Products

1 AIRPLANES If an airplane flies about 13 hours each day, about how many hours does it fly in 3 weeks?

There are 21 days in 3 weeks. So, estimate 21 × 13.

Step 1   Round each factor to the nearest ten.

$$\begin{array}{ccc} 21 & \longrightarrow & 20 \\ \times\ 13 & \longrightarrow & \times\ 10 \end{array}$$

21 rounds to 20.
13 rounds to 10.

Step 2   Multiply.

$$\begin{array}{r} 20 \\ \times\ 10 \\ \hline 200 \end{array}$$

Find the product of the basic fact.
2 × 1 = 2
Place the zeros.

So, one airplane flies about 200 hours in 21 days or 3 weeks. Since both factors were rounded down, the estimate is less than the actual product.

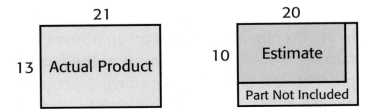

	21		20
13	Actual Product	10	Estimate
			Part Not Included

2 **MEASUREMENT** Marta's pet bullfrog jumped 77 centimeters in one leap. The frog continued to jump a total of 15 times. About how many centimeters did Marta's frog leap if all the jumps were the same distance as the first?

Remember

If one factor is rounded up and one factor is rounded down, it will not be obvious whether the estimate is greater or less than the actual product.

Step 1   Round each factor to its greatest place.

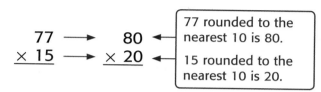

$$\begin{array}{r} 77 \\ \times\ 15 \end{array} \longrightarrow \begin{array}{r} 80 \\ \times\ 20 \end{array}$$

77 rounded to the nearest 10 is 80.

15 rounded to the nearest 10 is 20.

Step 2   Multiply.

$$\begin{array}{r} 80 \\ \times\ 20 \\ \hline 1{,}600 \end{array}$$

Find the product of the basic fact.
$8 \times 2 = 16$
Place the zeros.

So, Marta's frog leapt about 1,600 centimeters. Since both factors were rounded up, the estimate is greater than the actual product.

	77
15	Actual Product

	80
20	Actual Product
	Part is Included

✓ CHECK What You Know

Estimate. Tell whether the estimate is *greater than* or *less than* the actual product. See Examples 1 and 2 (pp. LA6–LA7)

1. $\begin{array}{r} 34 \\ \times\ 57 \end{array}$

2. $\begin{array}{r} 12 \\ \times\ 24 \end{array}$

3. $\$76 \times 17$

4. 52×43

5. A hamster sleeps about 14 hours a day. How many hours does it sleep in 28 days?

6. **Talk About It**   Explain how you know if an estimated product is more or less than the actual product.

Estimate. Tell whether the estimate is *greater than* or *less than* the actual product. See Examples 1 and 2 (pp. LA6–LA7)

7. 28
  × 25

8. 37
  × 14

9. $56
  × 16

10. 58
  × 29

11. 41
  × 64

12. 86
  × 55

13. $91
  × 94

14. 64
  × 82

15. $23 × 11

16. 35 × 37

17. 48 × 86

18. 35 × 42

19. 67 × 56

20. 73 × 84

21. 91 × 78

22. 19 × 92

23. A rabbit can run 35 miles per hour. About how many miles would it travel if it ran a total of 12 hours?

24. The food pyramid suggests children drink about 21 cups of milk a week. About how many cups of milk would that be in one year?

25. Measurement If 16 people stood with outstretched arms, fingertips touching, about how far would their arms extend?

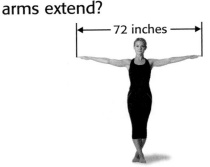

← 72 inches →

26. Measurement The world's largest pumpkin pie weighed 418 pounds. Suppose 12 pies of the same size were made. About how much of each ingredient would be needed?

Largest Pumpkin Pie	
Ingredients	**Amount**
Pumpkin	12 gallons
Flour	24 pounds
Sugar	56 pounds

Source: *Guinness World Records*

Real-World PROBLEM SOLVING

Sports Soccer is thought to be the most popular sport in the world.

27. Suppose a soccer player played 19 halves in one season, about how many minutes did he or she play that season?

28. If a players runs the length of the field 35 times, about how many meters would he or she have run?

Soccer Facts
- 11 players on each team
- two 45-minute halves
- 90 meter long field

H.O.T. Problems

29. OPEN ENDED Identify 2 two-digit factors that have an estimated product of 3,000.

30. NUMBER SENSE Estimate 41 × 29 and 82 × 45. Which is closer to its actual product? Explain your reasoning.

31. **WRITING IN ►MATH** Write a real-world problem that involves estimating the product of 2 two-digit numbers.

TEST Practice

32. What is the total length of 25 anacondas? (Lesson LA-2)

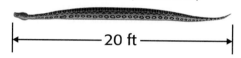

|← 20 ft →|

A 500 feet   **C** 800 feet

B 700 feet   **D** 900 feet

33. There are 24 hours in one day. Which is the best estimate of the number of hours in 61 days? (Lesson LA-1)

F 1,000   **H** 1,800

G 1,200   **J** 2,100

Spiral Review

Multiply. (Lesson 15-6)

34. 27
    × 3

35. 43
    × 5

36. $96
    × 7

Round to the nearest thousand. (Lesson 1-9)

37. 1,006

38. 4,650

39. 7,400

40. Find and extend the rule for the table. Then copy and complete. (Lesson 8-4)

Input	1	3	5	7	9	11
Output	4	12	20	▩	▩	▩

41. Arthur earns $20 for every lawn he mows. He mows 12 lawns twice a month. He has been mowing for 3 months. How much money does he make in 1 month? Identify any extra or missing information. Then solve. (Lesson 4-4)

Write the value of the underlined digit. (Lesson 1-4)

42. 9,3<u>9</u>7

43. <u>2</u>,670

44. <u>9</u>1,028

Mixed Numbers

MAIN IDEA

I will write mixed numbers.

New Vocabulary

mixed number

GET READY to Learn

Juanita made two peanut butter and jelly sandwiches for lunch. She cut the sandwiches into 4 pieces each. She has 5 pieces left. What fraction of the sandwiches are left?

Recall that a fraction is a number that names part of a whole. A **mixed number** has a whole number part and a fraction part.

Fractions	Mixed Numbers
$\frac{1}{2}$ $\frac{3}{4}$ $\frac{23}{6}$	$1\frac{1}{2}$ $2\frac{3}{4}$ $3\frac{5}{6}$

Real-World EXAMPLE   Mixed Numbers

① **FOOD What fraction of the sandwiches does Juanita have left?**

Each sandwich has 4 slices. There are 5 slices left.

Count the wholes and the parts.

$$\frac{4}{4} \quad + \quad \frac{1}{4} \quad = \quad 1\frac{1}{4}$$

whole   part

So, $1\frac{1}{4}$ of the sandwiches are left.

2 **Measurement** **Porcupine quills can be as short as 1 inch and as long as 12 inches.**

Write the length of this quill as a mixed number.

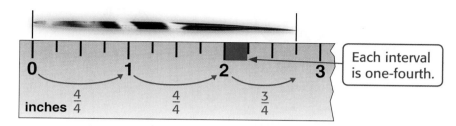

Each interval is one-fourth.

inches $\frac{4}{4}$ $\frac{4}{4}$ $\frac{3}{4}$

Count the wholes.

$\boxed{\frac{4}{4}} + \boxed{\frac{4}{4}} = 2$

Count the parts.

$\frac{1}{4} + \frac{1}{4} + \frac{1}{4} = \frac{3}{4}$

$2 + \frac{3}{4} = 2\frac{3}{4}$ inches

So, the quill is $2\frac{3}{4}$ inches long.

You can show mixed numbers on a number line.

EXAMPLE   Use a Number Line

3 **Identify point A as a mixed number.**

$\boxed{\frac{3}{3}} + \boxed{\frac{3}{3}} + \boxed{\frac{3}{3}} + \boxed{\frac{3}{3}} + \boxed{\frac{3}{3}} + \frac{1}{3} = 5\frac{1}{3}$

Each interval on the number line is one-third.
So, point A is $5\frac{1}{3}$.

CHECK What You Know

Write a mixed number for each model. See Example 1 and 2 (pp. LA10–LA11)

1.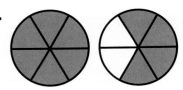

2.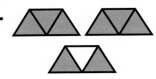

3.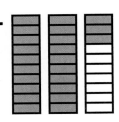

Identify each point. Write as a mixed number. See Example 3 (p. LA11)

4.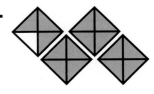

5.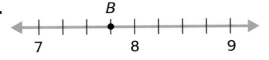

6. Richard has $1\frac{3}{8}$ of an apple sliced and Sabrina has $1\frac{2}{4}$ of an apple sliced. Who has more apples sliced? Explain.

7. **Talk About It** Explain how to compare $2\frac{3}{5}$ and $2\frac{8}{10}$.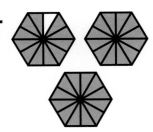

Practice and Problem Solving

Write a mixed number for each model. See Example 1 and 2 (pp. LA10–LA11)

8.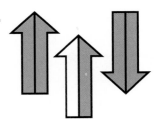

9.

10.

11.

12.

Identify each point. Write as a mixed number. See Example 3 (p. LA11)

13.

W   X

1   2   3

14.

M   N

4   5

15. Jon needs $1\frac{1}{2}$ cups of batter for pancakes and $1\frac{3}{4}$ cups of bananas for banana bread. Does Jon need more batter or more bananas? Explain.

16. Martin ate $2\frac{3}{5}$ cups of popcorn during the first half of the movie and $2\frac{4}{6}$ cups of popcorn during the second half. When did he eat more popcorn?

Real-World PROBLEM SOLVING

Tracking It is interesting to identify animals by the tracks they make.

17. Draw a model to represent the length of a black bear's track.

18. Is the length of a deer's track greater than the length of a lynx's track? Draw a model to explain.

Length of Animal Tracks	
Animal	Length (inches)
Black bear	$7\frac{1}{10}$
Deer	$3\frac{1}{2}$
Lynx	$3\frac{7}{10}$

H.O.T. Problems

19. OPEN ENDED Name two mixed numbers that name the same amount.

20. FIND THE ERROR Gwenith and Wesley are writing a mixed number to identify point A on the number line. Who is correct? Explain.

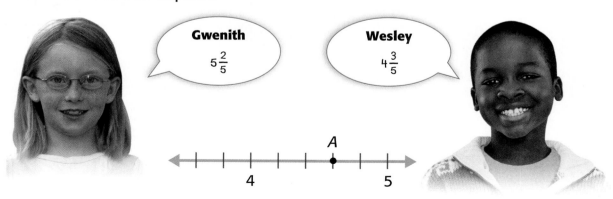

Gwenith
$5\frac{2}{5}$

Wesley
$4\frac{3}{5}$

A

4   5

21. WRITING IN ►MATH Compare a fraction and a mixed number.

Relate Mixed Numbers and Decimals

MAIN IDEA

I will identify, read, and write decimals greater than 1.

GET READY to Learn

Common iguanas live in trees and can be 3 to 6 feet long. Sydney has an iguana that is $3\frac{6}{10}$ feet in length from its head to the tip of its tail.

A mixed number like $3\frac{6}{10}$ is a fraction greater than one. You can write mixed numbers as decimals.

EXAMPLE Mixed Numbers as Decimals

1 **Write $3\frac{6}{10}$ as a decimal.**

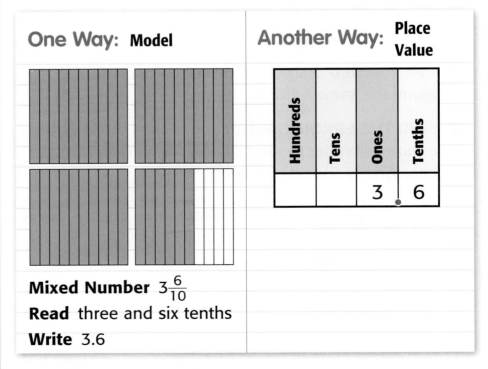

One Way: Model

Mixed Number $3\frac{6}{10}$

Read three and six tenths

Write 3.6

Another Way: Place Value

Hundreds	Tens	Ones	Tenths
		3	6

So, $3\frac{6}{10}$ as a decimal is 3.6.

Real-World EXAMPLE

2 **MEASUREMENT** Unionville, Maryland once recorded $1\frac{23}{100}$ inches of rain in one minute. Write $1\frac{23}{100}$ as a decimal.

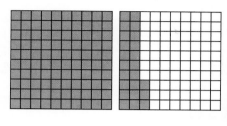

Remember

When reading a decimal, the word "and" represents the decimal.

Hundreds	Tens	Ones	Tenths	Hundredths
		1	2	3

Mixed Number $1\frac{23}{100}$

Read one and twenty-three hundredths

Write 1.23

CHECK What You Know

Write each as a mixed number and decimal. See Examples 1 and 2 (pp. LA14–LA15)

1.

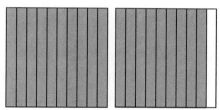

2.

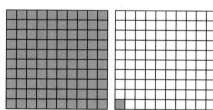

3. three and two tenths

4. twelve and eighteen hundredths

5. seven and seven tenths

6. three and two hundredths

Write each as a decimal. See Examples 1 and 2 (pp. LA14–LA15)

7. $5\frac{6}{10}$

8. $2\frac{5}{10}$

9. $16\frac{50}{100}$

10. $8\frac{24}{100}$

11. Jesse ran the 200-meter dash in 20.3 seconds. Merlene ran the 200-meter dash in 21.87 seconds. Write each runner's time as a mixed number.

12. **Talk About It** Do $8\frac{5}{10}$, $8\frac{1}{2}$, and 8.5 name the same amount? Explain.

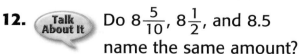

LA-4 Relate Mixed Numbers and Decimals **LA15**

Write each as a mixed number and decimal. See Examples 1 and 2 (pp. LA14–LA15)

13.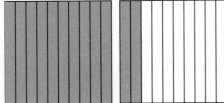

14.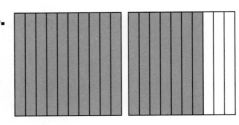

15.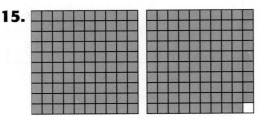

16.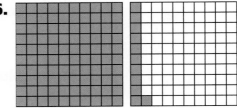

17. nine and five tenths

18. fifty-six and one tenth

19. nineteen and one hundred hundredths

20. fifty-six and one hundredth

Write each as a decimal. See Examples 1 and 2 (pp. LA14–LA15)

21. $6\frac{8}{10}$

22. $50\frac{5}{10}$

23. $78\frac{6}{10}$

24. $2\frac{1}{10}$

25. $22\frac{2}{100}$

26. $10\frac{25}{100}$

27. $60\frac{75}{100}$

28. $5\frac{16}{100}$

29. Measurement Jafari has grown $1\frac{4}{10}$ inches since last year. Write a decimal to show how many inches Jafari has grown.

30. Measurement Coastal Plains received 5.52 inches of rain. Write a mixed number to show the number of inches Coastal Plains received.

31. Measurement Ellin lives $2\frac{6}{10}$ miles from the mall. Write a decimal to show how many miles Ellin lives from the mall.

32. Measurement Gloria's paper airplane flew 3.05 yards. Write a mixed number to show how many yards the airplane flew.

33. Measurement A moose is one of the world's tallest mammals. Write a decimal to show how tall this moose is.

$6\frac{5}{10}$ feet

H.O.T. Problems

34. OPEN ENDED Write a mixed number and decimal that are less than seven and four tenths.

35. FIND THE ERROR Silvia and Matthew are writing $10\frac{3}{4}$ as a decimal. Who is correct? Explain your reasoning.

Silvia
$10\frac{3}{4} = 10.75$

Matthew
$10\frac{3}{4} = 10.34$

36. **WRITING IN** ►**MATH** Are $2\frac{3}{6}$ and 2.5 equivalent? Explain.

TEST Practice

37. Which number represents the shaded parts of the figure?
(Lesson 14-1)

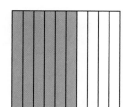

A 0.06

B 0.6

C 6.0

D 6.6

38. Which of the following is one and one hundredths? (Lesson 14-2)

F 0.11

G 1.01

H 1.1

J $1\frac{1}{10}$

Spiral Review

Write each as a fraction and as a decimal. (Lesson 14-2)

39. five tenths

40. fifty-six hundredths

Compare. Use >, <, or =. (Lesson 13-6)

41. $\frac{7}{12}$ ● $\frac{3}{12}$

42. $\frac{2}{4}$ ● $\frac{1}{4}$

43. $\frac{8}{10}$ ● $\frac{4}{5}$

44. $\frac{4}{16}$ ● $\frac{8}{16}$

45. Jack has read $\frac{4}{10}$ of a book. He then read $\frac{2}{10}$ more. Has he read more or less than half of the book? Explain.

Measurement: Area

MAIN IDEA

I will find the area of rectangles.

▶ **GET READY to Learn**

Two of Mrs. Alvarez's students washed the chalkboard after school. What is the area of the chalkboard?

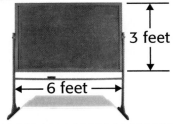

3 feet

← 6 feet →

Area is the number of square units needed to cover a figure without overlapping. Area is measured in square units.

1 square unit

Real-World EXAMPLE Area of a Rectangle

1 **CHALKBOARD** **Find the area of the chalkboard.**

You can use an area model to find the area of a figure.

6 ft

3 ft

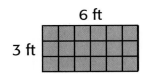

Add each row. 6 + 6 + 6 = 18

There are 18 square feet.

Check You can check by adding the columns.

3 + 3 + 3 + 3 + 3 + 3 = 18 ✔

So, the chalkboard has an area of 18 square feet.

Area of a Rectangle	Key Concept
Words	To find the area of a rectangle, multiply the length by the width.
Formula	$A = \ell \times w$

ℓ

w

You can also use multiplication to find the area of a rectangle.

Real-World EXAMPLE **Multiply to Find the Area**

2 **SIGNS** **What is the area of the rest area sign?**

Use the formula
$A = \ell \times w$.

$\ell = 4$ ft
$w = 3$ ft

$A = \ell \times w$

$A = 4$ ft $\times 3$ ft

$A = 12$ square feet

So, the area of the sign is 12 square feet.

Remember

If the unit of measurement is not given when determing area, use square units

4 ft

REST AREA
2 km

3 ft

CHECK **What You Know**

Find the area of each rectangle. See Examples 1 and 2 (pp. LA18–LA19)

1.

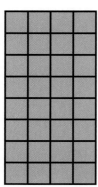

2.

9 yd

1 yd

3.

3 ft

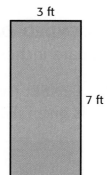

7 ft

4. The students in a third grade class have rectangular desks. Each desk measures 15 inches long by 30 inches wide. What is the area of each student's desk?

5. **Talk About It** Explain two ways to find the area of a rectangle.

Find the area of each rectangle. See Examples 1 and 2 (pp. LA18–LA19)

6.

7.

8.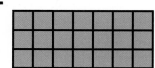

9.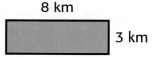
8 km
3 km

10.
8 km
8 km

11.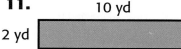
10 yd
2 yd

12. Measurement Chet is hanging a picture on a wall. How much wall space will the picture need if the frame has a length of 2 feet and a width of 9 inches?

13. Measurement Ricky watered his rectangular flower garden that has an area of 120 square feet. If the length of one side is 12 feet, what is the total perimeter of the garden?

14. Measurement A throw rug is 8 feet long and 4 feet wide. It is in a rectangular room with an area of 110 square feet. How much of the room is *not* covered by the rug?

15. Measurement A rectangular playground is 30 meters by 20 meters. Its area will be covered with shredded tires. Each bag of shredded tires covers 200 square meters and costs $30. Find the total cost for this project.

H.O.T. Problems

16. OPEN ENDED Draw three rectangles that each have an area of 24 square inches, but have different perimeters.

NUMBER SENSE The area and the measure of one side of each rectangle is given. Find the missing sides.

17.
4 in.
Area = 36 square inches

18.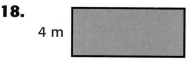
4 m
Area = 40 square meters

19.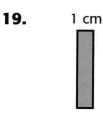
1 cm
Area = 7 square centimeters

20. WRITING IN ▶ MATH A rectangle has sides measuring 5 feet and 3 feet. If the sides of a rectangle are doubled, will the area also double? Explain.

21. Which number sentence below represents the area (A) of the rectangle in square feet? (Lesson LA-5)

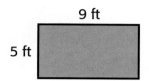

9 ft

5 ft

A $5 = A \times 9$

B $A = (5 + 9) + (5 + 9)$

C $A = 5 \times 9$

D $A = (5 \times 2) + (2 \times 9)$

22. Mrs. Morris plans to tile her front hallway shown below. If each tile is 1 foot long and 1 foot wide, how many tiles will she need? (Lesson LA-5)

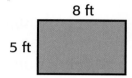

8 ft

5 ft

F 13 tiles

G 26 tiles

H 30 tiles

J 40 tiles

Spiral Review

Multiply. (Lesson 15-7)

23. 225
$\times 6$

24. 145
$\times 3$

25. 421
$\times 7$

26. 955
$\times 6$

Write each fraction as a decimal. (Lesson 14-1 and 14-2)

27. $\frac{3}{10}$

28. $\frac{1}{10}$

29. $\frac{71}{100}$

30. $\frac{5}{100}$

Describe each two-dimensional figure. Use the terms *sides* and *angles*. Then identify the figure. (Lesson 11-2)

31.

32.

33.

Solve.

34. Danielle put two textbooks, two story books, and her art project in her backpack. Are more or less than $\frac{3}{5}$ of the items in her backpack books? (Lesson 13-6)

Measurement: Area of Complex Figures

GET READY to Learn

MAIN IDEA

I will find the area of complex figures.

New Vocabulary

complex figure

Shrubs, trees, and plants can be bought at Mr. Ortega's Nursery. What is the area of the nursery's garden at the right?

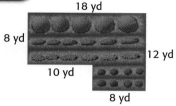

A **complex figure** is made up of two or more shapes. To find the area of a complex figure, break the figure into smaller parts.

Real-World EXAMPLE Area of a Complex Figure

① **NURSERY** Find the area of Mr. Ortega's Nursery garden.

Step 1 Break the figure into smaller parts. Look for rectangles.

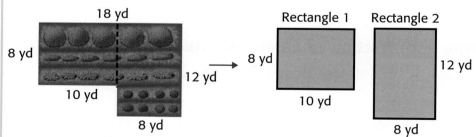

Step 2 Find the area of each part.

Rectangle 1
$A = \text{length} \times \text{width}$
$A = \ell \times w$
$A = 10 \text{ yd} \times 8 \text{ yd}$
$A = 80 \text{ square yards}$

Rectangle 2
$A = \text{length} \times \text{width}$
$A = \ell \times w$
$A = 8 \text{ yd} \times 12 \text{ yd}$
$A = 96 \text{ square yards}$

Step 3 Add the areas.

The area is $80 + 96$ or 176 square yards.

2 **Find the area of the complex figure.**

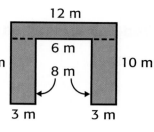

12 m

6 m

10 m   8 m   10 m

3 m   3 m

Step 1 Break the figure into smaller parts. Look for rectangles. This figure can be broken into 3 rectangles.

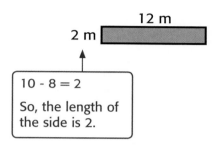

12 m

2 m

10 - 8 = 2

So, the length of the side is 2.

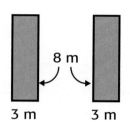

8 m

3 m   3 m

Step 2 Find the area of each part.

Rectangle 1
$A = \text{length} \times \text{width}$
$A = 12 \text{ m} \times 2 \text{ m}$
$A = 24$ square meters

Rectangles 2 and 3
$A = \text{length} \times \text{width}$
$A = 8 \text{ m} \times 3 \text{ m}$
$A = 24$ square meters

Step 3 Add the areas.
$24 + 24 + 24 = 72$

So, the area of the figure is 72 square meters.

✓ CHECK What You Know

Find the area of each figure. See Examples 1 and 2 (pp. LA22–LA23)

1.

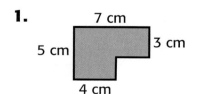

7 cm
5 cm   3 cm
4 cm

2.

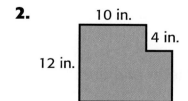

10 in.
4 in.
12 in.
14 in.

3.

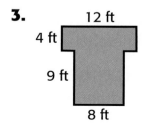

12 ft
4 ft
9 ft
8 ft

4. Talk About It Refer to Example 1. Find another way to break apart the total area.

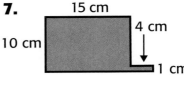

Practice and Problem Solving

Find the area of each figure. See Examples 1 and 2 (pp. LA 22–23)

5.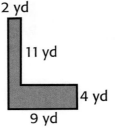
2 yd
11 yd
4 yd
9 yd

6.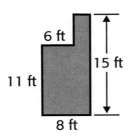
6 ft
15 ft
11 ft
8 ft

7.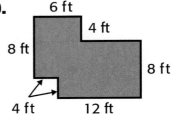
15 cm
4 cm
10 cm
1 cm

8.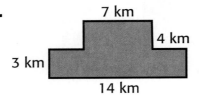
7 km
4 km
3 km
14 km

9.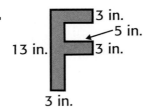
3 in.
5 in.
13 in.
3 in.
3 in.

10.
6 ft
4 ft
8 ft
8 ft
4 ft
12 ft

11. Courtney is playing miniature golf. What is the area of the entire figure?

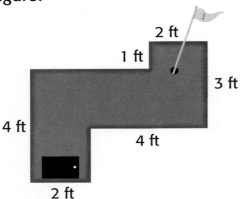
2 ft
1 ft
3 ft
4 ft
4 ft
2 ft

12. What is the area of the desk?

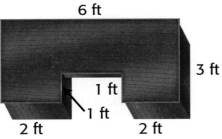
6 ft
3 ft
1 ft
1 ft
2 ft
2 ft

H.O.T. Problems

13. **OPEN ENDED** Draw and label two complex figures that have the same area but have different perimeters.

14. **CHALLENGE** Find the perimeter and area of the shaded figure in units.

15. **WRITING IN** ►**MATH** Create a word problem about a real-world situation involving perimeter and area of a complex figure.

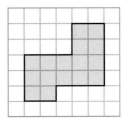

16. What is the area of the figure below? (Lesson LA-5)

7 in.

A 7 square inches

B 14 square inches

C 28 square inches

D 49 square inches

17. What is the total area of the figure below? (Lesson LA-6)

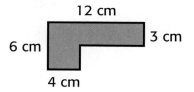

12 cm
3 cm
6 cm
4 cm

F 20 centimeters

G 25 centimeters

H 36 square centimeters

J 48 square centimeters

Compare. Use >, <, or = . (Lesson 13-6)

18. $\frac{1}{4}$ ● $\frac{3}{4}$

19. $\frac{1}{3}$ ● $\frac{1}{2}$

20. $\frac{2}{3}$ ● $\frac{4}{6}$

Spiral Review

21. The sum of two numbers is 26. One number is four more than the other. What are the two numbers? (Lesson 10-2)

22. Mr. Germaine is planting grass in part of his yard. The part is 9 yards long and 12 yards wide. What is the area of the newly planted part of the yard? (Lesson 9-6)

Find the perimeter of each figure. (Lesson 9-5)

23. 6 cm

6 cm

24. 12 m

5 m

Divide (Lesson 7-5)

25. $81 \div 9$

26. $63 \div 7$

27. $64 \div 8$

28. Algebra Copy and complete $6 \times (\blacksquare \times 5) = (6 \times 4) \times 5$. Identify the property used. (Lesson 5-9)

Problem-Solving Projects

Problem-Solving Projects

I Want to Go THERE

As a team member of a travel agency, you will create a travel brochure for a 5-day trip for 4 people. The trip should be at least 100 miles away. Your brochure should include how the customer will travel, the sites they will visit, hotels, food, and costs.

Getting Started

Day 1 The Best Place in the World

- Choose a location at least 100 miles away, and 3 sites to visit.
- Draw a thermometer that shows a temperature that is *likely* for this time of year.
- Estimate all costs for the trip.
- Create a table to organize and display this data.

Day 2 Travel and Sightseeing

- Create a bar graph comparing travel costs.
- Draw two analog clocks showing the time of departure and time of arrival.
- Determine round trip mileage in miles and kilometers.
- Construct a function table for admission cost for each site for 1, 2, 3, and 4 people.
- Create a pictograph comparing the total cost of visiting each site.

Day 3 Lodging and Food

- Construct a function table for lodging costs for 1, 2, 3, 4, and 5 nights.
- The budget for food is $500.
- Plan 3 meals a day for 4 people for 5 days.

Day 4 Advertise the Trip

- Collect and organize your data, graphs, and tables.
- Create a brochure to show your client. Record and display the data.

Day 5 Convince Your Client

- Present the brochure. Communicate why your trip is reasonably priced. Present your data.

Wrap Up

- Compare your estimated costs to your researched costs. What is the difference?
- Choose one part of the trip. Use what you have learned and tell how you could save money.
- Identify the mathematics from this project that could help you in everyday situations.

Basket-Golf

Have you ever wondered who made up your favorite sport? Or what would happen if two sports were combined? This week you will create a new sport. You can either invent a new sport or combine two sports that already exist.

Getting Started

Day 1 Create It
• Create a class table. List 6 sports and 5 characteristics of the game.
• Determine the characteristics for the new sport.

Day 2 Equipment, Playing Surface, and Area
• Design the equipment needed.
• Create a two-dimensional drawing of each piece of equipment.
• Label all measurements with standard units.

- Design the playing surface.
- Create a two-dimensional drawing. Describe the surface. Label its dimensions. Use standard units to find the perimeter.
- Decide whether the playing surface will be divided into halves, quarters, or another fractional part. Label this on your drawing.
- Use grid paper to find the area of the playing surface.

Day 3 Rules and Scoring

- Write the rules and procedures.
- Determine the playing time. Draw a clock for each playing period.
- Explain how to score and win. Make a function table to show the point system.

Day 4 Design the Uniforms

- Illustrate the uniform.
- Label each item with a reasonable price.
- Find the total cost of uniforms for a team. Write this as a number sentence.

Day 5 Introduce Your New Sport

- Present your new sport to the class.
- Describe one problem-solving strategy you used.

Wrap Up ·······

- Reflect on the math involved in playing sports. Give three examples of math connected to a sport(s).
- Describe parts of this project that remind you of an everyday situation in which you use math.

Crazy Cooks

Have you ever volunteered to help a charity? In this project, your class will form groups of 4 and make a recipe, sell the product, and then donate the proceeds to a local cause.

Getting Started

Day 1 Who Needs Your Help?

- Survey the class for charity ideas. Vote for one. Set a donation goal.
- Survey classmates about their favorite baked goods. Record in a tally chart. Display the top 5 in a bar graph.
- Make a table to show the servings needed and the price for each serving. Find the total that can be earned from each baked good.

Day 2 How Hungry Are You?

- How many of each baked good do you need?
- How much of each ingredient will you need?

- How much will the ingredients cost?
- How much should you charge?
- Find the total amount of each ingredient needed.
- Use grocery ads or the Internet to research the cost of each ingredient. Calculate the cost of each ingredient for multiple batches. Find the total cost.

Day 2 and 3 Plan and Prepare

- Find the number of servings your recipe makes. Determine how to divide the whole into equal servings.
- Determine the number of batches needed. Create a function table to show the number of servings for 1, 2, 3, 4, and 5 batches.
- Make the recipe.

Day 4 Sell

- Each group decides how much change is needed to start with and in what denominations.
- Sell the food at lunch time. Make the change. Count the change back.

Day 5 Display Results

- Create a bar graph to show how many of each baked good was sold. Analyze the data.
- Create another bar graph to show the money earned from each item.
- Use the bar graph to find the profit. Did you meet your goal? Explain.
- Write a letter to the leaders of the cause and explain what you did. Send the letters with the money you made.

Wrap Up

- Identify other ways you could make money to support a cause.
- Identify several steps you took that were similar to steps you take when solving word problems.
- Identify the mathematics from this project that may help you in everyday experiences.

PROJECT 4

I Have Always Wanted a Llama

You will collect data needed to choose your class pet. Research the needs and costs of taking care of a class pet. Design a habitat for your pet. Then, present your research to your class.

Getting Started

Day 1 So Many to Choose From

- Survey the class for ideas for a class pet. Record the data in a tally chart and display the data in a bar graph.
- Choose one pet to research. Make a table with two columns. Label one *Estimate* and the other *Actual.* Estimate its length and weight in standard units. Find the actual length and weight. Find the difference.

Day 2 and 3 Food and Home

- Create a function table to show what it eats and how much it eats for 1–7 days. Draw clocks to show when it should be fed.
- Design the pet's home. Make a three-dimensional design. Label the dimensions. Find the volume.
- Determine what accessories will be needed. Incude pictures of at least one three-dimensional object, one symmetrical object, and two congruent objects.

Day 4 Costs

- Suppose you have $50 to spend on what the pet needs. Decide from Day 3 what you must have and what is extra. Create a table to show what must be bought and what you can buy with any leftover money.
- Create a poster to show the most important parts of your research.

Day 5 Presentation

- Present your poster to the class.
- Describe one way math helped you with this project.

Wrap Up ··

- Did you have enough money on Day 4? How did you decide how to spend the money? Explain.
- How did this project help you make future decisions about money?

Student Handbook

Built-In Workbooks

Reference

How to Use the Student Handbook

The Student Handbook is the additional skill and reference material found at the end of books. The Student Handbook can help answer these questions.

What If I Need More Practice?

You, or your teacher, may decide that working through some additional problems would be helpful. The **Extra Practice** section provides these problems for each lesson so you have ample opportunity to practice new skills.

What If I Need to Prepare for a Standardized Test?

The **Preparing for Standardized Tests** section provides worked-out examples and practice problems for multiple-choice, short-response, and extended response questions.

What if I Want to Learn Additional Concepts and Skills?

Use the Concepts and Skills Bank section to either refresh your memory about topics you have learned in other math classes or learn new math concepts and skills.

What If I Forgot a Vocabulary Word?

The **English-Spanish Glossary** provides a list of important, or difficult, words used throughout the textbook. It provides a definition in English and Spanish as well as the page number(s) where the word can be found.

What If I Need to Find Something Quickly?

The **Index** alphabetically lists the subjects covered throughout the entire textbook and the pages on which each subject can be found.

What If I Forget Multiplication Facts or Measurement Conversions?

Inside the back cover of your math book is a multiplication table. You will also find a list of measurement conversions inside the back cover.

Extra Practice

Lesson 1-1

Pages 17–19

Identify a pattern. Then find the missing numbers.

1. 9, 12, 15, ■, 21, 24

2. 26, 31, ■, 41, 46, 51

3. 77, 71, 65, ■, 53, 47

4. 11, 15, ■, 23, 27, 31

5. 99, ■, 85, 78, 71, ■

6. 55, ■ 45, ■, 35, 30

7. 20, ■, 60, 80, ■

8. 86, 94, 102, ■, 118, ■

9. 700, 600, ■, 400, ■, 200

10. 85, ■, ■, 112, 121, 130

Lesson 1-2

Pages 20–21

Solve. Use the four-step plan.

1. Maria ate 7 grapes and 5 strawberries. How many pieces of fruit did she eat?

2. There were 22 people at the park. 7 people left. How many people were at the park then?

3. John has 18 more toy airplanes than Soto does. Soto has 13 toy airplanes. How many toy airplanes does John have?

4. During a treasure hunt, Megan walked 12 yards left, 6 yards forward, and 7 yards left. How many yards did Megan walk altogether?

Lesson 1-3

Pages 24–27

Write the place of the underlined digit. Then write the value of the digit.

1. 7<u>0</u>6

2. 2,<u>4</u>32

3. 5,68<u>2</u>

4. <u>6</u>,734

5. 8,<u>0</u>98

6. <u>3</u>,365

Write each number in expanded form and word form.

7. 4,371

8. 2,988

9. 5,654

10. 7,702

11. 6,520

12. 8,906

Lesson 1-4

Pages 28–30

Write the place of each underlined digit. Then write its value.

1. 4,3<u>2</u>2

2. <u>8</u>0,761

3. 3,<u>0</u>00

4. 67,02<u>3</u>

5. 5<u>1</u>,089

6. <u>2</u>7,055

Write each number in expanded form and word form.

7. 8,954

8. 14,523

9. 81,306

10. 27,621

11. 9,909

12. 50,345

Lesson 1-5

Pages 32–33

Use the four-step plan to solve each problem.

1. Tony has 7 games, Allison has 9 games, and Jarrod has 12 games. They each bought 3 more games. How many does each child have now?

2. Carlos jogged for 30 minutes today. Tomorrow he plans to jog 3 times as long as he did today. How long does he plan to jog tomorrow?

3. Gina bought a sweater for $14. She paid with a $20 bill. How much change did Gina get?

4. Sara picked 48 cherries. She ate 9 cherries. Her sister ate 12 cherries. How many cherries are left?

Lesson 1-6

Pages 34–37

Compare. Use >, <, or =.

1. 77 ● 67

2. $45 ● $54

3. 610 ● 610

4. 234 ● 342

5. 404 ● 440

6. 908 ● 889

7. 56 ● 65

8. 3,576 ● 3,567

9. 222 ● 232

10. 45 ● 450

11. 57 ● 57

12. 787 ● 878

Lesson 1-7

Order the numbers from least to greatest.

1. 888; 8,008; 81

2. 46; 49; 43

3. 678; 768; 5,667

4. 1,790; 1,978; 1,843

5. 3,438; 896; 2,122

6. 1,222; 2,221; 1,022

Order the numbers from greatest to least.

7. 765; 7,650; 79

8. 999; 3,221; 4,000

9. 368; 386; 833

10. 2,567; 2,982; 2,199

11. 4,235; 4,325; 3,443

12. 616; 6,116; 6,611

Lesson 1-8

Round to the nearest ten.

1. 68

2. 23

3. 84

4. 233

5. 397

6. 408

7. 1,656

8. 2,492

Round to the nearest hundred.

9. 231

10. 778

11. 645

12. 1,282

13. 442

14. 581

15. 4,774

16. 987

Lesson 1-9

Round to the nearest thousand.

1. 3,810

2. 1,221

3. 5,989

4. 8,297

5. 3,099

6. 6,572

7. 1,100

8. 2,667

9. 1,589

10. 4,088

11. 7,476

12. 2,821

Find the value of the coins.

1.

2.

Find the value of the bills and coins.

3.

4.

Find each sum. Identify the property.

1. $8 + 0 = $ ■

2. $7 + 3 = $ ■
$3 + 7 = $ ■

3. $(4 + 5) + 3 = $ ■
$4 + (5 + 3) = $ ■

4. $6 + 5 = $ ■
$5 + 6 = $ ■

5. $0 + 5 = $ ■

6. $9 + (4 + 3) = $ ■
$(9 + 4) + 3 = $ ■

Algebra **Find each missing number. Identify the property.**

7. $0 + 7 = 7 + $ ■

8. $6 + (3 + 5) = (6 + ■) + 5$

9. $8 + 4 = ■ + 8$

10. $5 + ■ = 7 + ■$

11. $(9 + 3) + 4 = ■ + (3 + 4)$

12. $■ + 0 = 5 + ■$

Tell whether an estimate or an exact answer is needed. Then solve.

1. How many sandwiches can be made with 3 loaves of bread that have 18 slices each?

2. Carmen walked 12 blocks forward and 25 blocks to the left. How many blocks did she walk in all?

3. A box of cookies costs $1.85. Milk costs $2.10. About how much will 2 boxes of cookies and milk cost?

4. Maya has 90 beads. She uses about 25 beads to make a necklace. Does she have enough beads to make 4 necklaces?

Lesson 2-3

Estimate each sum using rounding.

1. 22
$+ 41$

2. 58
$+ 39$

3. 67
$+ 23$

4. 47 + 33

5. 72 + 28

6. 51 + 24

7. 38
$+ 21$

8. 45
$+ 33$

9. 52
$+ 12$

10. 24 + 51

11. 38 + 29

12. 83 + 17

Lesson 2-4

Add. Use models if needed. Check for reasonableness.

1. 36
$+ 8$

2. 24
$+ 5$

3. 53
$+ 7$

4. 48
$+ 7$

5. 36
$+ 15$

6. 43
$+ 32$

7. 64
$+ 29$

8. 54
$+ 34$

9. 33
$+ 5$

10. 46 + 4

11. 67 + 8

12. 41 + 7

Lesson 2-5

Add. Use models if needed.

1. 10¢
$+ 80¢$

2. 30¢
$+ 50¢$

3. $47
$+ 33

4. $52
$+ 23

5. $38
$+ 23

6. $59
$+ 26

7. $28
$+ 13

8. $35
$+ 48

9. $35
$+ 58

10. $43 + $27

11. $28 + $55

12. $37 + $16

Lesson 2-6

Use the four-step plan to solve each problem.

1. Tony jogs about 28 miles each week. Does he jog more than 110 miles each month? Explain.

2. Terry made 4 dozen muffins. Rob made 35 muffins. How many muffins did Terry and Rob make altogether?

3. Oranges cost 3 for 50¢. Apples cost 4 for 20¢. Peter bought 11 pieces of fruit for 90¢. How many oranges and apples did he buy?

4. There are 28 windows on the first floor of a library. There are 34 windows on the second floor. About how many windows are there in the library?

Lesson 2-7

Add. Check for reasonableness.

1. $35 + 46$

2. $53 + 38$

3. $124 + \$49$

4. $237 + 57$

5. $\$425 + \$272$

6. $436 + 288$

7. $\$719 + \$255$

8. $409 + 354$

9. $73 + 236$

10. $174 + 349$

11. $\$384 + \$567$

12. $439 + 211$

13. $\$563 + \$398$

14. $277 + 562$

15. $\$478 + \$335$

Lesson 2-8

Find each sum. Use estimation to check for reasonableness.

1. $298 + 367$

2. $245 + 107$

3. $\$366 + \$523$

4. $648 + 751$

5. $1{,}988 + 3{,}766$

6. $\$1{,}375 + \$817$

7. $4{,}543 + 2{,}376$

8. $\$2{,}640 + \$3{,}765$

9. $3{,}905 + 4{,}227$

10. $3{,}465 + 5{,}555$

11. $2{,}988 + 2{,}675$

12. $6{,}042 + 2{,}309$

13. $\$1{,}991 + \$2{,}685$

14. $4{,}768 + 2{,}644$

15. $1{,}548 + 5{,}673$

Lesson 3-1

Pages 111–113

Subtract. Use models if needed. Check your answer.

1. 38
− 9

2. 55
− 8

3. 73
− 7

4. 92
− 5

5. 46
− 18

6. 37
− 21

7. 84
− 36

8. 54
− 27

9. 45
− 9

10. 93 − 25

11. 68 − 34

12. 72 − 19

13. 53 − 29

14. 85 − 72

15. 59 − 28

Lesson 3-2

Pages 114–117

Estimate. Round to the nearest ten or use compatible numbers.

1. 73
− 28

2. 86
− 58

3. 243
− 89

4. 92 − 34

5. 137 − 48

6. 81 − 47

Estimate. Round to the nearest hundred.

7. 698
− 322

8. 781
− 273

9. 542
− 386

10. 291 − 77

11. 362 − 125

12. 976 − 529

Lesson 3-3

Pages 118–120

Subtract. Use models if needed. Check your answer.

1. $76
− $24

2. $38
− $19

3. $63
− $47

4. 80¢
− 30¢

5. $33
− $25

6. $87
− $26

7. $58 − $39

8. $92 − $66

9. $67 − $59

10. $51 − $28

11. $78 − $39

12. $75 − $67

Lesson 3-4

Solve.

1. Julie had a pizza party for 7 of her friends. She ordered 4 large pizzas. Each pizza was cut into 10 slices. Is it reasonable to say that each person got at least 6 slices of pizza? Explain.

2. Joey bought a notebook for $2 and paint set for $4. He paid for them with two $5-bills. Is it reasonable to say that Joey has enough money to buy 2 sets of paintbrushes for $3 each? Explain.

3. At soccer camp, there are 220 drinks in coolers. There are 3 kinds of drinks. There are 64 bottles of water and 78 bottles of sports drinks. Is it reasonable to say that there are 115 bottles of juice?

4. Juan walks to and from school every day. His school is 9 blocks from his house. Is it reasonable to say that Juan walks about 150 blocks to and from school every week?

Lesson 3-5

Subtract. Check your answer.

1.   267
   − 154

2.   498
   − 207

3. $634
   − $321

4.   867
   − 89

5. $576
   − $283

6.   755
   − 448

7. $234
   − $97

8.   923
   − 542

9. $744
   − $452

10. $353 − $86

11. 824 − 619

12. 563 − 227

Lesson 3-6

Solve. Tell what strategy you used.

1. There were 123 people in line to ride a roller-coaster. 48 people got on the roller-coaster for the next ride. How many people were left in line?

2. Jesse bought a baseball bat for $16 and a baseball for $5. He paid for them with a $20-bill and a $10-bill. How much change did he get back?

3. Mei collected 240 seashells. Of the shells, 128 were spiral-shaped. The rest were scallop-shaped. About how many scallop-shaped shells did Mei collect?

4. Jim had 583 baseball cards. He gave 212 of his cards to his brother. How many baseball cards does Jim have left?

Lesson 3-7

Pages 134–136

Subtract. Check your answer.

1. 2,453
    − 1,231

2. 5,691
    − 207

3. $8,732
    − $6,215

4. 4,863
    − 3,788

5. 7,239
    − 908

6. 9,000
    − 3,455

7. $4,091
    − $1,637

8. 2,472
    − 848

9. 3,643
    − 1,784

10. $7,208 − $3,495

11. 5,064 − 2,659

12. $7,000 − $4,833

Lesson 3-8

Pages 138–141

Subtract. Check your answer.

1. 306
    − 87

2. 703
    − 405

3. 500
    − 376

4. 205
    − 56

5. 600
    − 248

6. $308
    − $221

7. 407
    − 99

8. $903
    − $775

9. 700
    − 456

10. 902 − 543

11. $507 − $489

12. 807 − 256

Lesson 3-9

Pages 142–143

Select addition or subtraction to solve.

1. There are 47 cats and 29 dogs in the animal shelter. How many cats and dogs are there in all?

2. Gina had $78 in her savings account. She deposited $15 into her account. How much does Gina have in her savings account?

3. Alberto picked 86 apples. His mother used 49 of the apples to make pies. How many apples are left?

4. There are 116 girls in the park playing soccer and softball. 72 of the girls are playing softball. How many girls are playing soccer?

Lesson 4-1

Pages 159–161

Write a multiplication sentence for each array.

1.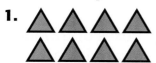

2. (array of circles)

3. (array of hearts)

Use the Commutative Property of Multiplication to find each missing number.

4. $4 \times 5 = 20$
   $\blacksquare \times 4 = 20$

5. $4 \times 9 = 36$
   $9 \times \blacksquare = 36$

6. $5 \times 7 = 35$
   $7 \times 5 = \blacksquare$

Lesson 4-2

Pages 162–164

Multiply.

1.

   4 groups of 2

2.

   7 groups of 2

3.

   8 groups of 2

Multiply. Use an array or draw a picture if needed.

4. $\begin{array}{r} 2 \\ \times\ 6 \\ \hline \end{array}$

5. $\begin{array}{r} 8 \\ \times\ 2 \\ \hline \end{array}$

6. $\begin{array}{r} 9 \\ \times\ 2 \\ \hline \end{array}$

7. $\begin{array}{r} 10 \\ \times\ 2 \\ \hline \end{array}$

8. $\begin{array}{r} 5 \\ \times\ 2 \\ \hline \end{array}$

9. $\begin{array}{r} 2 \\ \times\ 4 \\ \hline \end{array}$

10. $\begin{array}{r} 2 \\ \times\ 2 \\ \hline \end{array}$

11. $\begin{array}{r} 2 \\ \times\ 3 \\ \hline \end{array}$

Lesson 4-3

Pages 168–170

Multiply. Use models or draw a picture if needed.

1. $\begin{array}{r} 4 \\ \times\ 3 \\ \hline \end{array}$

2. $\begin{array}{r} 4 \\ \times\ 6 \\ \hline \end{array}$

3. $\begin{array}{r} 5 \\ \times\ 4 \\ \hline \end{array}$

4. $\begin{array}{r} 10 \\ \times\ 4 \\ \hline \end{array}$

5. $\begin{array}{r} 8 \\ \times\ 4 \\ \hline \end{array}$

6. $\begin{array}{r} 4 \\ \times\ 7 \\ \hline \end{array}$

7. $\begin{array}{r} 9 \\ \times\ 4 \\ \hline \end{array}$

8. $\begin{array}{r} 1 \\ \times\ 4 \\ \hline \end{array}$

9. 6×4

10. 4×0

11. 7×4

12. 4×8

13. 3×4

14. 4×9

Lesson 4-4

Pages 172–173

Solve. If there is missing information, tell what facts you need to solve the problem.

1. A vegetable garden has 4 rows of corn. There are 7 corn plants in each row. There are 5 rows of tomato plants next to the corn. How many corn plants are there?

2. Tony bought 5 boxes of crayons. There were 8 crayons in each box. Each box of crayons cost $2. How many crayons did Tony buy?

3. Zina played soccer for 30 minutes. Then she played basketball. How many minutes did Zina play sports?

4. Mark wants to buy a CD-player that costs $45. He has saved $20. How many hours will he have to work before he has enough money for the CD-player?

Lesson 4-5

Pages 174–176

Multiply. Use counters to model or draw a picture if needed.

1. $\begin{array}{r} 2 \\ \times 5 \\ \hline \end{array}$
2. $\begin{array}{r} 5 \\ \times 5 \\ \hline \end{array}$
3. $\begin{array}{r} 5 \\ \times 6 \\ \hline \end{array}$
4. $\begin{array}{r} 9 \\ \times 5 \\ \hline \end{array}$

5. $\begin{array}{r} 8 \\ \times 5 \\ \hline \end{array}$
6. $\begin{array}{r} 10 \\ \times 5 \\ \hline \end{array}$
7. $\begin{array}{r} 5 \\ \times 1 \\ \hline \end{array}$
8. $\begin{array}{r} 5 \\ \times 7 \\ \hline \end{array}$

9. 5×8
10. 10×5
11. 5×0

12. 5×9
13. 6×5
14. 7×5

Lesson 4-6

Pages 178–181

Multiply. Use patterns or models if needed.

1. $\begin{array}{r} 10 \\ \times 2 \\ \hline \end{array}$
2. $\begin{array}{r} 10 \\ \times 5 \\ \hline \end{array}$
3. $\begin{array}{r} 10 \\ \times 1 \\ \hline \end{array}$
4. $\begin{array}{r} 10 \\ \times 7 \\ \hline \end{array}$

5. $\begin{array}{r} 10 \\ \times 6 \\ \hline \end{array}$
6. $\begin{array}{r} 10 \\ \times 9 \\ \hline \end{array}$
7. $\begin{array}{r} 10 \\ \times 4 \\ \hline \end{array}$
8. $\begin{array}{r} 10 \\ \times 8 \\ \hline \end{array}$

9. 5×10
10. 10×7
11. 10×4

12. 3×10
13. 10×9
14. 8×10

Lesson 4-7

Pages 184–185

Use any strategy to solve. Tell what strategy you used.

1. Tennis balls are on sale for $3 a can or 4 cans for $10. If you buy 4 cans of tennis balls, how much will you save on each can?

2. Sara collected 3 colors of fall leaves to make a collage. She collected 9 orange leaves and 7 yellow leaves. How many red leaves did she collect?

3. There are about 18 soccer players on each team. There are 6 soccer teams in the league. About how many soccer players are in the league?

4. There are 7 spiders in a display at the zoo. Each spider has 8 legs. Each spider also has 2 body segments. How many spider legs are there altogether?

Lesson 4-8

Pages 186–188

Multiply.

1. $\begin{array}{r} 9 \\ \times\ 0 \\ \hline \end{array}$

2. $\begin{array}{r} 10 \\ \times\ 1 \\ \hline \end{array}$

3. $\begin{array}{r} 8 \\ \times\ 1 \\ \hline \end{array}$

4. $\begin{array}{r} 0 \\ \times\ 6 \\ \hline \end{array}$

5. $\begin{array}{r} 0 \\ \times\ 8 \\ \hline \end{array}$

6. $\begin{array}{r} 1 \\ \times\ 5 \\ \hline \end{array}$

7. $\begin{array}{r} 10 \\ \times\ 0 \\ \hline \end{array}$

8. $\begin{array}{r} 7 \\ \times\ 1 \\ \hline \end{array}$

9. 5×0

10. 0×4

11. 1×9

12. 0×7

13. 1×6

14. 3×0

Lesson 5-1

Pages 203–205

Multiply. Use models or draw a picture if needed.

1. $\begin{array}{r} 7 \\ \times\ 3 \\ \hline \end{array}$

2. $\begin{array}{r} 3 \\ \times\ 8 \\ \hline \end{array}$

3. $\begin{array}{r} 6 \\ \times\ 3 \\ \hline \end{array}$

4. $\begin{array}{r} 5 \\ \times\ 3 \\ \hline \end{array}$

5. $\begin{array}{r} 10 \\ \times\ 3 \\ \hline \end{array}$

6. $\begin{array}{r} 1 \\ \times\ 3 \\ \hline \end{array}$

7. $\begin{array}{r} 0 \\ \times\ 3 \\ \hline \end{array}$

8. $\begin{array}{r} 9 \\ \times\ 3 \\ \hline \end{array}$

9. 8×3

10. 3×2

11. 3×7

12. 4×3

13. 3×9

14. 3×0

Lesson 5-2

Pages 206–209

Multiply. Use models or draw a picture if needed.

1. $\begin{array}{r} 6 \\ \times\, 2 \\ \hline \end{array}$

2. $\begin{array}{r} 3 \\ \times\, 6 \\ \hline \end{array}$

3. $\begin{array}{r} 5 \\ \times\, 6 \\ \hline \end{array}$

4. $\begin{array}{r} 6 \\ \times\, 9 \\ \hline \end{array}$

5. $\begin{array}{r} 1 \\ \times\, 6 \\ \hline \end{array}$

6. $\begin{array}{r} 7 \\ \times\, 6 \\ \hline \end{array}$

7. $\begin{array}{r} 10 \\ \times\, 6 \\ \hline \end{array}$

8. $\begin{array}{r} 6 \\ \times\, 8 \\ \hline \end{array}$

9. 4×6

10. 0×6

11. 6×7

12. 6×8

13. 5×6

14. 6×6

Lesson 5-3

Pages 212–213

Solve. Use the *look for a pattern* strategy.

1. An oak tree was 3 feet tall at the end of the first year it was planted. By the end of the third year, it was 9 feet tall. If the tree grows 3 feet each year, how tall will the tree be at the end of the 5th year?

2. Carlos is making a castle. For each tower, he uses 6 triangle blocks and 8 square blocks. If Carlos makes 4 towers on his castle, how many more square blocks than triangle blocks will he use?

3. Suki ran one lap on a track in 84 seconds. The next week, she ran the same lap in 81 seconds. If she continues to decrease her time by 3 seconds how many seconds will it take her to run the lap on the 4th week?

4. The dancers in a play are arranged by height. The first student is 50 inches tall, the second is 47 inches tall, the third is 52 inches tall, and the fourth student is 49 inches tall. If the pattern continues, how tall is the fifth student?

Lesson 5-4

Pages 214–216

Multiply. Use models or draw a picture if needed.

1. $\begin{array}{r} 4 \\ \times\, 7 \\ \hline \end{array}$

2. $\begin{array}{r} 6 \\ \times\, 7 \\ \hline \end{array}$

3. $\begin{array}{r} 7 \\ \times\, 7 \\ \hline \end{array}$

4. $\begin{array}{r} 1 \\ \times\, 7 \\ \hline \end{array}$

5. $\begin{array}{r} 7 \\ \times\, 9 \\ \hline \end{array}$

6. $\begin{array}{r} 10 \\ \times\, 7 \\ \hline \end{array}$

7. $\begin{array}{r} 5 \\ \times\, 7 \\ \hline \end{array}$

8. $\begin{array}{r} 7 \\ \times\, 8 \\ \hline \end{array}$

9. 2×7

10. 7×6

11. 0×7

12. 7×10

13. 9×7

14. 7×3

Lesson 5-5

Pages 218–221

Multiply. Use models or a known fact if needed.

1. 8
    × 3

2. 8
    × 1

3. 7
    × 8

4. 8
    × 5

5. 8
    × 0

6. 10
    × 8

7. 8
    × 6

8. 9
    × 8

9. 2×8

10. 8×4

11. 8×7

12. 8×9

13. 5×8

14. 8×10

Lesson 5-6

Pages 222–224

Multiply. Use models or patterns if needed.

1. 9
    × 4

2. 3
    × 9

3. 9
    × 0

4. 5
    × 9

5. 7
    × 9

6. 9
    × 9

7. 9
    × 8

8. 10
    × 9

9. 6×9

10. 9×2

11. 0×9

12. 7×9

13. 1×9

14. 9×5

Lesson 5-7

Pages 228–229

Use any strategy to solve. Tell what strategy you used.

1. Rita made 6 paintings. Luis made 3 times as many paintings as Rita and 4 more paintings than Tara. How many paintings did Luis and Tara make?

2. In 1 week there are 7 days. In 2 weeks there are 14 days. There are 21 days in 3 weeks. How many days are in 4 and 5 weeks?

3. Freddie has 6 cats and 3 dogs. Each cat eats 4 ounces of food each day. How many ounces of food do Freddie's cats and dogs eat altogether?

4. Ivan unpacked16 bunches of grapes at the grocery store. He unpacked 13 bunches of bananas. How many bunches of fruit did Ivan unpack?

Lesson 5-8

Pages 230–233

Multiply. Use models or patterns if needed.

1. 12×7 **2.** 12×12 **3.** 5×11

4. 8×11 **5.** 12×8 **6.** 12×1

7. 11×4 **8.** 0×11 **9.** 11×6

Lesson 5-9

Pages 234–237

Find each product.

1. $2 \times 4 \times 3$ **2.** $4 \times 5 \times 1$ **3.** $4 \times 2 \times 6$

4. $3 \times 3 \times 3$ **5.** $2 \times 4 \times 7$ **6.** $5 \times 9 \times 1$

Algebra Find each missing factor.

7. $2 \times \blacksquare \times 3 = 18$ **8.** $5 \times \blacksquare \times 7 = 70$ **9.** $8 \times 2 \times \blacksquare = 48$

10. $5 \times \blacksquare \times 2 = 40$ **11.** $6 \times 1 \times \blacksquare = 54$ **12.** $5 \times 2 \times \blacksquare = 20$

Lesson 6-1

Pages 253–255

Use models to divide. Write a number sentence.

1. $12 \div 3$ **2.** $9 \div 3$ **3.** $16 \div 2$

4. $15 \div 5$ **5.** $24 \div 4$ **6.** $10 \div 2$

Use repeated subtraction to divide.

7. $21 \div 3$ **8.** $16 \div 4$ **9.** $8 \div 1$

10. $14 \div 2$ **11.** $20 \div 5$ **12.** $27 \div 9$

13. $18 \div 9$ **14.** $24 \div 3$ **15.** $12 \div 6$

Lesson 6-2

Pages 258–261

Draw an array to complete each pair of number sentences.

1. $2 \times 3 = \blacksquare$
$6 \div \blacksquare = 2$

2. $1 \times \blacksquare = 7$
$7 \div 7 = \blacksquare$

3. $4 \times \blacksquare = 16$
$\blacksquare \div 4 = 4$

4. $\blacksquare \times 4 = 12$
$12 \div \blacksquare = 3$

5. $5 \times 4 = \blacksquare$
$20 \div 4 = \blacksquare$

6. $8 \times \blacksquare = 24$
$24 \div \blacksquare = 3$

Write the fact family for each set of numbers.

7. 2, 4, 8

8. 3, 7, 21

9. 1, 5, 5

10. 2, 9, 18

11. 4, 3, 12

12. 4, 5, 20

Lesson 6-3

Pages 262–263

Choose an operation to solve.

1. Sung made 3 sets of frozen lemon treats and 2 sets of frozen cherry treats. There made 25 treats altogether. How many treats were in each set?

2. Jaime ordered 4 pizzas for himself and 5 of his friends. Each pizza had 6 slices. How many slices of pizza did each boy have?

3. Each mother duck has 4 ducklings. There are 4 mother ducks. How many ducklings are there?

4. Mrs. Diaz has five rows of peppers in her garden. There are 30 pepper plants in all. How many pepper plants are in each row?

Lesson 6-4

Pages 264–266

Divide. Write a related multiplication fact.

1. $18 \div 2$

2. $4 \div 2$

3. $8 \div 2$

4. $24 \div 2$

5. $14 \div 2$

6. $10 \div 2$

7. $22 \div 2$

8. $16 \div 2$

9. $20 \div 2$

10. $6 \div 2$

11. $2 \div 2$

12. $12 \div 2$

Lesson 6-5

Pages 270–273

Divide. Use models or related facts.

1. 35 ÷ 5   **2.** 20 ÷ 5   **3.** 55 ÷ 5

4. 40 ÷ 5   **5.** 25 ÷ 5   **6.** 45 ÷ 5

Algebra **Copy and complete each table.**

7.

Rule: − 5	
Input	**Output**
45	■
35	■
20	■
■	10

8.

Rule: ÷ 5	
Input	**Output**
50	■
30	■
■	4
■	1

9.

Rule: × 5	
Input	**Output**
9	■
8	■
6	■
■	15

Lesson 6-6

Pages 276–277

Use any strategy to solve. Tell what strategy you used.

1. Mike jogged for 15 minutes on Monday. He jogged twice as long on Tuesday. How long did Mike jog on Tuesday?

2. Jose had 18 baseball cards. He gave 4 to his sister. He bought 7 more. How many baseball cards does Jose have now?

3. Wanda is putting 4 colored candles in each of the cupcakes she made. She puts the candles in the order blue, green, red, green. If Wanda make 12 cupcakes, how many green candles will she use?

4. The Jackson family has $35 to spend for dinner. There are 2 adults and 3 children in the family. If they divide the money evenly, how much will each person spend on dinner?

Lesson 6-7

Pages 278–280

Divide.

1. 40 ÷ 10   **2.** 110 ÷ 10   **3.** 70 ÷ 10

4. 120 ÷ 10   **5.** 100 ÷ 10   **6.** 90 ÷ 10

Solve. Write a number sentence.

7. ■ ÷ 10 = 6   **8.** 50 ÷ 10 = ■   **9.** ■ ÷ 10 = 9

10. 80 ÷ 10 = ■   **11.** ■ ÷ 4 = 10   **12.** ■ ÷ 10 = 3

Lesson 6-8

Pages 282–283

Divide.

1. $9 \div 1$ **2.** $0 \div 6$ **3.** $10 \div 10$ **4.** $8 \div 1$

5. $5 \div 5$ **6.** $4 \div 1$ **7.** $0 \div 8$ **8.** $2 \div 1$

9. $10\overline{)0}$ **10.** $8\overline{)0}$ **11.** $9\overline{)9}$ **12.** $1\overline{)7}$

13. $2\overline{)2}$ **14.** $1\overline{)1}$ **15.** $4\overline{)0}$ **16.** $1\overline{)5}$

Lesson 7-1

Pages 297–299

Divide. Use models or related facts.

1. $9 \div 3$ **2.** $12 \div 3$ **3.** $0 \div 3$

4. $15 \div 3$ **5.** $3 \div 3$ **6.** $21 \div 3$

7. $3\overline{)24}$ **8.** $3\overline{)6}$ **9.** $3\overline{)30}$

Compare. Use $>$, $<$, or $=$.

10. $15 \div 3 \,\bullet\, 7$ **11.** $12 \div 2 \,\bullet\, 6$ **12.** $21 \div 3 \,\bullet\, 5$

Lesson 7-2

Pages 300–303

Divide. Use models or related facts.

1. $16 \div 4$ **2.** $8 \div 4$ **3.** $20 \div 4$

4. $24 \div 4$ **5.** $0 \div 4$ **6.** $32 \div 4$

7. $4\overline{)12}$ **8.** $4\overline{)28}$ **9.** $4\overline{)40}$

Algebra Find each missing number.

10. $20 \div \blacksquare = 4$ **11.** $\blacksquare \div 4 = 10$ **12.** $4 \times \blacksquare = 28$

Extra Practice

Solve. Use the *make a table* strategy.

1. Diego is picking lemons and oranges. For every 2 lemons he picks, he picks 4 oranges. When Diego has picked 6 lemons, how many oranges will he have picked?

2. Scott earns $4 per hour doing yard work. He works 2 hours a day. How many days will it take him to earn $32?

3. Jared is putting cans of food on a shelf. He can fit 6 large cans on each shelf. How many cans could he fit on 4 shelves?

4. Colored markers are on sale at a store for $0.20 each or a set of 6 for $1. How much would you save if you bought 5 sets of colored markers instead of the number of individual markers?

Divide. Use models or repeated subtraction.

1. $12 \div 6$
2. $28 \div 7$
3. $24 \div 6$
4. $35 \div 7$
5. $0 \div 7$
6. $21 \div 7$
7. $42 \div 6$
8. $30 \div 6$
9. $36 \div 6$
10. $6\overline{)6}$
11. $7\overline{)14}$
12. $6\overline{)18}$
13. $7\overline{)63}$
14. $6\overline{)48}$
15. $7\overline{)56}$

Divide. Use related facts or repeated subtraction.

1. $32 \div 8$
2. $27 \div 9$
3. $45 \div 9$
4. $40 \div 8$
5. $24 \div 8$
6. $36 \div 9$
7. $0 \div 9$
8. $16 \div 8$
9. $63 \div 9$
10. $9\overline{)27}$
11. $9\overline{)81}$
12. $8\overline{)80}$
13. $9\overline{)72}$
14. $8\overline{)64}$
15. $9\overline{)54}$

Lesson 7-6

Pages 316–319

Divide.

1. 77 ÷ 11

2. 24 ÷ 12

3. 55 ÷ 11

4. 36 ÷ 12

5. 33 ÷ 11

6. 99 ÷ 11

7. 60 ÷ 12

8. 144 ÷ 12

9. 11 ÷ 11

Find each missing number.

10. 120 ÷ ■ = 10

11. ■ ÷ 12 = 12

12. ■ ÷ 12 = 11

Lesson 7-7

Pages 320–321

Use any strategy to solve. Tell what strategy you used.

1. A swim team has 9 members. 5 members of the team each swam 3 laps. The other members each swam twice as many laps as the first 5. How many laps did the team members swim altogether?

2. A camp leader packed 8 sandwiches and 6 apples in each picnic basket. She packed 36 apples altogether. How many sandwiches did she pack? How many picnic baskets did she use?

3. At a flower shop, roses cost 6 for $5, and tulips cost 9 for $4. How much more would it cost to buy 24 roses than 36 tulips?

4. Rick is making a square pen for his turtle. Each side of the pen will be 3 feet long. Rick will use 4 posts for each foot of the pen. How many posts will he need?

Lesson 8-1

Pages 333–335

Model each problem. Use a number sentence.

1. Last year, Pedro was 48 inches tall. Now he is 53 inches tall. How many inches did Pedro grow since last year?

2. How much money does Anwon need to buy two packs of baseball cards for $1.95 each and a toy car for $4.75?

Model each number sentence. Use pictures and words.

3. 21 + 40 = ■

4. 27 − ■ = 9

5. 19 + 16 + 5 = ■

6. 17 + ■ = 64

7. 35 − 7 = ■

8. 37 + 24 = ■

Lesson 8-2

Pages 338–341

Write an expression and a number sentence for each problem.

1. A pet store has 32 hamsters and 17 turtles. How many more hamsters than turtles are there?

2. Thomas had $97 in his piggy bank. His uncle put $15 more into the piggy bank. How much does Thomas have in his piggy bank?

3. Jocelyn's mother has 75 apples. She uses 23 of the apples to make pies. How many apples are left?

4. There are 116 dogs in a park. Of these dogs, 28 are Labrador Retrievers. How many are not Labrador Retrievers?

Lesson 8-3

Pages 342–343

Solve. Use the *act it out* strategy.

1. A book is the middle book on a bookshelf. There are 7 books to the right of this book. How many books are on the shelf in all?

2. Janet made 2 out of every 5 basketball shots she took. She made 12 shots. How many shots did she take?

3. Freddie ate 3 pieces of pizza. Tim ate 2 more than Freddie. Half of the pizza is left. How many pieces were there to begin with?

4. Three friends have 8 video games each. If each friend sells 2, how many video games are left altogether?

Lesson 8-4

Pages 344–347

Find and extend the rule for each table. Then copy and complete.

1.

Rule: ☐	
Input	**Output**
4	20
6	▦
8	40
9	▦

2.

Rule: ☐	
Input	**Output**
3	21
▦	28
5	▦
6	42

3.

Rule: ☐	
Input	**Output**
4	▦
▦	20
6	24
7	28

Lesson 8-5

Pages 348–351

Copy each function table and extend the pattern.

1.

Rule: x + 9	
Input (x)	Output (y)
3	▨
4	▨
5	▨
6	▨

2.

Rule: x − 6	
Input (x)	Output (y)
20	▨
18	▨
16	▨
14	▨

3.

Rule: x + 11	
Input (x)	Output (y)
9	▨
10	▨
11	▨
12	▨

Make a function table for each situation. Write the function rule.

4. Hannah is 10 years old. Her brother is 3 years younger than her. How old will her brother be when Hannah is 11, 12, 13, and 14 years old?

5. Each book Quincy writes has 5 extra pages for drawings. How many pages will a book have that has 8, 9, 10, and 11 pages of writing?

Lesson 8-6

Pages 354–355

Use any strategy to solve. Tell what strategy you used.

1. Rafi has 8 nickels, 3 dimes, and 2 quarters. Does he have enough to buy a beach ball that costs $1.25? Explain.

2. Sonya made a batch of 22 brownies. She will give an equal number to each of 5 friends. How many brownies will each friend get? How many will be left over?

3. A waiting room has 6 rows of seats. There are 10 seats in each row. There are people sitting in half of the seats. How many seats are empty?

4. Sam is shorter than Rory. Ben is taller than Lee. Rory is shorter than Lee. Name the boys from tallest to shortest.

Lesson 8-7

Pages 356–359

Copy each function table and extend the pattern.

1.

Rule: x × 6	
Input (x)	Output (y)
7	▨
8	▨
9	▨
10	▨

2.

Rule: x ÷ 8	
Input (x)	Output (y)
40	▨
32	▨
24	▨
16	▨

3.

Rule: x × 3	
Input (x)	Output (y)
9	▨
10	▨
11	▨
12	▨

Estimate each length. Then measure each to the nearest half inch.

1.

2.

Choose the most appropriate unit to measure each length. Write *inch, foot, yard,* or *mile*.

1. the length of a bicycle

2. The distance from Texas to Georgia

3. the width of a book

4. The length of a truck

Choose the better estimate.

5. width of a TV screen
   19 yards or 19 inches

6. length of a slide
   4 yards or 4 miles

7. length of a fork
   8 feet or 8 inches

8. distance to the grocery store
   3 inches or 3 miles

Solve. Use the *work backward* strategy.

1. Jenna began a star map on Monday night. On Tuesday, she mapped 12 stars. On Wednesday night, she mapped 14 more stars. By then, she had mapped 33 stars in all. How many stars did she map on Monday?

2. Gustavo walked his dog for an hour. Next, he played at the park for 2 hours. Then, he did chores for 2 hours. He finished at 3:00 in the afternoon. At what time did he start walking his dog?

3. Petra ate some strawberries for breakfast. She ate 3 times as many strawberries for lunch. She ate 20 strawberries in all that day. How many strawberries did she eat for breakfast?

4. Sam added 8 shells to his collection. His sister gave him 9 shells. Sam found 5 more shells. Now he has 43 shells. How many shells did he have at first?

Lesson 9-4

Pages 386–389

Choose the most appropriate unit to measure each length. Write *millimeter*, *centimeter*, *meter*, or *kilometer*.

1. length of a crayon

2. length of a rake

3. length of a soccer field

4. Length of an ocean shoreline

Choose the better estimate.

5. length of an alligator
2 m or 2 cm

6. length of a pencil
13 km or 13 cm

7. width of California
480 cm or 480 km

8. length of a jet
62 km or 62 m

Lesson 9-5

Pages 392–395

Find the perimeter of each figure.

1.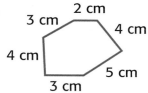

2.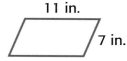

3.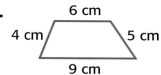

4.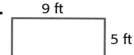

5.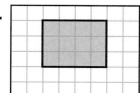

6.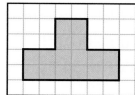

Lesson 9-6

Pages 398–401

Find the area of each figure.

1.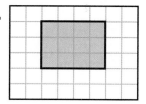

2.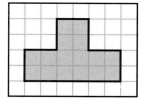

3.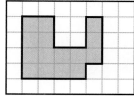

4.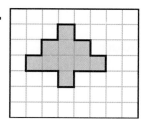

5.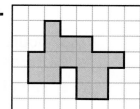

6.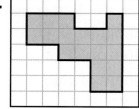

Extra Practice

Lesson 9-7

Pages 402–403

Use any strategy to solve. Tell what strategy you used.

1. A mall has 3 levels. Each level has 15 stores. Each store has 4 windows. How many windows are in the mall?

2. Carmen and Rob played 60 games of checkers. Rob won twice as many games as Carmen did. How many games did each person win?

3. For every $5 that Henry saves, his dad gives him $3. When Henry has saved $100, how much will he have in all?

4. Stella bought 12 stickers and 4 posters. The stickers cost 3 for $0.25. The posters cost 2 for $6. How much did Stella spend?

Lesson 9-8

Pages 408–411

Write each temperature in degrees Farenheit (°F).

1.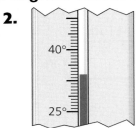

2.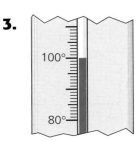

3.

Tell which temperature is greater.

4. 100°F or 10°F

5. 32°F or 68°F

6. 45°F or 54°F

7. 65°F or 56°F

Lesson 10-1

Pages 425–428

Choose the most appropriate unit to measure each capacity. Write *cup, pint, quart,* or *gallon*.

1. dog dish

2. milk bottle

3. cocoa

Choose the better estimate.

4. bathtub
50 c or 50 gal

5. bottle of shampoo
2 c or 2 qt

6. bottle of glue
1 gal or 1 c

7. can of wall paint
1 gallon or 1 cup

Lesson 10-2

Pages 430–431

Solve. Use the *guess and check* strategy.

1. Sara made 8 putts at the miniature golf course. Each putt was about 50 cm long. About how many meters was the total length of all the putts?

2. Each car on an amusement park ride holds 15 people. There are 20 cars on the ride. How many people can the ride hold?

3. David practiced playing his violin for $4\frac{1}{2}$ hours during 1 week. He practiced the same number of minutes from Monday through Friday. He practiced twice as long on Saturday and Sunday. How long did he practice each day?

4. There are 12 dog pens in an animal shelter. Each pen can hold about 6 large dogs or 10 small dogs. Is there enough room for 90 small dogs and 20 large dogs?

Lesson 10-3

Pages 432–434

Choose the most appropriate unit to measure each capacity. Write *milliliter* or *liter*.

1. a washing machine

2. a test tube

3. a drinking glass

4. a shark aquarium

5. a watering can

6. a juice box

Choose the better estimate.

7. bathroom sink
8 mL or 8 L

8. tablespoon
15 mL or 15 L

9. large bottle of soda
2 mL or 2 L

Lesson 10-4

Pages 436–437

Use any strategy to solve. Tell what strategy you used.

1. Kiri picks 500 g of blueberries every $\frac{1}{2}$ hour. How many hours will it take her to pick 3,000 g of blueberries?

2. Ana is 120 cm tall. Her father is 200 cm tall. How much taller is her father than Ana?

3. Deanna walked 7 blocks to get to school. She took a different route on the way home that was 3 times as many blocks. How many blocks did Deanna walk that day?

4. Potatoes cost $1.25 for 3 kg at the grocery store. How much will 12 kg of potatoes cost?

Lesson 10-5

Pages 438–441

Choose the most appropriate unit to measure the weight of each object. Write *ounce, pound,* or *ton*.

1. a tennis ball

2. a tiger

3. a cell phone

4. a pencil

5. a television

6. whale

Choose the better estimate.

7. baseball bat
3 oz or 3 lb

8. frisbee
12 oz or 12 lb

9. elephants
8 tons or 8 lb

Lesson 10-6

Pages 444–447

Choose the most appropriate unit to measure each mass. Write *gram* or *kilogram*.

1. a bowling ball

2. a grasshopper

3. a wheelbarrow

4. a golf ball

5. a bag of pretzels

6. A piece of chalk

Choose the better estimate.

7. zebra
300 g or 300 kg

8. penny
3 g or 3 kg

9. pumpkin
10 g or 10 kg

10. baseball
145 g or 145 kg

11. CD
15 g or 15 kg

12. microwave
20 g or 20 kg

Lesson 10-7

Pages 450–453

Use concrete models to find the volume of each figure.

1.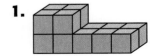

2.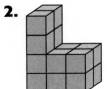

3.

4.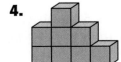

5.

6.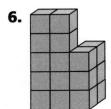

Lesson 10-8

Pages 454–455

Write the time shown on each analog or digital clock.

1.

2.

3.

4.

5.

6.

Lesson 11-1

Pages 467–471

Identify each three-dimensional figure.

1.

2.

3.

Classify each three-dimensional figure.

4. 1 edge, 1 face, and 1 vertex

5. 6 square faces

Lesson 11-2

Pages 472–475

Describe each two-dimensional figure. Use the terms *sides* and *angles*. Then identify the figure.

1.

2.

3.

4.

5.

6.

Classify each two-dimensional figure.

7. A closed figure has straight sides.

8. A polygon has 5 sides and 5 angles.

Extra Practice

Lesson 11-3

Pages 476–477

Solve. Use the *solve a simpler problem* strategy.

1. A chess board is a square with eight squares on each side. Each player begins with a playing piece on each square in the two rows closest to him or her. How many pieces are in a chess set?

2. There are 16 peaches, 8 plums, and 24 carrot sticks in the refrigerator. Toni took out 2 peaches, 3 plums, and 10 carrot sticks. Ali took out 2 plums, 8 carrot sticks, and one peach. How many of each is left?

3. Ben has 50 toy airplanes. Half of them are silver. There are 4 times as many blue planes as white planes. How many blue and white planes does Ben have?

4. Diego is using bricks to make a border around a flowerbed. His border is a rectangle with 6 bricks each on two sides and 8 bricks each on the other sides. How many bricks will he use in all if he stacks two more bricks on top of each brick?

Lesson 11-4

Pages 478–481

Identify and extend each pattern.

1.

2.

Apply each pattern.

3. Glenda makes a tile pattern. The first row has 2 blue tiles. The second row has 4 blue tiles and the third row has 8 blue tiles. How many tiles are in each of the next two rows?

4. If the pattern below continues until there are 20 polygons, how many squares will there be?

Lesson 11-5

Pages 484–485

Tell whether each pair of figures is congruent. Write *yes* or *no*.

1.

2.

3.

4. Each side of a hexagon measures 6 inches. What will each side of a congruent hexagon measure?

5. Explain why the figures below are not congruent.

Lesson 11-6

Pages 486–487

Use any strategy to solve. Tell what strategy you used.

1. Louis stopped working at 5:00. He pulled weeds for 2 hours, and he took a $\frac{1}{2}$-hour break. He raked leaves for $1\frac{1}{2}$ hours. What time did he start working?

2. Mei has 7 green marbles, 8 blue marbles, and 3 times as many red marbles as green ones. How many marbles does she have altogether?

Lesson 11-7

Pages 488–490

Tell whether each figure has line symmetry. Write *yes* or *no*. If yes, tell how many lines of symmetry the figure has.

1.

2.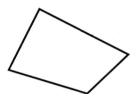

3.

4.

5.

6.

Lesson 11-8

Pages 492–493

Tell what point on the number line represents each number.

1. 496

2. 77

Tell what number on the number line each letter represents.

3. Point $A =$

4. Point $B =$

Lesson 11-9

Pages 494–497

Write the ordered pair for the location of each item on the grid.

1. plane

2. car

3. bike

4. train

Name the vehicle at each location.

5. (3, 4)

6. (9, 8)

7. (6, 8)

8. (7, 2)

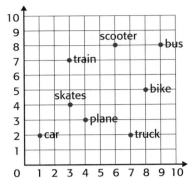

Lesson 12-1

Pages 515–517

Display each set of data in a pictograph.

1.

Instruments Played by Third Graders	
Instrument	Students
Piano	10
Flute	3
Guitar	6
Drums	7

2.

Favorite Vegetable	
Vegetables	Students
Green beans	12
Carrots	10
Corn	16
Potatoes	14
Broccoli	10

Lesson 12-2

Pages 518–519

For Exercises 1–4, refer to the pictograph.

Favorite Pets	
Snake	🐾
Dog	🐾 🐾 🐾
Cat	🐾 🐾
Bird	🐾 🐾
Fish	🐾
Key: 🐾 = 4	

1. How many students does the pictograph represent?

2. How many students liked the least favorite pet?

3. What are the two favorite pets?

4. What is the difference in the number of students who like the cat the best and those who like the bird best?

Lesson 12-3

Pages 522–523

Solve. Use the *make a list* strategy.

1. Marta's pocket has coins in it that add up to 15¢. How many combinations of coins could there be in her pocket?

2. Nita has a white scarf, a blue scarf, and a pink scarf. She also has purple mittens, pink mittens, and green mittens. How many combinations of scarves and mittens can Nita make?

Lesson 12-4

Pages 528–531

1. Display the set of data in a vertical bar graph.

Average Length of Adult Male Animals	
Animal	**Length**
Alligator	11 feet
Hippopotamus	15 feet
Otter	3 feet
Tiger	9 feet

Source: Smithsonian National Zoo

2. Display the set of data in a horizontal bar graph.

Most Popular Pets	
Animal	**Length**
Bird	卌 ‖
Cat	卌 卌 卌	
Dog	卌 卌 卌 卌 卌
Snake	‖

3. If you lined up the animals end to end how far would they stretch?

4. Can you tell what the most popular pet is without counting tallies? Explain.

Lesson 12-5

Pages 532–535

For Exercises 1–3, refer to the bar graph.

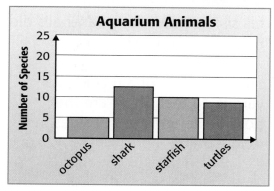

1. There are twice as many starfish species as what sea animal at the aquarium?

2. How would you find the total number of species represented by the graph?

3. Which two sea animals have close to the same number of species at the aquarium?

Extra Practice

Display each set of data in a line plot.

1.

Books Read in One Week	
Number of Books	Students
0	0
1	4
2	9
3	5
4	4
5	2

2.

Hours Spent on the Computer on the Weekend	
Hours	Students
0	6
1	5
2	8
3	2
4	4
5	0

Describe the probability of landing on each number or color. Write *certain, likely, unlikely,* or *impossible.*

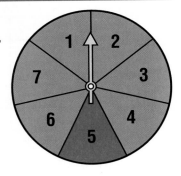

1. a number that is less than 5
2. a number that is between 0 and 8
3. an odd number
4. a number that is less than 2
5. orange
6. red
7. blue
8. not yellow

Use any strategy to solve. Tell what strategy you used.

1. Four teams in a tournament have to play every other team once. The top two teams then play each other. How many games will the top two teams play?

2. Hugo has 8 baseballs. He gave 3 baseballs to each of his 5 brothers. He gave 4 baseballs to his sister. How many baseballs did Hugo have at first?

3. Shani is making a bracelet. She will use 3 different colored beads. She can choose from a blue, green, yellow, purple, and pink bead. How many different combinations could she choose?

4. The perimeter of a rectangular yard is 24 feet. What are the possible lengths of the sides in whole units?

Lesson 13-1

Pages 561–563

Write the fraction for the part that is blue. Then write the fraction for the part that is *not* blue. Label your answers.

1.

2.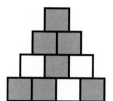

3.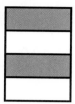

4.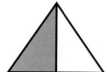

5.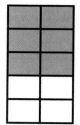

6.

Lesson 13-2

Pages 564–567

Write the fraction for the part of the set that is yellow. Then write the fraction of the set that is *not* yellow.

1.

2.

Lesson 13-3

Pages 568–569

Use any strategy to solve. Tell what strategy you used.

1. Josh rolled two number cubes at once. The sum of the numbers was 10. The difference was 2. What two numbers did Josh roll?

2. Lille bought a salad for $1.75 and a bowl of soup for $1.50. She has $2.25 left. How much did she have before she bought lunch?

3. For every blue headband Kitty has, she has 3 white ones and 2 red ones. Kitty has 3 blue headbands. How many white and red headbands does she have?

4. Alex delivered 26 papers on one block. He delivered half as many on another block. How many papers did Alex deliver?

Lesson 13-4

Pages 572–574

Complete each number sentence to find equivalent fractions.

1.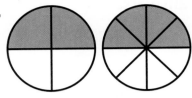

2.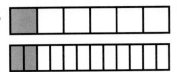

3.

$$\frac{2}{4} = \frac{\blacksquare}{8}$$

$$\frac{1}{6} = \frac{\blacksquare}{12}$$

$$\frac{\blacksquare}{3} = \frac{\blacksquare}{6}$$

Algebra Find each missing value. Use models if needed.

4. $\frac{2}{3} = \frac{\blacksquare}{12}$

5. $\frac{3}{5} = \frac{\blacksquare}{10}$

6. $\frac{2}{6} = \frac{\blacksquare}{3}$

Lesson 13-5

Pages 578–579

Solve. Use the *draw a picture* strategy.

1. John and Ben played 8 games of checkers. John won 2 more games than Ben did. What fraction of the games did each boy win?

2. Alisha sliced a pizza into 10 pieces. She ate $\frac{3}{5}$ of the pizza. How many slices are left?

3. Carli practiced dancing for $\frac{3}{6}$ of an hour. Lauren practiced for $\frac{3}{4}$ of an hour. Who practiced longer?

4. There were 24 cherries in a bowl. Tara took $\frac{1}{4}$ of them. Yuri took $\frac{1}{8}$ of them. How many cherries were left in the bowl?

Lesson 13-6

Pages 580–581

Compare. Use >, <, or =.

1.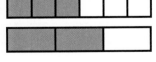

$$\frac{3}{6} \bullet \frac{2}{3}$$

2.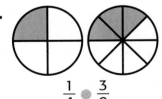

$$\frac{1}{4} \bullet \frac{3}{8}$$

3.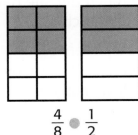

$$\frac{4}{8} \bullet \frac{1}{2}$$

4.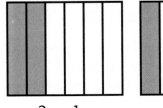

$$\frac{2}{6} \bullet \frac{1}{5}$$

Lesson 13-7

Pages 582–585

Tell what point each fraction represents.

1. $\frac{3}{4}$

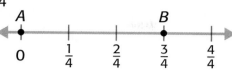

2. $\frac{2}{8}$

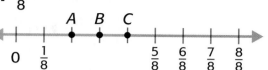

Lesson 14-1

Pages 601–604

Write a fraction and a decimal for the part that is shaded.

1.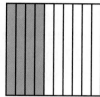

2.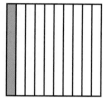

3.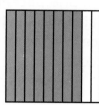

4.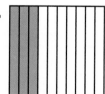

5.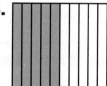

6.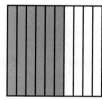

Lesson 14-2

Pages 606–608

Write each fraction as a decimal.

1. $\frac{18}{100}$ **2.** $\frac{36}{100}$ **3.** $\frac{88}{100}$

4. $\frac{20}{100}$ **5.** $\frac{48}{100}$ **6.** $\frac{57}{100}$

Write each decimal as a fraction.

7. 0.45   **8.** 0.10   **9.** 0.75

10. 0.14   **11.** 0.54   **12.** 0.23

13. 0.37   **14.** 0.98   **15.** 0.42

Lesson 14-3

Pages 612–613

Solve. Use the *work backward* strategy.

1. Tara played at the park for $1\frac{1}{2}$ hours. She ate lunch for $\frac{1}{2}$ hour. Then she spent $\frac{1}{2}$ hour walking home. She got home at 6:30. What time did she get to the park?

2. Jamal received 9 CDs for his birthday. Then, he bought 7 more CDs. Now, he has 37 CDs. How many CDs did he have before his birthday?

3. Ms. McCoy packed 6 coolers with equal numbers of juice and water bottles. She packed 36 bottles of juice altogether. How many bottles of juice and water did she pack in each cooler?

4. Tom had some money in his piggy bank. Last week he put twice as much money in his bank as he already had. This week he put 3 times as much in as he did last week. Now he has $27 in his bank. How much did he start with?

Lesson 14-4

Pages 616–619

Write the part of a dollar each amount represents.

1. 18¢

2. 40¢

3. 72¢

4. 85¢

5. 64¢

6. 55¢

7. 75¢

8. 14¢

9. 48¢

10. 57¢

11. 6¢

12. 30¢

Lesson 14-5

Pages 620–621

Use any strategy to solve. Tell what strategy you used.

1. A box contains 8 individual packages of 12 crackers each. How many crackers are in 3 boxes?

2. The temperature was 34 degrees at 5:00 P.M. The temperature fell 4 degrees every hour for the next 5 hours. What was the temperature at 10:00 P.M.?

3. There are 448 markers in a box to be shared by 56 students. Will each student get 7, 8, or 9 markers?

4. Dan's yard is 35 feet long and 20 feet wide. He has 135 feet of fencing. How much will be left after he fences his yard?

Multiply. Use basic facts and patterns.

1. $3 \times 7 =$ ▨
$3 \times 70 =$ ▨
$3 \times 700 =$ ▨
$3 \times 7{,}000 =$ ▨

2. $5 \times 9 =$ ▨
$5 \times 90 =$ ▨
$5 \times 900 =$ ▨
$5 \times 9{,}000 =$ ▨

3. $8 \times 6 =$ ▨
$8 \times 60 =$ ▨
$8 \times 600 =$ ▨
$8 \times 6{,}000 =$ ▨

4. $9 \times 4 =$ ▨
$9 \times 40 =$ ▨
$9 \times 400 =$ ▨
$9 \times 4{,}000 =$ ▨

5. $7 \times 7 =$ ▨
$7 \times 70 =$ ▨
$7 \times 700 =$ ▨
$7 \times 7{,}000 =$ ▨

6. $6 \times 5 =$ ▨
$6 \times 50 =$ ▨
$6 \times 500 =$ ▨
$6 \times 5{,}000 =$ ▨

Solve. Use *logical reasoning*.

1. Four friends have different types of pets. Tina has a cat. Suni does not have a turtle. Ed does not have a dog. Mark does not have a dog or a snake. What type of pet does each person have?

2. Cameron got on the elevator and rode down 5 floors. Then she rode up 7 floors. She rode down 6 floors and got off on the 10th floor. What floor did she start on?

3. Curtis and Anna have change that has the same value. The value is under $1. Curtis has 7 coins. Anna has 4 coins. What two possible combinations could they have?

4. A baby stacked 4 different shapes. The rectangle is on top. A triangle is on top of a square. A circle is not at the bottom. What is the order of the shapes from top to bottom?

Estimate. Round to the nearest ten.

1. $\begin{array}{r} 28 \\ \times\ 5 \\ \hline \end{array}$

2. $\begin{array}{r} 42 \\ \times 4 \\ \hline \end{array}$

3. $\begin{array}{r} 37 \\ \times 7 \\ \hline \end{array}$

Estimate. Round to the nearest hundred.

4. $\begin{array}{r} 170 \\ \times\ 6 \\ \hline \end{array}$

5. $\begin{array}{r} 210 \\ \times\ 8 \\ \hline \end{array}$

6. $\begin{array}{r} 390 \\ \times\ 5 \\ \hline \end{array}$

7. $\begin{array}{r} 289 \\ \times\ 3 \\ \hline \end{array}$

8. $\begin{array}{r} 113 \\ \times\ 9 \\ \hline \end{array}$

9. $\begin{array}{r} 274 \\ \times\ 6 \\ \hline \end{array}$

Lesson 15-4

Pages 642–645

Multiply. Use estimation to check.

1. 26
   ×3

2. 12
   ×4

3. 32
   ×3

4. 21
   ×4

5. 43
   ×2

6. 42
   ×4

7. 2 × 34

8. 2 × 41

9. 3 × 28

10. 3 × 31

11. 2 × 32

12. 3 × 21

Lesson 15-5

Pages 646–647

Use any strategy to solve. Tell what strategy you used.

1. Diana put 7 plums on one side of a scale. To balance the scale, she put 4 plums and 1 apple on the other side of the scale. Each plum weighs 2 ounces. How much does the apple weigh?

2. Troy, Andre, and Ana gathered 450 acorns. Troy gathered 5 times as many acorns as Ana did. Andre gathered 3 times as many as Ana did. How many acorns did each one collect?

3. Ms. McCoy has 3 pitchers of tea. Each pitcher holds 1 gallon of tea. How many cups of tea can she fill with the pitchers?

4. Alicia, Katie, and Megan each play a sport. Megan does not play soccer, Katie does not play basketball or tennis, and Alicia plays basketball. Which sport does each girl play?

Lesson 15-6

Pages 650–653

Multiply. Use models if needed.

1. 29
    ×5

2. 33
    ×6

3. 51
    ×9

4. 25
    ×4

5. 13
    ×8

6. 23
    ×5

7. 52
    ×3

8. 24
    ×7

9. 23
    ×4

Lesson 15-7

Pages 654–656

Multiply.

1. 214
    × 2

2. 128
    × 3

3. 405
    × 8

4. 160
    × 4

5. 391
    × 7

6. 286
    × 3

7. 622 × 3

8. 153 × 5

9. 112 × 9

10. 203 × 8

11. 1,089 × 3

12. 393 × 5

Facts Practice

Addition

1. $\begin{array}{r} 2 \\ +\,4 \\ \hline \end{array}$

2. $\begin{array}{r} 6 \\ +\,6 \\ \hline \end{array}$

3. $\begin{array}{r} 1 \\ +\,8 \\ \hline \end{array}$

4. $\begin{array}{r} 9 \\ +\,7 \\ \hline \end{array}$

5. $\begin{array}{r} 3 \\ +\,5 \\ \hline \end{array}$

6. $\begin{array}{r} 4 \\ +\,9 \\ \hline \end{array}$

7. $\begin{array}{r} 7 \\ +\,6 \\ \hline \end{array}$

8. $\begin{array}{r} 10 \\ +\,1 \\ \hline \end{array}$

9. $\begin{array}{r} 7 \\ +\,0 \\ \hline \end{array}$

10. $\begin{array}{r} 2 \\ +\,9 \\ \hline \end{array}$

11. $\begin{array}{r} 3 \\ +\,4 \\ \hline \end{array}$

12. $\begin{array}{r} 6 \\ +\,4 \\ \hline \end{array}$

13. $\begin{array}{r} 5 \\ +\,8 \\ \hline \end{array}$

14. $\begin{array}{r} 1 \\ +\,4 \\ \hline \end{array}$

15. $\begin{array}{r} 2 \\ +\,7 \\ \hline \end{array}$

16. $\begin{array}{r} 9 \\ +\,3 \\ \hline \end{array}$

17. $\begin{array}{r} 5 \\ +\,7 \\ \hline \end{array}$

18. $\begin{array}{r} 9 \\ +\,0 \\ \hline \end{array}$

19. $\begin{array}{r} 10 \\ +\,6 \\ \hline \end{array}$

20. $\begin{array}{r} 8 \\ +\,8 \\ \hline \end{array}$

21. $5 + 4$

22. $1 + 3$

23. $6 + 5$

24. $1 + 2$

25. $10 + 3$

26. $8 + 3$

27. $4 + 7$

28. $0 + 8$

29. $6 + 8$

30. $1 + 5$

31. $3 + 6$

32. $7 + 3$

33. $0 + 4$

34. $5 + 2$

35. $7 + 8$

36. $9 + 6$

37. $0 + 3$

38. $8 + 4$

39. $10 + 7$

40. $8 + 9$

Addition

1. $\begin{array}{r} 6 \\ + 3 \\ \hline \end{array}$

2. $\begin{array}{r} 1 \\ + 9 \\ \hline \end{array}$

3. $\begin{array}{r} 2 \\ + 5 \\ \hline \end{array}$

4. $\begin{array}{r} 0 \\ + 10 \\ \hline \end{array}$

5. $\begin{array}{r} 1 \\ + 7 \\ \hline \end{array}$

6. $\begin{array}{r} 3 \\ + 7 \\ \hline \end{array}$

7. $\begin{array}{r} 3 \\ + 9 \\ \hline \end{array}$

8. $\begin{array}{r} 9 \\ + 9 \\ \hline \end{array}$

9. $\begin{array}{r} 5 \\ + 3 \\ \hline \end{array}$

10. $\begin{array}{r} 2 \\ + 10 \\ \hline \end{array}$

11. $\begin{array}{r} 4 \\ + 1 \\ \hline \end{array}$

12. $\begin{array}{r} 6 \\ + 7 \\ \hline \end{array}$

13. $\begin{array}{r} 9 \\ + 5 \\ \hline \end{array}$

14. $\begin{array}{r} 8 \\ + 1 \\ \hline \end{array}$

15. $\begin{array}{r} 4 \\ + 4 \\ \hline \end{array}$

16. $\begin{array}{r} 3 \\ + 8 \\ \hline \end{array}$

17. $\begin{array}{r} 10 \\ + 10 \\ \hline \end{array}$

18. $\begin{array}{r} 4 \\ + 3 \\ \hline \end{array}$

19. $\begin{array}{r} 7 \\ + 2 \\ \hline \end{array}$

20. $\begin{array}{r} 6 \\ + 0 \\ \hline \end{array}$

21. 5 + 1

22. 4 + 6

23. 3 + 3

24. 8 + 7

25. 2 + 6

26. 0 + 5

27. 10 + 4

28. 9 + 8

29. 7 + 7

30. 8 + 5

31. 9 + 10

32. 3 + 2

33. 10 + 8

34. 1 + 6

35. 8 + 10

36. 5 + 5

37. 2 + 2

38. 7 + 9

39. 10 + 5

40. 8 + 6

Facts Practice

Subtraction

1. 4
   − 2

2. 11
   − 2

3. 8
   − 0

4. 12
   − 9

5. 10
   − 5

6. 8
   − 3

7. 13
   − 6

8. 5
   − 1

9. 12
   − 3

10. 7
   − 1

11. 10
   − 9

12. 6
   − 4

13. 10
   − 1

14. 14
   − 5

15. 7
   − 7

16. 5
   − 2

17. 15
   − 9

18. 8
   − 2

19. 6
   − 3

20. 13
   − 9

21. 7 − 2

22. 17 − 8

23. 10 − 3

24. 6 − 2

25. 13 − 7

26. 15 − 6

27. 8 − 5

28. 5 − 3

29. 11 − 4

30. 9 − 0

31. 12 − 8

32. 10 − 7

33. 13 − 5

34. 7 − 3

35. 10 − 4

36. 6 − 0

37. 10 − 2

38. 18 − 9

39. 14 − 7

40. 16 − 9

Facts Practice

Subtraction

1. $\begin{array}{r} 3 \\ -2 \\ \hline \end{array}$ **2.** $\begin{array}{r} 19 \\ -10 \\ \hline \end{array}$ **3.** $\begin{array}{r} 13 \\ -4 \\ \hline \end{array}$ **4.** $\begin{array}{r} 9 \\ -7 \\ \hline \end{array}$

5. $\begin{array}{r} 15 \\ -5 \\ \hline \end{array}$ **6.** $\begin{array}{r} 5 \\ -5 \\ \hline \end{array}$ **7.** $\begin{array}{r} 16 \\ -8 \\ \hline \end{array}$ **8.** $\begin{array}{r} 7 \\ -5 \\ \hline \end{array}$

9. $\begin{array}{r} 6 \\ -1 \\ \hline \end{array}$ **10.** $\begin{array}{r} 18 \\ -10 \\ \hline \end{array}$ **11.** $\begin{array}{r} 9 \\ -6 \\ \hline \end{array}$ **12.** $\begin{array}{r} 17 \\ -9 \\ \hline \end{array}$

13. $\begin{array}{r} 8 \\ -4 \\ \hline \end{array}$ **14.** $\begin{array}{r} 9 \\ -1 \\ \hline \end{array}$ **15.** $\begin{array}{r} 20 \\ -10 \\ \hline \end{array}$ **16.** $\begin{array}{r} 14 \\ -6 \\ \hline \end{array}$

17. $\begin{array}{r} 11 \\ -3 \\ \hline \end{array}$ **18.** $\begin{array}{r} 4 \\ -3 \\ \hline \end{array}$ **19.** $\begin{array}{r} 12 \\ -7 \\ \hline \end{array}$ **20.** $\begin{array}{r} 10 \\ -8 \\ \hline \end{array}$

21. $7 - 6$ **22.** $19 - 9$ **23.** $16 - 7$ **24.** $9 - 4$

25. $17 - 7$ **26.** $11 - 5$ **27.** $6 - 6$ **28.** $8 - 1$

29. $5 - 0$ **30.** $15 - 8$ **31.** $10 - 6$ **32.** $14 - 9$

33. $12 - 5$ **34.** $10 - 0$ **35.** $9 - 8$ **36.** $6 - 5$

37. $7 - 0$ **38.** $8 - 6$ **39.** $14 - 7$ **40.** $12 - 6$

Facts Practice

Multiplication

1. $\begin{array}{r} 2 \\ \times\ 2 \\ \hline \end{array}$

2. $\begin{array}{r} 4 \\ \times\ 3 \\ \hline \end{array}$

3. $\begin{array}{r} 5 \\ \times\ 8 \\ \hline \end{array}$

4. $\begin{array}{r} 4 \\ \times\ 0 \\ \hline \end{array}$

5. $\begin{array}{r} 8 \\ \times\ 8 \\ \hline \end{array}$

6. $\begin{array}{r} 7 \\ \times\ 8 \\ \hline \end{array}$

7. $\begin{array}{r} 9 \\ \times\ 7 \\ \hline \end{array}$

8. $\begin{array}{r} 1 \\ \times\ 6 \\ \hline \end{array}$

9. $\begin{array}{r} 4 \\ \times\ 10 \\ \hline \end{array}$

10. $\begin{array}{r} 6 \\ \times\ 8 \\ \hline \end{array}$

11. $\begin{array}{r} 5 \\ \times\ 3 \\ \hline \end{array}$

12. $\begin{array}{r} 0 \\ \times\ 2 \\ \hline \end{array}$

13. $\begin{array}{r} 9 \\ \times\ 9 \\ \hline \end{array}$

14. $\begin{array}{r} 5 \\ \times\ 1 \\ \hline \end{array}$

15. $\begin{array}{r} 8 \\ \times\ 3 \\ \hline \end{array}$

16. $\begin{array}{r} 5 \\ \times\ 7 \\ \hline \end{array}$

17. $\begin{array}{r} 0 \\ \times\ 5 \\ \hline \end{array}$

18. $\begin{array}{r} 6 \\ \times\ 3 \\ \hline \end{array}$

19. $\begin{array}{r} 10 \\ \times\ 1 \\ \hline \end{array}$

20. $\begin{array}{r} 9 \\ \times\ 6 \\ \hline \end{array}$

21. 4×7

22. 3×1

23. 2×8

24. 6×7

25. 8×4

26. 3×3

27. 6×0

28. 2×5

29. 5×6

30. 4×6

31. 3×9

32. 7×10

33. 9×2

34. 4×1

35. 0×10

36. 4×5

37. 1×7

38. 8×9

39. 6×6

40. 10×9

Multiplication

1. 1
   × 1

2. 10
   × 2

3. 5
   × 5

4. 8
   × 6

5. 7
   × 4

6. 3
   × 0

7. 9
   × 8

8. 5
   × 2

9. 8
   × 1

10. 6
    × 5

11. 7
    × 2

12. 9
    × 3

13. 5
    × 9

14. 6
    × 10

15. 7
    × 7

16. 0
    × 6

17. 4
    × 2

18. 2
    × 9

19. 10
    × 0

20. 7
    × 5

21. 1 × 9

22. 3 × 6

23. 4 × 8

24. 5 × 10

25. 2 × 3

26. 10 × 8

27. 7 × 0

28. 6 × 4

29. 4 × 10

30. 6 × 2

31. 0 × 4

32. 4 × 4

33. 1 × 0

34. 7 × 6

35. 9 × 4

36. 3 × 5

37. 8 × 0

38. 10 × 7

39. 8 × 7

40. 7 × 3

Facts Practice

Division

1. $5\overline{)5}$ **2.** $10\overline{)20}$ **3.** $7\overline{)28}$ **4.** $9\overline{)18}$

5. $3\overline{)21}$ **6.** $6\overline{)60}$ **7.** $5\overline{)0}$ **8.** $8\overline{)56}$

9. $4\overline{)24}$ **10.** $10\overline{)30}$ **11.** $5\overline{)45}$ **12.** $1\overline{)3}$

13. $8\overline{)80}$ **14.** $10\overline{)0}$ **15.** $2\overline{)2}$ **16.** $5\overline{)30}$

17. $2\overline{)20}$ **18.** $6\overline{)18}$ **19.** $3\overline{)27}$ **20.** $5\overline{)35}$

21. $16 \div 2$ **22.** $72 \div 9$ **23.** $3 \div 3$ **24.** $48 \div 8$

25. $9 \div 1$ **26.** $12 \div 3$ **27.** $8 \div 4$ **28.** $2 \div 1$

29. $40 \div 4$ **30.** $27 \div 9$ **31.** $0 \div 9$ **32.** $6 \div 2$

33. $54 \div 6$ **34.** $63 \div 7$ **35.** $36 \div 4$ **36.** $15 \div 5$

37. $32 \div 8$ **38.** $36 \div 4$ **39.** $6 \div 3$ **40.** $35 \div 7$

Division

1. $4\overline{)16}$ **2.** $10\overline{)50}$ **3.** $1\overline{)1}$ **4.** $6\overline{)48}$

5. $3\overline{)0}$ **6.** $7\overline{)56}$ **7.** $9\overline{)27}$ **8.** $5\overline{)25}$

9. $9\overline{)90}$ **10.** $6\overline{)36}$ **11.** $2\overline{)14}$ **12.** $4\overline{)32}$

13. $1\overline{)5}$ **14.** $8\overline{)72}$ **15.** $3\overline{)15}$ **16.** $7\overline{)0}$

17. $10\overline{)70}$ **18.** $2\overline{)18}$ **19.** $8\overline{)64}$ **20.** $1\overline{)8}$

21. $12 \div 4$ **22.** $10 \div 2$ **23.** $45 \div 9$ **24.** $36 \div 9$

25. $60 \div 10$ **26.** $10 \div 1$ **27.** $42 \div 7$ **28.** $9 \div 3$

29. $8 \div 8$ **30.** $81 \div 9$ **31.** $36 \div 6$ **32.** $20 \div 4$

33. $10 \div 5$ **34.** $54 \div 9$ **35.** $8 \div 2$ **36.** $49 \div 7$

37. $4 \div 1$ **38.** $40 \div 8$ **39.** $0 \div 6$ **40.** $100 \div 10$

Facts Practice

Preparing for Standardized Tests

Throughout the school year, you may be required to take several tests, and you may have many questions about them. Here are some answers to help you get ready.

How Should I Study?

The good news is that you've been studying all along—a little bit every day. Here are some of the ways your textbook has been preparing you.

- **Every Day** The lessons had multiple-choice practice questions.
- **Every Week** The Mid-Chapter Check and Chapter Test also had several multiple-choice practice questions.
- **Every Month** The Test Practice pages at the end of each chapter had even more questions, including short-response and extended-response questions.

Are There Other Ways to Review?

Absolutely! The following pages contain even more practice for standardized tests.

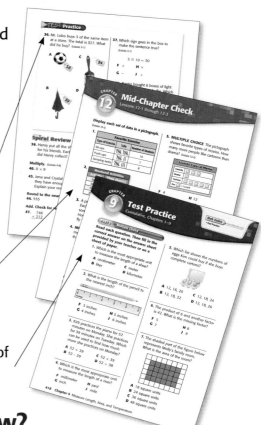

Tips for SUCCESS

Before the Test

- Go to bed early the night before the test. You will think more clearly after a good night's rest.
- Become familiar with common measurement units and when they should be used.
- Think positively.

During the Test

- Read each problem carefully. Underline key words and think about different ways to solve the problem.
- Watch for key words like *not.* Also look for order words like *least, greatest, first,* and *last.*
- Answer questions you are sure about first. If you do not know the answer to a question, skip it and go back to that question later.
- Check your answer to make sure that it is reasonable.
- Make sure that the number of the question on the answer sheet matches the number of the question on which you are working in your test booklet.

Whatever you do...

- Don't try to do it all in your head. If no figure is provided, draw one.
- Don't rush. Try to work at a steady pace.
- Don't give up. Some problems may seem hard to you, but you may be able to figure out what to do if you read each question carefully or try another strategy.

Relax: Just do your best!

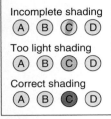

Incomplete shading
Ⓐ Ⓑ Ⓒ Ⓓ

Too light shading
Ⓐ Ⓑ Ⓒ Ⓓ

Correct shading
Ⓐ Ⓑ ● Ⓓ

Multiple-Choice Questions

Multiple-choice questions are the most common type of questions on standardized tests. You are asked to choose the best answer from four possible answers.

To record a multiple choice answer, you may be asked to shade in a bubble that is a circle or an oval. Always make sure that your shading is dark enough and completely covers the bubble.

Example

1 **The table shows the price of bread per loaf at a grocery store.**

Amount (loaf)	Price
2	$5
4	$10
6	$15

If the price increases at the same rate, how much would you pay for 10 loaves of bread?

A $2.50   **B** $5   **C** $20   **D** $25

STRATEGY

Patterns Can you find a pattern to solve the problem?

Read the Problem Carefully You know the price of two loaves of bread in dollars. Find how much you will pay if you buy 10 loaves of bread.

Solve the Problem Look for a pattern. Two loaves of bread cost $5. Four loaves cost $10. Six loaves cost $15. So, for every two loaves of bread, the price increases by $5.

Extend the pattern to find the price of ten loaves of bread.

8 pounds ⟶ $15 + $5 or $20

10 pounds ⟶ $20 + $5 or $25

So, five loaves of bread will cost $25.

The correct choice is D.

Example

2 The shaded part of the figure below represents the fraction $\frac{5}{8}$.

Which fraction represents the part of the figure that is not shaded?

F $\frac{1}{2}$ **G** $\frac{3}{5}$ **H** $\frac{3}{8}$ **J** $\frac{8}{3}$

Read the Problem Carefully You are asked to use the diagram to find which fraction represents the part of the figure that is not shaded.

Solve the Problem There are 3 parts of the whole that are *not* shaded. Since there are 8 parts in all, then three of the eight, or $\frac{3}{8}$, of the figure is not shaded.

The correct choice is H.

> **STRATEGY**
>
> **Key Words** When reading a question, look for words such as *not* or *both.*

Example

3 **Half of the students in a school play are in fourth or fifth grade. There are 7 fourth grade students and 11 fifth grade students in the play. How many students are in the play altogether?**

A 18   **B** 22   **C** 32   **D** 36

Read the Problem Carefully You are asked to find the total number of students in the play. You know how many students there are in fourth and fifth grade. You also know that half of the students in the play are in fourth and fifth grade.

Solve the Problem You need to find the total number of students. To find the total number, you need to add the number of fourth and fifth grade students and multiply that number by two.

> **STRATEGY**
>
> **Work Backward** Can you work backward to find the total number of students?

$$
\begin{array}{r}
7 \text{ fourth grade students} \\
+ 11 \text{ fifth grade students} \\
\hline
18 \text{ students in all}
\end{array}
\qquad
\begin{array}{r}
18 \text{ students} \\
\times 2 \\
\hline
36 \text{ students in the play}
\end{array}
$$

The correct choice is B.

Multiple-Choice Practice

DIRECTIONS
Read each question. Choose the best answer.

1. Which expression models the total number of seashells?

A 5 + 4

B 6 + 4

C 5 + 3

D 6 + 3

2. There are 364 students at Mountain Elementary School. What is this number rounded to the nearest hundred?

F 300   **H** 370

G 360   **J** 400

3. Frannie collects stickers in a book. The table shows the total number of stickers.

Sticker Collection	
Number of Pages	Number of Stickers
1	20
2	40
3	60

If each page has the same number of stickers, how many stickers are there on 8 pages?

A 120   **C** 150

B 140   **D** 160

4. Amy reads 3 pages of her book each minute. How many pages will she read in 9 minutes?

F 12   **H** 30

G 27   **J** 32

5. What is the perimeter of the figure?

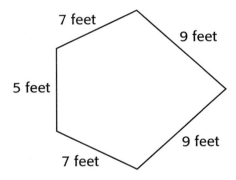

A 37 feet

B 39 feet

C 43 feet

D 47 feet

6. About how tall is a book?

F 1 inch   **H** 1 centimeter

G 1 yard   **J** 1 foot

7. How many angles does a triangle have?

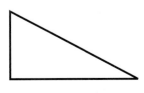

A 0   **C** 2

B 1   **D** 3

8. What is the shape of a soup can?

F circle

G cone

H cylinder

J sphere

9. Mrs. Redstone has 4 red pens, 2 green pens, and 3 blue pens in her desk drawer. Which bar graph best shows this information?

A

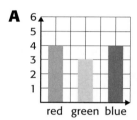

red green blue

B

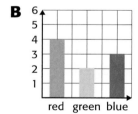

red green blue

C

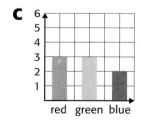

red green blue

D

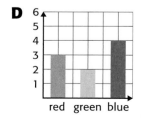

red green blue

10. Shannon has 12 white socks, 4 blue socks, and 2 brown socks in her dresser. She picks one sock without looking. Is it likely, unlikely, certain, or impossible that the sock will be white?

F likely

G unlikely

H certain

J impossible

11. The graph shows information about a summer reading program. Which student read 8 books?

Summer Reading	
Student	**Books Read**
Pamela	📖 📖 📖 📖
Rachel	📖 📖 📖
Orlando	📖 📖 📖 📖 📖
Key: Each 📖 means 2 books.	

A Orlando

B Rachel

C Pamela

D none of the above

12. Use the picture in Exercise 11. How many books did Pamela and Orlando read combined?

F 18

G 16

H 9

J 5

Short-Response Questions

Short-response questions ask you to find the answer to the problem as well as any method, explanation, and/or justification you used to arrive at the solution. You are asked to solve the problem, showing your work.

The following is a sample rubric, or scoring guide, for scoring short-response questions.

Credit	Score	Criteria
Full	2	Full Credit: The answer is correct and a full explanation is given that shows each step in finding the answer.
Partial	1	Partial Credit: There are two different ways to receive partial credit. • The answer is correct, but the explanation given is incomplete or incorrect. • The answer is incorrect, but the explanation and method of solving the problem is correct.
None	0	No credit: Either an answer is not provided or the answer does not make sense.

STRATEGY

Find the Operation
Which operation can be used to perform repeated addition?

Example

① **There are 3 tennis balls in each can. During practice, 6 cans are used. How many tennis balls are there altogether?**

Full Credit Solution

First, I will decide which operation to use. Since each can has the same number of tennis balls, I can use repeated addition or multiplication. I will use multiplication to find 6 × 3.

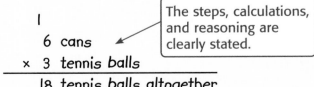

$$\begin{array}{r} 1 \\ 6 \text{ cans} \\ \times\ 3 \text{ tennis balls} \\ \hline 18 \text{ tennis balls altogether} \end{array}$$

The steps, calculations, and reasoning are clearly stated.

There are 18 tennis balls altogether.

The correct answer is given.

Partial Credit Solution

In this sample solution, the answer is correct. However, there is no explanation for any of the calculations.

6 cans, 3 tennis balls

There are 18 tennis balls altogether.

> There is no explanation of how the problem was solved.

Partial Credit Solution

In this sample solution, the answer is incorrect. However, the operation and explanation are correct.

Each can has the same number of tennis balls, so I can use repeated addition or multiplication. I will use multiplication to find 6 × 3.

$$\begin{array}{r} 6 \text{ cans} \\ \times\ 3 \text{ tennis balls} \\ \hline 16 \text{ tennis balls altogether} \end{array}$$

> The student did not multiply correctly.

There are 16 tennis balls altogether.

No Credit Solution

In this sample solution, the answer is incorrect and the student did not understand the problem.

6 + 3 = 9

There are 9 cans.

> The student does not understand the problem and adds 6 and 3.

Short-Response Practice

DIRECTIONS
Solve each problem.

1. Lyle has 52 blue marbles and 27 red marbles. He gives away 21 marbles. How many marbles does Lyle have left?

2. How is three thousand, sixty-two written in standard form?

3. What fraction of the figure below is shaded?

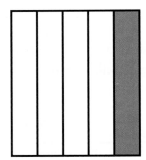

4. The table shows the number of toothpicks needed for different numbers of students in an art class.

Art Project	
Number of Students	**Toothpicks Needed**
1	8
2	16
3	24

If each student gets the same number of toothpicks, how many toothpicks are needed for 9 students?

5. Robert did the multiplication problem below. Write a number sentence that he can use to check his answer.

$$6 \times 9 = 54$$

6. What is the perimeter of the swimming pool shown below?

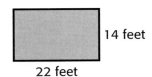

14 feet

22 feet

7. What time is shown on the clock below?

8. Ryan is designing a model rocket on grid paper. Each grid represents 1 square inch. What is the area of the design?

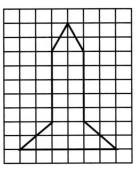

9. How many angles are there in a square?

10. Draw a figure that shows symmetry. Draw the line of symmetry.

11. The students at Brookville Elementary are collecting canned goods for a food drive. The bar graph shows the number of canned goods collected by each grade on the first day.

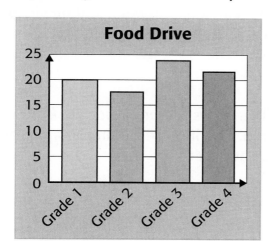

How many canned goods were collected by the students in the third grade?

12. A bag has 8 green marbles. There are a total of 9 marbles. Is it likely or unlikely to pick a green marble from the bag.

13. The graph shows the cars sold by five salespeople this month.

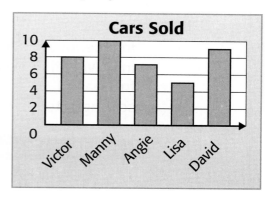

How many more cars did Manny sell than Lisa?

14. A pictograph shows the results of a baseball game. Each baseball on the pictograph represents two points. Team A scored 4 points. Team B scored 3 points. How many baseballs were shown on the graph?

15. The bar graph shows the number of students who chose different types of animals to write their animal reports.

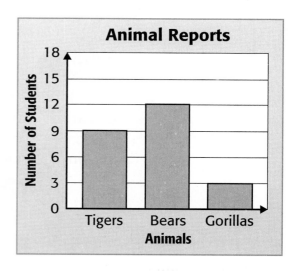

How many reports were written about gorillas or bears?

Extended-Response Questions

Most extended-response questions have multiple parts. You must answer all parts to receive full credit.

In extended-response questions, you must show all of your work in solving the problem. A rubric is used to determine if you receive full, partial, or no credit. The following is a sample rubric, for scoring extended-response questions.

Credit	Score	Criteria
Full	4	Full Credit: The answer is correct and a full explanation is given that shows each step in finding the answer.
Partial	3, 2, 1	Partial Credit: Most of the solution is correct, but it may have some mistakes in the explanation or solution. The more correct the solution, the greater the score.
None	0	No credit: Either an answer is not provided or the answer does not make sense.

Make sure that when the problem says to *Show your work*, you show every part of your solution. This includes figures, graphs, and any explanations for your calculations.

Example

1 **The table shows the number of hours it took four students to read two different books.**
Which student spent a total of 48 hours reading both books? Explain how you found your answer.

Student	Charlotte's Web (hours to read)	Sounder (hours to read)
Lisa	9	27
Jason	15	45
Torres	6	18
Monique	36	12

Full Credit Solution

In this sample answer, the student explains what calculations need to be done, and finds the correct solution.

First, I will list each student's name and write the expressions that will show the sums. Then I will add to see who spent 48 hours reading.

Lisa	Jason	Torres	Monique
27	45	18	36
+ 9	+ 15	+ 6	+ 12
36	60	24	48

Monique spent 48 hours reading both books.

> The steps, calculations and reasoning are clearly stated.

> The correct answer is given.

Partial Credit Solution

In this sample solution, only the calculations are shown.

27	45	18	36
+ 9	+ 15	+ 6	+ 12
36	60	24	48

No Credit Solution

A solution for this problem that will receive no credit may include incorrect answers and an inaccurate or incomplete explanation.

First, I will write the numbers. Then I will a few together.

27		45		18		36
	12		9		15	6

15	18	36	27
+ 9	+ 15	+ 6	+ 12
26	23	46	39

The answers are 26, 23, 46, and 39.

Extended-Response Practice

DIRECTIONS
Solve each problem. Show all your work.

1. What is the total amount of money shown below? Explain how you solved the problem.

2. Valeria has 21 books and 37 CDs on her bedroom shelf. How many books and CDs does she have altogether? Explain how you know which operation to use.

3. The table shows how many markers are needed if there are 3, 5, or 7 groups of students. Copy the table below, and extend it to show how many markers would be needed for 9 groups of students.

Number of Groups	Markers Needed
3	36
5	60
7	84

4. There are 5 rows of singers in the choir. In each row, there 7 singers. Explain how an array can help you find the total number of singers.

5. Carla is harvesting apples from her tree. She will put the same number of apples shown below in each basket.

How many baskets will she need for 30 apples? Draw a picture to help you solve the problem.

6. Eddie has soccer practice starting at the time on the clock. Practice lasts for 1 hour and 30 minutes. Describe the location of both hands of the clock when practice ends.

7. The shaded part of the figure below has a perimeter of 20 feet. Draw a rectangle that has the same perimeter but a different area as the figure shown. Label each side figure.

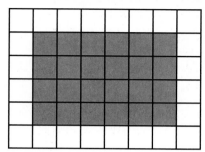

8. Name a polygon that could be made from tracing three triangles. Draw a picture to justify your answer.

9. The thermometer shows the temperature at 4:00. Suppose the temperature rises 2°F every 10 minutes. Explain how to find the temperature an hour later.

10. Explain why the line drawn is not the line of symmetry. Draw the arrow with the correct line of symmetry.

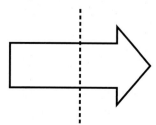

11. Carrie is 3 years older than her brother Michael. Make a table to show how old Michael is when Carrie is 7, 9, and 11.

12. The pictograph shows information about points scored in a basketball game.

Basketball Stats	
Player	**Points Scored**
Sven	🏀🏀🏀🏀🏀
Terrance	🏀🏀🏀
Alberto	🏀🏀🏀🏀

Make a key for the pictograph. Use the key to find the total number of points scored.

13. Use the spinner below to write a problem about probability. Find the solution.

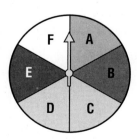

14. Describe how a bar graph and a pictograph are the same. Describe how they are different.

Concepts and Skills Bank

 ## Use Operations with Even and Odd Numbers

Even numbers are whole numbers that have 0, 2, 4, 6, or 8 in the ones place.
Whole numbers that have a 1, 3, 5, 7, or 9 in the ones place are **odd numbers**.

Even and Odd Numbers	Key Concept
Even numbers	**Odd numbers**
0, 2, 4, 6, 8, 10, 12…	1, 3, 5, 7, 9, 11, 13…
192, 194, 196…	281, 283, 285…
2,930, 2,932…	45,933, 45,935…

The expressions below show different combinations of even and odd numbers.

Addition	even + even	odd + odd	even + odd	odd + even
	4 + 6	3 + 9	2 + 3	7 + 2
Subtraction	even − even	odd − odd	even − odd	odd − even
	8 − 4	7 − 5	8 − 7	9 − 6
Multiplication	even × even	odd × odd	even × odd	odd × even
	6 × 2	5 × 3	4 × 7	1 × 8
Division	even ÷ even	odd ÷ odd	even ÷ odd	odd ÷ even
	8 ÷ 2	9 ÷ 3	6 ÷ 3	5 ÷ 2

Exercises

Solve each equation. Then use *odd* or *even* to complete the sentence.

1. 32 + 12 = ▦
even + even = ▦

2. 55 + 27 = ▦
odd + odd = ▦

3. 38 + 83 = ▦
even + odd = ▦

4. 72 − 25 = ▦
even − odd = ▦

5. 62 − 18 = ▦
even − even = ▦

6. 87 − 29 = ▦
odd − odd = ▦

Use *odd* or *even* to complete each sentence.

7. odd × odd = ▦

8. even × even = ▦

9. odd × even = ▦

10. even ÷ even = ▦

11. odd ÷ odd = ▦

12. even ÷ odd = ▦

Add and Subtract Money with Decimals

The steps to follow when adding or subtracting money with decimals are similar.

EXAMPLE Add Money with Decimals

1 **$3.93 + $5.25**

Step 1 Line up the decimal points.

$3.93
+ $5.25

Step 2 Before adding, place a decimal point for the answer. Line it up under the decimal points.

.

Step 3 Add the numbers as you would if you were adding whole numbers. Place a dollar sign ($) in front of the sum.

1
$3.93
+ $5.25
$9.18

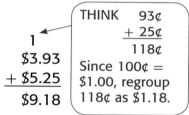

THINK   93¢
      + 25¢
       118¢
Since 100¢ = $1.00, regroup 118¢ as $1.18.

So, $3.93 + $5.25 = $9.18.

EXAMPLE Subtract Money with Decimals

2 **$15.28 − $8.12**

Step 1 Line up the decimal points. Place the decimal point.

$15.28
− $8.12

.

Step 2 Subtract. Place a dollar sign ($) in front of the difference.

$15.28
− $ 8.12
$ 7.16

So, $15.28 − $8.12 = $7.16.

Exercises

Add or subtract.

1. $7.22 + $5.37

2. $8.04 + $4.63

3. $13.44 + $35.52

4. $9.68 − $7.35

5. $7.42 − $3.31

6. $12.83 − 11.52

③ Multiply Money

Multiplying money can be used to find the total cost of many items that have the same price.

EXAMPLE **Multiplying Money**

① **Suppose a notebook costs $2.15.
How much would 6 notebooks cost?**

Step 1 Write the problem as shown.

$$\begin{array}{r} \$2.15 \\ \times\ \ \ 6 \\ \hline \end{array}$$

Step 2 Multiply the hundredths place.

$$\begin{array}{r} 3 \\ \$2.15 \\ \times\ \ \ 6 \\ \hline 0 \end{array}$$

Step 3 Multiply the tenths place.

$$\begin{array}{r} 0\ 3 \\ \$2.15 \\ \times\ \ \ 6 \\ \hline 90 \end{array}$$

Step 4 Multiply the ones.

$$\begin{array}{r} 0\ 3 \\ \$2.15 \\ \times\ \ \ 6 \\ \hline 1290 \end{array}$$

Step 5 Since there are two digits behind the decimal point, place the decimal point two places in front of the last digit of the answer. Write a dollar sign ($) in front of the product.

$$\begin{array}{r} 0\ 3 \\ \$\ 2.15 \\ \times\ \ \ 6 \\ \hline \$12.90 \end{array}$$

So, 6 notebooks cost $12.90.

Exercises

Multiply.

1.
$$\begin{array}{r} \$5.24 \\ \times\ \ \ \ 3 \\ \hline \end{array}$$

2.
$$\begin{array}{r} \$3.74 \\ \times\ \ \ \ 5 \\ \hline \end{array}$$

3.
$$\begin{array}{r} \$4.98 \\ \times\ \ \ \ 7 \\ \hline \end{array}$$

4.
$$\begin{array}{r} \$7.02 \\ \times\ \ \ \ 4 \\ \hline \end{array}$$

5.
$$\begin{array}{r} \$2.33 \\ \times\ \ \ \ 8 \\ \hline \end{array}$$

6.
$$\begin{array}{r} \$6.02 \\ \times\ \ \ \ 9 \\ \hline \end{array}$$

Relate Fractions to a Clock

Clocks can be used to show fractions. The hours and minutes divide a clock into equal parts.

The clock shows 4:15. The minute hand started at the 12 on the clock at the beginning of the 4:00 hour. It has moved around $\frac{1}{4}$ of the clock since 4:00. Compare the clock to the fraction model.

 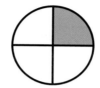

EXAMPLE

 Write the fraction of the clock that the minute hand has traveled since 6:00.

The minute hand started at the 12 and has traveled half way around the clock. So, the minute hand has traveled $\frac{1}{2}$ of the clock.

Exercises

Write the fraction of the clock that the minute hand has traveled since the beginning of the hour.

1.

2.

3.

Use the clock at the right for Exercises 4–7. Write the fraction of the clock that the minute hand will travel for each given time.

4. 3:15

5. 3:20

6. 3:45

7. 3:50

⑤ Points, Lines, Line Segments, Rays, and Angles

A **point** is shown by a dot. A **line** is a set of points that form a straight path that goes in opposite directions without ending. A **ray** is a line that has an endpoint and goes on forever in one direction. A **line segment** is a part of a line between two endpoints. An **angle** is made up of two rays that meet at the endpoints.

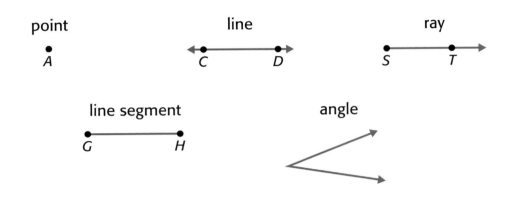

point

A

line

C   D

ray

S   T

line segment

G   H

angle

EXAMPLE   Identify a Figure

① **Describe the figure at the right.**

The figure has one endpoint. The arrow indicates that it goes on forever in one direction.

So, it is a ray.

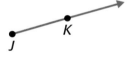

J   K

Exercises

Describe each figure as a *point, line, ray, line segment,* or *angle*.

1.

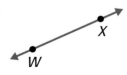

W   X

2.

C   D

3.

E   F

4.

P

5.

R   Q

6.

⑥ Attributes of Circles

A **circle** is a two-dimensional closed figure in which all points are the same distance from a fixed point, called the **center**. The distance around the circle is called the **circumference**. The parts of a circle are below.

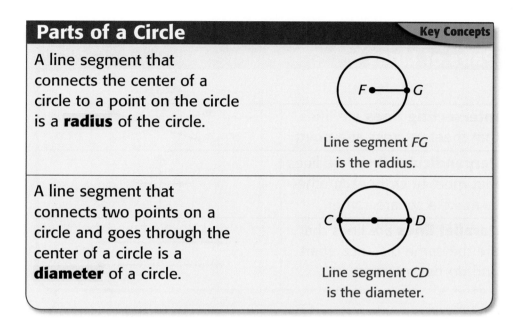

Parts of a Circle	Key Concepts
A line segment that connects the center of a circle to a point on the circle is a **radius** of the circle.	Line segment *FG* is the radius.
A line segment that connects two points on a circle and goes through the center of a circle is a **diameter** of a circle.	Line segment *CD* is the diameter.

EXAMPLE

① **Identify point *R* and line segment *TU*.**

Point *R* is in the middle of the circle. Line segment *TU* connects two points on a circle and goes through the center of the circle.

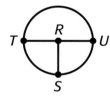

So, point *R* is the center, and line segment *TU* is the diameter.

Exercises

Identify the given points or line segments for each circle.

1. line segment *JK*   **2.** line segment *LN*   **3.** point *V*

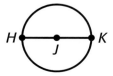   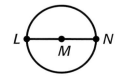   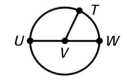

7 Intersecting, Perpendicular, and Parallel Lines

Two lines can be related in three ways. They can be intersecting, perpendicular, or parallel.

Pairs of Lines	Key Concept
Definition	**Model**
Intersecting lines are lines that meet or cross at a point.	
Perpendicular lines are lines that meet or cross each other to make a square corner.	
Parallel lines are lines that are the same distance apart and do not intersect.	

EXAMPLE   Describe a Pair of Lines

1 **Classify the lines at the right as *intersecting*, *perpendicular*, or *parallel*.**

The lines cross at one point, so they are intersecting. Since they do not form a square corner, they are not perpendicular lines.

Exercises

Classify each pair of lines as *intersecting*, *perpendicular*, or *parallel*.

1.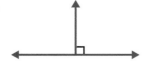

2.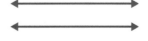

3.

4.

5.

6.

 8 Classify Angles

Angles can be classified by how far the rays spread apart. An angle that has the type of corner found on a square is called a **right angle**. Any angle that is smaller than a right angle is called an **acute angle**. An angle that is larger than a right angle but smaller than a straight line is called an **obtuse angle**.

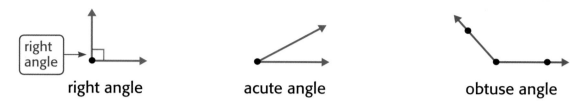

right angle   acute angle   obtuse angle

EXAMPLES

Classify the angle as *right, acute,* or *obtuse*.

1

The angle shown looks like the corner of a square. So, the angle is a right angle.

2

The angle shown is smaller than a right angle. So, it is an acute angle.

Exercises

Classify the angle at the right as *right, acute,* or *obtuse*. Use a square pattern block if needed.

1.

2.

3.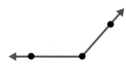

4.

5.

6.

7. Give a real-world example of each type of angle.

9 Triangles

You can classify triangles by the lengths of their sides or by their angles.

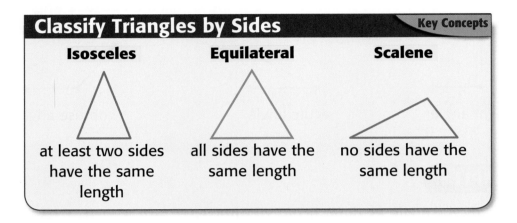

Classify Triangles by Sides — Key Concepts

Isosceles	Equilateral	Scalene
at least two sides have the same length	all sides have the same length	no sides have the same length

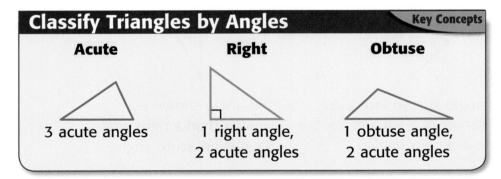

Classify Triangles by Angles — Key Concepts

Acute	Right	Obtuse
3 acute angles	1 right angle, 2 acute angles	1 obtuse angle, 2 acute angles

EXAMPLE

1. **Classify the triangle by its sides and by its angles.**

   None of the sides have the same length. There is one obtuse angle and 2 acute angles.

   So, the figure shown is a scalene triangle and an obtuse triangle.

Exercises

Classify each triangle by its sides and by its angles.

1.

2.

3.

4.

10 Similar and Congruent Figures

Two figures that have the same shape but different sizes are called **similar**. When figures have the same size and the same shape, they are **congruent**.

similar

same shape
different size

congruent

same shape
same size

EXAMPLE

1 **Classify the set of figures as *similar* or *congruent*.**

The figures are the same shape and same size.

So, the figures are congruent.

Exercises

Classify each set of figures as *similar* or *congruent*.

1.

2.

3.

4.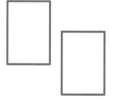

5.

6.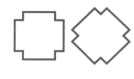

7. Write a real-world example of a pair of objects that are similar.

⑪ Transformations

When a geometric figure changes location, it has gone through a **transformation**. Three types of transformations are **turn**, **flip**, and **slide**.

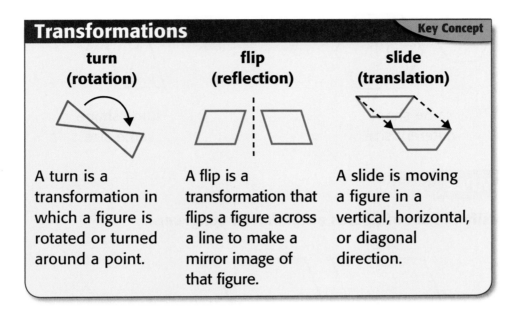

Transformations   **Key Concept**

turn (rotation)	flip (reflection)	slide (translation)
A turn is a transformation in which a figure is rotated or turned around a point.	A flip is a transformation that flips a figure across a line to make a mirror image of that figure.	A slide is moving a figure in a vertical, horizontal, or diagonal direction.

EXAMPLE

① **Identify the transformation. Write *turn*, *flip*, or *slide*.**

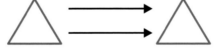

The triangle above has moved sideways. It has not turned or flipped.

So, the transformation of the triangle is a slide.

Exercises

Identify each transformation. Write *turn*, *flip*, or *slide*.

1.

2.

3.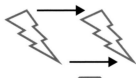

4.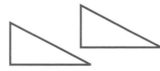

5.

6.

⑫ Minimum, Maximum, Mode, and Range

Data are pieces of information that often use numbers. One way to describe a set of data is to use the mode. The **mode** of a set of data is the number that occurs most often. The **range** is the difference between the least (**minimum**) and the greatest (**maximum**) numbers.

EXAMPLE

❶ Find the mode, minimum, maximum, and range in the set of data to the right.

To find the mode, find the most common numbers. Five is the number that occurs most often.

To find the range, subtract the minimum from the maximum.

So, 5 is the mode in the set of data. The minimum is 2, the maximum is 9, and the range is 7.

5, 2, 3, 9, 5, 5, 2

2, 2, 3, 5, 5, 5, 9

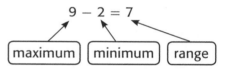
$9 - 2 = 7$
maximum   minimum   range

Exercises

For each set of data, find the mode, minimum, maximum, and range.

1. 6, 1, 4, 5, 4

2. 8, 1, 9, 3, 1

3. 3, 3, 3, 1, 5, 10

4. 4, 2, 4, 2, 4, 8

5. 12, 11, 4, 33, 3, 12

6. 15, 3, 5, 3, 51

7. SCHOOL Five students took a spelling test. The scores were 90, 88, 99, 100, and 99. Find the range of the scores.

8. TRUCKS A delivery truck company has seven trucks. Two of the trucks drove 25 miles. Three of the trucks drove 39 miles. The rest of the trucks drove 40 miles. Write the number of miles each truck drove to show the set of data. Find the mode.

13 Circle Graphs

Circle Graphs are used to compare parts of a whole.

EXAMPLE

1 **What is the most common way students arrive at school?**

The circle is divided into three parts. The parts are not equally divided. The sector or part that is the largest represents students that walk to school.

So, the most common way students arrive at a school is by walking to school.

How Students Arrive at School

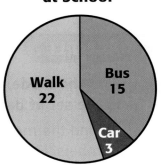

Exercises

For Exercises 1–3, use the circle graph at the right.

1. Are there more boys or more girls in the class?

2. How many students are in the class altogether?

3. What would happen to the graph if there were 10 boys and 10 girls?

Third Grade Class

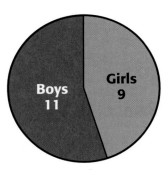

For Exercises 4–6, use the circle graph at the right.

4. Which fruit is liked the least?

5. How many students like oranges?

6. How many more students like apples than pears?

7. Create a circle graph with 3 sectors. Which part represents the most?

Favorite Fruits

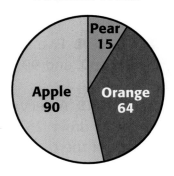

Acknowledgements

Unless otherwise credited, all currency courtesy of the US Mint. **v** Thomas Barwick/Getty Images; **vi** Doug Martin; **vii** (br)courtesy Dinah Zike, (others)Doug Martin; **x-xi** Arthur Morris/CORBIS; **xii-xiii** CORBIS; **xiv-xv** BRIAN P. KENNEY/Animals Animals; **xvi-xvii** Index Stock; **xviii-xix** CORBIS; **xx-xxi** PETE OXFORD/Minden Pictures; **xxii-xxiii** CORBIS; **xxiv-xxv** Joseph Sohm-Visions of America/Getty Images; **xxix** Ed-Imaging; **xxvii** Alaska Stock LLC/Alamy Images; **xxviii** Tim Fuller; **1** Comstock/PunchStock; **2** Transparencies; **3** J. Mallwitz/Peter Arnold, Inc.; **4** David Brody/epa/CORBIS; **5** (tr)Daniel Dempster Photography/Alamy Images, (b) Adam Jones/Photo Researchers; **6** Andrew Lichtenstein/CORBIS; **7** (tr)Photodisc/Getty Images, (b) AP Images; **8** Hutchings Photography; **9** (tr)Bettmann/CORBIS, (b)Jon Arnold Images/Alamy Images; **10** (t)Mike Grandmaison/CORBIS, (c)Creatas/Punchstock, (bc)Photodisc/Alamy Images; **11** Hutchings Photography; **12** Joel Rogers/CoasterGallery.com; **13** IndexStock; **14-15** Getty Images; **17** S. Wanke/PhotoLink/Getty Images; **20** David Kjaer/Nature Picture Library; **22** Ed-Imaging; **24** Image Source/Getty Images; **27** Ed-Imaging; **28** John Giustina/Getty Images; **32** CORBIS; **40** (l to r)Getty Images, (2)C Squared Studios/Getty Images, (3)Getty Images, (4-6)G.K. & Vikki Hart/Getty Images; **41** Ed-Imaging; **42-43** CORBIS; **44** Bob Torrez/PhotoEdit; **47** Ed-Imaging; **49** David Crausby/Alamy Images; **51** (bl)Jupiterimages/photos.com, (bcl)Getty Images, (bcr)Siede Preis/Getty Images, (br)Ryan McVay/Getty Images; **53 through 65** Michael Houghton/StudiOhio; **66-67** Joe McBride/Getty Images; **72** Lori Adamski Peek/Getty Images; **77** Ed-Imaging; **81** (cr)CORBIS, (r)The McGraw-Hill Companies; **83** (tr)Photodisc/Getty Images, (kites)C Squared Studios/Getty Images, (robot)Getty Images, (car)Ken Karp/The McGraw-Hill Companies, (money)Michael Houghton/StudiOhio, (yo-yo)Siede Preis/Getty Images; **85** (c)Photodisc/Getty Images, (r)Ryan McVay/Getty Images; **86-87** CORBIS; **88** Marilyn Conway/Getty Images; **90** Ed-Imaging; **92** David De Lossy/Getty Images; **93** Comstock Images/Alamy Images; **95** Ed-Imaging; **96** Tim Jones/Getty Images; **97** Michael Newman/PhotoEdit; **99** (tl)Ed-Imaging, (tr)Stockbyte/Getty Images, (cl)2006 Photos To Go, (c)PhotoObjects.net/Jupiter Images, (cr)Siede Preis/Getty Images; **108-109** Getty Images; **114** Photodisc/Getty Images; **115** (l)Mira/Alamy Images, (r)Daniel Dempster Photography/Alamy Images; **116** William Leaman/Alamy Images; **117** (l to r, t to b)Richard Hutchings, (2)Getty Images, (3)CORBIS, (4)The McGraw-Hill Companies, (5)2006 Photos To Go, (6)Photodisc/Getty Images; **122** (inset)Courtesy of New England Conservatory; **122-123** (bkgd)Culliganphoto/Alamy Images; **124** Ed-Imaging; **128** David Young-Wolff/PhotoEdit; **129** (tl)Michael Pole/CORBIS, (tr)Robert McGouey/Alamy Images, (bcr)Punchstock, (br)CORBIS; **132** PunchStock; **137** Ed-Imaging; **139** Siede Preis/Getty Images; **142** Roger Harris/Photo Researchers; **150** C Squared Studios/Getty Images; **154-155** BRIAN P. KENNEY/Animals Animals; **157** Ed-Imaging; **161** (bl)Ed-Imaging, (br)Getty Images; **162** BananaStock/Alamy Images; **163** Brand X Pictures/PunchStock, (c)Stockdisc/PunchStock; **164** (tl)Getty Images, (r)CORBIS; **166** (bl)Renee Morris/Alamy Images,(br)CORBIS; **167** (l)Jupiterimages, (tcr)Photodisc/Getty Images, (tr)Getty Images; **169** Comstock Images/Alamy Images; **170 - 171** Ed-Imaging; **172** Chip Henderson/Index Stock; **174** Inga Spence/IndexStock; **176** Brand X Pictures/PunchStock; **180** Ellen McKnight/Alamy Images; **181** (r)Stockdisc/PunchStock, (bl)Ken Cavanagh/The McGraw-Hill Companies, (br)Stockdisc/PunchStock; **182-183** Jeffrey L. Rotman/CORBIS; **184** image100 Ltd; **194** Jupiterimages/photos.com; **198-199** Michael Newman/PhotoEdit; **203** Stockdisc/PunchStock; **211–212** Ed-Imaging; **214** CORBIS; **215** Photos.com/jupiterimages; **216** Ryan McVay/Getty Images; **218** Getty Images; **219** G.K. & Vikki Hart/Getty Images; **223** CORBIS; **224** Getty Images; **226-227** (bkgd)John Warden/SuperStock; **227** (inset)Museum of Fine Arts, Houston, Texas, USA, The Bayou Bend Collection/Bridgeman Art Library; **228** Ed-Imaging; **235** C Squared Studios/Getty Images; **240** Stockdisc/Getty Images/PunchStock; **245** Photos.com/Jupiterimages; **248-249** CORBIS; **250** (l)C Squared Studios/Getty Images, (r)Photodisc/Getty Images; **251** Ed-Imaging; **253** The McGraw-Hill Companies; **254** Don Smetzer/PhotoEdit; **256** Ed-Imaging; **258** David Young-Wolff/Photoedit; **259** Jupiterimages; **262** Jose Luis Pelaez, Inc /jupiterimages; **266** (tr)Photos.com/jupiterimages, (cr)Jupiter Images, (bl)BananaStock/Alamy Images, (br)Richard Hutchings; **271** PhotoLink/Getty Images; **274-275** (bkgd)David Young Wolff/PhotoEdit, **275** (inset)Jim West/The Image Works; **276** Ed-Imaging; **277** Stockdisc/PunchStock; **281** Ed-Imaging; **282** (l to r)2006 Photos To Go/Index Open, (2)Index Stock/Photolibrary, (3)Photolibrary; **283** Photos.com/Jupiter Images; **292-293** Index Stock;

295 Ed-Imaging; **297** Getty Images; **298** 2006 Photos To Go/Index Open; **302** (tl)2006 Photos To Go, (bl)Getty Images, **303** (tl)Blend Images/Alamy Images, (tr)Ed-Imaging; **304** (tc)Ingram Publishing/SuperStock, (tr)C Squared Studios/Getty Images; **310-311** (inset)Dennis Hallinan/Alamy Images, (bkgd)CORBIS; **315** Ed-Imaging; **316** Image Source/Getty Images; **319-320** Ed-Imaging; **321** Getty Images; **330-331** Gandee Vasan/Getty Images; **332** (bc)C Squared Studios/Getty Images, (br)Getty Images; **333** Earth Scenes; **335** (bl)Blend Images/Alamy Images, (br)PunchStock; **342** digitalvision/PunchStock; **344** Ed-Imaging; **350** Stockdisc/PunchStock; **351-352** Ed-Imaging; **354** PunchStock; **355** (tl-tc)Ken Cavanagh/The McGraw-Hill Companies, (tr)The McGraw-Hill Companies; **356** G.K. & Vikki Hart/Getty Images; **357** Stockdisc/CORBIS; **358** Stockbyte/Getty Images; **360-361** Ciaran Griffin/Stockbyte Platinum/Getty Images; **363** (tl)Getty Images, (tc)Siede Preis/Getty Images; **364** C Squared Studios/Getty Images; **370-371** CORBIS; **372** (tc)Getty Images, (tr)C Squared Studios/Getty Images; **373** (c)C Squared Studios/Getty Images, (c)PhotoLink/Getty Images, (bc)Ed-Imaging; **374** (tr cr)PhotoObjects/jupiterimages, (bl)Getty Images; **375** Jupiter Images; **376** (tc)The McGraw-Hill Companies, (bl)Image Source, (br)Stockbyte/PictureQuest; **377** (l to r, t to b)PhotoObjects/jupiterimages, (2)The McGraw-Hill Companies, (3)Max Delson/istockphoto, (4)Jackie DesJarlais/istockphoto, (5)The McGraw-Hill Companies, (6)Gerry Ellis/Minden Pictures; **378** (tr)Ed-Imaging, (cl)C Squared Studios/Getty Images, (bl)2006 Photos To Go; **379** C Squared Studios/Getty Images; **381** Getty Images; **382** Doug Menuez/Getty Images; **383** (t)Joe Polillio/The McGraw-Hill Companies, (b)The McGraw-Hill Companies/Ken Cavanagh; **385** (tl)C Squared Studios/Getty Images, (cl)CORBIS; **386** (tr) Michael Wong/CORBIS, (cl)Comstock/jupiterimages, (cr)Ed-Imaging, (bl)Dorling Kindersley/Getty Images, (br)CORBIS; **387** Dorling Kindersley/Getty Images; **390** Ed-Imaging; **391** PhotoLink/Getty Images; **393** Brand X Pictures/PunchStock; **396** Ed-Imaging; **401** Brand X Pictures/PunchStock; **402** Ed-Imaging; **404-405** (bkgd)Joel A. Rogers/CoasterGallery.com; **406** Ed-Imaging; **413** (tr)C Squared Studios/Getty Images, (bl)Dorling Kindersley/Getty Images; **414** Photodisc/Getty Images; **417** (cl)David Toase/Getty Images, (cl)Jupiter Images/Photos.com; **420-421** Hutchings Photography; **422** (l to r, t to b)Ingram Publishing/Superstock, (1)Ken Cavanagh/The McGraw-Hill Companies, (2-4)C Squared Studios/Getty Images, (5)Photodisc/Getty Images, (6-7)Getty Images, (8)C Squared Studios/Getty Images; **423** (tl)Handke-Neu/CORBIS, (tcl)Marielle/photocuisine/CORBIS, (tcr)Envision/CORBIS, (tr)Ryan McVay/Getty Images; **425** Nick Daly/Getty Images; **426** (tr)Amy Etra/PhotoEdit, (cl)D. Hurst/Alamy Images, (c) www.comstock.com, (cr)Spencer Grant/PhotoEdit, (bl)Hirdes/f1online/Alamy Images, (br)Ton Kinsbergen/Beateworks/CORBIS; **427** (tl)Jupiter Images, (tc)Getty Images, (tr)CORBIS, (cl)2006 Photos To Go/Index Open, (c)G.K. & Vikki Hart/Getty Images, (cr)Brian Leatart/Jupiter Images; **428** (t to b)Judith Collins/Alamy Images, (2-3) Jupiterimages, (4) Jupiter Images/Photos.com; **429** Ed-Imaging; **430** Hutchings Photography; **432** (cr)Burke/Triolo Productions/Brand X/CORBIS, (bl)Adrianna Williams/zefa/CORBIS, (br)Amon/PhotoCuisine/CORBIS; **433** (bl)Jupiter Images, (bc)Didier Robcis/CORBIS, (br) The McGraw-Hill Companies; **434** (tl) Mick Broughton/Alamy Images, (tc) Comstock Images/Alamy Images, (tr) Jupiter Images, (cl) Jupiter Images/Brand X/Alamy, (c)Judith Collins/Alamy Images, (cr)Ingram Publishing/Superstock; **435** (tl)Ryan McVay/Getty Images, (tc)Siede Preis/Getty Images, (cl)FoodCollection/SuperStock, (c)Ford Smith/CORBIS; **436** (tr)Geoff du Feu/Alamy Images, (others)Jupiterimages; **437** Stockdisc/PunchStock; **438** Mark Karrass/CORBIS; **439** (l to r, t to b)C Squared Studios/Getty Images, (2)Siede Preis/Getty Images, (3)Ryan McVay/Photodisc/Getty Images, (4)Phototdisc/Getty Images, (5)Comstock Images/Alamy Images, (6)Getty Images, (7)Image Source/Jupiter Images; **440** (tl)Ryan McVay/Getty Images, (tc)Martin Ruegner/jupiterimages, (tc)Burke/Triolo/Brand X Pictures/jupiterimages, (cl)C Squared Studios/Getty Images, (c)Getty Images, (cr)Photodisc Collection/Getty Images, (br)PhotoDisc/Getty Images; **441** (tc)Stockdisc/PunchStock, (tr)Siede Preis/Getty Images, (c)CORBIS; **442** (t to b)D. Robert & Lorri Franz/CORBIS, (2)RAOUL SLATER/WWI/Peter Arnold, Inc., (3)Punchstock, (4)Michael Durham/Minden Pictures; **442-443** (bkgd)Michael Hagedorn/zefa/CORBIS; **444** (tr-cr)2006 Photos To Go, (cl)C Squared Studios/Getty Images, (bc)CORBIS, (br)Photodisc/Getty Images; **445** (bl)Stockdisc, (bc)C Squared Studios/Getty Images, (br)2006 Photos To Go; **446** (tl)Photos.com/Jupiter Images, (tc)CORBIS, (tr)DK Limited/CORBIS, (bl)CORBIS, (bc)Michael Newman/PhotoEdit; **448** (5)Ed-Imaging, (others)The McGraw-Hill Companies; **453** (tl)Getty Images, (tr)Ed-Imaging;

457 (tl)Handke-Neu/CORBIS, (tc)Ryan McVay/Getty Images, (tr)C Squared Studios/Getty Images; 458 (tl)Adrianna Williams/zefa/CORBIS, (tcl)Amon/PhotoCuisine/CORBIS, (tcr)Foodfolio/Alamy Images, (tr)Brian Hagiwara/jupiterimages; 459 (tc)Getty Images, (tr)2006 Photos To Go/Index Open, (cr)Digital Vision/Getty Images; 461 (tc)G.K. & Vikki Hart/Getty Images, (tr)Lew Robertson/jupiterimages, (c)2006 Photos To Go/Index Open, (bc)CMCD/Getty Images, (br)Getty Images; 464-465 Index Stock; 467 (tl)Klaus Hackenberg/zefa/CORBIS, (tc)Stockbyte Platinum/Alamy Images, (tr)Michael Newman/PhotoEdit, (br)The McGraw-Hill Companies; 469 (tr)Getty Images, (cr)C Squared Studios/Photodisc/Getty Images, (bc)Ryan McVay/Getty Images, (br)Hemera Technologies/jupiterimages; 472 (br)CORBIS, (others)Ryan McVay/Getty Images; 473 (bl)Dynamicgraphics/InMagine Images, (bc)AgeFotostock/SuperStock, (br)Image Source/Jupiter Images; 474 (tl)Ken Cavanagh/The McGraw-Hill Companies, (tc)Burke/Triolo Productions/Brand X/CORBIS, (tr)Stockdisc/CORBIS, (cl)2006 Photos To Go/Index Open, (c)Comstock Images/Alamy Images, (cr)2006 Photos to Go/Index Open; 475 (tr)Digital Vision Ltd./SuperStock, (bl)Ed-Imaging, (br) Brad Wilson/Getty Images; 481 (bl)Steve Hamblin/Alamy Images, (br)Ryan McVay/Getty Images; 482 Ed-Imaging; 484 Ryan McVay/Getty Images; 485 (tl)David Spindel/SuperStock, (tc)2006 Photos to Go/Index Open, (tc)Siede Preis/Getty Images, (tr)2006 Photos to Go/Index Open, (c)Siede Preis/Getty Images; 486 Ed-Imaging; 488 The McGraw-Hill Companies Inc.; 489 (br)Siede Preis/Getty Images, (others)Jupiter Images; 490 (tl tc)Getty Images, (tr)C Squared Studios/Getty Images, (br)Brand X Pictures/PunchStock; 495 2006 Photos To Go/Index Open; 496 (bl)Comstock/jupiterimages, (br)Ed-Imaging; 498-499 Michael Ventura/Alamy; 501 (tl)C Squared Studios/Getty Images, (tr)2006 Photos To Go/Index Open, (tr)Brand X Pictures/Getty Images, (br)Comstock Images/Alamy Images, (br)Ken Cavanagh/The McGraw-Hill Companies; 504 (br)The McGraw-Hill Companies, (others)Ken Karp/The McGraw-Hill Companies; 505 (l)Stockbyte, (tr)Getty Images, (tr)Siede Preis/Getty Images; 507 (l to r, t to b)Brian Leatart /jupiterimages, (2)Hemera Technologies/jupiterimages, (3)Jupiter Images/Photos.com, (4)Getty Images; 510-511 Bob Krist/CORBIS; 512 (t to b)Photodisc/Getty Images, (2)Siede Preis/Getty Images, (3)Getty Images, (4)David Toase/Getty Images; 513 Ed-Imaging; 520 (l to r, t to b)G.K. & Vikki Hart/Getty Images, (2)Getty Images, (3)C Squared Studios/Getty Images, (4)CMCD/Getty Images, (5)Stockdisc/PunchStock, (6)Getty Images, (7)CORBIS; 521 G.K. & Vikki Hart/Getty Images; 522 Ed-Imaging; 524 Ed-Imaging; 528 (tc)CORBIS, (tr)Getty Images; 536 2006 Photos To Go/Index Open; 540-541 (bkgd) Frans Lanting/CORBIS; 541 (inset)Siede Preis/Getty Images; 542 Ed-Imaging; 544 (br)Joseph Sohm/Visions of America/CORBIS; 545 (tl)PunchStock, (tr)Ed-Imaging; 546 CORBIS; 547 Ken Cavanagh/The McGraw-Hill Companies; 556-557 Ed-Imaging; 559 Ed-Imaging; 565 (tl tc tr)G.K. & Vikki Hart/Getty Images, (t cr)Getty Images, (bc)jupiterimages/Photos.com, (br)Getty Images; 566 (tc)Getty Images, (tr)Jupiter Images/Photos.com, (br)Stephen Mallon/Getty Images; 568 CORBIS; 574 (bl)2006 Photos To Go/Index Open, (br 2)Getty Images, (others)Siede Preis/Getty Images; 576 Ed-Imaging; 578 Burke/Triolo Productions/Brand X/CORBIS; 579 Getty Images; 580 Siede Preis/Getty Images; 584 Getty Images; 587 (tl)Thomas Northcut/Getty Images, (tr)Digital Vision/Getty Images; 588 (l to r, t to b)Dorling Kindersley/Getty Images, (2)James Cotier/Getty Images, (3)CORBIS, (4)Getty Images, (5-6)John T. Fowler/Alamy Images, (7)CORBIS, (8)Dorling Kindersley/Getty Images, (9)Getty Images, (10)James Cotier/Getty Images; 588-589 (bkgd)CORBIS; 591 (br)Stockdisc/PunchStock, (others)CMCD/Getty Images; 598-599 Ed-Imaging; 605 Getty Images; 607 Ed-Imaging; 612-613 PunchStock/CORBIS; 614 Steve Smith/SuperStock; 616 Ed-Imaging; 619 Stockdisc/PunchStock; 622 David Young-Wolff/PhotoEdit, Inc.; 626 Stockdisc/PunchStock; 629 (br)2006 Photos To Go/Index Open, (br)PhotoObjects.net/Jupiter Images, (br)Photodisc/Getty Images; 632-633 Jeff Vanuga/CORBIS; 636 Stockdisc Classic/Alamy Images; 640 Jeff Greenberg/Alamy Images; 642 (bl)Getty Images, (br)Ed-Imaging; 644 (tr)Ken Cavanagh/The McGraw-Hill Companies, (others)The McGraw-Hill Companies; 648 PunchStock; 652 David Young-Wolff/PhotoEdit; 654 (c)Hemera Technologies/jupiterimages, (r)Cre8tive Studios/Alamy Images, (br)CORBIS; 656 CORBIS; 657 Getty Images; 659 Ed-imaging; 660-661 (bkgd)Sergio Delle Vedove/Alamy Images; 661 (c)FARINA CHRISTOPHER/CORBIS SYGMA, (bl)The British Library/HIP/The Image Works, (bc)PhotoObjects.net/Jupiter Images, (br)The Granger Collection, New York; 669 The McGraw-Hill Companies; LA-0 Mark Steinmetz; LA-1 Tim Fuller; LA-2 Food Image Source/Frank Rogozienski/StockFood; LA-3 INTERFOTO Pressebildagentur/Alamy Images; LA-6 David R. Frazier Photolibrary/Alamy Images; LA-8 (cl)Rubberball/Alamy Images, (br)Aflo Foto Agency/Alamy Images; LA-10 Thom DeSanto/Stock Food; LA-11 (tc)AfriPics.com/Alamy Images, (cl)Alaska Stock LLC/Alamy Images; LA-13 (l)David Young-Wolff/PhotoEdit, (r)Digital Vision/Getty Images; LA-14 Martin Harvey/CORBIS; LA-16 Kennan Ward/CORBIS; LA-17 (tl)Ed-Imaging, (tr)Getty Images; LA-18 Image Ideas Inc./Index Stock; LA-19 AgeFotostock/SuperStock; P0 (tl)Getty Images, (cr)Punchstock, (bl)Bob Daemmrich/PhotoEdit; P1 Tim Fuller; P2 (t)The Plane Picture Company/jupiterimages, (br)Amy Etra/PhotoEdit; P3 Ed-Imaging; P4 (br)Ed-Imaging, (bkgd)Zak Waters/Alamy Images; P5 (tr)Ed-Imaging, (br)Tim Pannell/CORBIS; P6 Photo Network/Alamy images; P7 (tr)Ed-Imaging, (br)Michael Newman/PhotoEdit; P8 (t)Ross Couper-Johnston/npl/Minden Pictures, (br)Helene Rogers/Alamy Images; P9 (cr)Ed-Imaging, (others)Michael Houghton/StudiOhio; R0 Ed-Imaging; R5 Michael Houghton/StudiOhio; R11 (cl)Jupiter Images, (c)Ian Cartwright/Getty Images, (cr)Ken Cavanagh/The McGraw-Hill Companies; R24 (tl)Getty Images, (tr)C Squared Studios/Getty Images; R26 (bl)Comstock Images/Alamy Images, (br)Photos.com/Jupiterimages; R29 (l)Photos.com/Jupiterimages, (c)Ryan McVay/Getty Images, (cr)Getty Images; R31 CORBIS; R51 Ed-Imaging; R62 (r)Stockdisc/PunchStock, (tl)Michael Houghton/StudiOhio; R84 Michael Houghton/StudiOhio; R88 Getty Images.

McGraw-Hill would like to acknowledge the artists and agencies who contributed to illustrating this program: Cover Mick McGinty represented by Mendola Artists; Argosy Publishing; Gary Ciccarelli, Shawn McKelvey represented by AA Reps. Inc.; Richard Carbajal represented by Deborah Wolfe Ltd.; Trevor Moo-Young/Black Dot Group.

Acknowledgements

Glossary/Glosario

A mathematics multilingual glossary is available at www.math.glencoe.com/multilingual_glossary. The glossary includes the following languages.

Arabic	Cantonese	Korean	Tagalog
Bengali	English	Russian	Urdu
Brazilian	Haitian Creole	Spanish	Vietnamese
Portuguese	Hmong		

Cómo usar el glosario en español:
1. Busca el término en inglés que desces encontrar.
2. El término en español, junto con la definición, se encuentran en la columna de la derecha.

English   Ⓐ   Español

add (adding, addition) An operation on two or more *addends* that results in a *sum*.

$$9 + 3 = 12$$

suma (sumar) Operación que se realiza en dos o más *sumandos* y que resulta en una *suma*.

$$9 + 3 = 12$$

addend (p. 69) Any numbers being added together.

sumando Cualquier número que se le suma a otro.

analog clock (p. 454) A clock that has an **hour hand** and a **minute hand**.

reloj analógico Reloj que tiene un horario y un minutero.

area (p. 396) The number of *square units* needed to cover the inside of a region or plane figure.

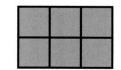

area = 6 square units

área Número de *unidades cuadradas* necesarias para cubrir el interior de una región o figura plana.

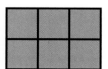

área = 6 unidades cuadradas

array (p. 159)   Objects or symbols displayed in rows of the same length and columns of the same length.

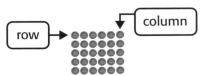

arreglo   Objetos o símbolos representados en filas de la misma longitud y columnas de la misma longitud.

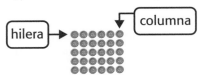

Associative Property of Addition (p. 69)   The property which states that the grouping of the *addends* does not change the *sum*.

$$(4 + 5) + 2 = 4 + (5 + 2)$$

propiedad asociativa de la adición   Propiedad que establece que la agrupación de los *sumandos* no altera la *suma*.

$$(4 + 5) + 2 = 4 + (5 + 2)$$

Associative Property of Multiplication (p. 234)   The property which states that the grouping of the factors does not change the *product*.

$$3 \times (6 \times 2) = (3 \times 6) \times 2$$

propiedad asociativa de la multiplicación   Propiedad que establece que la agrupación de los factores no altera el *producto*.

$$3 \times (6 \times 2) = (3 \times 6) \times 2$$

B

bar graph (p. 528)   A graph that compares *data* by using bars of different lengths or heights to show the values.

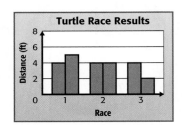

gráfica de barras   Gráfica que compara los datos usando barras de distintas longitudes o alturas para mostrar los valores.

C

capacity (p. 425)   The amount a container can hold, measured in units of dry or liquid measure.

capacidad   Cantidad que puede contener un envase, medida en unidades liquídas o secas.

centimeter (cm) (p. 384) A *metric unit* for measuring *length and height.*

100 centimeters = 1 meter

centímetro (cm) *Unidad métrica* para medir de longitud y altura.

100 centímetros = 1 metro

Commutative Property of Addition (p. 69) The property that states that the order in which two numbers are added does not change the *sum.*

$$12 + 15 = 15 + 12$$

propiedad conmutativa de la adición Propiedad que establece que el orden en el cual se suman dos o más números no altera la *suma.*

$$12 + 15 = 15 + 12$$

Commutative Property of Multiplication (p. 160) The property that states that the order in which two numbers are multiplied does not change the *product.*

$$7 \times 2 = 2 \times 7$$

propiedad conmutativa de la multiplicación Propiedad que establece que el orden en el cual se multiplican dos o más números no altera el *producto.*

$$7 \times 2 = 2 \times 7$$

compatible numbers (p. 74) Numbers in a problem or related numbers that are easy to work with mentally.

720 and 90 are compatible numbers for division because 72 ÷ 9 = 8.

números compatibles Números en un problema o números relacionados con los cuales es fácil trabajar mentalmente.

720 y 90 son números compatibles en la división porque 72 ÷ 9 = 8.

complex figure (p. LA 22) A shape that is made up of two or more shapes.

figura compleja Figura compuesta por dos o más formas.

cone (p. 467) A *three-dimensional figure* with a curved surface and a circular *base* which comes to a point called the *vertex.*

cono *Figura tridimensional* con una superficie curva y una la *base* circular que terminan en un punto llamado *vértice.*

congruent figures (p. 484) Two figures having the same size and the same shape.

congruentes figuras Dos figuras con la misma forma y el mismo tamaño.

coordinate plane or grid (p. 494) A graph that displays a set of points and gives the position of a point on a line.

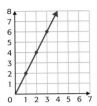

gráfica de coordenadas o cuadriculado Gráfica que representa un conjunto de puntos y da, en términos numéricos, la posición de un punto sobre una recta.

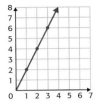

cube (p. 467) A *three-dimensional* figure with six faces that are congruent.

cubo Figura *tridimensional* con seis caras cuadradas *congruentes*.

cubic unit (p. 448) A unit for measuring *volume*, such as a cubic inch or a cubic centimeter.

unidad cúbica Unidad para medir *volumen*, como la pulgada cúbica o el centímetro cúbico.

cup (c) (p. 425) A customary unit for measuring *capacity*.

1 cup = 8 ounces
16 cups = 1 gallon

taza (c) Unidad inglesa de *capacidad*.

1 taza = 8 onzas
16 tazas = 1 galón

customary system The measurement system that includes units such as foot, pound, quart, and degrees Fahrenheit. Also called *standard measurement*.

sistema inglés Sistema de medición que incluye unidades como el pie, la libra, el cuarto de galón y los grados Fahrenheit. También llamado *medición estándar*.

Glossary/Glosario

cylinder (p. 467)   A *three-dimensional figure* having two circular *bases* and a curved surface connecting the two *bases*.

cilindro   *Figura tridimensional* que tiene dos *bases* circulares y una superficie curva que une las dos *bases*.

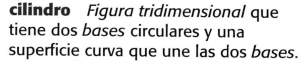

data (p. 510)   Numbers or symbols sometimes collected from a *survey* or experiment to show information. Datum is singular; data is plural.

datos   Números o símbolos, algunas veces recolectados de una *encuesta* o un experimento, para mostrar información.

decimal (p. 601)   A number with one or more digits to the right of the decimal point, such as 8.37 or 0.05.

número decimal   Número con uno o más dígitos a la derecha del punto decimal, como 8.37 ó 0.05.

decimal point (p. 601)   A period separating the ones and the *tenths* in a decimal number.

0.8 or $3.77

punto decimal   Punto que separa las unidades de las *décimas* en un número decimal.

0.8 ó $3.77

degree (°) (p. 408)   A unit of measure used to describe temperature.

grado (°)   Unidad de medida que se usa para describir la temperatura.

denominator (p. 561)   The bottom number in a *fraction*.
   In $\frac{5}{6}$, 6 is the denominator.

denominador   El número inferior en una *fracción*.
   En $\frac{5}{6}$, 6 es el denominador.

difference (p. 111)   The answer to a *subtraction* problem.

diferencia   Respuesta a un problema de *resta*.

digit (p. 24)   A symbol used to write numbers. The ten digits are 0, 1, 2, 3, 4, 5, 6, 7, 8, and 9.

dígito   Símbolo que se usa para escribir números. Los diez dígitos son 0, 1, 2, 3, 4, 5, 6, 7, 8 y 9.

digital clock (p. 454)   A clock that uses only numbers to show time.

reloj digital   Reloj que sólo utiliza números para mostrar la hora.

Distributive Property of Multiplication (p. LA 6)   To multiply a *sum* by a number, multiply each *addend* by the number and add the *products*.

$$4 \times (1 + 3) = (4 \times 1) + (4 \times 3)$$

propiedad distributiva de la multiplicación   Para multiplicar una *suma* por un número, puedes multiplicar cada *sumando* por el número y sumar los *productos*.

$$4 \times (1 + 3) = (4 \times 1) + (4 \times 3)$$

divide (division) (p. 251)   To separate into equal groups.

dividir (división)   Separar en grupos iguales.

dividend (p. 256)   A number that is being divided.

$3\overline{)9}$ 9 is the dividend

dividendo   El número que se divide.

$3\overline{)9}$ 9 es el dividendo

divisor (p. 256)   The number by which the dividend is being divided.

$3\overline{)19}$ 3 is the divisor

divisor   Número entre el cual se divide el dividendo.

$3\overline{)19}$ 3 es el divisor

dollar ($) (p. 53)   One dollar = 100¢ or 100 cents. Also written as $1.00.

front   back

dólar ($)   Un dólar = 100¢ ó 100 centavos. También se escribe como $1.00.

frente   revés

E

edge (p. 468)   The line segment where two faces of a solid figure meet.

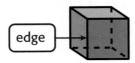

edge

arista   Segmento de recta donde concurren dos caras de una figura sólida.

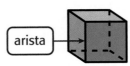

arista

equals sign (=) (p. 34)   A symbol of equality.

signo de igualdad (=)   Símbolo que muestra igualdad.

equation   A sentence that contains an equals sign (=) showing that two *expressions* are equal.

$$5 + 7 = 2$$

ecuación   Expresión que contiene un signo de igualdad y que muestra que dos expresiones son iquales.

$$5 + 7 = 2$$

equivalent fractions (p. 570) *Fractions* that have the same value.

$$\frac{2}{4} = \frac{1}{2}$$

fracciones equivalentes   *Fracciones* que tienen el mismo valor.

$$\frac{2}{4} \text{ y } \frac{1}{2}$$

estimate (p. 74)   A number close to an exact value. An estimate indicates *about* how much.

47 + 22 is about 70.

estimación   Número cercano a un valor exacto. Una estimación indica aproximadamente cuánto.

47 + 22 es aproximadamente 70.

expanded form/expanded notation (p. 25)   The representation of a number as a sum that shows the value of each digit.

536 is written as 500 + 30 + 6.

forma desarrollada/notación desarrollada   Representación de un número como suma que muestra el valor de cada dígito.

536 se escribe como 500 + 30 + 6.

expression (p. 338)   A combination of numbers and operations.

5 + 7

expresión   Combinación de números y símbolos.

5 + 7

F

face (p. 468)   The flat part of a *three-dimensional figure.*
   A square is a face of a cube.

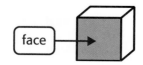

cara   La parte llana de una figura tridimensional.
   Un cuadrado es una cara de un cubo.

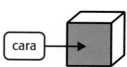

fact family (p. 259)   A group of *related facts* using the same numbers.

$$5 + 3 = 8 \quad | \quad 5 \times 3 = 15$$
$$3 + 5 = 8 \quad | \quad 3 \times 5 = 15$$
$$8 - 3 = 5 \quad | \quad 15 \div 5 = 3$$
$$8 - 5 = 3 \quad | \quad 15 \div 3 = 5$$

familia de operaciones   Grupo de *operaciones relacionadas* que usan los mismos números.

$$5 + 3 = 8 \quad | \quad 5 \times 3 = 15$$
$$3 + 5 = 8 \quad | \quad 3 \times 5 = 15$$
$$8 - 3 = 5 \quad | \quad 15 \div 5 = 3$$
$$8 - 5 = 3 \quad | \quad 15 \div 3 = 5$$

factor (p. 159)   A number that divides a whole number evenly. Also a number that is multiplied by another number.

factor   Número que divide exactamente a otro número entero. También es un número multiplicado por otro número.

foot (ft) (p. 378)   A *customary unit* for measuring *length*. Plural is feet.
1 foot = 12 inches

pie (pie)   *Unidad inglesa* para medir *longitud*.
1 pie = 12 pulgadas

fraction (p. 559)   A number that represents part of a whole or part of a set.

$$\frac{1}{2}, \frac{1}{3}, \frac{1}{4}, \frac{3}{4}$$

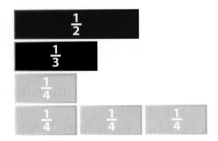

fracción   Número que representa parte de un todo o parte de un conjunto.

$$\frac{1}{2}, \frac{1}{3}, \frac{1}{4}, \frac{3}{4}$$

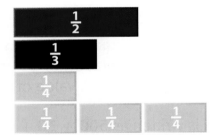

function (p. 348)   A relationship in which one quantity depends upon another quantity.

función   Relación en que una cantidad depende de otra cantidad.

G

gallon (gal) (p. 425)   A *customary unit* for measuring *capacity* for liquids.
1 gallon = 4 quarts

galón (gal)   *Unidad de medida inglesa* para medir la *capacidad* líquida.
1 galón = 4 cuartos de galón

gram (g) (p. 444)   A *metric unit* for measuring *mass*.

gramo (g)   *Unidad métrica* para medir la *masa*.

graph (p. 526)   An organized drawing that shows sets of data and how they are related to each other. Also a type of chart.

a bar graph

gráfica   Dibujo organizado que muestra conjuntos de datos y cómo se relacionan. También, un tipo de diagrama.

una gráfica de barras

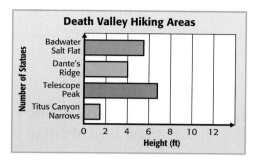

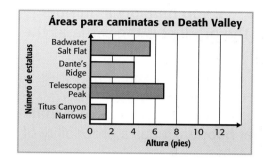

H

hexagon (p. 472)   A *polygon* with six *sides* and six *angles*.

hexágono   *Polígono* con seis *lados* y seis *ángulos*.

hour (h) (p. 454)   A unit of time equal to 60 *minutes*.

1 hour = 60 minutes

hora (h)   Unidad de tiempo igual a 60 *minutos*.

1 hora = 60 minutos

hundredths (p. 608)   A place value position. One of one hundred equal parts. In the number 0.05, the number 5 is in the hundredths place.

centésima   Valor de posición. Una de cien partes iguales.
En el número 0.05, 5 está en el lugar de las centésimas.

Glossary/Glosario

Glossary/Glosario

Identity Property of Addition (p. 69)
If you add zero to a number, the sum is the same as the given number.
$$3 + 0 = 3 \text{ or } 0 + 3 = 3$$

propiedad de identidad de la suma
Si sumas cero a un número, la suma es igual al número dado.
$$3 + 0 = 3 \text{ ó } 0 + 3 = 3$$

Identity Property of Multiplication
(p. 186)   If you multiply a number by 1, the product is the same as the given number.
$$8 \times 1 = 8 = 8 \times 1$$

propiedad de identidad de la multiplicación   Si multiplicas un número por 1, el producto es igual al número dado.
$$8 \times 1 = 8 = 1 \times 8$$

impossible (p. 542)   An event that cannot happen. It has a probability of zero.

It is impossible to choose red.

imposible   Evento que no puede suceder, el cual tiene probabilidad cero.

Es imposible elegir rojo.

inch (in.) (p. 375)   A *customary unit* for measuring *length*. Plural is inches.

pulgada (pulg)   *Unidad inglesa* para medir la *longitud*.

inverse operation   Operations that undo each other.
Addition and subtraction are inverse or opposite operations.
Multiplication and division are also inverse operations.

operación inversa   Operaciones que se anulan entre sí.
La suma y la resta son operaciones inversas u opuestas.
La multiplicación y la división también son operaciones inversas.

is equal to (=) (p. 34)

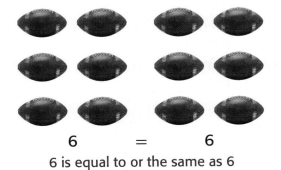

6 = 6
6 is equal to or the same as 6

es igual a (=)

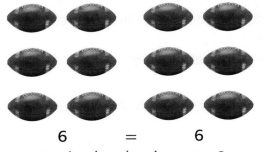

6 = 6
6 es igual a o lo mismo que 6

is greater than > (p. 34)   An inequality relationship showing that the value on the left of the symbol is greater than the value on the right.

5 > 3 5 is greater than 3

es mayor que >   Relación de desigualdad que muestra que el valor a la izquierda del símbolo es mayor que el valor a la derecha.

5 > 3 5 es mayor que 3

is less than < (p. 34)   The value on the left side of the symbol is smaller than the value on the right side.

4 < 7 4 is less than 7

es menor que <   El valor a la izquierda del símbolo es más pequeño que el valor a la derecha.

4 < 7 4 es menor que 7

K

key (p. 514)   Tells what or how many each symbol stands for.

clave   Indica qué significa o cuánto vale cada símbolo.

kilogram (kg) (p. 444)   A *metric unit* for measuring *mass*.

kilogramo (kg)   *Unidad métrica* para medir la *masa*.

kilometer (km) (p. 386)   A *metric unit* for measuring length.

kilómetro (km)   *Unidad métrica* para medir la *longitud*.

L

length (p. 375)   Measurement of the distance between two points.

longitud   Medida de la distancia entre dos puntos.

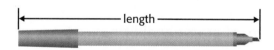

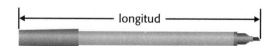

likely (p. 542)   An event that will probably happen.
It is likely you will choose a red tile.

posible   Evento que probablemente sucederá.
Es posible que elijas un cubo rojo.

line of symmetry (p. 488)   A *line* on which a figure can be folded so that its two halves match exactly.

eje de simetría   *Recta* sobre la cual se puede doblar una figura de manera que sus mitades se correspondan exactamente.

Glossary/Glosario

line plot (p. 476)   A graph that uses columns of Xs above a *number line* to show frequency of data.

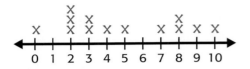

esquema lineal   Gráfica que usa columnas de X sobre una *recta numérica* para representar frecuencias de datos

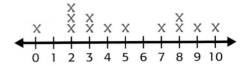

liter (L) (p. 432)   A *metric unit* for measuring *volume* or *capacity*.
1 liter = 1000 milliliters

litro (L)   *Unidad métrica* para medir *volumen* o *capacidad*.
1 litro = 1000 mililitros

M

mass (p. 444)   The amount of matter in an object. Two examples of units of measure would be gram and kilogram.

masa   Cantidad de materia de un cuerpo. Dos ejemplos de unidades de medida son el gramo y el kilogramo.

meter (m) (p. 386)   A *metric unit* for measuring *length*.
1 meter = 100 centimeters

metro (m)   *Unidad métrica* para medir la *longitud*.
1 metro = 100 centímetros

metric system (SI)   The measurement system based on powers of 10 that includes units such as meter, gram, and liter.

sistema métrico (SI)   Sistema de medición que se basa en potencias de 10 y que incluye unidades como el metro, el gramo y el litro.

mile (mi) (p. 378)   A *customary unit* of measure for distance.
1 mile = 5,280 feet

milla (mi)   *Unidad inglesa* para medir la distancia.
1 milla = 5,280 pies

milliliter (mL) (p. 432)   A *metric unit* used for measuring *capacity*.
1,000 milliliters = 1 liter

mililitro (ml)   *Unidad métrica* para medir la *capacidad*.
1,000 mililitros = 1 litro

millimeter (mm) (p. 385)   A *metric unit* used for measuring *length*.
1,000 millimeters = 1 meter

milímetro (mm)   *Unidad métrica* que se usa para medir la *longitud*.
1,000 milímetros = 1 metro

minute (min) (p. 454)   A unit used to measure time.

1 minute = 60 seconds

minuto (min)   Unidad que se usa para medir el tiempo.

1 minuto = 60 segundos

mixed number (p. LA 10)   A number that has a *whole number* part and a *fraction* part.

$6\frac{3}{4}$

número mixto   Número compuesto por una *parte entera* y una parte *fraccionaria*.

$6\frac{3}{4}$

multiplication (p. 157)   An operation on two numbers to find their product. It can be thought of as repeated *addition*.

$3 \times 4 = 12$
$4 + 4 + 4 = 12$

multiplicación   Operación en dos números para hallar su producto. Se puede considerar una suma repetida.

$3 \times 4 = 12$
$4 + 4 + 4 = 12$

multiple (p. 636)   A multiple of a number is the *product* of that number and any whole number.

15 is a multiple of 5 because $3 \times 5 = 15$.

múltiplo   Un múltiplo de un número es el *producto* de ese número y cualquier otro número entero.

15 es múltiplo de 5 porque $3 \times 5 = 15$.

multiply (p. 159)   To find the product of 2 or more numbers.

multiplicar (multiplicación)   Calcular el producto de 2 o más números.

N

number line (p. 492)   A line with numbers on it in order at regular intervals.

0 0.1 0.2 0.3 0.4 0.5 0.6 0.7 0.8 0.9 1.0

recta numérica   Recta que representa números como puntos.

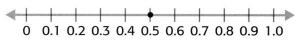

0 0.1 0.2 0.3 0.4 0.5 0.6 0.7 0.8 0.9 1.0

number sentence (p. 362)   An expression using numbers and the = sign, or the < or > sign.

$5 + 4 = 9; 8 > 5$

expresión numérico   Expresión que usa números y el signo = o los signos < o >.

$5 + 4 = 9; 8 > 5$

Glossary/Glosario

numerator (p. 561) The number above the bar in a *fraction*; the part of the fraction that tells how many of the equal parts are being used.

numerador Número que está encima de la barra de *fracción*; la parte de la fracción que indica cuántas partes iguales se están usando.

O

octagon (p. 472) A *polygon* with eight *sides*.

octágono *Polígono* de ocho *lados*.

operation A mathematical process such as addition (+), subtraction (−), multiplication (×), and division (÷).

operación Proceso matemático como la suma (+), la resta (−), la multiplicación (×), y la división.

ordered pair (p. 494) A pair of numbers that are the *coordinates* of a point in a coordinate plane or grid in this order (horizontal coordinate, vertical coordinate).

par ordenado Par de números que son *coordenadas* de un punto en un plano de coordenadas o cuadriculado, en este orden (coordenada horizontal, coordenada vertical).

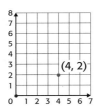

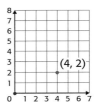

ounce (oz) ounces (p. 438) A *customary unit* for measuring *weight or capacity*.

onza (oz) *Unidad inglesa* para medir *peso* o *capacidad*.

P

pattern (p. 17) A sequence of numbers, figures, or symbols that follows a rule or design.

2, 4, 6, 8, 10

patrón Sucesión de números, figuras o símbolos que sigue una regla o un diseño.

2, 4, 6, 8, 10

pentagon (p. 472)   A *polygon* with five *sides*.

pentágono   *Polígono* de cinco *lados*.

perimeter (p. 392)   The *distance* around a shape or region.

perímetro   *Distancia* alrededor de una figura o región.

period (p. 28)   The name given to each group of three digits on a place-value chart.

período   Nombre dado a cada grupo de tres dígitos en una tabla de valores de posición.

pictograph (p. 515)   A graph that compares *data* by using picture symbols.

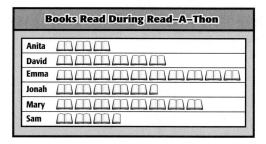

pictograma   Gráfica que compara *datos* usando figuras.

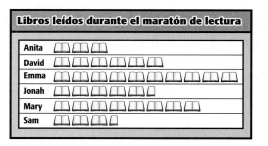

pint (pt) (p. 425)   A *customary unit* for measuring *capacity*.

1 pint = 2 cups

pinta (pt)   *Unidad inglesa* para medir la *capacidad*.

1 pinta = 2 tazas

place value (p. 24)   The value given to a *digit* by its place in a number.

valor de posición   El valor de un *dígito* según su lugar en el número.

plane figure (p. 472)   A *two-dimensional figure* that lies entirely within one *plane* such as a triangle or square.

figura plana   *Figura bidimensional* que yace completamente en un *plano*, como un triángulo o un cuadrado.

polygon (p. 472)   A closed plane figure formed using line segments that meet only at their endpoints.

polígono   Figura plana cerrada formada por segmentos de recta que sólo concurren en sus extremos.

pound (lb) (p. 438)   A *customary unit* for measuring *weight*.

1 pound = 16 ounces.

libra (lb)   *Unidad inglesa* para medir el *peso*.

1 libra = 16 onzas

prediction   Something you think will happen such as a specific outcome of an experiment.

predicción   Algo que crees que sucederá, como un resultado específico de un experimento.

prism (p. 467)   A *three-dimensional figure* with two parallel, congruent polygons as bases and parallelograms for faces.

prisma   *Figura tridimensional* con dos polígonos paralelos y congruentes como bases y paralelogramos como caras.

probability (p. 542)   The chance that an event will happen.

probabilidad   La posibilidad de que ocurra un evento.

product (p. 159)   The answer to a multiplication problem.

producto   Respuesta a un problema de multiplicación.

pyramid (p. 467)   A *three-dimensional figure* with a *polygon* as a base and triangular shaped *faces* that share a common *vertex.*

pirámide   Figura sólida con un *polígono* como base y *caras* triangulares que comparten un *vértice* común.

Glossary/Glosario

Q

quadrilateral (p. 472) A shape that has 4 sides and 4 angles.

square   rectangle   parallelogram

cuadrilátero Figura con 4 lados y 4 ángulos.

cuadrado rectángulo paralelogramo

quart (qt) (p. 425) A *customary unit* for measuring *capacity*.

1 quart = 4 cups

cuarto de galón (ct) *Unidad inglesa* de galón para mdir la *capacidad*.

1 cuarto de galón = 4 tazas

quotient (p. 256) The answer to a *division problem*.

15 ÷ 3 = 5 ← [5 is the quotient.]

cociente Respuesta a un *problema de división*.

15 ÷ 3 = 5 ← [5 es el cociente.]

R

rectangle A *quadrilateral* with four *right angles*; opposite *sides* are equal length and *parallel*.

rectángulo *Cuadrilátero* con cuatro *ángulos rectos*; los *lados* opuestos son de igual longitud y *paralelos*.

rectangular solid (p. 467)
A *three-dimensional figure* with six faces that are rectangles.

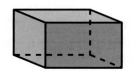

sólido rectangular *Figura tridimensional* con seis caras rectangulares.

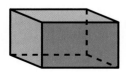

regroup (p. 78)   To use place value to exchange equal amounts when renaming a number.

reagrupar   Usar el valor de posición para intercambiar cantidades iguales cuando se convierte un número.

related fact(s) (p. 259)   Basic facts using the same numbers. Sometimes called a fact family.

operación (u operaciones relacionada(s)   Operaciones básicas que usan los mismos números. A veces llamadas familia de operaciones.

4 + 1 = 5	5 × 6 = 30
1 + 4 = 5	6 × 5 = 30
5 − 4 = 1	30 ÷ 5 = 6
5 − 1 = 4	30 ÷ 6 = 5

4 + 1 = 5	5 × 6 = 30
1 + 4 = 5	6 × 5 = 30
5 − 4 = 1	30 ÷ 5 = 6
5 − 1 = 4	30 ÷ 6 = 5

rhombus   A *parallelogram* with four *sides* of the same *length*.

rombo   *Paralelogramo* con cuatro *lados* del mismo *largo*.

round (p. 44)   To change the *value* of a number to one that is easier to work with. To find the nearest value of a number based on a given place value. 27 rounded to the nearest 10 is 30.

redondear   Cambiar el *valor* de un número por uno con el que es más fácil trabajar. Calcular el valor más cercano de un número en base a un valor de posición dado. 27 redondeado a la décima más cercana es 30.

S

scale   Equally spaced marks along an axis of a graph.

escala   Marcas igualmente separadas a lo largo del eje de una gráfica.

simplest form   A fraction in which the numerator and the denominator have no common factor greater than 1. $\frac{3}{5}$ is the simplest form of $\frac{6}{10}$.

forma reducida   Fracción en que el numerador y el denominador no tienen un factor común mayor que 1. $\frac{3}{5}$ es la forma reducida de $\frac{6}{10}$.

skip count (p. 163)   To count forward or backward by a given number or in intervals of a number. 3, 6, 9, 12 …

conteo salteado   Contar hacia adelante o hacia atrás por un número dado o en intervalos de un número. Ejemplo: 3, 6, 9, 12, …

solid figure (p. 467)   A solid figure having the three dimensions: length, width, and height.

figura sólida   Figura sólida tridimensional: largo, ancho y alto.

sphere (p. 467)   A *three-dimensional figure* that has the shape of a round ball.

esfera   *Figura tridimensional* con forma de pelota redonda.

square   A rectangle with four *congruent sides*.

cuadrado   Rectángulo con cuatro *lados congruentes*.

standard form/standard notation (p. 25)   The usual way of writing a number that shows only its *digits*, no words.

537   89   1642

forma estándar/notación estándar   La manera habitual de escribir un número que sólo muestra sus *dígitos*, sin palabras.

537   89   1642

standard units   Measuring units from the customary or metric system.

unidades estándar   Unidades de medida del sistema inglés o del métrico.

subtraction (subtract) (p. 111)   An operation that tells the difference, when some or all are taken away.

$$9 - 4 = 5$$

resta (restar)   Operación que indica la diferencia cuando se elimina algo o todo.

$$9 - 4 = 5$$

sum   The answer to an addition problem.

$$8 + 5 = 13$$

suma   Respuesta a un problema de suma.

$$8 + 5 = 13$$

survey (p. 528)   A method of collecting data.

encuesta   Un método para reunir datos.

T

table A way to organize and display data with rows and columns.

tabla Manera de organizar y representar datos con filas y columnas.

tally chart (p. 515) A way to keep track of data using tally marks to record the results.

tabla de conteo Una manera de llevar la cuenta de los datos usando marcas de conteo para anotar los resultados.

What is Your Favorite Color?	
Color	Tally
Blue	卌 |||
Green	||||

¿Cuál es tu color favorito?	
Color	Conteo
Azul	卌 |||
Verde	||||

tally mark(s) (p. 515) A mark made to keep track and display data recorded from a survey.

marcas(s) de conteo Marca hecha para llevar la cuenta y presentar datos reunidos con una encuesta.

tenth (p. 603) One of ten equal parts or $\frac{1}{10}$.

décima Una de diez partes iguales ó $\frac{1}{10}$.

thousand(s) (p. 24) A place value of a number.
In 1,253, the **1** is in the thousands place.

millares Valor de posición de un número.
En 1,253, el **1** está en el lugar de las unidades de millar.

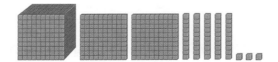

three-dimensional figure (p. 467) A solid figure that has *length*, *width*, and *height*.

figura tridimensional Figura sólida que tiene *largo*, *ancho* y *alto*.

ton (T) (p. 438) A customary unit to measure weight.

1 ton = 2,000 pounds

tonelada (T) Unidad inglesa de peso

1 tonelada = 2,000 libras

triangle (p. 472) A *polygon* with three sides and three angles.

triángulo *Polígono* con tres lados y tres ángulos.

Glossary/Glosario

two-dimensional figure (p. 472)
The outline of a shape, such as a
triangle, square, or rectangle, which
has only *length*, *width*, and *area*.
Also called a plane figure.

figura bidimensional   El contorno de
una figura, como un triángulo, un
cuadrado o un rectángulo, que sólo
tiene *largo*, *ancho* y *área*. También se
llama figura plana.

unlikely (p. 542)   An event that will
probably *not* happen.

It is unlikely you will
choose a yellow tile.

improbable   Evento que
probablemente *no* sucederá.

Es improbable que elijas
una baldosa amarilla.

volume (p. 450)   The number of cubic
units needed to fill a *three-dimensional
figure* or solid figure.

volumen   Número de unidades cúbicas
necesarias para llenar una *figura
tridimensional* o sólida.

weight (p. 438)   A measurement that
tells how heavy an object is.

peso   Medida que indica la pesadez de
un cuerpo.

whole number   The numbers 0, 1, 2,
3, 4 …

número entero   Los números
0, 1, 2, 3, 4 …

yard (yd) (p. 478)   A *customary unit*
for measuring *length*.
    1 yard = 3 feet or 36 inches

yarda (yd)   *Medida inglesa* para medir
la *longitud*.
    1 yarda = 3 pies eo 36 pulgadas

Zero Property of Multiplication
(p. 186)   The property that states any
number multiplied by zero is zero.

$$0 \times 5 = 0 \qquad 5 \times 0 = 0$$

**propiedad del producto nulo de la
multiplicación**   Propiedad que
establece que cualquier número
multiplicado por cero es igual a cero.
$$0 \times 5 = 0 \qquad 5 \times 0 = 0$$

Glossary/Glosario

Glossary/Glosario   R99

Index

Index

Index

Index

Index

Index

Index

Measurement Conversions

	Metric	**Customary**
Length	1 kilometer (km) = 1,000 meters (m) 1 meter = 100 centimeters (cm)	1 mile (mi) = 1,760 yards (yd) 1 mile = 5,280 feet (ft) 1 yard = 3 feet 1 foot = 12 inches (in.)
Volume and Capacity	1 liter (L) = 1,000 milliliters (mL)	1 gallon (gal) = 4 quarts (qt) 1 gallon = 128 ounces (oz) 1 quart = 2 pints (pt) 1 pint = 2 cups (c) 1 cup = 8 ounces
Weight and Mass	1 kilogram (kg) = 1,000 grams (g)	1 pound (lb) = 16 ounces (oz)

Time	1 year (yr) = 365 days (d) 1 year = 12 months (mo) 1 year = 52 weeks (wk) 1 week = 7 days 1 day = 24 hours (h) 1 hour = 60 minutes (min) 1 minute = 60 seconds (s)

Multiplication Table

×	1	2	3	4	5	6	7	8	9	10	11	12
1	1	2	3	4	5	6	7	8	9	10	11	12
2	2	4	6	8	10	12	14	16	18	20	22	24
3	3	6	9	12	15	18	21	24	27	30	33	36
4	4	8	12	16	20	24	28	32	36	40	44	48
5	5	10	15	20	25	30	35	40	45	50	55	60
6	6	12	18	24	30	36	42	48	54	60	66	72
7	7	14	21	28	35	42	49	56	63	70	77	84
8	8	16	24	32	40	48	56	64	72	80	88	96
9	9	18	27	36	45	54	63	72	81	90	99	108
10	10	20	30	40	50	60	70	80	90	100	110	120
11	11	22	33	44	55	66	77	88	99	110	121	132
12	12	24	36	48	60	72	84	96	108	120	132	144

Formulas

	Perimeter	Area
Square	$P = 4s$	$A = s \times s$ $P = \ell + \ell + w + w$ or $P = 2\ell + 2w$
Rectangle	$P = \ell + \ell + w + w$ $P = 2\ell + 2w$	$A = \ell \times w$